17th International Symposium Symmetries in Science (SIS XVII)

Journal of Physics: Conference Series
Volume 1071

Bregenz, Austria
30 July – 4 August 2017

ISBN: 978-1-5108-7029-1
ISSN: 1742-6588

Printed from e-media with permission by:

Curran Associates, Inc.
57 Morehouse Lane
Red Hook, NY 12571

Printed by Curran Associates, Inc. (2018)

For permission requests, please contact the Institute of Physics at the address below.

Institute of Physics
Dirac House, Temple Back
Bristol BS1 6BE UK

Phone: 44 1 17 929 7481
Fax: 44 1 17 920 0979

techtracking@iop.org

Additional copies of this publication are available from:

Curran Associates, Inc.
57 Morehouse Lane
Red Hook, NY 12571 USA
Phone: 845-758-0400
Fax: 845-758-2633
Email: curran@proceedings.com
Web: www.proceedings.com

Table of contents

Volume 1071

Symmetries in Science XVII

30 July to 4 August 2017, Bregenz

Accepted papers received: 27 July 2018
Published online: 3 September 2018

Preface

Papers

IOP Conf. Series: Journal of Physics: Conf. Series **1071** (2018) 011001 doi:10.1088/1742-6596/1071/1/011001

Preface

The international symposium "Symmetries in Science" (initiated by Bruno Gruber in 1979) continued with SIS XVII, July 30 - August 4, 2017 in Bregenz, Austria, following a break in 2015. Thirty-three scientists from 13 countries attended, again almost half of them from countries outside the European Union (from Mexico to South Korea, Brazil to Russia).

For the first time since being in our charge, the conference took place at Gasthof Hotel Lamm, a location well-known to former participants and just across the road from the conference site of previous years. Gasthof Lamm was always a favourite meeting place for after-lecture and evening discussions (prolonged and inspired by one's drink of choice) that sometimes went till well after midnight; this year was no exception.

The symbiosis of experts (old and young), upstarts, advocates of well-established theories and proponents of more speculative concepts provided the basis for stimulating debate, lively interaction and constructive exchange of ideas (in and outside the lecture rooms), hallmarks of the symposium series.

Wide-ranging topics from theoretical chemistry, fundamental problems in quantum theory, nuclear and mathematical physics, cosmology to history of physics and philosophy were explored, setting enthusiastic discussions in motion. Several talks can also be viewed on the conference website.

Most of the contributions are included in these Proceedings; a few were however previously submitted or already published elsewhere. To make these contributions also available to readers of our Proceedings, the respective references are given as follow:

- Goldin G, Sharp D H 2017, Diffeomorphism Group Representations in Relativistic Quantum Field Theory. In Piotr Kielanowski, Anatol Odzijewicz, & Emma Previato (Eds.), *"Geometric Methods in Physics XXXVI - Workshop and Summer School, Białowieża, Poland, 2017"*
- Kiefer C, Branislav N 2017, Conformal and Weyl–Einstein gravity: Classical geometrodynamics. *Phys. Rev.* D **95** 084018
- Kiefer C 2012, *"Quantum Gravity"* (3rd edition, Oxford University Press)
- Giulini D, Kiefer C 2007, The Canonical approach to quantum gravity: General ideas and geometrodynamics. *Lect. Notes Phys.* **721** 131-150.

We are especially grateful to Daniel and the Family Schenk, owners and operators of Gasthof Hotel Lamm, an institution in its own right. Their altruistic and generous support over the years meant we were able to enjoy yet another lavish and exquisite conference dinner (free of charge!). More importantly, they provided (practically impromptu) a wonderful and well-equipped setting to host and accommodate our symposium and its participants.

Last, but not least, our sincere thanks to Yvette (who does not like being mentioned); but these Proceedings would certainly not have materialized without her persistence and support.

Dieter Schuch, Frankfurt am Main, Germany
Michael Ramek, Graz, Austria
June 2018

fohrenburger
fohrenburger
fohrenburger
fohrenburger

IOP Conf. Series: Journal of Physics: Conf. Series **1071** (2018) 011003 doi:10.1088/1742-6596/1071/1/011003

Peer review statement

All papers published in this volume of *Journal of Physics: Conference Series* have been peer reviewed through processes administered by the proceedings Editors. Reviews were conducted by expert referees to the professional and scientific standards expected of a proceedings journal published by IOP Publishing.

IOP Conf. Series: Journal of Physics: Conf. Series **1071** (2018) 012001 doi:10.1088/1742-6596/1071/1/012001

Effects of Lorentz boosts on Dirac bispinor entanglement

Victor A. S. V. Bittencourt

E-mail: vbittencourt@df.ufscar.br
Departamento de Física, Universidade Federal de São Carlos, PO Box 676, 13565-905, São Carlos, SP, Brasil.

Alex E. Bernardini

E-mail: alexeb@ufscar.br
Departamento de Física, Universidade Federal de São Carlos, PO Box 676, 13565-905, São Carlos, SP, Brasil.

Massimo Blasone

E-mail: blasone@sa.infn.it
Dipartimento di Fisica, Università degli studi di Salerno, Via Giovanni Paolo II, 132 84084 Fisciano, Italy
Also at: INFN Sezione di Napoli, Gruppo collegato di Salerno, Italy

Abstract. In this paper we describe the transformation properties of quantum entanglement encoded in a pair of spin 1/2 particles described via Dirac bispinors. Due to the intrinsic parity-spin internal structure of the bispinors, the joint state is a four-qubit state exhibiting multipartite entanglement, and to compute global correlation properties we consider the averaged negativities over each possible bi-partition. We also consider specific bipartitions, such as the spin-spin and the particle-particle bipartitions. The particle-particle entanglement, between all degrees of freedom of one particle and all degrees of freedom of the other particle, is invariant under boosts if each particle has a definite momentum, although the spin-spin entanglement is degraded for high speed boosts. Correspondingly, the mean negativities are not invariant since the boost drives changes into correlations encoded in specific bipartitions. Finally, the results presented in the literature about spin-momentum entanglement are recovered by considering the projection of bispinorial states into positive intrinsic parity, and some striking differences between the appropriate approach for this case and the one usually treated in the literature are discussed.

1. Introduction

In the last decades many researches have been devoted to describe and contextualize quantum entanglement in relativistic setups [1, 2, 3, 4, 5, 6, 7, 8, 9]. In particular, the implementation of information protocols, such as clock synchronization [10], requires the description of the effects of frame transformations in quantum correlations, and since the beginning of the 2000s many studies described how entanglement encoded in a pair of spins changes under Lorentz boosts [1, 2, 3, 4, 5, 6, 7, 8, 9]. The construction of spin states in this context, in its majority, follows

IOP Conf. Series: Journal of Physics: Conf. Series **1071** (2018) 012001 doi:10.1088/1742-6596/1071/1/012001

the classification of the irreducible representations (irreps) of the Poincaré group [11, 12], and the effect of a Lorentz boost is given by a momentum dependent rotation of the spin [13], which is the basis of some of the most iconic results in the field, such as the non-invariance of the reduced spin entropy of a single particle [2].

Although the setup usually adopted to describe transformation properties of quantum entanglement has given some interesting insights into the physics of relativistic quantum information, when massive charged fermions are considered as the physical carriers of spin 1/2, a more complete description of the problem is required. The physical particles, such as electrons and muons, are described by quantum electrodynamics, a theory which, apart from the usual Poincaré symmetry, also exhibit invariance under parity transformation [12, 14]. This last symmetry operation exchange two irreps of the Poincaré group, and a proper description of such carriers of spin is given in terms of the irreps of the so called complete Lorentz group [12]. The state of the particles are then described by four component objects, the Dirac bispinors, which satisfy the Dirac equation.

The solutions of Dirac equation were previously considered in the information theory framework to discuss the definition of spin operators in relativisitc quantum mechanics [15], and the effects of Lorentz boosts on the entanglement encoded in superposition of Dirac equation solutions were considered for some specific states in connection with Wigner rotations [16] and in discussing spin-spin entanglement in the context of the Fouldy-Wouthuysen spin operator [17]. However, due to the group structure intrinsic to the bispinors, such type of states carries two quantum bits [18], spin and intrinsic parity, and a given two particle state in this context is thus a four-qubit state.

In this paper we provide a general description of the changes driven by Lorentz boosts on quantum entanglement encoded in superpositions of two particle states described in terms of Dirac bispinors. We consider the effects of Lorentz boosts on the averaged entanglement encoded in each type of possible bi-partition of the system as well as in the entanglement shared among specific degrees of freedom (DoFs) of the system, for instance only between the spins or between the parities. In a first approach, we consider superpositions between spinor states where each particle has a definite momentum while the spins are superposed, the momenta are supposed to be (anti)parallel, and we consider both parallel and perpendicular boosts. For this scenario, we prove that the entanglement shared between all DoFs of one particle and all DoFs of the other particle is Lorentz invariant, despite the overall entanglement encoded in other types of bi-partitions is non-trivially affected by boosts. In a second scenario, we describe the effects of boosts on states with momentum superposition on a simplified framework and recover previous results of the literature about spin-momentum entanglement by considering projections into definite parities. We shown that boosts cannot create spin-momentum entanglement, a striking difference between our approach and previous ones based on the irreps of the Lorentz group.

The paper is organized as follows. In Sec. II the basic properties of Dirac bispinors, including intrinsic spin-parity entanglement, is reviewed. Section III introduces the two particle state without momentum superposition, and presents the different boost scenarios which will be considered. It is proved that the particle-particle entanglement is invariant under Lorentz Boosts although entanglement in other bi-partitions are not invariant. In Sec. IV momentum superposition is introduced and the connection with results derived in the literature. Additionally, differences between the proper approach and the one adopted in the literature are discussed. To end up, Sec. VI presents our last conclusion and future perspectives.

2. Spin-parity entanglement

Dirac equation was proposed as a wave equation invariant under Poincaré transformations and through which is possible to define a non-negative conserved probability current [19]. In its

Hamiltonian form, in the momentum space, Dirac equation reads

$$H\,\psi = (\boldsymbol{p}\cdot\hat{\boldsymbol{\alpha}} + m\hat{\beta})\psi = E_p\psi, \tag{1}$$

where bold variables represent vector quantities with modulus denoted by $a = |\boldsymbol{a}| = \sqrt{\boldsymbol{a}\cdot\boldsymbol{a}}$, and $\hat{\alpha}_i$ and $\hat{\beta}$ (hats "^" denoting operators from hereafter) are anticommuting $n\times n$ matrices satisfying the relations

$$\hat{\alpha}_i\hat{\alpha}_j + \hat{\alpha}_j\hat{\alpha}_i = 2\delta_{ij}\hat{I}_4, \qquad \hat{\alpha}_i\hat{\beta} + \hat{\beta}\hat{\alpha}_i = 0,$$
$$\hat{\beta}^2 = \hat{I}, \tag{2}$$

with $\hat{I}$ is the identity operator. The matrices $\hat{\alpha}_i$ and $\hat{\beta}$ have different representations, interconnect via unitary transformations, and we shall adopt the Dirac representation in which

$$\hat{\alpha}_i = \begin{bmatrix} 0 & \hat{\sigma}_i \\ \hat{\sigma}_i & 0 \end{bmatrix} \quad \hat{\beta} = \begin{bmatrix} \hat{I} & 0 \\ 0 & -\hat{I} \end{bmatrix}. \tag{3}$$

The eigenstates of the Dirac Hamiltonian $\hat{H} = \boldsymbol{p}\cdot\hat{\boldsymbol{\alpha}} + m\hat{\beta}$, $u(p,s)$, with eigen-energy $+E_p$, and $v(p,s)$ with eigen-energy $-E_p$, are the 4-component *Dirac bispinors* given by

$$u(p,s) = \begin{bmatrix} \sqrt{\frac{E_p+m}{2E_p}}\chi_s \\ \frac{\boldsymbol{p}\cdot\boldsymbol{\sigma}}{\sqrt{2E_p(E_p+m)}}\chi_s \end{bmatrix}, \qquad v(p,s) = \begin{bmatrix} \frac{\boldsymbol{p}\cdot\boldsymbol{\sigma}}{\sqrt{2E_p(E_p+m)}}\chi_s \\ \sqrt{\frac{E_p+m}{2E_p}}\chi_s \end{bmatrix}. \tag{4}$$

The two-component spinors χ_s ($s=\pm$) are related with the spin of the particle and their explicit form depends on the specific polarization adopted. In this paper we describe the bispinors as eigenstates of the Helicity operator

$$\hat{h} = \frac{\boldsymbol{p}\cdot\hat{\boldsymbol{\Sigma}}}{|\boldsymbol{p}|},$$

with

$$\hat{\boldsymbol{\Sigma}} = \begin{bmatrix} \hat{\boldsymbol{\sigma}} & 0 \\ 0 & \hat{\boldsymbol{\sigma}} \end{bmatrix}$$

the Pauli-Dirac spin operator. For such choice, $|\chi_s\rangle$ are the eigenstates of the operator $\boldsymbol{p}\cdot\hat{\boldsymbol{\sigma}}$ which, in terms of the eigenstates $|\pm\rangle$ of the $\hat{\sigma}_z$ operators, are given by

$$|\chi_\pm\rangle = \frac{1}{\sqrt{2}}\left(\hat{I} \pm \frac{\boldsymbol{p}}{|\boldsymbol{p}|}\cdot\hat{\boldsymbol{\sigma}}\right)|\pm\rangle, \tag{5}$$

and are orthonormalized $\chi_s^\dagger\chi_l = \delta_{s,l}$. The orthogonality relations satisfied by u and v are

$$u^\dagger(p,s)u(p,r) = v^\dagger(p,s)v(p,r) = \delta_{sr}, \qquad u^\dagger(p,s)v(p,r) = v^\dagger(p,s)u(p,r) = 0, \tag{6}$$

and the completeness relation reads

$$\sum_s\left[u(p,s)u^\dagger(p,s) - v(p,s)v^\dagger(p,s)\right] = \frac{E_p+m}{E_p}\,\hat{I}.$$

The quantum information framework of Dirac equation can be set by noticing that in the representation (3), Dirac matrices are given in the form of tensor products, as to have

$$\hat{\alpha}_i = \hat{\sigma}_x^{(P)}\otimes\hat{\sigma}_i^{(S)}, \qquad \hat{\beta} = \hat{\sigma}_z^{(P)}\otimes\hat{I}^{(S)},$$

IOP Conf. Series: Journal of Physics: Conf. Series **1071** (2018) 012001 doi:10.1088/1742-6596/1071/1/012001

such that the Dirac Hamiltonian reads

$$H = \boldsymbol{p} \cdot (\hat{\sigma}_x^{(P)} \otimes \hat{\boldsymbol{\sigma}}^{(S)}) + m(\hat{\sigma}_z^{(P)} \otimes \hat{I}^{(S)}). \tag{7}$$

Therefore, relativistic quantum mechanics can be interpreted as a two-qubit information theory associated with the discrete DoFs of the system: P, associated with the Hilbert space $\mathcal{H}_P$ describing the intrinsic parity DoF, and S, associated with the Hilbert space $\mathcal{H}_S$ describing the spin DoF [18]. This two DoFs are related with the underlying group structure of the Dirac bispinors. The invariance under proper Lorentz transformations and space inversion, a symmetry present for example in quantum electrodynamics, requires the description in terms of the irreps of the complete Lorentz group [12], which contains, additionally to the spin, the intrinsic parity quantum number. In the case of spin 1/2 particles (electrons, protons, muons, etc) such irreps belongs to $SU(2) \otimes SU(2)$ [11, 12], as is explicitly shown in Eq. (7). In this context the bispinors (4) are rewritten as two-qubit states. For instance, the positive energy bispinor reads

$$|\, u(p,s)\,\rangle = \sqrt{\frac{E_p + m}{2E_p}} |+\rangle_P \otimes |\chi_s\rangle_S + \frac{1}{\sqrt{2E_p(E_p+m)}} |-\rangle_P \otimes (\,\boldsymbol{p} \cdot \hat{\boldsymbol{\sigma}}^{(S)}\, |\chi_s\rangle_S\,), \tag{8}$$

where $|\chi_s\rangle_S$ are the helicity eigenstates spinors (5), and we have introduced the subscripts P and S to indicate the intrinsic parity and spin spaces.

The generic form (8) is *spin-parity* entangled. Entanglement is defined by means of the separability concept: A state with density matrix ρ describing a system composed of two subsystems A and B, each associated to its corresponding Hilbert space $\mathcal{H}_{A(B)}$, is separable if it can be written as [20]

$$\rho = \sum_i c_i \, \rho_i^{(A)} \otimes \rho_i^{(B)}, \tag{9}$$

with $c_i \geq 0$, $\sum_i c_i = 1$, $\rho_i^{A(B)} \in \mathcal{H}_{A(B)}$. If ρ is not a separable state, then it is entangled. Separability can also be defined in terms of the Peres criterion [21], which establishes that ρ is separable iff the partial transpose density matrix $\rho^{A(B)}$, with respect to the A subsystem, has only positive eigenvalues. With respect to a fixed basis on the composite Hilbert space $\{|\mu_i\rangle \otimes |\nu_j\rangle\}$ (with $|\mu_i\rangle \in \mathcal{H}_A$ and $|\nu_i\rangle \in \mathcal{H}_B$), the matrix elements of the partial transpose with respect to the first subsystem ρ^A are given by

$$\langle\mu_i| \otimes \langle\nu_j|\rho^A|\mu_k\rangle \otimes |\nu_l\rangle = \langle\mu_k| \otimes \langle\nu_j|\,\rho\,|\mu_i\rangle \otimes |\nu_l\rangle. \tag{10}$$

With Peres criterion one defines the bi-partite entanglement quantifier called Negativity $\mathcal{N}$ [22], given by

$$\mathcal{N}^{A;\,B}[\rho] = \frac{1}{d_i - 1}(\,\textstyle\sum_i |\mu_i| - 1\,), \tag{11}$$

where μ_i are the eigenvalues of the matrix ρ^A, and $d_i = \dim[\,\mathcal{H}_A\,]$. Although many other entanglement quantifiers can be defined, negativity has the advantage to be computable without requiring any extremization process, and is the entanglement measurement adopted in this paper.

The intrinsic entanglement of Dirac bispinors was previously described for a plethora of scenarios [25, 26, 23, 24]. For example, a simple bidimensional scattering process of a plane wave by a step barrier can create entanglement between spin and parity DoFs in both reflected and transmitted waves [23]. A more complete description of the influence of external fields on spin-parity entanglement was also studied [24], and its subsequent translation to Dirac-like systems was considered [25, 26]. For example, the tight binding model for bilayer graphene is formally equivalent to a modified Dirac Hamiltonian including external fields, and the single-particle excitations of the material were shown to exhibit a lattice-layer entanglement, analogous to the spin-parity entanglement of its corresponding relativistic bispinor description [26].

IOP Conf. Series: Journal of Physics: Conf. Series **1071** (2018) 012001 doi:10.1088/1742-6596/1071/1/012001

Different from the single particle spin-parity entanglement encoded in a single-particle, here we consider entanglement encoded among the DoFs of two bispinorial particles constructed in terms of tensor products of bispinors of the generic form

$$\frac{1}{\sqrt{N}}\sum_{i}^{n} c_i \,|\, u_A(p_i, \alpha_i)\rangle \otimes |\, u_B(q_i, \beta_i)\,\rangle, \tag{12}$$

where N is a normalization factor. Due to the intrinsic structure of the bispinors (7,8), the superposition (12) is a 4-qubit state and although many effort has been devoted to devise how to proper characterize and quantify multipartite entanglement, up to now there is no precise method to accomplish such task [20]. A simplified picture of the global measure of entanglement shared among different partitions of the system can be compute through averages over the possible bi-partitions [27]. For the 4-qubit states (12) there are 4 types of bi-partitions:

- $\{i;\, j,\, k,\, l\}$, for example the partition $\{S_1;\, P_1,\, P_2,\, S_2\}$, which is used to compute the entanglement between one the DoF and all the others;
- $\{i,\, j;\, k,\, l\}$, which described the entanglement between pairs of DoFs. One type of such partition is $\{S_1,\, P_1;\, P_2,\, S_2\}$, which encodes the entanglement of all DoFs of one particle with all DoFs of the other;
- $\{i;\, j,\, k\}$, obtained by tracing one of the DoFs. For example in the partition $\{S_1;\, P_2,\, S_2\}$ is encoded correlation between the spin of particle 1 and all DoFs of particle 2;
- $\{i;\, j\}$, which encodes entanglement between only two DoFs. For example in the partition $\{P_1;\, P_2\}$ is encoded the entanglement between only the intrinsic parities.

In particular, to quantify the entanglement between spin of particle 1 and spin of particle 2 in a joint state φ, the partition $\{S_1;\, S_2\}$ should be considered, and entanglement would be given by $\mathcal{N}^{S_1;S_2}$, calculated with the reduced density matrix $\rho_{S_1,\,S_2} = \mathrm{Tr}_{P_1,\,P_2}[\varphi\varphi^{\dagger}]$. Another possible approach to quantify the multipartite entanglement shared among the different DoFs of the system is to consider the geometry of the Hilbert space itself and compute different distance between the state and the set of so called K-separable states [28]. Such procedure requires an involving extremization procedure and it will be postponed to future investigations.

As global measures of entanglement encoded in a state φ, one adopts the mean negativity in each bipartition as to have 4 quantities defined as

$$\begin{aligned} N^{(1)} &= \bar{\mathcal{N}}^{i;\,j\,k\,l}[\varphi] \quad N^{(2)} = \bar{\mathcal{N}}^{i;\,j}[\varphi] \\ N^{(3)} &= \bar{\mathcal{N}}^{i;\,j\,k}[\varphi] \quad N^{(4)} = \bar{\mathcal{N}}^{i\,j;\,k\,l}[\varphi]. \end{aligned} \tag{13}$$

For example, $N^{(1)} = (\mathcal{N}^{P_1;\,P_2\,S_1\,S_2} + \mathcal{N}^{P_2;\,P_1\,S_1\,S_2} + \mathcal{N}^{S_1;\,P_1\,P_2\,S_2} + \mathcal{N}^{S_2;\,P_1\,P_2\,S_1})/4$. In fact, the relation between linear entropies and negativities for pure states sets an equivalence between $N^{(1)}$ and $N^{(4)}$, and the global measures of entanglement defined in [27].

One of the simplest superpositions of the form (12) is the Bell-like state

$$\Psi = \cos(\theta)\, u_A(p,+) \otimes u_B(q,-) + \sin(\theta)\, u_A(p,-) \otimes u_B(q,+), \tag{14}$$

where through this paper we describe the bispinors $u(p,s)$ polarized in the e_z direction, such that $|\chi_{\pm}\rangle_S = |\pm\rangle_S$ in Eq. (8). The measures (13) for the state Ψ are depicted in Fig. 1 for particles with opposite momentum and rapidity ξ_0 in the center of momentum (CoM) frame, i.e. $\boldsymbol{p} = -\boldsymbol{q} = (0, 0, m\sinh(\xi_0))$, as function of the superposition parameter θ. The entanglement measures are maximed for $\theta = \pi/4$, and $3\pi/4$, which corresponds to a maximal superposition.

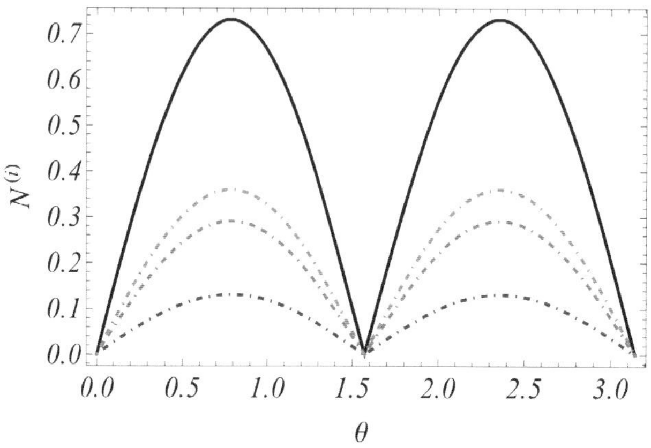

Figure 1. Mean negativities $N^{(i)}$ (13) of the state (14) in the CoM frame $\boldsymbol{p} = -\boldsymbol{q}$ as function of the superposition angle θ. The initial rapidity of the particles is $\xi_0 = \text{arctanh}(E_p/|\boldsymbol{p}|) = 1/2$ and the curves correspond to $N^{(1)}$ (black solid line), $N^{(2)}$ (red dashed line), $N^{(3)}$ (blue dot-dashed line) and $N^{(4)}$ (magenta dotted line). All quantifiers are maximed for maximal superpositions, for $\theta = \pi/4$ and $3\pi/4$

3. Lorentz boost of entangled bispinors

Once the general framework is set, we now describe the effects of frame transformations on the entanglement content of a two bispinoral state. For a Lorentz transformation Λ relating two frames $\mathcal{S}$ and $\mathcal{S}'$

$$(x')^\mu = \Lambda^\mu_{\ \nu} x^\nu, \tag{15}$$

the covariant form of (1) is invariant if (for $\hat{\boldsymbol{\gamma}} = \hat{\beta}\hat{\boldsymbol{\alpha}}$ and $\hat{\gamma}^0 = \hat{\beta}$)

$$(\hat{\gamma}^\mu p_\mu - m\hat{I})\psi(x) = 0 \rightarrow ((\hat{\gamma}')^\mu p'_\mu - m\hat{I})\psi'(x') = 0, \tag{16}$$

and thus the bispinor $\psi(x)$ transform as

$$\psi(x) \rightarrow \psi'(x') = \hat{S}[\,\Lambda\,]\psi(\Lambda^{-1}x'), \tag{17}$$

where $\hat{S}[\,\Lambda\,]$ is the representation of the Lorentz transformation acting on the bispinor space. For two inertial frames moving with respect to each other at a constant speed $\boldsymbol{v} = |\,\boldsymbol{v}\,|\,(e_x, e_y, e_z)$, the transformation Λ, a Lorentz boost, is given by

$$[\Lambda(\omega)]_{ij} = \delta_{ij} + (\cosh(\omega) - 1)\, n_i\, n_j, \quad [\Lambda(\omega)]_{i0} = [\Lambda(\omega)]_{0i} = \sinh(\omega)\, n_i, \quad [\Lambda(\omega)]_{00} = \cosh(\omega),$$

where $\boldsymbol{n} = (n_x, n_y, n_z)$ is the unity vector specifying the direction of the boost and ω is the *rapidity*

$$\omega = \text{arctanh}\left[\frac{|\boldsymbol{v}|}{\sqrt{1 - |\boldsymbol{v}|^2}}\right]. \tag{18}$$

The representation of a boost in the bispinor space $\hat{S}[\,\Lambda(\omega)\,]$ is explicitly given by [11, 19]

$$\hat{S}[\Lambda(\omega)] = \exp\left[-\frac{\omega}{2}\boldsymbol{n}\cdot\hat{\boldsymbol{\alpha}}\right] = \cosh\left(\frac{\omega}{2}\right)\hat{I} - \sinh\left(\frac{\omega}{2}\right)\boldsymbol{n}\cdot\hat{\boldsymbol{\alpha}}, \tag{19}$$

and bispinor u therefore transforms as

$$u(p,s) \quad \rightarrow \quad \frac{1}{\sqrt{\cosh(\omega)}}\hat{S}[\Lambda(\omega)]u(p,s). \tag{20}$$

In particular, for two successive boosts with rapidities ω_1 and ω_2 in the same direction one has

$$\hat{S}[\Lambda(\omega_2)]\hat{S}[\Lambda(\omega_1)] = \hat{S}[\Lambda(\omega_1+\omega_2)], \tag{21}$$

which can be explored to construct the general solutions (4) by performing a Lorentz boost from particles in the rest frame to the frame where the particle has momentum $\boldsymbol{p}$. In this case, the rapidity of the boost to construct the solutions from the rest frame is given by $\omega = |\boldsymbol{p}|/m$. For boosts parallel to the particles momentum, the transformation between $\mathcal{S}$ and $\mathcal{S}'$ corresponds to just a change on the momentum of the state.

The action of $\hat{S}$ on superpositions of bispinors were previously considered mainly to discuss the definition of spin operators in relativistic quantum mechanics [15]. Due to the form of $\mathcal{S}$, some possible definitions of spin operators, apart from the usual Pauli-Dirac operator $1/2\,\hat{\Sigma}$, other equally meaningful spin operators can be proposed. In particular, the Fouldy-Wouthuysen (FW) spin operator was considered to define a covariant spin reduced density matrix for Dirac bispinors as well as to define a proper position operator [17, 29]. Additionally, considering superpositions of bispinors, as those of generic form (12), the effects of using different spin operators, with particular focus on eigenstates of the FW operator, were considered in the context of Bell's inequality [17].

On the other hand, we shall consider the effects of boosts (19) on the entanglement content of two bispinoral particles. In terms of two-qubit operators (19) reads

$$\hat{S}[\Lambda(\omega)] = \cosh\left(\frac{\omega}{2}\right)\hat{I}^{(P)}\otimes\hat{I}^{(S)} - \sinh\left(\frac{\omega}{2}\right)\boldsymbol{n}\cdot(\,\hat{\sigma}_x^{(P)}\otimes\hat{\boldsymbol{\sigma}}^{(S)}\,), \tag{22}$$

and , from now on, we consider a simplified analysis in which, in the unboosted frame $\mathcal{S}$, the joint state is prepared as Ψ (14) and address individually the effects of boosts parallel and perpendicular to the particle momenta.

When a Lorentz transformation Λ is performed, the state (14) transforms as

$$\begin{aligned}\Psi \rightarrow \Psi' \;&=\; (\hat{S}[\Lambda]\otimes\hat{S}[\Lambda])\Psi\\ &=\; \cos(\theta)\,(\,\hat{S}[\Lambda]u(p,+)\,)\otimes(\,\hat{S}[\Lambda]u(q,-)\,)+\sin(\theta)\,(\,\hat{S}[\Lambda]u(p,-)\,)\otimes(\,\hat{S}[\Lambda]u(q,+)\,),\end{aligned}$$

and the changing on entanglement between different partitions driven by the boost can be described by considering different partial transpositions and partial traces of the Lorentz transformed density matrix $\rho' = \Psi'(\,\Psi'\,)^{\dagger}$. In particular because the state is pure, negativities of the type $\mathcal{N}^{\,i;\,j\,k\,l}$ and $\mathcal{N}^{\,i\,j;\,k\,l}$ shall exhibit the same behavior as the linear entropies $1-\mathrm{Tr}_i[\rho_i^2]$ and $1-\mathrm{Tr}_{i,\,j}[\rho_{i\,j}^2]$. For example, the entanglement in the bi-partition $\{P_1,\,S_1;\,P_2,\,S_2\}$, which shall be called hereafter **particle-particle** entanglement, has the same behavior of the linear entropy $E_L = 2(1-\mathrm{Tr}[\rho_1^2])$, where

$$\rho_1 = \mathrm{Tr}_{P_2,\,S_2}[\Psi\,\Psi^{\dagger}] = \cos^2(\theta)\,u(p,+)u^{\dagger}(p,+)+\sin^2(\theta)\,u(p,-)u^{\dagger}(p,-), \tag{23}$$

and through the orthogonality relations (6)

$$E_L = \sin^2(\,2\theta\,). \tag{24}$$

Since

$$(\,\hat{S}[\Lambda]u(p,\alpha)\,)^{\dagger}\,(\,\hat{S}[\Lambda]u(p,\beta)\,) = \delta_{\alpha\,\beta}, \tag{25}$$

with respect to $\mathcal{S}'$ one has the reduced density operator to the particle 1 given by

$$\begin{aligned} \rho_1' &= \mathrm{Tr}_{P_2,\,S_2}[\Psi'\,(\,\Psi'\,)^\dagger] \\ &= \cos^2(\theta)\,(\,\hat{S}[\Lambda]u(p,+)\,)\,(\,\hat{S}[\Lambda]u(p,+)\,)^\dagger + \sin^2(\theta)\,(\,\hat{S}[\Lambda]u(p,-)\,)(\,\hat{S}[\Lambda]u(p,-)\,)^\dagger, \end{aligned} \tag{26}$$

and, due to (25), the same result (24) holds. Therefore entanglement between all DoFs of particle 1 and all DoFs of particle 2 is invariant under Lorentz boosts. We point here that, although a specific state was considered, the invariance property derived here holds for states with no momentum superposition, i.e. the bispinors on the superposition associated with a given particle have all the same momentum, which have the general form

$$\frac{1}{\sqrt{N}} \sum_{\alpha=\pm,\beta=\pm} c_i\, u(p,\alpha) \otimes u(q,\beta). \tag{27}$$

Despite the invariance of particle-particle entanglement, other bi-partitions may have non-invariant entanglement, as is now addressed for boosts parallel and perpendicular to the $\boldsymbol{e}_z$ direction.

3.1. Parallel boost

The simplest boost framework can be constructed by considering the reference frame $\mathcal{S}$ as the rest frame of particle 1 while particle 2 is moving with rapidity $\xi_0 = \mathrm{arcsinh}\,(|\boldsymbol{q}|/m)$ in the $-e_z$ direction. The 4-momenta p and q of particle 1 and 2 with respect to $\mathcal{S}$ are given by:

$$p = (m, 0, 0, 0) \qquad q = (m\cosh{(-\xi_0)}, 0, 0, m\sinh{(-\xi_0)}). \tag{28}$$

The frame $\mathcal{S}'$ in this setup moves with rapidity ω in the direction $-e_z$, such that the momenta with respect to $\mathcal{S}$ reads

$$p' = (m\cosh{(\omega)}, 0, 0, m\sinh{(\omega)}) \qquad q' = (m\cosh{(\omega-\xi_0)}, 0, 0, m\sinh{(\omega-\xi_0)}), \tag{29}$$

and one notices that for a boost $\omega = \xi_0$, $\mathcal{S}$ is the rest frame of particle 2, while for a boost $\omega = \xi_0/2$, $\mathcal{S}'$ is the CoM of the system. Additionaly, if $\xi_0 = 0$ then the particles are in rest with respect to each other. Figure 2 depicts pictorically the framework described.

The boost $\hat{S}[\Lambda]$ relating $\mathcal{S}$ to $\mathcal{S}'$ is in the same direction of $\boldsymbol{q}$, and therefore it only changes the momentum of the corresponding bispinor, i.e.

$$u'(q', s) = u(q', s). \tag{30}$$

In this framework, one can explicitly compute the following negativities

$$\begin{aligned} \mathcal{N}^{S_1;P_1,\,P_2,S_2}[\Psi'] &= \mathcal{N}^{S_2;P_1,\,P_2,S_1}[\Psi'] = |\sin 2\theta|, \\ \mathcal{N}^{P_1\,P_2;S_1\,S_2}[\Psi'] &= \sqrt{1-\mathrm{sech}^2\,\omega\,\mathrm{sech}^2\,(\,\omega-\xi_0\,)}\,\frac{|\sin 2\theta|}{3}, \quad \mathcal{N}^{P_1;P_2\,S_1\,S_2}[\Psi'] = |\,\tanh\omega\,\sin 2\theta\,|, \\ \mathcal{N}^{S_1;S_2}[\Psi'] &= \mathrm{sech}\,\omega\,\mathrm{sech}\,(\,\omega-\xi_0\,)\,|\sin 2\theta|, \quad \mathcal{N}^{P_2;P_1\,S_1\,S_2}[\Psi'] = |\,\tanh{(\,\omega-\xi_0\,)}\,\sin 2\theta\,|, \end{aligned} \tag{31}$$

The partitions $\{S_1; P_1,\, P_2, S_2\}$ and $\{S_2; P_1,\, P_2,\, S_1\}$ exhibit invariant entanglement. Other bi-partitions are non-invariant as shown in Fig. 3, which depicts the negativities (31) for a maximal superposition $\theta = \pi/4$ as function of the boost rapidity ω for $\xi_0 = 0$ (left plot) and $\xi_0 = 1$ (right plot). While entanglement between parity 1 (2) and the rest of the subsystems increases with the

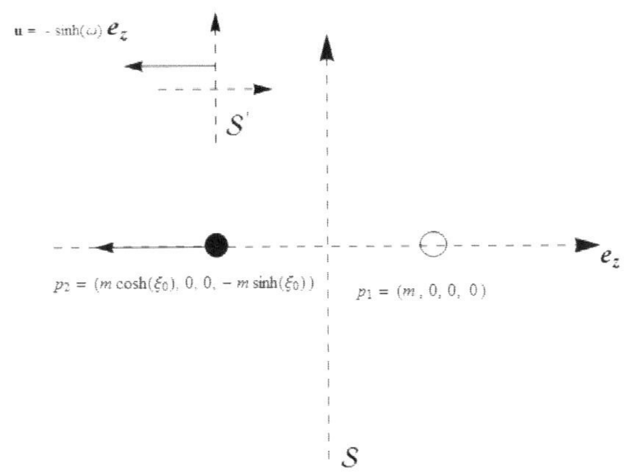

Figure 2. In the parallel Boost framework $\mathcal{S}$ corresponds to the rest frame of particle 1, and in which particle 2 moves with rapidity ξ_0 in the direction $-e_z$. The frame $\mathcal{S}'$ moves with respect to $\mathcal{S}$ with rapidity ω in the $-e_z$ direction.

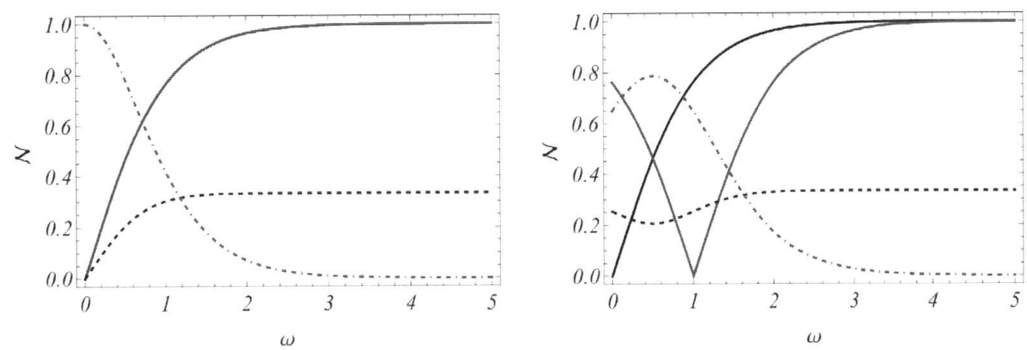

Figure 3. Negativities (31) - $\mathcal{N}^{P_1;P_2\,S_1\,S_2}$ (black solid line), $\mathcal{N}^{P_2;P_1\,S_1\,S_2}$ (blue solid line), $\mathcal{N}^{S_1;S_2}$ (blue dot-dashed line) and $\mathcal{N}^{P_1\,P_2;S_1\,S_2}$ (black dashed line) as function of the boosts rapidity ω, for $\xi_0 = 0$ (left plot) and $\xi_0 = 1$ (right plot), and for $\theta = \pi/4$. In the situation where both particles are at rest with respect to each other, entanglement exhibit a monotonous behavior. While entanglement between parities and the rest of subsystems, and between both parities and both spins increases with the rapidity, the spins DoFs disentangle for high speed boosts. A similar behavior is shown when one of the particles moves with respect to the other in $\mathcal{S}$, although in this case entanglement is not a monotonous function of the rapidity. In particular, for $\omega = \xi_0$, parity 2 disentangles for the rest of the system as for such boost $\mathcal{S}'$ is the rest frame of particle 2. Additionally, spin-spin entanglement is maximum at the CoM $\omega = \xi_0/2$, for which the entanglement between both parities and both spins reaches its minimum.

boost rapidity, spin-spin entanglement vanishes in the limit $\omega \to \infty$, being a non-monotonous function of ω if the particles are not in rest with respect to each other, and exhibiting a maximum in the CoM. Additionally $\mathcal{N}^{P_2;P_1\,S_1\,S_2}$ exhibit a zero at $\omega = \xi_0$, which is expected since this point corresponds to the rest frame of particle 2. In this framework parity-parity entanglement vanishes.

The other partitions also have non-invariant negativities. For instance the negativity $\mathcal{N}^{P_i;S_1\,S_2}$, quantifying the entanglement between parity i and both spins is a non-monotonous function of the boost rapidity vanishing for $\omega \to \infty$ and in the frame where the i-th particle is at rest. The mean partition entanglements are thus not invariant, as shown if Fig. 4 which depicts the variations of the mean negativites defined in (13), $\Delta N^{(i)} = N^{(i)}[\Psi'] - N^{(i)}[\Psi]$, for the different types of bi-partitions and for the same set of parameters of Fig. 3. The mean negativity over the partitions of the type $\{i;\ j,\ k,\ l\}$ always increases under boosts. For boosts with $\omega < \xi_0$, the mean entanglement in partitions of the type $\{i;\ j,\ k,\ l\}$ and$\{i;\ j\}$ while the mean entanglement in the partition $\{i;\ j,\ k\}$ decreases. When the boost has a rapidity bigger than the rapidity of particle 2, one observe an increasing on the entanglement of partition $\{i;\ j,\ k,\ l\}$ and $\{i;\ j,\ k\}$, while the other two partitions exhibit a degradation in the mean entanglement. Among all partitions, $\{i;\ j,\ k\}$ has the smaller entanglement variation due to the boost.

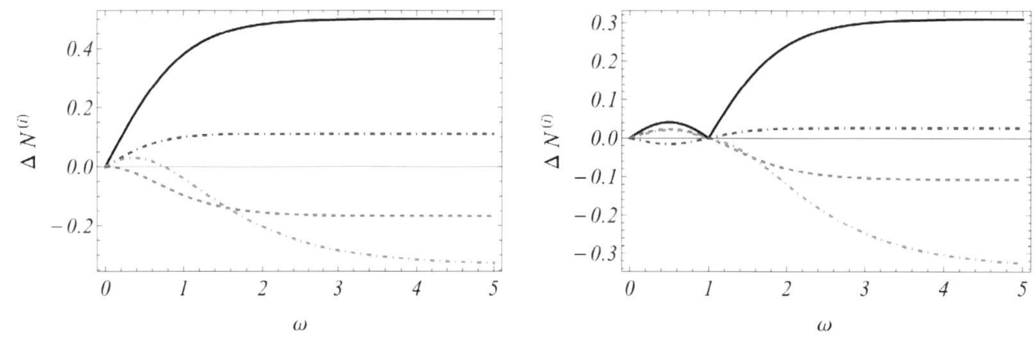

Figure 4. Variation of mean Negativities $\Delta N^{(i)} = N^{(i)}[\Psi'] - N^{(i)}[\Psi]$ as function of the boost rapidity ω, for $\xi_0 = 0$ (left plot) and $\xi_0 = 1$ (right plot), and for $\theta = \pi/4$. The curves correspond to $\Delta N^{(1)}$ (black solid line), $N^{(2)}$ (red dashed line), $N^{(3)}$ (blue dot-dashed line) and $N^{(4)}$ (magenta dotted line). One notice that all differences vanishes for $\omega = \xi_0$, when a Boost to the reference frame of particle 2 is considered. Moreover, the average $N^{(4)}$ exhibit an invariance point even if both particles are at rest with respect to each other, and $N^{(1)}$ always increases with the boost.

3.2. Perpendicular boost

When a boost is performed in a direction different from the particles momentum, the components of the bispinor get mixed. In the perpendicular boost framework, $\mathcal{S}$ corresponds to the CoM in which the 4-momenta p and q reads

$$p = (m\cosh(\xi_0), 0, 0, m\sinh(\xi_0)) \qquad q = (m\cosh(-\xi_0), 0, 0, m\sinh(-\xi_0)). \tag{32}$$

The inertial frame $\mathcal{S}'$ moves in the e_x direction with rapidity ω with respect to $\mathcal{S}$ and the transformed 4-momenta are given by

$$\begin{aligned} p' &= (m\cosh(\xi_0)\cosh(\omega), m\cosh(\xi_0)\sinh(\omega), 0, m\sinh(\xi_0)) \\ q' &= (m\cosh(-\xi_0)\cosh(\omega), m\cosh(-\xi_0)\sinh(\omega), 0, m\sinh(-\xi_0)). \end{aligned} \tag{33}$$

The bispinors transforms as

$$\begin{aligned} u(p,\pm) \to u' &= \frac{1}{\sqrt{\cosh\omega}}\left[\cosh\left(\frac{\omega}{2}\right)\hat{I} - \sinh\left(\frac{\omega}{2}\right)\hat{\alpha}_x\right] u(p,\pm) \\ &= \frac{1}{\sqrt{\cosh\omega}}\begin{bmatrix} \cosh\left(\frac{\omega}{2}\right) f(p)\chi_\pm \mp \sinh\left(\left(\frac{\omega}{2}\right)\right) g(p)\chi_\mp \\ \pm\cosh\left(\frac{\omega}{2}\right) g(p)\chi_\pm - \sinh\left(\frac{\omega}{2}\right) f(p)\chi_\mp \end{bmatrix}, \end{aligned} \tag{34}$$

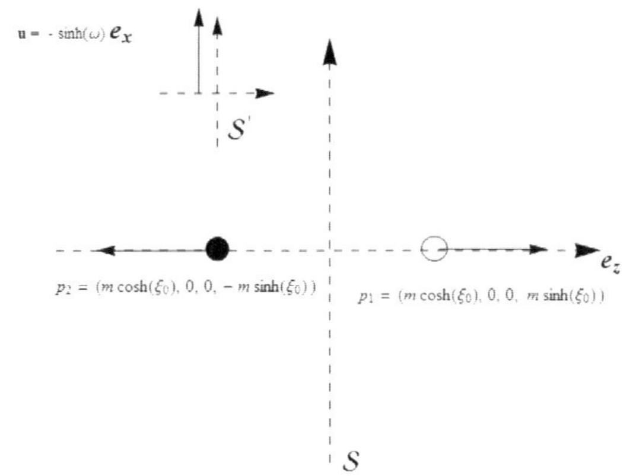

Figure 5. Perpendicular Boost framework. From the CoM frame $\mathcal{S}$, one performs a Lorentz boost with rapidity ω in the direction e_x, perpendicular to the particles momenta. Different from the previous scenario, the boost not only changes the momentum of the bispinors but also mixed its components, changing the entanglement between the different DoFs of the system.

where the shorthand notation $f(p) = \sqrt{\frac{E_p+m}{2E_p}}$, $g(p) = \frac{p}{\sqrt{2E_p(E_p+m)}}$ was adopted. Figure 5 depicts schematically the Boost scenario.

Figure 6 shows the same negativities depicted in Fig. 3 but for the boost from the CoM in the e_x direction . The qualitative behavior of the functions are the same as those depicted in the right plot of Fig. 3. Similar to the parallel boost from the common rest frame, the entanglement between spins is degraded by the boost and for boosts with high rapidity the spins of the particles are completely separable. On the other hand, the entanglement between one intrinsic parity and the other DoFs and between both parities and both spins are increasing functions of the boost rapidity, suggesting that the boost distribute the spin-spin entanglement among other partitions. Additionally, the correlations between one spin and the other DoFs are invariant under such boosts.

Different from the parallel boost, the mean negativities exhibit a monotonous behavior as function of the rapidity for a perpendicular boost, as depicted in Fig. 7 that shows the variations of the mean negativities $\Delta N^{(i)}$ as function of ω for $\theta = \pi/4$ considering that $\mathcal{S}_0$ is the CoM frame for which is supposed that $\xi_0 = 1/2$. The partitions $\{i;\, j,\, k,\, l\}$ and $\{i;\, j,\, k\}$ have increasing mean entanglement, while the other two partitions lost entanglement due to the boost. The behaviors exhibit in Figs. 4 and 7 suggest that a boost has an overall effect of increasing the mean entanglement in the partition $\{i;\, j,\, k,\, l\}$, that can be used as a global measure of entanglement for such pure states [27]. This general behavior is also observed for anti-symmetric states and more general boosts [30]. A part from this general behavior, there is no compensation between degradation and increasing of the mean partition entanglement considered.

4. Momentum superposition - Recovering the results in the Literature

The behavior of entanglement under Lorentz boost was also described for spin states in momentum superposition [2, 6, 7]. In the framework of the irreps of the Lorentz group, the simplest separable two particle state in a momentum superposition is given by [7]

$$|\phi\rangle = (\ \cos(\alpha)\,|\boldsymbol{p}\rangle_1 \otimes |\boldsymbol{q}\rangle_2 + \sin(\alpha)\,|\boldsymbol{q}\rangle_1 \otimes |\boldsymbol{p}\rangle_2\,) \otimes |\phi_{spin}\rangle, \tag{35}$$

where $|\phi_{spin}\rangle$ is a joint spin state, and $|\boldsymbol{p}\rangle_i$ is the momentum state of the particle i. The state (35) is separable between spin and the momenta, and it was shown that a Lorentz boost entangles

IOP Conf. Series: Journal of Physics: Conf. Series **1071** (2018) 012001 doi:10.1088/1742-6596/1071/1/012001

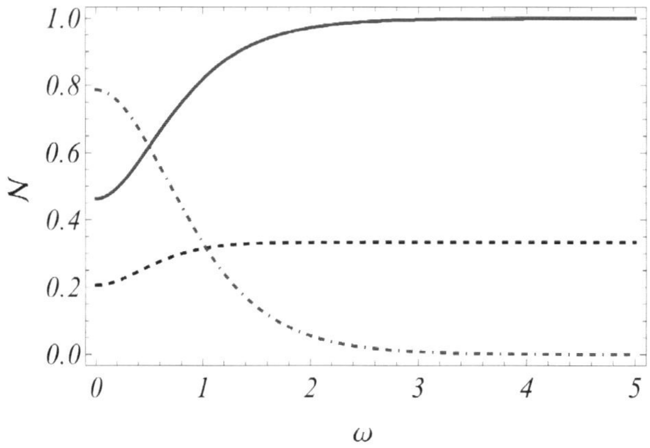

Figure 6. Negativities as function of the boost rapidity ω in the direction perpendicular to the momenta of the particles. The plot-styles are in correspondence with Fig. 3 and $\mathcal{S}$ corresponds to the CoM of the particles, in which $\xi_0 = 1/2$. Entanglement exhibit a monotonous behavior as function of ω, and while $\mathcal{N}^{P_{1(2)};P_{2(1)},S_1,S_2}$ and $\mathcal{N}^{P_1,P_2;S_1,S_2}$ increase with the boost rapidity, spin-spin entanglement is degraded. In the limit of high speed boost, S_1 and S_2 are separable, while the entanglement between parities and the other subsystems is maximum.

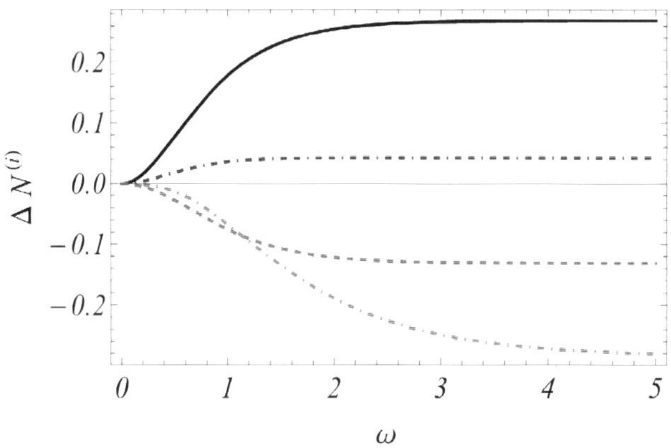

Figure 7. Variations of mean negativities $N^{(i)}$ as function of the boost rapidity ω for the perpendicular boost. The plot styles are in correspondence with those of Fig. 5. The variations on entanglement behave similar to those exhibited on parallel boost framework (see the left plot of Fig. 5). While the mean entanglement on the partitions $\{i;\, k,\, k,\, l\}$ and $\{i,\, j,\, k\}$ increase, the other two partitions have the entanglement degraded by the boost.

the spins and the momenta DoFs [2, 7] depending on the value of the Wigner rotation angle δ, which for boosts from the CoM and perpendicular to the particle momenta is given in terms of the initial rapidity and the boost rapidity by

$$\tan(\delta) = \frac{\sinh(\xi_0)\sinh(\omega)}{\cosh(\xi_0) + \cosh(\omega)}.$$

The momentum eigenstates $|\boldsymbol{p}\rangle$ have normalization given by

$$\int \frac{d^3p}{2E_p}\langle \boldsymbol{p}|\boldsymbol{q}\rangle = 1, \tag{36}$$

and, if one considers a particle with highly concentrate momentum distribution, for instance if the state $|\boldsymbol{p}\rangle$ is associated with a gaussian distribution centered around $\boldsymbol{p}$ with a very small width, the distributions of the opposite momenta $\boldsymbol{p} = -\boldsymbol{q}$ will have a negligible overlap. With this assumption, the momentum DoF introduced in (35) can be effectively described as dichotomic and treated as an additional qubit: $|\boldsymbol{p}\rangle \equiv |1\rangle$ and $|\boldsymbol{q}\rangle \equiv |0\rangle$. This simplified approximation shall be adopted in this section.

To describe transformation properties of quantum entanglement carried by pairs of physical particles in the context of a parity symmetric theory, such as the electron in QED, the Dirac bispinors must be used. For example, the Bell-like spin state with momentum superposition constructed with vectors belonging to the irreps of the Lorentz group

$$|\Phi\rangle = (\,\cos(\alpha)\,|\boldsymbol{p}\rangle_1 \otimes |\boldsymbol{q}\rangle_2 + \sin(\alpha)\,|\boldsymbol{q}\rangle_1 \otimes |\boldsymbol{p}\rangle_2\,) \otimes (\,\cos(\theta)\,|+\rangle \otimes |-\rangle + \sin(\theta)|-\rangle \otimes |+\rangle\,) \tag{37}$$

has its correspondent on the bispinor level constructed as the superposition

$$\begin{aligned}\Phi \;=\; & \frac{1}{N}\Big[\,\cos(\alpha)(\,\cos(\theta)\,u(p,+)\otimes(u(q,-)+\sin(\theta)\,u(p,-)\otimes u(q,+)\,)\otimes|\boldsymbol{p}\rangle_1\otimes|\boldsymbol{q}\rangle_2 \\ & +\;\sin(\alpha)(\,\cos(\theta)\,u(q,+)\otimes u(p,-)+\sin(\theta)\,u(q,-)\otimes u(p,+)\,)\otimes|\boldsymbol{q}\rangle_1\otimes|\boldsymbol{p}\rangle_2\Big]\end{aligned} \tag{38}$$

where N is the normalization factor, α is the superposition angle between the momenta and θ is the superposition angle between the spins. We emphasize that such state is a 6-qubit state as each particle carries 3-qubits: spin, intrinsic parity and (the discrete) momentum, and the description of its entanglement properties is even more involving than those presented for the state (14). We thus focus on recovering previous results presented on the literature and pointing out some striking differences arising from our treatment.

The behavior of spin-momentum entanglement for (37) under boosts can be reproduced by considering the projection of (38) into positive parity states. The consideration of such projections discards the contribution of negative parity components of the bispinors, reducing the state vector to two non-vanishing components which, when the parity DoF is traced out, corresponds to the framework usually adopted the above mentioned transformation properties of entanglement. Starting from the full density matrix $\rho = \Phi\Phi^\dagger$, the projection into positive parity is obtained by

$$\varrho = \frac{\hat{\Pi}^{(P)}_{+}\,\rho\,\hat{\Pi}^{(P)}_{+}}{\mathrm{Tr}[\hat{\Pi}^{(P)}_{+}\rho]}, \tag{39}$$

where $\hat{\Pi}^{(P)}_{+} = |+\rangle\langle+|_1 \otimes |+\rangle\langle+|_2 \otimes \hat{I}$ (with $\hat{I}$ the identity operator on the spins and momenta spaces). The entanglement between the spins and the momenta are evaluated through the reduced density matrix $\varrho_{S_1,p_1,S_2,p_2} = \mathrm{Tr}_{P_1,P_2}[\varrho]$. For example, to compute entanglement between the spin of particle 1 and all other 3 DoFs, one compute the negativity $\mathcal{N}^{S_1;\,p_1,\,S_2,\,p_2}$ by the partial transposition of $\varrho_{S_1,p_1,S_2,p_2}$ with respect to S_1. In the same fashion, the entanglement between all spins and all momenta is evaluated by $\mathcal{N}^{S_1,\,S_2;\,p_1,\,p_2}$. The variation of $\mathcal{N}^{S_1;\,p_1,\,S_2,\,p_2}$ and $\mathcal{N}^{S_1,\,S_2;\,p_1,\,p_2}$ under boosts perpendicular to the momenta, from the CoM frame, is depicted in Fig. 8 in function of the superposition parameters α and θ and for parameters such that $\delta = \pi/2$ (left plot) and $\delta = \pi/4$ (right plot). The behavior of spin-momentum entanglement under boosts obtained with the above procedure is exactly the same reported in the literature [7], with a characteristic egg-tray behavior.

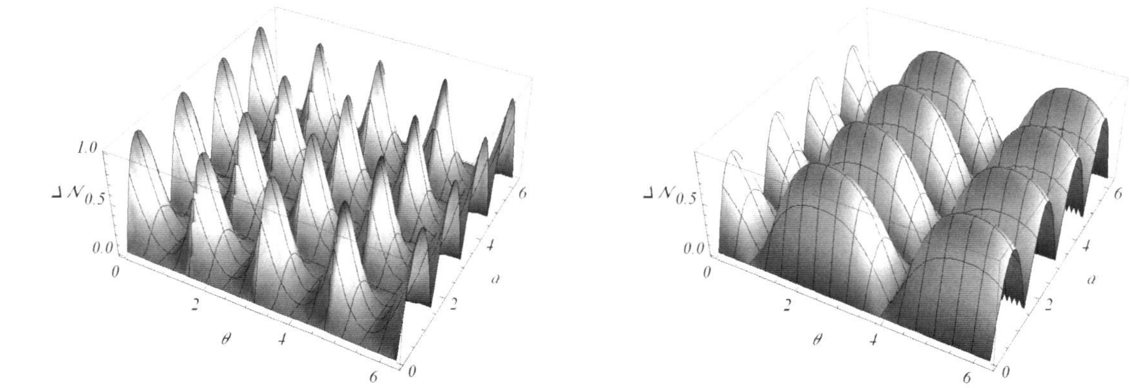

Figure 8. Entanglement in the partitions $\{S_1;\, p_1,\, S_2,\, p_2\}$ (left plot) and $\{S_1,\, S_2;\, p_1,\, p_2\}$ (right plot) for the positive parity projection (39) of the state (38) under a Lorentz boost from the CoM perpendicular to the momenta. The initial and the boost rapidities are chosen to correspond to Wigner angles $\delta = \pi/2$ (left plot) and $\delta = \pi/4$. The variation of quantum entanglement in this partitions is exactly the same reported in the literature for spin-momentum entanglement encoded in Bell-like states (37) belonging to the irreps of the Lorentz group under boosts [7].

Although the projection into positive parity states reproduces previous results, a correct characterization of spin-momentum entanglement in states such as (38) is obtained by tracing out the parity DoFs, that is, by calculating the appropriate negativity of the density matrix

$$\rho_{S_1,p_1,S_2,p_2} = \mathrm{Tr}_{P_1,P_2}[\rho]. \tag{40}$$

By construction, the state (38) is momentum-spin separable, and thus $\mathcal{N}^{S_1;\, p_1,\, S_2,\, p_2}[\rho_{S_1,p_1,S_2,p_2}] = \mathcal{N}^{S_1,\, S_2;\, p_1,\, p_2}[\rho_{S_1,p_1,S_2,p_2}] = 0$, but different from its counterpart (37), a boost does not create entanglement in such partitions for any perpendicular boost from the CoM. This is a striking difference from the usual treatment that need to be addressed if one wants to describe entanglement encoded in the DoFs of a physical particle, such as an electron, a proton, a muon etc. A boost does not create spin-momentum entanglement in a separable Bell-like state and, if one wants to use correlations of such DoFs for some practical purpose, such as cryptography, such correlations need to be present from the beginning, for example, they must have been created by some scattering or decay process (see for example [31]). The significant difference between the results obtained by the procedures set by (39) and (40) are also exhibited in other bipartitions of the system. For example, the entanglement encoded only between the spin DoFs, quantified by $\mathcal{N}^{S_1;\, S_2}$ depicted in Fig. 9, is invariant under boosts for (39), while for (40) the boost generates a degradation in spin-spin entanglement similar to those reported in Figs. 6,3.

5. Conclusions and perspectives

The behavior of entanglement encoded in spin states under Lorentz boosts has been the focus of recent research due to its importance in the implementation and characterization of physical protocols in setups involving reference frames relativistically moving with respect to each other [1, 2, 3, 4, 5, 6, 7, 8]. Usually, the quantum state considered is constructed with vectors belonging to the irreps of the Lorentz group, and the effects of Lorentz boosts are characterized by a rotation of the spin which depends on the momentum of the state [13] – a Wigner rotation. Although many interesting and insightful results were derived in this setup, when physical

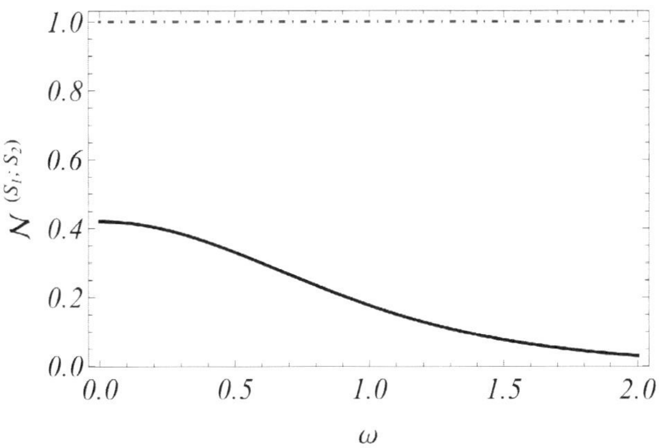

Figure 9. Variation of quantum entanglement between the spins of state (38), for $\theta = \pi/4$ and $\alpha = \pi/4$, for a boost with rapidity ω perpendicular to the CoM frame. The solid line corresponds to the proper procedure to compute spin-spin entanglement, via the density matrix (40) obtained by tracing the parity DoF. The dot-dashed line corresponds to the projection into positive parity (39) and corresponds to the framework usually considered in the literature. While by (40) one predicts a degradation of the spin-spin entanglement, the consideration of only the posity parity component of the bispinorial state would lead to the conclusion that spin spin is invariant.

particles created by QED processes, such as electrons and protons, are considered, the states must be described in terms of the irreps of the complete Lorentz group [12], the Dirac bispinors.

This paper was concerned with the behavior of quantum entanglement encoded in superpositions of Dirac bispinors under Lorentz boost. The transformation law of bispinors is given by the exponential of the generator of the boost and it was supposed that, in the unboosted frame particles have momentum in the e_z direction, and that the bispinors were helicity eigenstates. Due to the intrinsic spin-parity structure of Dirac bispinors [18], two particle states corresponds to four-qubit states, which can be entangled in several non equivalent ways. The entanglement between different bi-partitions of the systems was calculated in terms of the appropriate negativity, and as a global measures of entanglement it was adopted the average negativity in each type of bi-partition. For example, the spin-spin entanglement, which can be used for quantum information purposes, was calculated through the negativity of the density matrix reduced to the spin-spin subsystem, obtained by tracing all other subsystems. In such multipartite set, the focus was on the study of the effects of Lorentz boosts on the different entanglements to investigate how this type of transformation redistribute the entanglement initially encoded in a set of bipartition among the other possible types, which can be directly related, for example, to the non-locality of the correlations encoded on the bispinors [3, 4] and as a characterization of the informational content of fermionic particles in a full relativistic setup [16, 29]. It is important to stress that the averages negativities provides a qualitative approach to the multipartite entanglement encoded among the different partitions of the systems once the quantification of multipartite entanglement measure is still an open problem. Two different boosts scenarios were considered: boosts parallel to the particle momenta, and boosts from the CoM frame perpendicular to the particle momenta.

The first approach does not take into account momentum superposition. In this scenario each particle has a definite momentum and particle-particle entanglement, i.e. entanglement between all DoFs of one particle and all DoFs of the other particle, was shown to be invariant. On the other hand, other bi-partitions exhibit non-invariant behaviors. Spin-spin entanglement is degraded by the boost, being a non-monotonous function of the boost rapidity and exhibiting a maximum value at the CoM reference frame, in which entanglement between all intrinsic parities and all spins is minimal. As expected the entanglement between a given parity and all other DoFs vanish in the rest frame of one of the particles. For boosts perpendicular to the momenta, a similar behavior was observed for the negativities. The mean entanglement in bipartitions $\{i;\, j,\, k,\, l\}$ and $\{i;\, j,\, k\}$ increases under boosts, while $\{i,\, j;\, k,\, l\}$ and $\{i;\, j\}$ bi-partitions have a decreasing entanglement under Boosts. The bi-partition $\{i,\, j;\, k,\, l\}$ exhibit a non-monotonous behavior under boosts, decreasing for high-rapidity boosts.

Momentum superposition was also considered. Additionally to the superposition of the bispinors, it was supposed that the momenta of the particle were also superposed, and considering the simplified hypothesis of highly concentrated momenta distribution, allowing the effective description of the momentum DoF as an additional qubit, it was shown that, by considering the projection of the state into positive parity, i.e. by disregarding the negative parity components of the bispinors, it is possible to recover the egg-tray behavior of spin-momentum entanglement under boosts quoted in the literature [7]. Although the consistence with this previous known results, the proper way to evaluate quantum entanglement in such scenario is by tracing out the parity DoF without any projection, and we showed that, in this case, the spin-momentum entanglement is an invariant quantity. Such striking difference between both methods is also present in the spin-spin entanglement which, for the correct method of tracing the parity DoFs, exhibit the same degradation observed previously, while for the projection method, this correlation is invariant.

Considering that in the present literature quantum entanglement in relativistic scenarios is mostly described in the light of the irreps of the Poincaré group, our results set a new framework to discuss the transformation properties of quantum correlations in a complete covariant scenario, and are a fruitful and interesting addition to previous constructions that consider Dirac bispinors [15, 16, 17]. Furthermore, given that bispinors naturally appear in QED processes, such as scatterings and creation/annihilation processes [14, 32], the framework developed in our paper can be further extended to the description of quantum entanglement in such context, which has also attracted recent attention [33].

Acknowledgments - The work of AEB is supported by the Brazilian Agencies CAPES (grant 15/05903-4), FAPESP (grant 15/05903-4) and CNPq (grant 300809/2013-1). The work of VASVB is supported by the Brazilian Agency CAPES (grant 88881.132389/2016-1).

References

[1] Gingrich R M and Adami C 2002 Phys. Rev. Lett. **89** 270402.
[2] Peres A, Scudo P F and Terno D R 2002 Phys. Rev. Lett. **88** 230402.
[3] Ahn D, Lee H J, Moon Y H and Hwang S W 2003 Phys. Rev. A **67** 012103.
[4] Terashima H and Ueda M 2003 Int. J. Quant. Inf. **1** 93.
[5] Bartlett S D and Terno D R 2005 Phys. Rev. A **71** 012302.
[6] Jordan T F, Shaji A and Sudarshan E C G 2007 Phys. Rev. A **75** 022101.
[7] Friis N, Bertlmann R A, Huber M and Hiesmayr B C 2010 Phys. Rev. A **81** 042114.
[8] Palge V and Dunningham J 2012 Phys. Rev. A **85** 042322.
[9] Dunningham J, Palge V and Vedral V 2009 Phys. Rev. A **80** 044302; Palge V, Vedral V and Dunningham J. A. 2011 Phys. Rev. A **84** 044303.
[10] Jozsa R, Abrams D S, Dowling J P and Williams C P 2000 Phys. Rev. Lett. **85** 2010; Giovannetti V, Lloyd S and Maccone L 2001 Nature **412** 417; Yurtsever U and Dowling J P 2002 Phys. Rev. A **65** 052317.
[11] Fonda L and Ghirardi G C 1970 *Symmetry Principles in Quantum Physics* (New York: Marcel Dekker INC).
[12] Tung W K 2003 *Group Theory* (London: World Scientific Publishing).
[13] Wigner E 1939 Ann. Math. **40** 149
[14] Weinberg S 1995 *Quantum theory of Fields vol. 1* (New York: Cambridge University Press).
[15] Caban P, Rembielinski J and Wlodarczyk M 2013 Phys. Rev. A **88** 022119; Bauke H, Ahrens S, Keitel C H

and Grobe R 2014 New Journal of Physics **16** 043012; Saldanha P L and Vedral V 2012 New Journal of Physics **14** 023041.
[16] Alsing P M and Milburn G J [arXiv:quant-ph/020305].
[17] Choi T, Hur J and Kim J 2011 Phys. Rev. A **84** 012334; Choi T 2013 Journal of the Korean Physical Society **62** 1085.
[18] Mizrahi S S 2009 Physica Scripta **T135** 014007; Bernardini A E and Mizrahi S S 2014 Physica Scripta **89** 075105.
[19] Greiner W 2000 *Relativistic Quantum Mechanics: Wave equations* (Berlin: Springer).
[20] Horodecki R, Horodecki P, Horodecki M. and Horodecki K. 2009 Rev. Mod. Phys. **81** 865.
[21] Peres A 1996 Phys. Rev. Lett. **77** 1413.
[22] Vidal G and Werner R F 2002 Phys. Rev. A **65** 032314.
[23] Bittencourt V A S V, Mizrahi S S and Bernardini A E 2015 Annals of Physics **355** 35-47.
[24] Bittencourt V A S V and Bernardini A E 2016 Annals of Physics **364** 182.
[25] Bittencourt V A S V, Bernardini A E and Blasone M 2016 Phys. Rev. A **93** 053823.
[26] Bittencourt V A S V and Bernardini A E 2017 Phys. Rev. B **95** 195145.
[27] Rigolin G, de Oliveira T R and de Oliveira M C 2006 Phys. Rev. A **74** 022314.
[28] Blasone M, Dell'Anno F, De Siena S and Illuminati F 2008 Phys. Rev. A **77**, 062304.
[29] Céleri L C, Kiosses V and Terno D R 2016 Phys. Rev. A **94** 062115.
[30] Bittencourt V A S V, Bernardini A E and Blasone M 2018 arXiv:1801.00758[quant-ph].
[31] Pachos J and Solano E 2003 Quant. Info. Comm. **3** 115.
[32] Mandl F and Shaw G 2010 *Quantum Field Theory* (West Sussex: John Wiley & Sons)
[33] Peschanski R and Seki S 2016 Phys. Lett. B **758** 89; Ratzel D, Wilkens M and Menzel R 2017 Phys. Rev. A **95** 012101; Fan J, Deng Y and Huang Y-C 2017 Phys. Rev. D **95** 065017.

IOP Conf. Series: Journal of Physics: Conf. Series **1071** (2018) 012002 doi:10.1088/1742-6596/1071/1/012002

On the canonical quantization of the electromagnetic field and the emergence of gauge invariance

M. Blasone[1], E. Celeghini[2], P. Jizba[3], F. Scardigli[4,5] and G. Vitiello[1]

[1] Dipartimento di Fisica, Università di Salerno, Fisciano (SA) - 84084 Italy & INFN, Gruppo Collegato di Salerno
[2]Dipartimento di Fisica, Università di Firenze and INFN-Firenze, 50019 Sesto Fiorentino, Italy
[3]FNSPE, Czech Technical University in Prague, Břehová 7, 115 19 Praha 1, Czech Republic
[4]Dipartimento di Matematica, Politecnico di Milano, Piazza L.da Vinci 32, 20133 Milano, Italy
[5]Institute-Lorentz for Theoretical Physics, Leiden University, P.O. Box 9506, Leiden, The Netherlands

Abstract. In the framework of the canonical quantization of the electromagnetic field, we impose as primary condition on the dynamics the positive definiteness of the energy spectrum. This implies that (Glauber) coherent states have to be considered for the longitudinal and the scalar photon fields. As a result we obtain that the relation holds which in the traditional approach is called the Gupta-Bleuler condition. Gauge invariance emerges as a property of the physical states. The group structure of the theory is recognized to be the one of $SU(2)\otimes SU(1,1)$.

1. Introduction

The quest for a deterministic basis of present physical theories has led 't Hooft to propose a cellular automaton interpretation of Quantum Mechanics [1]: here the central concept is that of a complete set of operators commuting at all times, the *beables*, representing the true (ontological) variables of the system and allowing for a deterministic description. One possibility in such a framework, is represented by loss of information: it may happen that a deterministic theory has a dissipative character, so that, on a given time scale, the dynamics leads to a reduction of the number of degrees of freedom of the system, and the result is the emergence of ordinary quantum states with non-local character. Several examples of such a scheme have been discussed [1]–[6], in particular the system of two coupled damped–amplified harmonic oscillators (Bateman system)[7]–[11], turned out to be a paradigmatic one, in the sense that the quantum harmonic oscillator here emerges out of the dissipative dynamics of the original (deterministic) system [2, 4]. A common feature of these models, is the fact that the Hamiltonian of the (primordial) deterministic system is not positive–definite implying a spectrum unbounded from below: by imposing the constraint of positive definiteness, one then obtains a genuine quantum system with unitary evolution and a stable groundstate.

Motivated by such ideas, we consider in this paper the quantization procedure for the electromagnetic field in Lorenz gauge in the operatorial version, which historically was obtained in the 1950's by Gupta and Bleuler. We show that, by only resorting to the condition of positive definiteness of the Hamiltonian, i.e. without assuming the gauge invariance as a primary ingredient of the theory, we arrive at a definition of the physical states of the theory in terms of coherent states of longitudinal and scalar photons, which turn out to be equivalent to the traditional Gupta–Bleuler states. We also show that the theory has the group structure of $SU(2) \otimes SU(1,1)$.

The structure of the paper is as follows. In Section 2, we review the traditional quantization procedure for the electromagnetic field, in Section 3 we present our results. Section 4 is devoted to conclusions and perspectives.

2. Canonical quantization of electromagnetic field

In this Section we summarize briefly the main steps in the canonical quantization of the electromagnetic field in the traditional approach. In our presentation we closely follow Ref.[12].

Canonical quantization of the Maxwell field in the Lorenz gauge requires the introduction of a gauge fixing term leading to the so-called Fermi Lagrangian density [12]

$$\mathcal{L} \;=\; -\frac{1}{4}F^{\mu\nu}F_{\mu\nu} - \frac{1}{2}\zeta\left(\partial_\mu A^\mu\right)^2 \tag{1}$$

Equations of motions following from such Lagrangian are

$$\Box A^\mu - (1-\zeta)\partial_\mu\left(\partial_\sigma A^\sigma\right) \;=\; 0 \tag{2}$$

If we restrict to the case $\zeta = 1$ (Fermi-Feynman gauge), Lagrangian and equations of motion assume the simple form:

$$\mathcal{L} \;=\; -\frac{1}{2}\partial_\mu A_\nu \partial^\mu A^\nu \tag{3}$$

$$\Box A^\mu \;=\; 0 \tag{4}$$

The Fourier expansion of the A^μ field can be written as

$$A^\mu(x) \;=\; \int \frac{d^3k}{\sqrt{2\omega_k(2\pi)^3}} \sum_{\lambda=0}^{3}\left(a_{\mathbf{k},\lambda}\,\epsilon^\mu(\mathbf{k},\lambda)e^{-ik\cdot x} + a^*_{\mathbf{k},\lambda}\,\epsilon^\mu(\mathbf{k},\lambda)e^{ik\cdot x}\right), \tag{5}$$

where $\omega_k = k_0 = |\mathbf{k}|$ because of the vanishing photon mass. Moreover we have the orthogonality four-dimensional relation

$$\epsilon_\mu(\mathbf{k},\lambda)\epsilon^\mu(\mathbf{k},\lambda') = g_{\lambda\lambda'}\,, \tag{6}$$

and the completeness relation

$$\sum_{\lambda=0}^{3} g_{\lambda\lambda}\epsilon_\mu(\mathbf{k},\lambda)\epsilon_\nu(\mathbf{k},\lambda) = g_{\mu\nu}\,. \tag{7}$$

By introducing the conjugate momenta

$$\pi_\mu \;=\; \frac{\partial\mathcal{L}}{\partial(\partial_0 A^\mu)} = -\partial^0 A_\mu \tag{8}$$

we obtain the Hamiltonian density

$$\mathcal{H} \;=\; -\frac{1}{2}\pi_\mu\pi^\mu + \frac{1}{2}\partial_k A_\nu \partial^k A^\nu \tag{9}$$

IOP Conf. Series: Journal of Physics: Conf. Series **1071** (2018) 012002 doi:10.1088/1742-6596/1071/1/012002

which can be rewritten as

$$\mathcal{H} = \frac{1}{2}\sum_{k=1}^{2}\left[(\dot{A}^k)^2 + (\nabla A^k)^2\right] - \frac{1}{2}\left\{\left[(\dot{A}^0)^2 + (\nabla A^0)^2\right] - \left[(\dot{A}^3)^2 + (\nabla A^3)^2\right]\right\}. \tag{10}$$

Quantization is achieved by imposing commutation relations for the operators A^μ and π^μ:

$$[A^\mu(\mathbf{x},t), \pi^\nu(\mathbf{y},t)] = ig^{\mu\nu}\delta^3(\mathbf{x}-\mathbf{y}) \tag{11}$$

or

$$\left[A^\mu(\mathbf{x},t), \dot{A}^\nu(\mathbf{y},t)\right] = ig^{\mu\nu}\delta^3(\mathbf{x}-\mathbf{y}) \tag{12}$$

By use of the field expansion (5), we obtain the commutation relations for the ladder operators:

$$\left[a_{\mathbf{k}',\lambda'}, a^\dagger_{\mathbf{k},\lambda}\right] = -g_{\lambda\lambda'}\delta^3(\mathbf{k}'-\mathbf{k}) \tag{13}$$

and the Hamiltonian becomes:

$$H = \int d^3\mathbf{k}\,\omega_k\left(\sum_{\lambda=1}^{3} a^\dagger_{\mathbf{k},\lambda}a_{\mathbf{k},\lambda} - a^\dagger_{\mathbf{k},0}a_{\mathbf{k},0}\right) \tag{14}$$

At this point, the Lorenz gauge is implemented as a condition (the Gupta–Bleuler condition) on the Hilbert space, defining the physical states $|\Phi\rangle$ as those for which

$$\partial^\mu A^{(+)}_\mu|\Phi\rangle = 0 \tag{15}$$

where the "+" denotes the positive frequency part of the field. The condition (15) gives

$$(a_{\mathbf{k},0} - a_{\mathbf{k},3})|\Phi\rangle = 0 \tag{16}$$

implying that

$$\langle\Phi|\left(a^\dagger_{\mathbf{k},0}a_{\mathbf{k},0} - a^\dagger_{\mathbf{k},3}a_{\mathbf{k},3}\right)|\Phi\rangle = 0. \tag{17}$$

It can be also shown [12] that the physical states can be generated from the purely transverse states $|\Phi_T\rangle$ in the following way:

$$|\Phi\rangle = R|\Phi_T\rangle \tag{18}$$

where

$$R = 1 + \int d^3k c(\mathbf{k})L^\dagger_{\mathbf{k}} + \int d^3k d^3k' c(\mathbf{k})c(\mathbf{k}')L^\dagger_{\mathbf{k}}L^\dagger_{\mathbf{k}'} + \dots, \tag{19}$$

$L_{\mathbf{k}} \equiv a_{\mathbf{k},0} - a_{\mathbf{k},3}$ with $[L_k, R] = 0$, and in Eq.(19) the $c(\mathbf{k})$ coefficients can be freely chosen (see Ref.[12], Sec.7.4). The purely transversal states $|\Phi_T\rangle$ are those which do not contain any longitudinal or scalar photons:

$$a_{\mathbf{k},0}|\Phi_T\rangle = a_{\mathbf{k},3}|\Phi_T\rangle = 0. \tag{20}$$

Thus, for such states, the condition (16) is trivially satisfied. The physical states $|\Phi\rangle$ defined by the Gupta-Bleuler condition $L_{\mathbf{k}}|\Phi\rangle = 0$, are related to the transverse states $|\Phi_T\rangle$ through a gauge transformation:

$$\langle\Phi|A_\mu(x)|\Phi\rangle = \langle\Phi_T|A_\mu(x) + \partial_\mu\Lambda(x)|\Phi_T\rangle = \langle\Phi_T|A_\mu(x)|\Phi_T\rangle + \partial_\mu\Lambda(x). \tag{21}$$

where $\Lambda(x)$ is a c-number function satisfying $\Box\Lambda(x) = 0$ [12].

3. Algebraic structure of the Hamiltonian and coherent states

We take as starting point the Lagrangian density Eq.(1) with $\zeta = 1$ (Fermi-Feynman gauge):

$$\mathcal{L} \;=\; -\frac{1}{4}F^{\mu\nu}F_{\mu\nu} \;-\; \frac{1}{2}\left(\partial_\mu A^\mu\right)^2 . \tag{22}$$

We stress that we assume now the theory described by Lagrangian (22) as fundamental, contrarily to the usual pattern described in previous Section, in which one starts from the Maxwell theory exhibiting gauge invariance and then introduces the Lagrangian (1) in order to implement canonical quantization. In such a way, we privilege the possibility of a canonical formalism, with respect to the gauge invariance, which we do not assume as a fundamental concept.

The Hamiltonian density following from (22) is the one given in Eq.(9), leading, upon quantization, to the Hamiltonian operator (14).

Let us focus on the algebraic structure of the Hamiltonian (14). To this end, we define the following operators:

$$J_+ \,\equiv\, a^\dagger_{\mathbf{k},1}a_{\mathbf{k},2}\,, \quad J_- \,\equiv\, a^\dagger_{\mathbf{k},2}a_{\mathbf{k},1}\,, \quad J_3 \,\equiv\, \frac{1}{2}\left(a^\dagger_{\mathbf{k},1}a_{\mathbf{k},1} - a^\dagger_{\mathbf{k},2}a_{\mathbf{k},2}\right) \tag{23}$$

$$[J_+, J_-] \,=\, 2\,J_3\,, \quad [J_3, J_+] \,=\, +\,J_+\,, \quad [J_3, J_-] \,=\, -\,J_- \tag{24}$$

and

$$K_+ \,\equiv\, a^\dagger_{\mathbf{k},3}a_{\mathbf{k},0}\,, \quad K_- \,\equiv\, a^\dagger_{\mathbf{k},0}a_{\mathbf{k},3}\,, \quad K_3 \,\equiv\, \frac{1}{2}\left(a^\dagger_{\mathbf{k},0}a_{\mathbf{k},0} + a^\dagger_{\mathbf{k},3}a_{\mathbf{k},3}\right) \tag{25}$$

$$[K_+, K_-] \,=\, -2\,K_3\,, \quad [K_3, K_+] \,=\, +\,K_+\,, \quad [K_3, K_-] \,=\, -\,K_- \tag{26}$$

We see that the minus sign in the commutation relations for the a_0 operator reflects into the different algebraic structure for the K operators, closing the $su(1,1)$ algebra, with respect to the J operators, which close the $su(2)$ algebra.

The Casimir operators are $J_0 = \frac{1}{2}\left(a^\dagger_{\mathbf{k},1}a_{\mathbf{k},1} + a^\dagger_{\mathbf{k},2}a_{\mathbf{k},2}\right)$ and $K_0 = \frac{1}{2}\left(a^\dagger_{\mathbf{k},0}a_{\mathbf{k},0} - a^\dagger_{\mathbf{k},3}a_{\mathbf{k},3}\right)$.

In the above definitions, for notational simplicity, we have omitted the k index on the generators J and K.

Thus the hamiltonian (14) can be written as

$$H \;=\; \int d^3k\,\omega_k\,(J_0 \,-\, K_0)\,. \tag{27}$$

The fact that the Hamiltonian is written in terms of Casimir operators is remarkable [13] and guarantees the conservation of energy under transformations induced by the algebraic generators J and K above defined.

The Hamiltonian (27) is clearly not positive definite, therefore we define the physical states as those for which the Hamiltonian reduces to J_0, namely:

$$_{phys}\langle\psi|K_0|\psi\rangle_{phys} \;=\; 0. \tag{28}$$

Our main task is to characterize these physical states and analyze if they are equivalent to the physical states defined via the GB condition.

The problem is to find the explicit form of such physical states. We start by assuming them to be of the form:

$$|\psi\rangle_{phys} \;=\; \prod_{\mathbf{k}} |n_{\mathbf{k},1}\rangle_1 \otimes |n_{\mathbf{k},2}\rangle_2 \otimes |\alpha_{\mathbf{k}}\rangle \tag{29}$$

IOP Conf. Series: Journal of Physics: Conf. Series **1071** (2018) 012002 doi:10.1088/1742-6596/1071/1/012002

where we denote with subscript 1 and 2 the states for the transverse photons and with $|\alpha_{\mathbf{k}}\rangle$ a generic state (to be determined) for the longitudinal and scalar photons. Although for notational simplicity we do not make it explicit, it is understood that the tensorial product notation used in Eq. (29) is also adopted for the operators, e.g. $a^\dagger_{\mathbf{k},1} \equiv a^\dagger_{\mathbf{k},1} \otimes 1 \otimes 1$, etc..

We now concentrate on the state $|\alpha\rangle \equiv \prod_{\mathbf{k}} |\alpha_{\mathbf{k}}\rangle$, which is required to satisfy the above condition (28), i.e.

$$\langle\alpha| \left(a^\dagger_{\mathbf{k},0} a_{\mathbf{k},0} - a^\dagger_{\mathbf{k},3} a_{\mathbf{k},3}\right) |\alpha\rangle \;=\; 0. \tag{30}$$

We also require such a state to be different from the vacuum state $|0\rangle \equiv |0\rangle_3 \otimes |0\rangle_0$, which trivially fulfills such condition. Furthermore, we restrict to states of the form $|\alpha\rangle = |\alpha\rangle_3 \otimes |\alpha\rangle_0$ where $|\alpha\rangle_3$ and $|\alpha\rangle_0$ denote (Glauber) coherent states for a_3 and a_0:

$$a_{\mathbf{k},3}|\alpha\rangle_3 \;=\; \alpha_k|\alpha\rangle_3 \tag{31}$$

$$a_{\mathbf{k},0}|\alpha\rangle_0 \;=\; \alpha_k|\alpha\rangle_0 \tag{32}$$

with the same α_k, for any $\mathbf{k}$. Note that the choice of the coherent states defined in (31), (32) is dictated by the fact they are the only states (different from the vacuum $|0\rangle$) that satisfy Eq.(30). The coherent state generators are

$$G_3(\alpha) = \exp \sum_{\mathbf{k}} \left(\alpha_k^*\, a_{\mathbf{k},3} - \alpha_k\, a^\dagger_{\mathbf{k},3}\right) \tag{33}$$

$$|\alpha\rangle_3 \;=\; G_3^{-1}(\alpha)|0\rangle \tag{34}$$

$$a_{\mathbf{k},3}(\alpha) \equiv G_3^{-1}(\alpha) a_{\mathbf{k},3} G_3(\alpha) \;=\; a_{\mathbf{k},3} - \alpha_k \tag{35}$$

and

$$G_0(\alpha) = \exp \sum_{\mathbf{k}} \left(-\alpha_k^*\, a_{\mathbf{k},0} + \alpha_k\, a^\dagger_{\mathbf{k},0}\right) \tag{36}$$

$$|\alpha\rangle_0 \;=\; G_0^{-1}(\alpha)|0\rangle \tag{37}$$

$$a_{\mathbf{k},0}(\alpha) G_0^{-1}(\alpha) a_{\mathbf{k},0} G_\alpha \;=\; a_{\mathbf{k},0} - \alpha_k\,. \tag{38}$$

The sign difference in the commutator for a_0 and $a_0^\dagger$ has dictated the choice of the sign in the definition of the G_0 generator. We thus obtain

$$\left(a_{\mathbf{k},0} - a_{\mathbf{k},3}\right)|\alpha\rangle \;=\; 0 \tag{39}$$

$$\langle\alpha| \left(a^\dagger_{\mathbf{k},0} - a^\dagger_{\mathbf{k},3}\right) \;=\; 0\,, \tag{40}$$

which immediately extends to the physical states $|\psi\rangle_{phys}$. The condition (28) easily follows from these relations. Note that Eqs.(39), (40) are actually the Gupta-Bleuler condition Eq.(16) and its hermitan conjugate.

By resorting to the $L_{\mathbf{k}}$ and $L^\dagger_{\mathbf{k}}$ operators above defined, we note that they are invariant under α-translation: $G_0^{-1}(\alpha)G_3^{-1}(\alpha)\left(a_{\mathbf{k},0} - a_{\mathbf{k},3}\right)G_3(\alpha)G(\alpha) = \left(a_{\mathbf{k},0} - a_{\mathbf{k},3}\right)$ and similar for $L^\dagger_{\mathbf{k}}$.

Let us now consider the explicit form of the coherent states $|\alpha\rangle_0$ and $|\alpha\rangle_3$. By use of the Campbell-Baker-Hausdorff relation and the commutation relations Eq.(13), we obtain

$$|\alpha\rangle_3 \;=\; \exp\left(-\frac{1}{2}\int d^3k\, |\alpha_k|^2\right) \exp\left(\int d^3k\, \alpha_k\, a^\dagger_{\mathbf{k},3}\right)|0\rangle_3 \tag{41}$$

$$|\alpha\rangle_0 \;=\; \exp\left(\frac{1}{2}\int d^3k\, |\alpha_k|^2\right) \exp\left(-\int d^3k\, \alpha_k\, a^\dagger_{\mathbf{k},0}\right)|0\rangle_0 \tag{42}$$

and

$$|\alpha\rangle \equiv |\alpha\rangle_3 \otimes |\alpha\rangle_0 = \exp\left(\int d^3k\,\alpha_k\left(a^\dagger_{\mathbf{k},3} - a^\dagger_{\mathbf{k},0}\right)\right)|0\rangle \tag{43}$$

$$= \left(1 + \int d^3k\,(-\alpha_k)\,L^\dagger_{\mathbf{k}} + \int d^3k d^3k'\,\frac{(-\alpha_k)(-\alpha_{k'})}{2!}L^\dagger_{\mathbf{k}}L^\dagger_{\mathbf{k}'} + ...\right)|0\rangle \tag{44}$$

Thus a one-to-one correspondence exists between the coherent states above defined and those used in the Gupta-Bleuler quantization. We can therefore identify the physical states of the Gupta-Bleuler condition with the ones defined by the condition Eq.(28).

We can thus write

$$_{phys}\langle\psi|A_\mu(x)|\psi\rangle_{phys} = \langle\phi_T|A_\mu(x)|\phi_T\rangle + \partial_\mu\Lambda(x) \tag{45}$$

with $\Lambda(x)$ satisfying $\Box\Lambda = 0$.

In conclusion, only the transverse photons contribute to the energy in the coherent state representation $\{|\psi\rangle_{phys}\}$:

$$_{phys}\langle\psi|H|\psi\rangle_{phys} = \int d^3k\,\omega_k \sum_{\lambda=1,2} n_{\mathbf{k},\lambda}. \tag{46}$$

We conclude that, by choosing the coherent state representation $|\alpha\rangle$, we obtain Eq.(30) which in the Gupta-Bleuler method is instead postulated in order to impose the Lorenz gauge condition. In our case, we do not assume any gauge invariance from the beginning, rather we obtain it as an emergent property of the physical states.

4. Conclusions

In this paper we have explored the possibility to consider as fundamental the theory described by the Lagrangian density Eq.(22). On the other hand, the traditional approach is to start with the gauge invariant Maxwell theory and then proceed to the canonical quantization, the gauge invariance being assumed to be a fundamental ingredient of the theory. In our approach the canonical structure, namely the operatorial algebraic structure of the theory, is instead assumed to be fundamental, the gauge invariance being implied by the positive definiteness of the Hamiltonian operator. In short, we show that the primary condition that the energy spectrum has to be definite positive fully determines the structure of the physical states of the theory. Thus, instead of postulating it, we obtain as a result what in the traditional approach is called the Gupta-Bleuler condition.

Remarkably, the Hamiltonian for the electromagnetic field is shown to be given by the Casimir operators of the $su(2) \otimes su(1,1)$ algebras defining the theory. Such a result is a general feature of the dynamics, holding in the traditional approach and in our treatment, and makes evident the invariance of the spectrum under the of $SU(2) \otimes SU(1,1)$ group transformations.

We remark that, although we discussed only free modes, our conclusions are valid also in the case of interacting fields, because in perturbation theory one still works with annihilation and creation operators of free fields. The conclusions can be also extended to non abelian gauge theories (see Ref.[17]).

Finally, note that by acting with the generator G_3 and G_0 on the K_0 term in the Hamiltonian in Eq.(27), instead than on the states $|0\rangle_3$ and $|0\rangle_0$, we have

$$a^\dagger_{\mathbf{k},3}a_{\mathbf{k},3} - a^\dagger_{\mathbf{k},0}a_{\mathbf{k},0} \rightarrow \left(a^\dagger_{\mathbf{k},3}a_{\mathbf{k},3} - \alpha^*_k a_{\mathbf{k},3} - \alpha_k a^\dagger_{\mathbf{k},3}\right) - \left(a^\dagger_{\mathbf{k},0}a_{\mathbf{k},0} - \alpha^*_k a_{\mathbf{k},0} - \alpha_k a^\dagger_{\mathbf{k},0}\right), \tag{47}$$

i.e., under the shift transformations Eqs. (35) and (38), the longitudinal and the scalar photon modes are submitted to external forces (represented by the linear operator terms $a_{\mathbf{k},3}$ and $a_{\mathbf{k},0}$ and their hermitian conjugates, respectively). This shows that the two sets of longitudinal and scalar modes, each constitutes actually an open dissipative system on which an external force acts. The energy of each of the modes, for given $\mathbf{k}$, is proportional to $|\alpha_k|^2$ (cf. Eq. (30)). Both sets, combined together, form a closed system whose total energy is conserved and vanishes.

It is in our future plans to extend the analysis here presented in the direction of the 't Hooft scheme for quantization of deterministic systems [1]. In this context, the emergence of various symmetries, including local gauge symmetry, has been discussed in Refs.[14, 15]. Another development will be the formulation of these topics in the path-integral formalism, in the line of Ref.[16].

Acknowledgements

P.J. was supported by the Czech Science Foundation Grant No. 17-33812L. M.B., E.C. and G.V. were supported by INFN and Miur. F.S. was supported by a Delta-ITP (Leiden) visiting scholar fellowship.

References

[1] 't Hooft G 2016 *The Cellular Automaton Interpretation of Quantum Mechanics* (Springer Int. Publ.)
[2] Blasone M, Jizba P and Vitiello G 2001 *Phys. Lett. A* **287** 205
[3] Blasone M, Celeghini E, Jizba P and Vitiello G 2003 *Phys. Lett. A* **310** 393.
[4] Scardigli F 2007 *Found. of Phys.*, **37**, 1278.
[5] Blasone M, Jizba P, Scardigli F 2009 *J.Phys.Conf.Ser.* 174, 012034, arXiv:0901.3907 [quant-ph]
[6] Blasone M, Scardigli F, Jizba P and Vitiello G 2009 *Phys. Lett. A* **373** 4106.
[7] Bateman H 1931 *Phys. Rev.* **38** 815
[8] Celeghini E, Rasetti M and Vitiello G 1992 *Ann. Phys.* **215** 156.
[9] Blasone M and Jizba P 2004 *Ann. Phys.* **312** 354. Blasone M, Jizba P, Vitiello G 2003 *J. Phys. Soc. Jap.* **72** (suppl. C) 50; Blasone M, Jizba P 2002 *Can. J. Phys.* **80** 645.
[10] Chruscinski D 2006 *Ann. Phys.* **321** 840; Chruscinski D and Jurkowski J 2006 *Ann. Phys.* **321** 854.
[11] Schuch D and Blasone M 2017 *J. Phys. Conf. Ser.* **880** 012050.
[12] Greiner W and Reinhardt J 1996 *Field Quantization* (Springer-Verlag, Berlin)
[13] Iachello F 2015 *Lie Algebras and Applications* (Lecture Notes in Physics, Springer)
[14] 't Hooft G. 2007 *AIP Conf. Proc.* **957** 154
[15] Elze H T 2009 *J. Phys. Conf. Ser.* **171** 012034.
[16] Blasone M, Jizba P and Kleinert H 2005 *Phys. Rev.* A **71** 052507; 2005 *Ann. Phys.* **320** 468.
[17] Blasone M, Jizba P, Vitiello G 2011 *Quantum Field Thoery and its Macroscopic Manifestations*, Imperial College Press, London.

IOP Conf. Series: Journal of Physics: Conf. Series **1071** (2018) 012003 doi:10.1088/1742-6596/1071/1/012003

Schwinger-Dyson equations and flavor mixing

M Blasone[1], P Jizba[2] and L Smaldone[1]

[1] Dipartimento di Fisica, Università di Salerno, Via Giovanni Paolo II, 132 84084 Fisciano, Italy & INFN Sezione di Napoli, Gruppo collegato di Salerno, Italy
[2] FNSPE, Czech Technical University in Prague, Břehová 7, 115 19 Praha 1, Czech Republic

E-mail: blasone@sa.infn.it

E-mail: p.jizba@fjfi.cvut.cz

E-mail: lsmaldone@sa.infn.it

Abstract. We briefly review some results on Green's functions for mixed fermion fields and reinterpret them in terms of Schwinger–Dyson equations. Working with retarded propagators, we identify the self-energy operator for the mixing interaction, and find an exactly solvable expansion for the complete-fermion propagator. Finally, we show that an equation à la Bethe–Salpeter can be derived from Schwinger–Dyson equations, leading to the formal concept of single-particle bound state.

1. Introduction

The idea of flavor oscillations was first conceived by Bruno Pontecorvo [1] in order to quantitatively explain the results of the Homestake experiment [2]. While phenomena such as neutrino or K^0–$\overline{K}^0$ oscillations are currently experimentally well established [3], there are still open issues and loopholes in theoretical description of fermion mixing. The standard theoretical approach to this subject is based on quantum mechanics (QM) or on perturbative quantum field theory (QFT) [4]. A problem with this approach is that flavor-states cannot be described as standard free asymptotic $in-out$ states and hence one cannot use the LSZ formalism to directly relate Green functions with the S matrix [5].

A non-perturbative QFT approach to mixing was first developed in Ref. [6] where the non-trivial structure of the mixing transformation generator was discovered. As consequence, corrections to the standard Pontecorvo oscillation formula were found [7]. This result was then extended to the case of many flavors in Ref. [8]. Other developments can be found, e.g. in Ref. [9, 10]. A pedagogical introduction to this subject can be found in Ref. [11].

In this work we re-interpret the known results on fermion mixing in QFT, in a new and quite illuminating way. In particular, with the help of Schwinger–Dyson (SD) equations, we find a formal analogy with the description of bound states via Bethe–Salpeter equation. This in turn will allow us to introduce the formal concept of *single particle bound-state*. Aforementioned analogy with the bound-state description will be instrumental in providing a new fresh view in the description of flavor fields. Although in the present paper we deal with Dirac fermions, we expect that similar results also holds for the case of Majorana fermions, see Ref. [12] for a treatment of mixing of Majorana neutrinos in QFT.

IOP Conf. Series: Journal of Physics: Conf. Series **1071** (2018) 012003 doi:10.1088/1742-6596/1071/1/012003

The paper is organized as follows: in Section 2 we review some fundamentals on fermion mixing transformation in QFT and related flavor Fock space. In Section 3, following Ref. [7] and Ref. [8] we discuss the problem related to the definition of Green's functions for flavor fields. Finally, in Section. 4 we reinterpret previous results by means of SD equations and we introduce the concept of single-particle bound states.

2. Fermion mixing and flavor vacuum

Let us consider the fermion mixing Lagrangian:

$$\mathcal{L}(x) \;=\; \overline{\nu}(x)\;(i\gamma^\mu\partial_\mu \;-\; M)\;\nu(x)\,, \tag{1}$$

with

$$\nu(x) \;=\; \begin{bmatrix}\nu_e(x)\\ \nu_\mu(x)\end{bmatrix}, \tag{2}$$

and

$$M \;=\; \begin{bmatrix} m_e & m_{e\mu}\\ m_{e\mu} & m_\mu\end{bmatrix}. \tag{3}$$

The field equations are then:

$$(i\gamma^\mu\partial_\mu \;-\; M)\,\nu(x) \;=\; 0\,. \tag{4}$$

These can be rewritten as two independent Dirac equations thanks to the Pontecorvo mixing transformation:

$$\nu_e(x) \;=\; \nu_1(x)\cos\theta + \nu_2(x)\sin\theta\,, \tag{5}$$

$$\nu_\mu(x) \;=\; -\nu_1(x)\sin\theta + \nu_2(x)\cos\theta\,, \tag{6}$$

with $\tan 2\theta = 2m_{e\mu}/(m_\mu - m_e)$. ν_e and ν_μ are called flavor fields while ν_1 and ν_2 are the mass fields, satisfying free Dirac equations:

$$(i\gamma^\mu\partial_\mu \;-\; m_j)\,\nu_j(x) \;=\; 0\,, \qquad j=1,2\,. \tag{7}$$

The masses m_1 and m_2 are related to the original parameters as:

$$m_e \;=\; m_1\,\cos^2\theta \;+\; m_2\,\sin^2\theta\,, \tag{8}$$

$$m_\mu \;=\; m_1\,\sin^2\theta \;+\; m_2\,\cos^2\theta\,. \tag{9}$$

Now ν_1 and ν_2 can be expanded as[1]

$$\nu_i(x) = \frac{1}{\sqrt{V}}\sum_{\mathbf{k},r} e^{i\mathbf{k}\cdot\mathbf{x}}\,[u^r_{\mathbf{k},i}(t)\,\alpha^r_{\mathbf{k},i} \;+\; v^r_{-\mathbf{k},i}(t)\,\beta^{r\dagger}_{-\mathbf{k},i}]\,, \qquad i=1,2\,. \tag{10}$$

These are truly asymptotic fields and the (mass) vacuum satisfies:

$$\alpha^r_{\mathbf{k},i}|0\rangle_{1,2} \;=\; \beta^r_{\mathbf{k},i}|0\rangle_{1,2} \;=\; 0\,. \tag{11}$$

[1] Here we consider the system in a box of volume V. This regularization scheme permits to avoid problems with inequivalent representation of field algebra [13, 14, 15]. One has to perform the limit $V\to\infty$ at the the end of calculations.

Now we rewrite the Pontecorvo relations (5),(6) as:

$$\nu_e(x) = G_\theta^{-1}(t)\,\nu_1(x)\,G_\theta(t)\,, \tag{12}$$

$$\nu_\mu(x) = G_\theta^{-1}(t)\,\nu_2(x)\,G_\theta(t)\,, \tag{13}$$

where $G_\theta(t)$, the *mixing generator*, is given by

$$G_\theta(t) = \exp\left[\theta \int d^3x \left(\nu_1^\dagger(x)\nu_2(x) - \nu_2^\dagger(x)\nu_1(x)\right)\right], \tag{14}$$

and in a finite volume limit is a unitary operator, preserving the canonical anticommutation relations. The equivalence between Eqs.(12),(13) and the Pontecorvo relations (5),(6) can be verified by using the Baker-Campbell-Hausdorff formula. The action of the mixing generator does not leave the vacuum $|0\rangle_{1,2}$ invariant:

$$|0(t)\rangle_{e,\mu} = G_\theta^{-1}(t)\;|0\rangle_{1,2}\,. \tag{15}$$

$|0(t)\rangle_{e,\mu}$ is the vacuum for the *flavor Fock space* $\mathcal{H}_{e,\mu}$ and is called flavor vacuum. Thanks to $G_\theta(t)$ we can write at every time the annihilation operators of the flavor vacuum as

$$\alpha^r_{\mathbf{k},\sigma}(t) \equiv G_\theta^{-1}(t)\;\alpha^r_{\mathbf{k},j}\;G_\theta(t)\,, \tag{16}$$

$$\beta^r_{\mathbf{k},\sigma}(t) \equiv G_\theta^{-1}(t)\;\beta^r_{\mathbf{k},j}\;G_\theta(t)\,, \quad (\sigma, i) = (e,1), (\mu,2)\,. \tag{17}$$

In terms of these operators we can expand the flavor fields ν_e and ν_μ:

$$\nu_e(x) = \frac{1}{\sqrt{V}}\sum_{\mathbf{k},r} e^{i\mathbf{k}\cdot\mathbf{x}}\left[u^r_{\mathbf{k},1}(t)\,\alpha^r_{\mathbf{k},e}(t) + v^r_{-\mathbf{k},1}(t)\,\beta^{r\dagger}_{-\mathbf{k},e}(t)\right], \tag{18}$$

$$\nu_\mu(x) = \frac{1}{\sqrt{V}}\sum_{\mathbf{k},r} e^{i\mathbf{k}\cdot\mathbf{x}}\left[u^r_{\mathbf{k},2}(t)\,\alpha^r_{\mathbf{k},\mu}(t) + v^r_{-\mathbf{k},2}(t)\,\beta^{r\dagger}_{-\mathbf{k},\mu}(t)\right]. \tag{19}$$

Choosing a Lorentz frame so that $\mathbf{k} = (0,0,|\mathbf{k}|)$, Eqs.(16),(17) explicitly read:

$$\alpha^r_{\mathbf{k},e}(t) = \cos\theta\;\alpha^r_{\mathbf{k},1} + \sin\theta\left(U^*_{\mathbf{k}}(t)\;\alpha^r_{\mathbf{k},2} + \epsilon^r\;V_{\mathbf{k}}(t)\;\beta^{r\dagger}_{-\mathbf{k},2}\right), \tag{20}$$

$$\alpha^r_{\mathbf{k},\mu}(t) = \cos\theta\;\alpha^r_{\mathbf{k},2} - \sin\theta\left(U_{\mathbf{k}}(t)\;\alpha^r_{\mathbf{k},1} - \epsilon^r\;V_{\mathbf{k}}(t)\;\beta^{r\dagger}_{-\mathbf{k},1}\right), \tag{21}$$

$$\beta^r_{-\mathbf{k},e}(t) = \cos\theta\;\beta^r_{-\mathbf{k},1} + \sin\theta\left(U^*_{\mathbf{k}}(t)\;\beta^r_{-\mathbf{k},2} - \epsilon^r\;V_{\mathbf{k}}(t)\;\alpha^{r\dagger}_{\mathbf{k},2}\right), \tag{22}$$

$$\beta^r_{-\mathbf{k},\mu}(t) = \cos\theta\;\beta^r_{-\mathbf{k},2} - \sin\theta\left(U_{\mathbf{k}}(t)\;\beta^r_{-\mathbf{k},1} + \epsilon^r\;V_{\mathbf{k}}(t)\;\alpha^{r\dagger}_{\mathbf{k},1}\right), \tag{23}$$

where $\epsilon^r = (-1)^r$ and we defined

$$U_{\mathbf{k}}(t) \equiv u^{r\dagger}_{\mathbf{k},2}(t)u^r_{\mathbf{k},1}(t) = v^{r\dagger}_{-\mathbf{k},1}(t)v^r_{-\mathbf{k},2}(t)\,,$$

$$V_{\mathbf{k}}(t) \equiv \epsilon^r\;u^{r\dagger}_{\mathbf{k},1}(t)v^r_{-\mathbf{k},2}(t) = -\epsilon^r\;u^{r\dagger}_{\mathbf{k},2}(t)v^r_{-\mathbf{k},1}(t)\,. \tag{24}$$

Now it is possible to define flavor states as excitations of the flavor vacuum. For example a flavor neutrino state will be

$$|\nu^r_{\mathbf{k},\sigma}(t)\rangle \equiv \alpha^{r\dagger}_{\mathbf{k},\sigma}(t)|0(t)\rangle_{e,\mu}\,, \quad \sigma = e,\mu\,. \tag{25}$$

This is an eigenstate of the (non-conserved) flavor charge:

$$Q_\sigma(t) = \int d^3x\,\nu_\sigma^\dagger(x)\nu_\sigma(x) = \sum_{\mathbf{k},r}\left(\alpha^{r\dagger}_{\mathbf{k},\sigma}(t)\alpha^r_{\mathbf{k},\sigma}(t) - \beta^{r\dagger}_{\mathbf{k},\sigma}(t)\beta^r_{\mathbf{k},\sigma}(t)\right), \quad \sigma = e,\mu\,, \tag{26}$$

as one easily check.

3. Green's functions for flavor fields and oscillation formula

Following Ref.[7] let us construct the propagator for flavor fields. The main problem resides in the choice of the vacuum. As shown in Ref.[16], this problem is equivalent to the choice of appropriate boundary conditions.

A first possibility is to consider the propagator constructed on the mass vacuum[2] $|0\rangle_{1,2}$:

$$S_f(x,y) = \begin{bmatrix} S^{\alpha\beta}_{ee}(x,y) & S^{\alpha\beta}_{e\mu}(x,y) \\ S^{\alpha\beta}_{\mu e}(x,y) & S^{\alpha\beta}_{\mu\mu}(x,y) \end{bmatrix} \tag{27}$$

$$= \begin{bmatrix} {}_{1,2}\langle 0|T\left[\nu^\alpha_e(x)\,\overline{\nu}^\beta_e(y)\right]|0\rangle_{1,2} & {}_{1,2}\langle 0|T\left[\nu^\alpha_e(x)\,\overline{\nu}^\beta_\mu(y)\right]|0\rangle_{1,2} \\ {}_{1,2}\langle 0|T\left[\nu^\alpha_\mu(x)\,\overline{\nu}^\beta_e(y)\right]|0\rangle_{1,2} & {}_{1,2}\langle 0|T\left[\nu^\alpha_\mu(x)\,\overline{\nu}^\beta_\mu(y)\right]|0\rangle_{1,2} \end{bmatrix}.$$

This can be explicitly written in terms of the propagators of mass fields:

$$S_f(x,y) = \begin{bmatrix} S^{\alpha\beta}_1(x,y)\cos^2\theta + S^{\alpha\beta}_2(x,y)\sin^2\theta & (S^{\alpha\beta}_2(x,y) - S^{\alpha\beta}_1(x,y))\cos\theta\sin\theta \\ (S^{\alpha\beta}_2(x,y) - S^{\alpha\beta}_1(x,y))\cos\theta\sin\theta & S^{\alpha\beta}_2(x,y)\cos^2\theta + S^{\alpha\beta}_1(x,y)\sin^2\theta \end{bmatrix}, \tag{28}$$

where

$$S^{\alpha\beta}_j(x,y) = i\int \frac{\mathrm{d}^4k}{(2\pi)^4}\, e^{-ik\cdot(x-y)}\, \frac{\not{k} + m_j}{k^2 - m_j^2 + i\varepsilon}, \qquad j = 1,2\,. \tag{29}$$

In Ref.[7] it was defined the amplitude of the process where a e-flavor particle was created at $t = 0$ and is observed unchanged at time $t > 0$, as:

$$\mathcal{P}^{>}_{ee}(\mathbf{k},t) = i\, u^{r\dagger}_{\mathbf{k},1}\, e^{i\omega_{\mathbf{k},1}t}\, S^{>}_{ee}(\mathbf{k},t)\,\gamma^0\, u^r_{\mathbf{k},1}\,, \tag{30}$$

where $S^{>}_{ee}(\mathbf{k},t)$ is the Fourier transform of the Wightman function

$$S^{>}_{ee}(t,\mathbf{x};0,\mathbf{y}) = {}_{1,2}\langle 0|\nu_e(t,\mathbf{x})\,\overline{\nu}_e(0,\mathbf{y})|0\rangle_{1,2}\,. \tag{31}$$

An explicit calculation gives us:

$$\mathcal{P}^{>}_{ee}(\mathbf{k},t) = \cos^2\theta \;+\; \sin^2\theta\,|U_{\mathbf{k}}|^2\, e^{-i(\omega_{\mathbf{k},2}-\omega_{\mathbf{k},1})t}\,. \tag{32}$$

However this result is unacceptable because

$$\mathcal{P}^{>}_{ee}(\mathbf{k},0^+) = \cos^2\theta \;+\; \sin^2\theta|U_{\mathbf{k}}|^2 \;<\; 1\,. \tag{33}$$

Therefore, in Ref.[7], it was proposed to consider the propagator

$$\mathcal{G}_f(x,y) = \begin{bmatrix} \mathcal{G}^{\alpha\beta}_{ee}(x,y) & \mathcal{G}^{\alpha\beta}_{e\mu}(x,y) \\ \mathcal{G}^{\alpha\beta}_{\mu e}(x,y) & \mathcal{G}^{\alpha\beta}_{\mu\mu}(x,y) \end{bmatrix} \tag{34}$$

$$= \begin{bmatrix} {}_{e,\mu}\langle 0(y_0)|T\left[\nu^\alpha_e(x)\,\overline{\nu}^\beta_e(y)\right]|0(y_0)\rangle_{e,\mu} & {}_{e,\mu}\langle 0(y_0)|T\left[\nu^\alpha_e(x)\,\overline{\nu}^\beta_\mu(y)\right]|0(y_0)\rangle_{e,\mu} \\ {}_{e,\mu}\langle 0(y_0)|T\left[\nu^\alpha_\mu(x)\,\overline{\nu}^\beta_e(y)\right]|0(y_0)\rangle_{e,\mu} & {}_{e,\mu}\langle 0(y_0)|T\left[\nu^\alpha_\mu(x)\,\overline{\nu}^\beta_\mu(y)\right]|0(y_0)\rangle_{e,\mu} \end{bmatrix},$$

[2] In this Section we use the same notation as in Ref.[7]. The propagator is thus defined as the vacuum expectation value of the time-ordered products of fields. Such a definition differs from the one of Ref. [17] just by a factor i. However, in the next Section it will be useful to follow the convention of Ref.[17].

on the flavor vacuum. As anticipated, $\mathcal{G}_f(x,y)$ differs from $S_f(x,y)$ just by boundary terms. For example, we consider the Fourier transform of $\mathcal{G}_{ee}(x,y)$:

$$\begin{aligned}\mathcal{G}_{ee}(\mathbf{k},t) &= S_{ee}(\mathbf{k},t) + 2\pi i \sin^2\theta \left[|V_\mathbf{k}|^2 (\not{k} + m_2)\delta(k^2 - m_2^2) \right. \\ &- \left. |U_\mathbf{k}||V_\mathbf{k}| \sum_r \left(\epsilon^r u^r_{\mathbf{k},2}\overline{v}^r_{-\mathbf{k},2}\delta(k_0 - \omega_2) + \epsilon^r v^r_{-\mathbf{k},2}\overline{u}^r_{\mathbf{k},2}\delta(k_0 + \omega_2)\right) \right] .\end{aligned} \tag{35}$$

By introducing, as before, the Wightman function $\mathcal{G}^>_{ee}(\mathbf{k},t)$, we define

$$\mathcal{P}^>_{ee}(\mathbf{k},t) = i\, u^{r\dagger}_{\mathbf{k},1}\, e^{i\omega_{\mathbf{k},1}t}\, \mathcal{G}^>_{ee}(\mathbf{k},t)\,\gamma^0\, u^r_{\mathbf{k},1} . \tag{36}$$

This can be evaluated explicitly:

$$\mathcal{P}^>_{ee}(\mathbf{k},t) = \cos^2\theta + \sin^2\theta \left(|U_\mathbf{k}|^2 e^{-i(\omega_{\mathbf{k},2}-\omega_{\mathbf{k},1})t} + |V_\mathbf{k}|^2 e^{i(\omega_{\mathbf{k},1}+\omega_{\mathbf{k},2})t} \right) . \tag{37}$$

Now $\mathcal{P}^>_{ee}(\mathbf{k},t)$ satisfies the right initial condition:

$$\mathcal{P}^>_{ee}(\mathbf{k},0^+) = 1 . \tag{38}$$

Moreover

$$|\mathcal{P}^>_{ee}(\mathbf{k},t)|^2 + |\mathcal{P}^>_{\mu\mu}(\mathbf{k},t)|^2 + |\mathcal{P}^>_{e\mu}(\mathbf{k},t)|^2 + |\mathcal{P}^>_{\mu e}(\mathbf{k},t)|^2 = 1 . \tag{39}$$

Here $\mathcal{P}^>_{\rho\sigma}$, $\rho,\sigma = e,\mu$ are defined in a similar way as $\mathcal{P}^>_{ee}$. We can thus identify the different pieces in the Eq.(39) as flavor (un)-changing probabilities. The QFT oscillation formula can be explicitly derived [7]:

$$P_{\nu_e\to\nu_\mu}(\mathbf{k},t) = \sin^2(2\theta)\left[|U_\mathbf{k}|^2 \sin^2\left(\frac{\omega_{\mathbf{k},1}-\omega_{\mathbf{k},2}}{2}t\right) + |V_\mathbf{k}|^2 \sin^2\left(\frac{\omega_{\mathbf{k},1}+\omega_{\mathbf{k},2}}{2}t\right)\right] . \tag{40}$$

Note that in the ultrarelativistic limit $|\mathbf{k}| \to \infty$, $|U_\mathbf{k}|^2 \to 1$ and $|V_\mathbf{k}|^2 \to 0$, and we thus recover the standard quantum mechanical oscillation formula [4].

Tough the result in Eq.(40) could seem related with the use of the flavor vacuum, in Ref. [8] it was shown that it can be re-obtained by using retarded propagators both on flavor and mass vacua:

$$S^{ret}(t,\mathbf{x};0,y) = \theta(t)\, {}_{1,2}\langle 0|\, \{\nu_\rho(t,\mathbf{x}), \overline{\nu}_\sigma(0,\mathbf{y})\}\, |0\rangle_{1,2} , \tag{41}$$

$$\mathcal{G}^{ret}(t,\mathbf{x};0,y) = \theta(t)\, {}_{e,\mu}\langle 0|\, \{\nu_\rho(t,\mathbf{x}), \overline{\nu}_\sigma(0,\mathbf{y})\}\, |0\rangle_{e,\mu} , \quad \rho,\sigma = e,\mu . \tag{42}$$

In fact these turns out to be equal and, passing to Fourier space, we get:

$$S^{ret}(\mathbf{k},t) = \mathcal{G}^{ret}(\mathbf{k},t) . \tag{43}$$

Defining

$$P_{\nu_\rho\to\nu_\sigma}(\mathbf{k},t) = \mathrm{Tr}\left[\mathcal{G}^{ret}_{\sigma\rho}(\mathbf{k},t)\mathcal{G}^{ret\dagger}_{\sigma\rho}(\mathbf{k},t)\right] , \tag{44}$$

one can check that the probability (44) gives us the result (40). The oscillation formula (40) has thus to be interpreted as the consequence of the QFT treatment of mixing and not as directly related with the use of the flavor vacuum.

4. Schwinger-Dyson equations and single particle bound state

The partition function can be written down from the Lagrangian (1) thanks to the functional integral formalism:

$$\mathcal{Z}[\eta,\overline{\eta}] \;=\; \int \mathcal{D}\overline{\nu}\,\mathcal{D}\nu\, e^{i\int \mathrm{d}^4x(\mathcal{L}(x)\,+\,\overline{\eta}(x)\nu(x)\,+\,\overline{\nu}(x)\eta(x))}\,. \tag{45}$$

Now the fields $\nu(x) = (\nu_e(x), \nu_\mu(x))$ and the currents $\eta(x) = (\eta_e(x), \eta_\mu(x))$ are Grassman variables. The partition function (45) can be explicitly written as:

$$\mathcal{Z}[\eta,\overline{\eta}] \;=\; \mathcal{N} e^{-i\int \mathrm{d}^4x \int \mathrm{d}^4y\left(\overline{\eta}(x)S^F(x,y)\eta(y)\right)}\,, \tag{46}$$

where $\mathcal{N}$ is a normalization factor, and $S_F(x,y)$ is a 2×2 matrix Green's function, satisfying the delta-source equation:

$$(i\partial\!\!\!/ - M)S^F(x,y) \;=\; \delta^4(x-y)\mathbb{I}\,. \tag{47}$$

Here $\mathbb{I}$ is the 2×2 identity matrix. It is well known that $S^F(x,y)$ is defined up to a solution of the homogeneous equation

$$(i\partial\!\!\!/ - M)S_0^F(x,y) \;=\; 0\,. \tag{48}$$

This is exactly the aforementioned freedom in the choice of boundary conditions. Moreover, from the observation that the functional integral of a functional derivative is zero

$$\int \mathcal{D}\overline{\nu}\,\mathcal{D}\nu\,\frac{\delta}{\delta\overline{\nu}(x)} e^{i\int \mathrm{d}^4x'(\mathcal{L}(x')\,+\,\overline{\eta}(x')\nu(x')\,+\,\overline{\nu}(x')\eta(x'))} \;=\; 0\,, \tag{49}$$

we find the SD equations for $\mathcal{Z}[\eta,\overline{\eta}]$:

$$\left(\eta(x) \,+\, (i\partial\!\!\!/ \,-\, M)\frac{\delta}{i\delta\overline{\eta}(x)}\right)\mathcal{Z}[\eta,\overline{\eta}] \;=\; 0\,, \tag{50}$$

$$\tag{51}$$

which explicitly read

$$\eta_e(x)\mathcal{Z}[\eta,\overline{\eta}] \;=\; -(i\partial\!\!\!/ \,-\, m_e)\frac{\delta}{i\delta\bar{\eta}_e(x)}\mathcal{Z}[\eta,\overline{\eta}] \,+\, m_{e\,\mu}\frac{\delta}{i\delta\bar{\eta}_\mu(x)}\mathcal{Z}[\eta,\overline{\eta}]\,, \tag{52}$$

$$\eta_\mu(x)\mathcal{Z}[\eta,\overline{\eta}] \;=\; -(i\partial\!\!\!/ \,-\, m_\mu)\frac{\delta}{i\delta\bar{\eta}_\mu(x)}\mathcal{Z}[\eta,\overline{\eta}] \,+\, m_{e\,\mu}\frac{\delta}{i\delta\bar{\eta}_e(x)}\mathcal{Z}[\eta,\overline{\eta}]\,. \tag{53}$$

Deriving respect to $\eta_e(y)$ and $\eta_\mu(y)$ and putting $\eta_\mu = \eta_e = \bar{\eta}_e = \bar{\eta}_\mu = 0$, we get:

$$\left(S^F_{e\,e}\right)^{-1}(x-y) \;=\; (i\partial\!\!\!/ \,-\, m_e)\,\delta^4(x-y) \,-\, m_{e\,\mu}\int \mathrm{d}^4z\, S^F_{e\,\mu}(x-z)\left(S^F_{e\,e}\right)^{-1}(z-y)\,, \tag{54}$$

$$\left(S^F_{\mu\,\mu}\right)^{-1}(x-y) \;=\; (i\partial\!\!\!/ \,-\, m_\mu)\,\delta^4(x-y) \,-\, m_{e\,\mu}\int \mathrm{d}^4z\, S^F_{\mu\,e}(x-z)\left(S^F_{\mu\,\mu}\right)^{-1}(z-y)\,, \tag{55}$$

$$m_{e\mu}S^F_{\mu\,\mu}(x-y) \;=\; (i\partial\!\!\!/ \,-\, m_e)S^F_{\mu\,e}(x-y)\,, \tag{56}$$

$$m_{e\mu}S^F_{e\,e}(x-y) \;=\; (i\partial\!\!\!/ \,-\, m_\mu)S^F_{e\,\mu}(x-y)\,. \tag{57}$$

It is straightforward to check that these equations are equivalent to the matrix equation (47). Therefore we can derive:

$$(i\partial\!\!\!/ \,-\, m_\mu)\left(S^F_{e\,e}\right)^{-1}(x-y) \;=\; (i\partial\!\!\!/ \,-\, m_\mu)(i\partial\!\!\!/ \,-\, m_e)\,\delta^4(x-y) \,-\, m^2_{e\,\mu}\,\delta^4(x-y)\,, \tag{58}$$

$$(i\partial\!\!\!/ \,-\, m_e)\left(S^F_{\mu\,\mu}\right)^{-1}(x-y) \;=\; (i\partial\!\!\!/ \,-\, m_e)(i\partial\!\!\!/ \,-\, m_\mu)\,\delta^4(x-y) \,-\, m^2_{e\,\mu}\,\delta^4(x-y)\,. \tag{59}$$

IOP Conf. Series: Journal of Physics: Conf. Series **1071** (2018) 012003 doi:10.1088/1742-6596/1071/1/012003

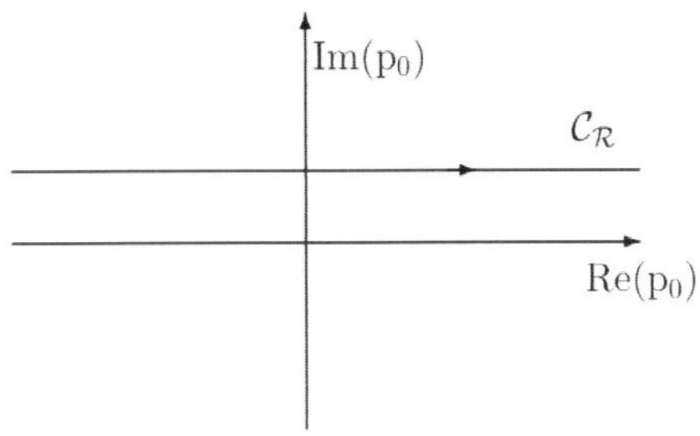

Figure 1. The complex contour $\mathcal{C}_{\mathcal{R}}$.

Passing to Fourier space we need to specify which propagator we are dealing with, i.e. the boundary conditions. These are determined by choosing a contour on the complex p_0-plane. Because of Eq.(43), we will deal with retarded propagators. Let us remember that, for free Dirac fields, the retarded propagator reads as:

$$S^{ret}_{\alpha\beta}(x-y) \;=\; (i\partial\!\!\!/ + m)_{\alpha\beta}\, \Delta^{ret}(x-y)\,, \tag{60}$$

where $\Delta^{ret}(x-y)$ is the Klein-Gordon retarded propagator:

$$\Delta^{ret}(x-y) \;=\; \frac{1}{(2\pi)^4}\int_{C_R} \mathrm{d}^4k\, \frac{e^{-ik\cdot(x-y)}}{k^2-m^2}\,, \tag{61}$$

and the contour is shown in figure 1. With this prescription in mind, we replace $\mathcal{G}^{ret}(x,y)$ to $S^F(x,y)$ and we write the SD equations in momentum space:

$$(k\!\!\!/ \,-\, m_\mu)\left(\mathcal{G}^{ret}_{ee}\right)^{-1}(k) \;=\; (k\!\!\!/ \,-\, m_\mu)(k\!\!\!/ \,-\, m_e) \,-\, m^2_{e\,\mu}\,, \tag{62}$$

$$(k\!\!\!/ \,-\, m_e)\left(\mathcal{G}^{ret}_{\mu\,\mu}\right)^{-1}(k) \;=\; (k\!\!\!/ \,-\, m_e)(k\!\!\!/ \,-\, m_\mu) \,-\, m^2_{e\,\mu}\,. \tag{63}$$

introducing $\mathcal{G}^{ret}_e(p) = (k\!\!\!/ - m_e)^{-1}$ and $\mathcal{G}^{ret}_\mu = (k\!\!\!/ - m_\mu)^{-1}$, we can rewrite Eqs.(62),(63) as:

$$\mathcal{G}^{ret}_{e\,e}(k) \;=\; \mathcal{G}^{ret}_e(k)\left(1 \,-\, m^2_{e\,\mu}\, \mathcal{G}^{ret}_e(k)\, \mathcal{G}^{ret}_\mu(k)\right)^{-1}\,, \tag{64}$$

$$\mathcal{G}^{ret}_{\mu\,\mu}(k) \;=\; \mathcal{G}^{ret}_\mu(k)\left(1 \,-\, m^2_{e\,\mu}\, \mathcal{G}^{ret}_e(k)\, \mathcal{G}^{ret}_\mu(k)\right)^{-1}\,. \tag{65}$$

These can be rewritten as sum of the series ($\hbar$-expansion):

$$\mathcal{G}^{ret}_{e\,e}(k) \;=\; \mathcal{G}^{ret}_e(k) \,+\, \mathcal{G}^{ret}_e(k)\,\Sigma_e(k)\,\mathcal{G}^{ret}_e(k) \,+\, \ldots\,, \tag{66}$$

$$\mathcal{G}^{ret}_{\mu\,\mu}(k) \;=\; \mathcal{G}^{ret}_\mu(k) \,+\, \mathcal{G}^{ret}_\mu(k)\,\Sigma_\mu(k)\,\mathcal{G}^{ret}_\mu(k) \,+\, \ldots\,, \tag{67}$$

where

$$\Sigma_e(k) \;=\; m^2_{e\,\mu}\,\mathcal{G}^{ret}_\mu(k)\,, \tag{68}$$

$$\Sigma_\mu(k) \;=\; m^2_{e\,\mu}\,\mathcal{G}^{ret}_e(k)\,. \tag{69}$$

These are the self-energy operators in the momentum space. In fact:

$$\Gamma_{e\,e}(k) \;=\; \Gamma_e(k) \;-\; m_{e\,\mu}^2\,\mathcal{G}_\mu^{ret}(k)\,, \tag{70}$$

$$\Gamma_{\mu\,\mu}(k) \;=\; \Gamma_\mu(k) \;-\; m_{e\,\mu}^2\,\mathcal{G}_e^{ret}(k)\,, \tag{71}$$

where the Γ's are two-point truncated functions (inverse propagators). These relations indicate that flavor-particles propagation is characterized by an infinite series of processes of flavor-oscillation in the vacuum.

Finally we introduce the formal but intriguing concept of *single-particle bound state*. The idea is to derive an equation à la Bethe–Salpeter for the flavor fields. Let us calculate:

$$\begin{aligned}(\not{k} - m_e - \Sigma_e)(\not{k} - m_\mu - \Sigma_\mu)\,\mathcal{G}_{e\,\mu}^{ret}(k) \;&=\; m_{e\,\mu}^2\,(\not{k} - m_e - \Sigma_e)\,\mathcal{G}_{e\,e}^{ret}(k) \\ &-\; (\not{k} - m_e - \Sigma_e)\,\Sigma_\mu\,\mathcal{G}_{e\,\mu}^{ret}(k)\,.\end{aligned} \tag{72}$$

By using that $\mathcal{G}_{e\,e}^{ret}(k) \;=\; (\not{k} - m_e - \Sigma_e)^{-1}$, we get:

$$(\not{k} - m_e - \Sigma_e)(\not{k} - m_\mu - \Sigma_\mu)\mathcal{G}_{e\,\mu}^{ret}(k) \;=\; m_{e\,\mu}^2 \;-\; (\not{k} - m_e - \Sigma_e)\,\Sigma_\mu\,\mathcal{G}_{e\,\mu}^{ret}(k)\,. \tag{73}$$

If we define the potential:

$$V_{e\,\mu}(k) \;=\; (\not{k} - m_e - \Sigma_e(k))\,\Sigma_\mu(k)\,, \tag{74}$$

we derive a Bethe–Salpeter-like equation in the flavor-space:

$$(\not{k} - m_e - \Sigma_e)(\not{k} - m_\mu - \Sigma_\mu)\mathcal{G}_{e\,\mu}^{ret}(k) \;=\; m_{e\,\mu}^2 \;-\; V_{e\,\mu}\,\mathcal{G}_{e\,\mu}^{ret}(k)\,. \tag{75}$$

This suggestive relation indicates that the flavor fields can be formally viewed as bound states in the abstract flavor-space, revealing their non-asymptotic nature. Therefore flavor particles cannot be described by LSZ asymptotic states. This is evident in the time-dependence of the flavor vacuum and flavor creation and annihilation operators (see Eqs.(15),(20)-(23)).

5. Conclusions

Structural richness of the phenomenon of fermion mixing in QFT can be appreciated by analyzing the propagator of flavor fields. It turns out that self-energy operator does not lead to a simple renormalization of the bare mass, but it regulates the phenomenon of oscillations in the vacuum (see Eq. (66)). Moreover, Schwinger–Dyson equations permit to derive a relation which in its structure resembles a Bethe–Salpeter equation describing bound states in the flavor space, rather than in spacetime as usual (cf. Eq. (75)). This non-perturbative nature of mixing is evidently at the origin of the time dependence of flavor annihilation and creation operators (cf. Eqs. (20)-(23)).

Generally the problem of the boundary conditions in the flavor-vacuum Green functions is rather complicated issue but fortunately in the present paper one could resort to a formal identity for retarded Green functions of mass and flavor vacua. There the ignorance of the flavor-vacuum boundary conditions was translated on the level of boundary conditions for mass vacuum where these are well understood. This fact was explicitly represented in the identity (43). Recently, in Ref. [16], a *generalized generating functional* for mixed-representations Green's functions (i.e. Green's functions on different vacua) was derived. Such functional permits to take into account automatically the boundary conditions thanks to a consistent extension of the standard generating functional for Green's functions. A treatment of mixing with this technique was done in Ref. [18].

IOP Conf. Series: Journal of Physics: Conf. Series **1071** (2018) 012003 doi:10.1088/1742-6596/1071/1/012003

Acknowledgments

P.J. was supported by the Czech Science Foundation Grant No. 17-33812L.

References

[1] Pontecorvo B 1957 *Sov. Phys. JETP* **6** 429; *Sov. Phys. JETP* **7** 172-02; Gribov V and Pontecorvo B 1969 *Physics Letters* B **28** 493; Bilenky S and Pontecorvo B 1978 *Phys. Rep.* **41** 225

[2] Davis R 1994 *Prog. Part. Nucl. Phys.* **32** 13

[3] Smy M B (Super-Kamiokande) 2003 *Nucl. Phys. Proc. Suppl.* **118** 25; Hallin A L et al. 2003 *Nucl. Phys. Proc. Suppl.* **118** 3; Eguchi K et al. (KamLAND) 2003 *Phys. Rev. Lett.* **90** 021802; Hampel W et al. (GALLEX) 1999 *Phys. Lett.* B **447** 127; Altmann M et al. (GNO) 2000 *Phys. Lett.* B **490** 16

[4] Giunti C 2002 *JHEP* **11** 017; Giunti C and Kim C W 2007 *Fundamentals of Neutrino Physics and Astrophysics* (Oxford Univ. Press); Cohen A G, Glashow S L and Ligeti Z 2009 *Phys. Lett.* B **678** 191; Akhmedov E K and A. Y. Smirnov 2009 *Phys. Atom. Nucl.* **72** 1363

[5] Lehmann H, Symanzik K and Zimmerman W 1955 *Nuovo Cimento* **1** 205

[6] Blasone M and Vitiello G 1995 *Ann. Phys.* **244** 283

[7] Blasone M, Henning P A and Vitiello G 1999 *Phys. Lett.* B **451** 140

[8] Fujii K, Habe C and Yabuki T 2001 *Phys. Rev.* D **64** 013011

[9] Blasone M, Gargiulo M V and Vitiello G 2016 *Phys. Lett.* B **716** 104

[10] Blasone M, Capolupo A, Romei O and Vitiello G 2001 *Phys. Rev.* D **63** 125015; Blasone M, Jizba P, Lambiase G and Mavromatos N E 2014 *J. Phys.* **538** 012003; Blasone M, Jizba P and Smaldone L 2015 *Nuovo Cimento* **38 C** 201 Blasone M, Lambiase G and Luciano G G 2017 *Phys. Rev.* D **96** 025023

[11] Blasone M, Jizba P and Vitiello G 2011 *Quantum Field Theory and its Macroscopic Manifestations* (London: World Scientific & ICP)

[12] Blasone M and Palmer J S 2004 *Phys. Rev.* D **69** 057301

[13] Umezawa H, Matsumoto H and Tachiki M 1982 *Thermo Field Dynamics And Condensed States* (Amsterdam: North-holland); Umezawa H 1993 *Advanced Field Theory: Micro,Macro and Thermal Physics* (New York: AIP)

[14] Haag R 1996 *Local Quantum Physics: Fields, Particles, Algebras,* (Berlin: Springer-Verlag)

[15] Blasone M, Jizba P and Smaldone L 2017 *Ann. Phys.* **383** 207–238; *J. Phys: Conf. Ser.* **804** 012006

[16] Blasone M, Jizba P and Smaldone L 2017 *Phys. Rev.* A **96** 052107

[17] Greiner W and Reinhardt J 1996 *Field Quantization* (Berlin Heidelberg: Springer-Verlag)

[18] Blasone M, Jizba P and Smaldone L 2017 *J. Phys: Conf. Ser.* **880** 012051

IOP Conf. Series: Journal of Physics: Conf. Series **1071** (2018) 012004 doi:10.1088/1742-6596/1071/1/012004

Interference in semiclassical electron dynamics

M. Buchholz

Institut für Theoretische Physik, Technische Universität Dresden, D-01062 Dresden, Germany

F. Grossmann

Institut für Theoretische Physik, Technische Universität Dresden, D-01062 Dresden, Germany

E-mail: frank@physik.tu-dresden.de

Abstract. We further strengthen the hypothesis that the difference between singlet and triplet initial states in the scattering of two electrons is an interference effect. This is achieved by comparing semiclassical results based on multiple classical trajectories à la Herman and Kluk published previously [Phys. Rev. A **89**, 032104 (2014)] with classical Wigner (LSC-IVR) results based on a similar set of trajectories but not taking into account their phases (classical action divided by $\hbar$).

1. Introduction

Electronic many particle dynamics in the time-domain in atoms and quantum dots with or without magnetic fields has been investigated semiclassically only recently [1, 2, 3]. In addition, the scattering of two electrons without any external potential has been studied in [4], to answer the question if the Pauli principle is accounted for semiclassically. It was found out that the multi-trajectory Herman-Kluk method does account for the differences in the triplet versus the singlet state *without* the need for additional, artificial terms in the Hamiltonian [5], enforcing this difference. Single trajectory-based semiclassical approaches like the thawed Gaussian approximation of Heller [6] are not capable of displaying this difference in the asymptotic limit, however. This fact has been taken as a first indication that interference effects between different trajectories might be responsible for the difference between singlet and triplet scattering.

Here we further strengthen this hypothesis, by contrasting again the Herman-Kluk results from [4], but this time against multi-trajectory results without interference effects that have been gained from an application of the linearized semiclassical initial value approach, which is also referred to as classical Wigner method and uses a similar set of trajectories as the true semiclassical method.

To this end, in Section 2, we first recapitulate the semiclassical initial value representation of the quantum propagator à la Herman and Kluk and then the linearized version of the semiclassical approximation in the case of arbitrary correlation functions. In Section 3, we review a fully quantum treatment with a focus on the difference between the initial triplet and singlet states. In Section 4 we then apply both trajectory based approaches to describe the dynamics of scattering of two electrons, initially in the two different states. Conclusions and an outlook are given in Section 5.

2. Semiclassical and classical initial value representations

The solution of the time-dependent Schrödinger equation in terms of classical trajectories has a long history and is intimately connected to the semiclassical approximation to the Feynman path integral representation of the propagator [7]. An early review of different semiclassical approximations with a focus on the Herman-Kluk (HK) propagator is given in [8].

The computational advantage of the semiclassical HK propagator is its dependence on the solution of classical initial value problems. In their terms and for two degrees of freedom (DOF), which we will focus on later, it is given by

$$K(\mathbf{x},t;\mathbf{x}',0) = \int \frac{\mathrm{d}^2p\mathrm{d}^2q}{(2\pi\hbar)^2} R(\mathbf{p},\mathbf{q},t)\mathrm{e}^{\mathrm{i}S(\mathbf{p},\mathbf{q},t)/\hbar}\langle\mathbf{x}|g_\gamma(\mathbf{p}_t,\mathbf{q}_t)\rangle\langle g_\gamma(\mathbf{p},\mathbf{q})|\mathbf{x}'\rangle, \tag{1}$$

with $\mathbf{x}^{\mathrm{T}} = (x,y), \mathbf{q}^{\mathrm{T}} = (q_x,q_y), \mathbf{p}^{\mathrm{T}} = (p_x,p_y)$, $\mathbf{p}_t = \mathbf{p}(\mathbf{p},\mathbf{q},t), \mathbf{q}_t = \mathbf{q}(\mathbf{p},\mathbf{q},t)$,

$$g_\gamma(\mathbf{q},\mathbf{p}) = \left(\frac{\det\gamma}{\pi^2}\right)^{1/4} \mathrm{e}^{-\frac{1}{2}(\mathbf{x}-\mathbf{q})^{\mathrm{T}}\gamma(\mathbf{x}-\mathbf{q})+\frac{\mathrm{i}}{\hbar}\mathbf{p}^{\mathrm{T}}(\mathbf{x}-\mathbf{q})} \tag{2}$$

denoting a normalized Gaussian wavefunction, and the classical action

$$S(\mathbf{p},\mathbf{q},t) = \int_0^t \mathrm{d}t'(T-V) \tag{3}$$

for the motion of a system governed by a potential energy V. Furthermore, it contains a prefactor

$$R(\mathbf{q},\mathbf{p},t) = \sqrt{\det\left[\frac{1}{2}\left(\mathbf{m}_{11}+\boldsymbol{\gamma}\mathbf{m}_{22}\boldsymbol{\gamma}^{-1}-\mathrm{i}\hbar\boldsymbol{\gamma}\mathbf{m}_{21}-\frac{1}{\mathrm{i}\hbar}\mathbf{m}_{12}\boldsymbol{\gamma}^{-1}\right)\right]}, \tag{4}$$

which is dependent on the so-called stability or monodromy matrix [9] and has been derived by Herman and Kluk [10] following up on earlier ground-breaking work by Heller [11]. The matrices $\boldsymbol{\gamma}$, appearing in the prefactor as well as in the Gaussians, are usually chosen diagonal. Over a wide range of reasonable parameters, the final result does not depend on their value. For diagonal elements of γ that go to infinity, the semiclassical van Vleck-Gutzwiller propagator is recovered [9]. This formulation is based on the solution of classical boundary value problems, however, and therefore has not found such a widespread acceptance in numerical work as the HK propagator that is based on classical initial value problems, i.e., is a semiclassical initial value representation of the propagator.

Due to the presence of the phase factor in the propagator, different trajectories can "interfere" constructively or destructively and a lot of quantum effects, like plateau formation in high-order harmonic generation [12, 13], energy quantization in Bose-Hubbard systems [14] and breather oscillation [15] can be described properly by the HK propagator. A review of semiclassical initial value representations with a focus on applications in chemistry can be found in [16] and a semiclassical hybrid approach combining Heller's thawed Gaussians [6] with Herman-Kluk propagation for a limited number of degrees of freedom is laid out in [17]. This latter approach allows the semiclassical treatment of dissipative quantum systems. Deep tunneling effects, however, seem to be out of reach in a formulation with real trajectories and in real time [18].

In correlation functions of the form

$$C_{AB}(t) = \mathrm{tr}[\hat{A}\mathrm{e}^{\mathrm{i}\hat{H}t/\hbar}\hat{B}\mathrm{e}^{-\mathrm{i}\hat{H}t/\hbar}], \tag{5}$$

two time-evolution operators do appear and therefore a double Herman-Kluk expression emerges in a semiclassical description. For a sizable number of degrees of freedom, this might impose an unsurmountable numerical wall for the evaluation of such expressions. Simplifications are thus sought for and a linear expansion of the action difference leads to a purely classical expression with no quantum interference effects. The final result

$$C_{AB}^{\mathrm{W}}(t) = \int \frac{\mathrm{d}^2p\mathrm{d}^2q}{(2\pi\hbar)^2} A^{\mathrm{W}}(\mathbf{q},\mathbf{p})\, B^{\mathrm{W}}(\mathbf{q}_t,\mathbf{p}_t) \tag{6}$$

is given in terms of a single (multi-dimensional) phase space integral, where A^{W} is the Wigner transform

$$A^{\mathrm{W}}(\mathbf{q},\mathbf{p}) = \int \mathrm{d}^2s\, \mathrm{e}^{-\mathrm{i}\mathbf{p}\cdot\mathbf{s}/\hbar} \left\langle \mathbf{q}+\frac{\mathbf{s}}{2}\middle|\hat{A}\middle|\mathbf{q}-\frac{\mathbf{s}}{2}\right\rangle \tag{7}$$

of the operator $\hat{A}$, and an analogous relation holds for the operator $\hat{B}$. Heller had written down the final expression for the correlation function intuitively [19]. Semiclassical derivations have been given by Miller and coworkers [20, 21] as well as by Herman and Coker [22] and for dipole-dipole correlation functions in [23, 24]. In the latter cases, the starting point was a double HK expression and the mathematics involved is straightforward but tedious.

Obviously the correlation function (6) does not contain time-dependent oscillatory integrands (initial states that are not single Gaussians may induce "static" negativities, however, as can be seen below) and interference effects are not described any more. The method is therefore also referred to as the classical Wigner method. It has a long history of analysis and applications [19, 25, 26, 27] and is known under different pseudonyms, as, e.g., linearized semiclassical initial value representation (LSC-IVR) or truncated Wigner approximation. The reason for the latter naming convention is that the approximation is analogous to the truncation of the Moyal bracket in the equation of motion for the Wigner function at the lowest (classical) order [28].

If one is interested in the time-dependent (pure state) density of the system of interest, $\hat{A}$ and $\hat{B}$ are the projection operators representing the initial state and position space, according to

$$\hat{A} = |\Psi_\alpha(0)\rangle\langle\Psi_\alpha(0)|, \tag{8}$$
$$\hat{B} = |\mathbf{x}\rangle\langle\mathbf{x}|. \tag{9}$$

For an initial wavefunction

$$\Psi_\alpha(\mathbf{x},0) = \exp\left[-\frac{1}{2}(\mathbf{x}-\mathbf{q}_\alpha)^{\mathrm{T}}\gamma(\mathbf{x}-\mathbf{q}_\alpha) + \frac{\mathrm{i}}{\hbar}\mathbf{p}_\alpha^{\mathrm{T}}(\mathbf{x}-\mathbf{q}_\alpha)\right], \tag{10}$$

centered around $(\mathbf{q}_\alpha, \mathbf{p}_\alpha)$, the diagonal elements of the density matrix in position representation in LSC-IVR thus are given by

$$\begin{aligned}\rho(\mathbf{x},t) &= \int \frac{\mathrm{d}^2p\mathrm{d}^2q}{(\pi\hbar)^2}\delta[\mathbf{q}_t-\mathbf{x}] \\ &\times \exp\left[-\frac{1}{\hbar^2}(\mathbf{p}-\mathbf{p}_\alpha)^{\mathrm{T}}\gamma^{-1}(\mathbf{p}-\mathbf{p}_\alpha) - (\mathbf{q}-\mathbf{q}_\alpha)^{\mathrm{T}}\gamma(\mathbf{q}-\mathbf{q}_\alpha)\right],\end{aligned} \tag{11}$$

which contains the classical trajectories $\mathbf{q}_t$ as the only dynamical input.

3. Scattering of two electrons

Depending on their spin, two electrons will approach each other to different degrees, thus revealing the effect of the Pauli principle. Our test system is described by a Hamiltonian of two interacting electrons that are not subject to any external potential in one dimension (atomic units (a.u.) will be used throughout the rest of this work)

$$\hat{H} = \frac{p_x^2}{2} + \frac{p_y^2}{2} + V = \frac{p_x^2}{2} + \frac{p_y^2}{2} + \frac{1}{\sqrt{(x-y)^2 + c}} \quad . \tag{12}$$

We use the smoothing parameter $c = 0.55$ for the repulsive Coulomb interaction that has often been used for electron dynamics before, because it closely reproduces the ground state of the 3D Helium atom [29].

In the following, we investigate the scattering of two electrons as a function of their total spin, comparing the semiclassical HK method with exact quantum mechanical propagation and the classical (truncated) Wigner approach. A similar comparison has been made before in [4], where HK results have been compared to the quantum ones and the ones from a thawed Gaussian approximation. Initially, the two electrons are assumed to be far apart, occupying the two spatial orbitals

$$\varphi_1(x) = \left(\frac{\gamma_x}{\pi}\right)^{1/4} \mathrm{e}^{-\frac{\gamma_x}{2}(x-q_{x,\alpha})^2 + \mathrm{i}p_{x,\alpha}(x-q_{x,\alpha})}, \tag{13}$$

$$\varphi_2(y) = \left(\frac{\gamma_y}{\pi}\right)^{1/4} \mathrm{e}^{-\frac{\gamma_y}{2}(y-q_{y,\alpha})^2 + \mathrm{i}p_{y,\alpha}(y-q_{y,\alpha})}, \tag{14}$$

with $q_{x,\alpha} = 0, p_{x,\alpha} = 0$ and $q_{y,\alpha} = 15, p_{y,\alpha} = -2$. The first orbital initially describes an electron at rest and at the origin, and the second a far away electron, moving towards the first electron. These spatial orbitals can be combined either to create a singlet

$$\begin{aligned}\Psi_\mathrm{s}(x,y) &= \frac{1}{\sqrt{2}}\left(\frac{1}{\sqrt{2}}\begin{vmatrix} \varphi_{1\uparrow} & \varphi_{1\uparrow} \\ \varphi_{2\downarrow} & \varphi_{2\downarrow}\end{vmatrix} - \frac{1}{\sqrt{2}}\begin{vmatrix} \varphi_{1\downarrow} & \varphi_{1\downarrow} \\ \varphi_{2\uparrow} & \varphi_{2\uparrow}\end{vmatrix}\right) \\ &= \frac{1}{2}\left(\varphi_1(x)\varphi_2(y) + \varphi_2(x)\varphi_1(y)\right)\left(|\uparrow\downarrow\rangle - |\downarrow\uparrow\rangle\right)\end{aligned} \tag{15}$$

or a triplet

$$\begin{aligned}\Psi_\mathrm{t}(x,y) &= \frac{1}{\sqrt{2}}\begin{vmatrix} \varphi_{1\uparrow} & \varphi_{1\uparrow} \\ \varphi_{2\uparrow} & \varphi_{2\uparrow}\end{vmatrix} \\ &= \frac{1}{\sqrt{2}}\left(\varphi_1(x)\varphi_2(y) - \varphi_2(x)\varphi_1(y)\right)|\uparrow\uparrow\rangle\end{aligned} \tag{16}$$

wave function. The energy of the two states is identical at all times, since at $t = 0$ the electrons are far apart, and the total energy is conserved. A graphical display of the real part of the spatial part of the two initial states is given in Fig. 1. The structure in the upper left corner of both figures is identical, whereas the one in the lower right corner has opposite sign.

An exact quantum mechanical propagation leads to a time-dependent one particle electron density

$$\rho_\mathrm{s}(x,t) = \int \mathrm{d}y \rho_\mathrm{s}(\mathbf{x},t) \tag{17}$$

for the singlet, which illustrates the scattering, see Fig. 2. If the same calculation is repeated for the triplet state, we obtain a different time-dependent density $\rho_\mathrm{t}(x,t)$, where the non-zero blue area at the end is notably wider. To emphasize the difference between the two, we show the difference density $\Delta\rho(x,t) = \rho_\mathrm{t}(x,t) - \rho_\mathrm{s}(x,t)$ as a function of time in the third panel of Fig. 2. In the magenta area, $\rho_\mathrm{s} > \rho_\mathrm{t}$ (Fermi hole), while in the green area $\rho_\mathrm{s} < \rho_\mathrm{t}$. This demonstrates the Pauli principle: Two electrons in a triplet cannot come as close to each other as two electrons in a singlet.

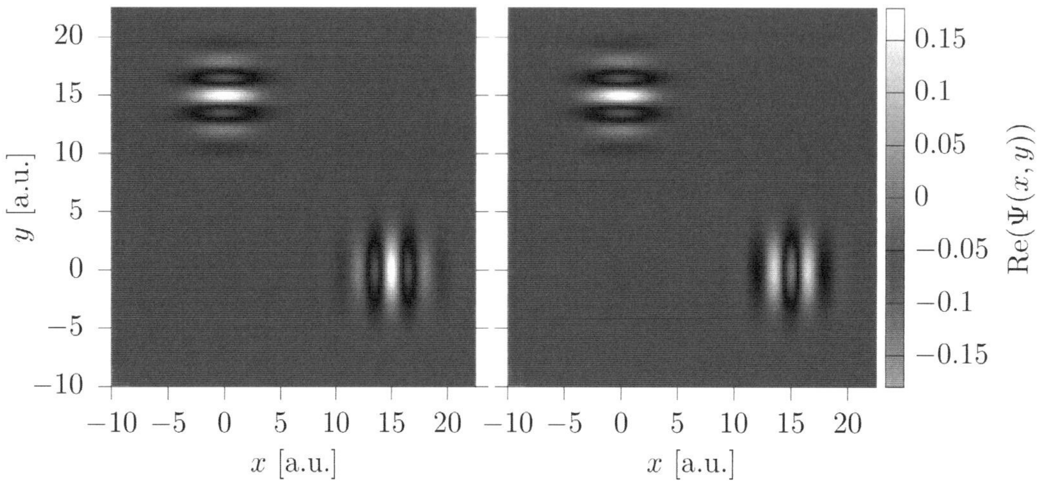

Figure 1. Real part of the initial spatial part of the two-electron wavefunction in the singlet (left panel) triplet (right panel) case.

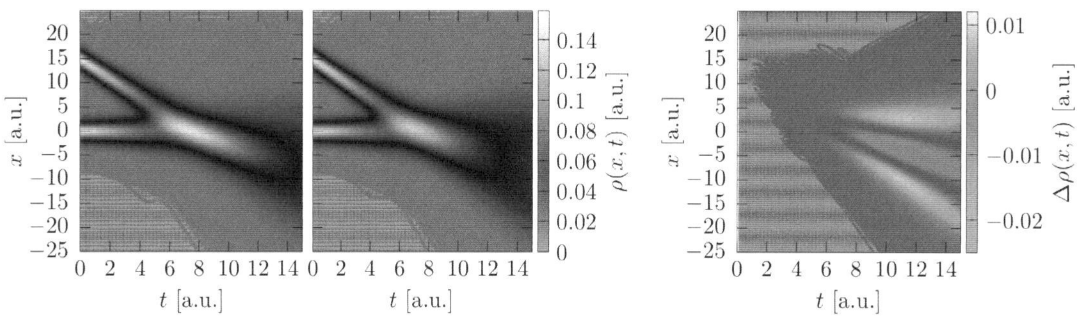

Figure 2. Time dependent singlet (left panel), triplet (middle panel), and difference (right panel) one particle electron density from an exact quantum dynamical calculation.

4. Semiclassical and classical numerical results

After having set the stage with the full quantum results, which can still be calculated easily for just two degrees of freedom, we now show the numerical results of a full semiclassical calculation as well as of an LSC-IVR calculation of the observable

$$\langle(\hat{x}-\hat{y})^2\rangle^{1/2} = \sqrt{\int \mathrm{d}^2x \Psi^*(\mathbf{x},t)(x-y)^2\Psi(\mathbf{x},t)} = \sqrt{\mathrm{Tr}[\hat{\rho}(\hat{x}-\hat{y})^2]}, \tag{18}$$

based on the Herman-Kluk propagator and its linearized approximation, respectively. Here the abbreviation $\mathrm{d}^2x = \mathrm{d}x\mathrm{d}y$ has been used. For the pure initial states that we propagate, the full semiclassical calculations have been performed on the wavefunction level, according to

$$\Psi(\mathbf{x},t) = \int \mathrm{d}^2x' K(\mathbf{x},t;\mathbf{x}',0)\Psi(\mathbf{x}',0). \tag{19}$$

The overlap between the initial wavefunction and the Gaussian that appears in the propagator is given by (see, e.g., [17] for the single Gaussian case)

$$
\begin{aligned}
\langle g_\gamma(\mathbf{q},\mathbf{p},0)|\Psi_s\rangle \;=\;& \Big[\mathrm{e}^{-\frac{1}{4}(\mathbf{q}-\mathbf{q}_\alpha)^{\mathrm{T}}\boldsymbol{\gamma}(\mathbf{q}-\mathbf{q}_\alpha)+\frac{\mathrm{i}}{2}(\mathbf{q}-\mathbf{q}_\alpha)^{\mathrm{T}}(\mathbf{p}+\mathbf{p}_\alpha)-\frac{1}{4}(\mathbf{p}-\mathbf{p}_\alpha)^{T}\boldsymbol{\gamma}^{-1}(\mathbf{p}-\mathbf{p}_\alpha)} \\
&+\mathrm{e}^{-\frac{1}{4}(\mathbf{q}-\tilde{\mathbf{q}}_\alpha)^{\mathrm{T}}\boldsymbol{\gamma}(\mathbf{q}-\tilde{\mathbf{q}}_\alpha)+\frac{\mathrm{i}}{2}(\mathbf{q}-\tilde{\mathbf{q}}_\alpha)^{\mathrm{T}}(\mathbf{p}+\tilde{\mathbf{p}}_\alpha)-\frac{1}{4}(\mathbf{p}-\tilde{\mathbf{p}}_\alpha)^{\mathrm{T}}\boldsymbol{\gamma}^{-1}(\mathbf{p}-\tilde{\mathbf{p}}_\alpha)}\Big] \\
&\frac{1}{2}(|\uparrow\downarrow\rangle - |\downarrow\uparrow\rangle) \qquad (20)
\end{aligned}
$$

for the singlet case and a corresponding formula with a relative minus sign in the first bracket for the triplet case. The vectors appearing in the exponent are defined by

$$
\mathbf{q}_\alpha^{\mathrm{T}} = (q_{x,\alpha}, q_{y,\alpha}); \qquad \tilde{\mathbf{q}}_\alpha^{\mathrm{T}} = (q_{y,\alpha}, q_{x,\alpha}) \qquad (21)
$$
$$
\mathbf{p}_\alpha^{\mathrm{T}} = (p_{x,\alpha}, p_{y,\alpha}); \qquad \tilde{\mathbf{p}}_\alpha^{\mathrm{T}} = (p_{y,\alpha}, p_{x,\alpha}). \qquad (22)
$$

As usual, the Gaussians in phase space in Eq. (20) serve as weight functions for a Monte-Carlo integration scheme with random initial conditions for the classical trajectories [30, 31]. Initially, the two distinct contributions to the wavefunction corresponding to the two classes of trajectories centered around $(\mathbf{q}_\alpha, \mathbf{p}_\alpha)$ and $(\tilde{\mathbf{q}}_\alpha, \tilde{\mathbf{p}}_\alpha)$ do not overlap and there will be only "intra-interference" of trajectories coming from the same class and no difference between singlet and triplet. As time increases, however, we expect a difference in the final result to occur, as soon as the two classes of trajectories approach each other and now also can "inter-interfere". The singlet and triplet results will be different due to the difference in sign of the two contributions.

In the classical case, a density matrix expression analogous to Eq. (11) will be used. To this end, the correlation function has first to be adapted to the case of the initial wavefunctions given in Eqs. (15,16), however. Because the initial density (looking only at the spatial part of the wavefunction)

$$
\begin{aligned}
\rho(x,y,0) &= \Psi^*(x,y)\Psi(x,y) \\
&= \frac{1}{4}\left[\varphi_1^*(x)\varphi_2^*(y) \pm \varphi_2^*(x)\varphi_1^*(y)\right]\left[\varphi_1(x)\varphi_2(y) \pm \varphi_2(x)\varphi_1(y)\right] \\
&= \frac{1}{4}\Big\{\varphi_1(x)\varphi_1^*(x)\varphi_2(y)\varphi_2^*(y) + \varphi_2(x)\varphi_2^*(x)\varphi_1(y)\varphi_1^*(y) \\
&\quad \pm\varphi_1(x)\varphi_2^*(x)\varphi_2(y)\varphi_1^*(y) \pm \varphi_2(x)\varphi_1^*(x)\varphi_1(y)\varphi_2^*(y)\Big\} \qquad (23)
\end{aligned}
$$

allows for different indices of the Gaussian wavepackets, $i, j = 1, 2$, we have to generalize the projection operator from (8). For a single one out of the four contributions to the dynamics of the initial density

$$
\rho_{i,i';j,j'}(x,y,0) = \varphi_i(x)\varphi_{i'}^*(x)\varphi_j(y)\varphi_{j'}^*(y), \qquad (24)
$$

we then get the final Wigner expression

$$
\begin{aligned}
\rho_{i,i';j,j'}(x,y,t) = & \frac{1}{(\pi\hbar)^2}\int \mathrm{d}p_x\mathrm{d}p_y\mathrm{d}q_x\mathrm{d}q_y\; \delta\left(x - q_{x,t}\right)\delta\left(y - q_{y,t}\right) \\
\times\quad & \mathrm{e}^{-\frac{\gamma_x}{2}\left[(q_x-(\mathbf{q}_\alpha)_i)^2+(q_x-(\mathbf{q}_\alpha)_{i'})^2\right]+\frac{\gamma_x}{4}((\mathbf{q}_\alpha)_i-(\mathbf{q}_\alpha)_{i'})^2-\frac{1}{\hbar^2\gamma_x}\left(p_x-\frac{(\mathbf{p}_\alpha)_i+(\mathbf{p}_\alpha)_{i'}}{2}\right)^2} \\
\times\quad & \mathrm{e}^{-\frac{\mathrm{i}}{\hbar}\left[\frac{1}{2}((\mathbf{q}_\alpha)_i+(\mathbf{q}_\alpha)_{i'})((\mathbf{p}_\alpha)_i-(\mathbf{p}_\alpha)_{i'})-q_x((\mathbf{p}_\alpha)_i-(\mathbf{p}_\alpha)_{i'})+p_x((\mathbf{q}_\alpha)_i-(\mathbf{q}_\alpha)_{i'})\right]} \\
\times\quad & \mathrm{e}^{-\frac{\gamma_y}{2}\left[(q_y-(\mathbf{q}_\alpha)_j)^2+\left(q_y-(\mathbf{q}_\alpha)_{j'}\right)^2\right]+\frac{\gamma_y}{4}\left((\mathbf{q}_\alpha)_j-(\mathbf{q}_\alpha)_{j'}\right)^2-\frac{1}{\hbar^2\gamma_y}\left(p_y-\frac{(\mathbf{p}_\alpha)_j+(\mathbf{p}_\alpha)_{j'}}{2}\right)^2} \\
\times\quad & \mathrm{e}^{-\frac{\mathrm{i}}{\hbar}\left[\frac{1}{2}\left((\mathbf{q}_\alpha)_j+(\mathbf{q}_\alpha)_{j'}\right)\left((\mathbf{p}_\alpha)_j-(\mathbf{p}_\alpha)_{j'}\right)-q_y\left((\mathbf{p}_\alpha)_j-(\mathbf{p}_\alpha)_{j'}\right)+p_y\left((\mathbf{q}_\alpha)_j-(\mathbf{q}_\alpha)_{j'}\right)\right]}, \qquad (25)
\end{aligned}
$$

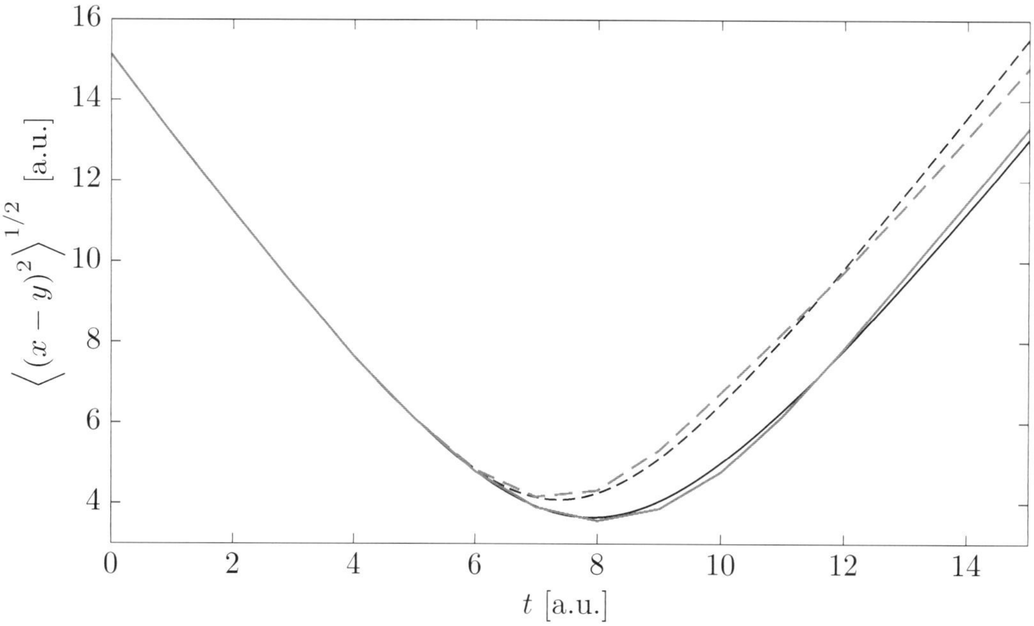

Figure 3. Multiple trajectory HK result for the distance between two electrons starting in a singlet state (solid red line) and a triplet state (dashed red line). For a comparison, full quantum results are displayed in black.

which (in principle) contains negativities due to the phase terms. The two "off-diagonal" terms with $i \neq i'$ and $j \neq j'$ that could in principle account for a difference between singlet and triplet, are vanishingly small, however. This can be seen as follows: Consider the case where $i = 1$ and $i' = 2$. Then in order for the exponent to be close to zero and therefore the contribution to the integral to be large, q_x would have to be around zero as well as around 15 at the same time, which is impossible. The off diagonal terms can therefore be safely neglected. This also immediately leads to the observation that there is no difference to be expected between singlet and triplet in the LSC-IVR results. We will have just two finite contributions to the LSC-IVR result, that are the same in both the singlet as well as the triplet case and each of these contributions is in the form given by (11), one with initial phase space center $(\mathbf{q}_\alpha, \mathbf{p}_\alpha)$ and the other one with the center $(\tilde{\mathbf{q}}_\alpha, \tilde{\mathbf{p}}_\alpha)$.

Let us now come to a comparison of the numerical results. The HK results have been obtained by propagating the wavefunction in grid representation, which emerges by analytical integration over d^2x' in Eq. (19). The remaining phase space integration was performed numerically, using 10^5 trajectories for converged results. The expectation values were then found in the usual way according to Eq. (18). In the LSC-IVR case, we employed the density from Eq. (28) and calculated the expectation value grid free, by performing the integration over d^2x analytically, which is very easy as it only amounts to evaluating the delta-functions. The phase space integration was then performed numerically, using 10^4 trajectories for converged results. As can be seen in Fig. 3, by a comparison to full quantum calculations, the HK results do account for the difference between singlet and triplet to a very high degree, see also [4]. The classical Wigner results for the same quantity, however, do not show any difference between the two cases [32]. They are both identical and lie right in-between the singlet and triplet quantum results as can be seen in Fig. 4. The set of trajectories used to calculate the (converged) classical results is a subset of the ones used for the HK results. Due to the neglect of the phase information

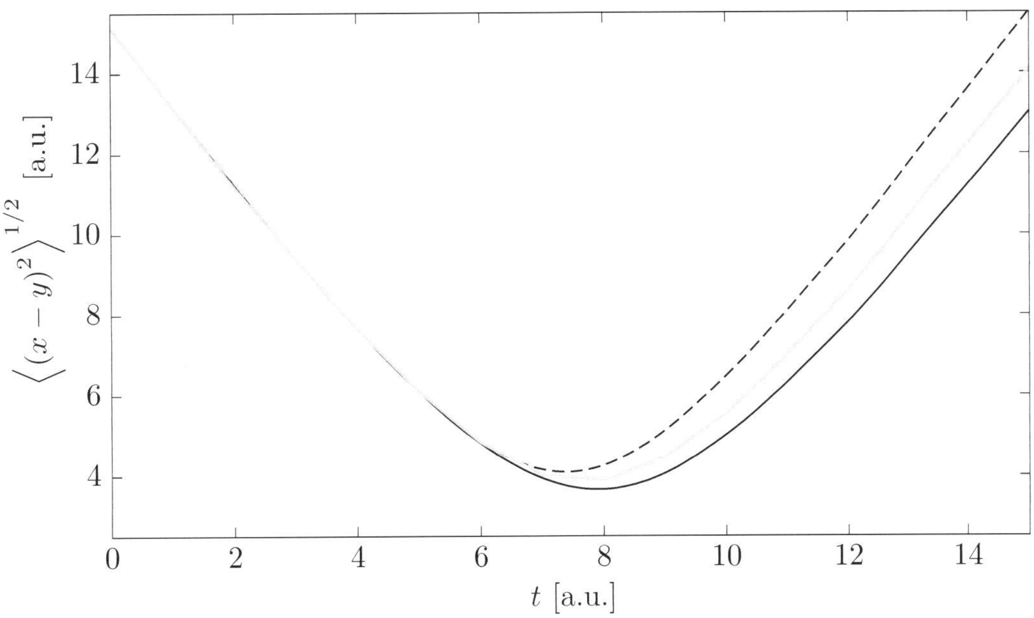

Figure 4. Multiple trajectory classical Wigner result for the distance between two electrons starting in a singlet state (solid green line) and a triplet state (dashed green line). The two results are on top of each other. For a comparison, full quantum results are displayed in black.

by working with the density, the classical results show no difference between singlet and triplet, however!

5. Conclusions and Outlook

We have worked out that the difference between singlet and triplet initial state scattering of two electrons cannot be accounted for by a classical (truncated) Wigner method based on multiple trajectories. Both cases, singlet and triplet, lead to the same result for all times, and especially to the same long-time limit. This is in stark contrast to the HK result presented previously in [4] and repeated here for convenience. We are therefore led to the conclusion that the difference between singlet and triplet scattering is an interference effect. This conclusion had been made before by a comparison to the single trajectory Gaussian wavepacket results, but is even more convincing now that we have used a similar set of trajectories as in the HK calculation, without taking into account the phase information. We also stress that there was no need in the semiclassical HK calculation to invoke a fermionic molecular dynamics extension [33] or use Grassmann variable path integrals [34]. The symmetrization/antisymmetrization of the wavefunction is at the heart of the difference in interference in the HK approach and was also used successfully for the extraction of spectra of the different "species" of Helium [35]. In the classical (truncated Wigner) approach it was shown that the symmetrization on the level of the density is not leading to any observable difference between singlet and triplet.

An interesting project for future investigations will be the semiclassical investigation of the dynamics of more than two electrons. A recent study related to this generalization deals with the leading correction due to symmetrization in a quantum (Bohmian) trajectory setup [36].

Acknowledgements
The authors gratefully acknowledge the collaboration with M. Nest and E. Pollak, from which parts of the reviewed results originate. FG would like to thank for the opportunity to present this work at the conference "Symmetries in Science XVII", held in Bregenz.

References
[1] Harabati C and Kay K G 2007 *J. Chem. Phys.* **127** 084104
[2] Elliott P and Maitra N T 2011 *J. Chem. Phys.* **135** 104110
[3] Grossmann F and Kramer T 2011 *J. Phys. A: Math. Theor.* **44** 445309
[4] Grossmann F, Buchholz M, Pollak E and Nest M 2014 *Phys. Rev. A* **89** 032104
[5] Wilets L, Henley E, Kraft M and MacKellar A 1977 *Nuc. Phys. A* **282** 341
[6] Heller E J 1975 *J. Chem. Phys.* **62** 1544
[7] Feynman R P and Hibbs A R 2010 *Quantum Mechanics and Path Integrals* emended ed (Mineola: Dover)
[8] Grossmann F 1999 *Comm. At. Mol. Phys.* **34** 141
[9] Grossmann F 2013 *Theoretical Femtosecond Physics: Atoms and Molecules in Strong Laser Fields* 2nd ed (Berlin, Heidelberg: Springer)
[10] Herman M F and Kluk E 1984 *Chem. Phys.* **91** 27
[11] Heller E J 1981 *J. Chem. Phys.* **75** 2923
[12] van de Sand G and Rost J M 1999 *Phys. Rev. Lett.* **83** 524
[13] Zagoya C, Goletz C M, Grossmann F and Rost J M 2012 *Phys. Rev. A* **85** 041401(R)
[14] Ray S, Ostmann P, Simon L, Grossmann F and Strunz W T 2016 *J. Phys. A* **49** 165303
[15] Zagoya C, Schulman L and Grossmann F 2014 *J. Phys. A* **47** 165102
[16] Miller W H 2001 *J. Phys. Chem. A* **105** 2942 and references therein
[17] Grossmann F 2006 *J. Chem. Phys.* **125** 014111
[18] Kay K G 2013 *Phys. Rev. A* **88**(1) 012122
[19] Heller E J 1976 *J. Chem. Phys.* **65** 1289
[20] Sun X, Wang H and Miller W H 1998 *J. Chem. Phys.* **109** 4190
[21] Wang H, Sun X and Miller W H 1998 *J. Chem. Phys.* **108** 9726
[22] Herman M F and Coker D F 1999 *J. Chem. Phys.* **111** 1801
[23] Noid W G, Ezra G S and Loring R F 2003 *J. Chem. Phys.* **119** 1003
[24] Grossmann F 2016 *Physica Scripta* **91** 044004
[25] Wigner E 1932 *Phys. Rev.* **40** 749
[26] Jaffé C and Brumer P 1985 *J. Chem. Phys.* **82** 2330
[27] Møller K B, Dahl J P and Henriksen N E 1994 *J. Phys. Chem.* **98** 3272
[28] Hillery M, O'Connell R F, Scully M O and Wigner E P 1984 *Phys. Rep.* **106** 121
[29] Bauer D 1997 *Phys. Rev. A* **56** 3028
[30] Kluk E, Herman M F and Davis H L 1986 *J. Chem. Phys.* **84** 326
[31] Kay K G 1994 *J. Chem. Phys.* **100** 4432
[32] Buchholz M 2017 *Semiclassical initial value representations for complex dynamics* Ph.D. thesis Technische Universität Dresden
[33] Kirrander A and Shalashilin D V 2011 *Phys. Rev. A* **84** 033406
[34] Engl T, Plössl P, Urbina J D and Richter K 2014 *Theor. Chem. Acc.* **133** 1563
[35] Tanner G, Richter K and Rost J M 2000 *Rev. Mod. Phys.* **72** 497
[36] Gu B, Rassolov V and Garashchuk S 2016 *Theor. Chem. Acc.* **135** 267

IOP Conf. Series: Journal of Physics: Conf. Series **1071** (2018) 012005 doi:10.1088/1742-6596/1071/1/012005

Invariant functions of vector field realizations of Lie algebras and some applications to representation theory and dynamical systems

Rutwig Campoamor-Stursberg
Instituto de Matemática Interdisciplinar IMI-UCM,
Plaza de Ciencias 3, E-28040 Madrid, Spain
E-mail: rutwig@ucm.es

Abstract. Realizations of Lie algebras by means vector fields associated to a linear representation and their corresponding invariant functions are inspected from the perspective of the embedding problem of Lie algebras and the branching rules for the subduction of representations. Some applications concerning the construction of dynamical systems with prescribed Lie algebras of point symmetries and the symmetry breaking with respect to embeddings of algebras into $\mathfrak{so}(N)$ are discussed.

To the memory of Sigitas Ališauskas

1. Introduction

The literature of realizations of Lie algebras as vector fields on manifolds is extensive, having many ramifications in different disciplines, ranging from the theory of differential equations and differential geometric problems to cohomology theories as well a s integrable systems and General Relativity (see e.g. [1, 2, 3, 4, 5] and references therein). Among the pioneering contributions to the field, we may refer to the work of Dickson on differential equations from the Lie group perspective [6], an approach that revitalized Lie's classical theory and motivated later studies on the Lie symmetry method [7, 8] from both the mathematical and physical perspective [9]. Special types of realizations, corresponding to the coadjoint representation of Lie algebras, have been shown to be an indispensable tool in the modelization and interpretation of physical phenomena [10, 11], such as the symmetry groups of physical systems, and constitutes nowadays a well-established research subject. A relevant research line that combines the analytical and geometrical properties of Lie groups is given by those systems of differential equations admitting a nonlinear superposition principle [12, 13, 14], an approach that has provided new tools for the elucidation of physical properties at the classical and quantum level.

In this work we analyze some features of realizations of Lie algebras by vector fields from the perspective of the representation theory and its relation to the branching rules of representations for embeddings of (simple) Lie algebras. It turns out that the analysis of the invariants of a given realization can provide useful information concerning the embedding of a Lie algebra $\mathfrak{g}'$ into another algebra $\mathfrak{g}$, allowing to determine the corresponding conjugacy class within $\mathfrak{g}$.

This can further be used to decide whether a subalgebra in a given realization corresponds to an irreducible embedding, or either determine the multiplicity of the trivial representation in the branching rules. An an application to second-order dynamical systems, it is shown that invariants of realizations can be used for the construction of (non-conservative) Lagrangian systems having an exact prescribed Lie algebra of Lie point and/or Noether symmetries.

Unless otherwise stated, any Lie algebra $\mathfrak{g}$ and any representation is considered over the field $\mathbb{R}$ of real numbers. The Einstein summation convention is used.

2. Realizations of Lie algebras by vector fields

Let $\{X_1, \cdots, X_r\}$ be a basis of a Lie algebra $\mathfrak{g}$ and consider a realization by vector fields in n-coordinates, i.e., a map $\Phi : \mathfrak{g} \rightarrow \mathfrak{X}(\mathbb{R}^n)$ such that $\Phi(X_\alpha) \neq 0$ holds for all α and satisfying the property $\Phi[X_\alpha, X_\beta] = [\Phi(X_\alpha), \Phi(X_\beta)]$ for all α, β. In matrix terms, the map Φ can be described as

$$(\Phi(X_\alpha))_\alpha = A_\Phi(\mathfrak{g}) \frac{\partial}{\partial \mathbf{x}}, \tag{1}$$

with $A_\Phi(\mathfrak{g}) = \left(\xi_i^j(\mathbf{x})\right) \in M_{r\times n}(C^\infty(\mathbb{R}^n))$ being a the functional matrix that specifies the components of $\mathbf{X}_\alpha = \Phi(X_\alpha)$ in the coordinates $\left\{x^1, \cdots, x^n\right\}$. We recall that $\{\mathbf{X}_1, \cdots, \mathbf{X}_s\}$ are linearly independent as first-order differential operators if a dependence relation

$$\lambda^1(\mathbf{x})\mathbf{X}_1 + \cdots + \lambda^s(\mathbf{x})\mathbf{X}_s = 0 \tag{2}$$

with $\lambda^1(\mathbf{x}) \in C^\infty(\mathbb{R}^n)$ implies that $\lambda^1(\mathbf{x}) = \cdots = \lambda^s(\mathbf{x}) = 0$. In this context, it is important to observe that albeit the generators $X_1, \ldots, X_r$ are independent, this does generally not imply that the vector fields $\mathbf{X}_1, \cdots, \mathbf{X}_r$ are linearly independent.[1] The number of linearly independent vector fields is given by the rank $r_\Phi(\mathfrak{g})$ of the coefficient matrix $A_\Phi(\mathfrak{g})$, from which $r_\Phi(\mathfrak{g}) \leq \dim \mathfrak{g} = r$ follows. As obviously the maximal possible number of linearly independent vector fields is given by the number of independent variables n, in any case the bound $r_\Phi(\mathfrak{g}) \leq \min(\dim \mathfrak{g}, n)$ holds. Dependence relations of the type (2) are necessarily given whenever the inequality $n < \dim \mathfrak{g}$ holds.

Invariant functions of a realization $\Phi : \mathfrak{g} \rightarrow \mathfrak{X}(\mathbb{R}^n)$ correspond to elements $F(\mathbf{x}) \in C^\infty(\mathbb{R}^n)$ such that the linear first-order system of PDEs

$$A_\Phi(\mathfrak{g}) \frac{\partial F}{\partial \mathbf{x}} = \begin{pmatrix} \xi_1^1 & \cdots & \xi_1^n \\ \vdots & & \vdots \\ \xi_r^1 & \cdots & \xi_r^n \end{pmatrix} \begin{pmatrix} \frac{\partial F}{\partial x^1} \\ \vdots \\ \frac{\partial F}{\partial x^n} \end{pmatrix} = 0 \tag{3}$$

is satisfied. As the vector fields span a finite-dimensional Lie algebra, the system (3) is always complete, and can thus be reduced to its Jacobian form [15]. A particularly relevant case is given when $A_\Phi(\mathfrak{g})$ is the coefficient matrix of the realization associated to the coadjoint representation of the Lie algebra $\mathfrak{g}$, where these functions coincide with the (generalized) Casimir invariants of $\mathfrak{g}$ [10]. In the general case, if the matrix $A_\Phi(\mathfrak{g})$ has rank $r_\Phi(\mathfrak{g})$, the solution of (3) is given in terms of $\chi_0 = n - r_\Phi(\mathfrak{g})$ independent functions J_k that can be taken as the fundamental invariants. We observe that if J is a solution of (3), then for any vector field $\mathbf{X}_\alpha \in \Phi(\mathfrak{g})$ the condition

$$\mathbf{X}_\alpha(J) = \xi_\alpha^l(\mathbf{x}) \frac{\partial J}{\partial x^l} = \xi_\alpha^l(\mathbf{x}) \frac{\partial J}{\partial x^k} dx^k \left(\frac{\partial}{\partial x^l}\right) = dJ(\mathbf{X}_\alpha) = 0 \tag{4}$$

[1] While for $\mathfrak{g}$ a dependence relation is always expressed by scalars, for $\Phi(\mathfrak{g})$ dependence relations are given in terms of functions.

IOP Conf. Series: Journal of Physics: Conf. Series **1071** (2018) 012005 doi:10.1088/1742-6596/1071/1/012005

is satisfied, showing that the total differential dJ of the scalar field J annihilates the realization [2].

Assuming that the rank of $A_{\Phi}(\mathfrak{g})$ is $k < n$, we can suppose without loss of generality that, after a reordering of the indices, the vector fields $\mathbf{X}_i$ with $1 \leq i \leq k$ are linearly independent. For the remaining generators we thus have dependence relations

$$\mathbf{X}_{k+s} = \lambda_s^1(\mathbf{x})\mathbf{X}_1 + \cdots \lambda_s^k(\mathbf{x})\mathbf{X}_k, \; 1 \leq s \leq r-k. \tag{5}$$

In these conditions, we can find linearly independent differential 1-forms $\omega^j \in \Omega^1(\mathbb{R}^n)$ such that $\omega^i(\mathbf{X}_j) = \delta_i^j$ for $1 \leq i,j \leq k$ and $\omega^i(\mathbf{X}_{k+s}) = \lambda_s^k(\mathbf{x})$ for $1 \leq s \leq r-k$. As $k < n$, the set $\{\omega^1, \cdots, \omega^k\}$ can be augmented with $(n-k)$ 1-forms θ^l $(1 \leq l \leq n-k)$ such that $\theta^l(\mathbf{X}_j) = 0$ for $1 \leq j \leq k$ and $1 \leq l \leq n-k$, implying that θ^l belongs to the annihilator $\mathrm{Ann}(\Phi(\mathfrak{g}))$ of the realization $\Phi(\mathfrak{g})$ [2]. Hence, if J is an invariant of the realization, then necessarily $dJ \in \mathrm{Ann}(\Phi(\mathfrak{g}))$. As the vector fields span the Lie algebra $\mathfrak{g}$, it follows that $\mathrm{Ann}(\Phi(\mathfrak{g}))$ is a differential ideal, meaning that for each $\theta \in \mathrm{Ann}(\Phi(\mathfrak{g}))$, the exterior derivative $d\theta$ also belongs to $\mathrm{Ann}(\Phi(\mathfrak{g}))$ [1]. We remark that this reformulation in terms of differential forms can be seen as an algebraic approach to find the solutions of the system (3), up to the determination of integrating factors. Indeed, if $\theta \in \mathrm{Ann}(\Phi(\mathfrak{g}))$, we can ask whether there exists some function $J \in C^{\infty}(\mathbf{R}^n)$ such that $dJ = \theta$. Although θ is not necessarily an exact 1-form, we can (at least locally, under certain natural assumptions) find functions f, J' such that $\theta = f dJ'$. The condition to be satisfied for this to happen (see e.g. [1]) is that

$$\theta \wedge d\theta = 0 \tag{6}$$

holds. This result can be easily extended to a (local) basis of independent one-forms ω^i $(1 \leq i \leq \chi_0)$ in the annihilator $\mathrm{Ann}(\Phi(\mathfrak{g}))$ of the realization, requiring that for the relative volume form $\Omega = \omega^1 \wedge \cdots \wedge \omega^{\chi_0}$, the constraints

$$d\omega^i \wedge \Omega = 0, \; 1 \leq i \leq \chi_0 \tag{7}$$

are satisfied. As a consequence, for any invariant J of the realization (3), it follows that the total differential is a combination of the ω^i:

$$dJ = \varphi_j(\mathbf{x})\omega^i, \; \varphi(\mathbf{x}) \in C^{\infty}(\mathbb{R}^n). \tag{8}$$

Letting $\omega = \omega_i(\mathbf{x})\, dx^i \in \Omega^1(\mathbb{R}^n)$, the action of ω on the (linearly independent) vector fields $\mathbf{X}_{\alpha}$ leads to relations of the type

$$\omega(\mathbf{X}_{\alpha}) = \omega_i(\mathbf{x})\, \xi_{\alpha}^i(\mathbf{x}) = 0, \; 1 \leq \alpha \leq r_{\Phi}(\mathfrak{g}) \tag{9}$$

that can formally be solved with respect to $\chi_0 = n - r_{\Phi}(\mathfrak{g})$ components $\omega_1, \cdots, \omega_{\chi_0}$ of ω, so that the latter is expressed as

$$\omega = \sum_{k=1}^{\chi_0} \omega_k(\mathbf{x})\, dx^k + \sum_{l=\chi_0+1}^{n} \Psi_l\left(\omega_1, \cdots, \omega_{\chi_0}, \xi_{\alpha}^1, \cdots, \xi_{\alpha}^n\right) dx^l. \tag{10}$$

Evaluating now the exterior derivative $d\omega = 0$ provides us with a system of conditions on the components ω_k that eventually allow us to derive independent scalar fields $\{\Theta_1(\mathbf{x}), \cdots, \Theta_{\chi_0}(\mathbf{x})\}$ such that the relations

$$\omega_k = \sum_{a}^{\chi_0} \frac{\partial \Theta_a}{\partial x^k}, \; \Psi_l = \sum_{a}^{\chi_0} \frac{\partial \Theta_a}{\partial x^l}, \; 1 \leq k \leq \chi_0, \; \chi_0 + 1 \leq l \leq n \tag{11}$$

IOP Conf. Series: Journal of Physics: Conf. Series **1071** (2018) 012005 doi:10.1088/1742-6596/1071/1/012005

are simultaneously satisfied. It then follows that $\omega = d\Theta_1 + \cdots + d\Theta_{\chi_0}$ is a total differential, and the solutions $\{\Theta_1(\mathbf{x}), \cdots, \Theta_{\chi_0}(\mathbf{x})\}$ can be taken as a set of fundamental invariants of the system. Although this procedure is certainly not simpler than solving directly the first-order system (3) by the usual methods [15], it gives some hints on the structure of the invariants, specially for nonlinear realizations.

As an illustrating example, consider the five-dimensional (solvable) Lie algebra $\mathfrak{g}$ generated by the vector fields in $\mathbb{R}^5$:

$$\begin{array}{l}\mathbf{X}_1 = x^1\frac{\partial}{\partial x^5}, \quad \mathbf{X}_2 = \left(x^1+x^2\right)\frac{\partial}{\partial x^5}, \quad \mathbf{X}_3 = \left(x^2+x^3\right)\frac{\partial}{\partial x^5}, \quad \mathbf{X}_4 = \left(x^3+x^4\right)\frac{\partial}{\partial x^5},\\ \mathbf{X}_5 = x^1\frac{\partial}{\partial x^1} + \left(x^1+x^2\right)\frac{\partial}{\partial x^2} + \left(x^2+x^3\right)\frac{\partial}{\partial x^3} + \left(x^3+x^4\right)\frac{\partial}{\partial x^4}.\end{array}$$

If $\omega = \omega_i(\mathbf{x})dx^i \in \Omega^1\left(\mathbb{R}^5\right)$ belongs to the annihilator, the condition $\omega\left(\mathbf{X}_1\right) = 0$ implies that $\omega_5\left(\mathbf{x}\right) = 0$. A dependence condition relating the remaining components $\omega_i(\mathbf{x})$ is obtained from the identity $\omega\left(\mathbf{X}_5\right) = 0$:

$$x^1\omega_1\left(x\right) + \left(x^1+x^2\right)\omega_2\left(x\right) + \left(x^2+x^3\right)\omega_3\left(x\right) + \left(x^3+x^4\right)\omega_4\left(x\right) = 0.$$

As at most three of these functions can be chosen freely, the realization admits three invariants (alternatively, the rank of the matrix $A_\Phi\left(\mathfrak{g}\right)$ is two). Taking e.g. $\omega_3\left(\mathbf{x}\right) = \omega_4\left(\mathbf{x}\right) = 0$ gives the 1-form

$$\theta_1 = -\frac{\left(x^1+x^2\right)\omega_2\left(x\right)}{x^1}dx^1 + \omega_2\left(x\right)dx^2.$$

Now θ_1 is closed ($d\theta_1 = 0$) whenever $\omega_2\left(\mathbf{x}\right) = \frac{1}{x^1}F\left(\frac{x^2}{x^1} - \ln x^1\right)$ holds. The simplest solution can thus be taken as $\omega_2\left(\mathbf{x}\right) = \exp\left(-\frac{x^2}{x^1}\right)$. A routine computation shows that

$$\theta_1 = \exp\left(-\frac{x^2}{x^1}\right)\left(-\frac{\left(x^1+x^2\right)}{x^1}dx^1 + dx^2\right) = d\left(x^1\exp\left(-\frac{x^2}{x^1}\right)\right),$$

from which we conclude that θ_1 is an exact form. In analogous manner, the remaining fundamental invariants

$$J_2 = \frac{2x^1x^3 - \left(x^2\right)^2}{\left(x^1\right)^2}, \quad J_3 = \frac{3\left(x^1\right)^2x^4 - 3x^1x^2x^3 + \left(x^2\right)^3}{\left(x^1\right)^3}$$

can be easily found starting from the choices

$$\omega_1(\mathbf{x}) = \frac{2((x^2)^2 - x^1x^3)}{(x^1)^3}, \quad \omega_2(\mathbf{x}) = -\frac{2x^2}{(x^1)^2}, \quad \omega_3(\mathbf{x}) = \frac{2}{x^1}, \quad \omega_4(\mathbf{x}) = 0$$

and

$$\omega_1(\mathbf{x}) = \frac{6x^1x^2x^3 - 3(x^2)^3 - 2(x^1)^2x^4}{(x^1)^4}, \ \omega_2(\mathbf{x}) = \frac{3((x^2)^2 - x^1x^3)}{(x^1)^3}, \ \omega_3(\mathbf{x}) = -\frac{3x^2}{(x^1)^2}, \ \omega_4(\mathbf{x}) = \frac{2}{x^1},$$

respectively.

IOP Conf. Series: Journal of Physics: Conf. Series **1071** (2018) 012005 doi:10.1088/1742-6596/1071/1/012005

2.1. First-order systems and realizations

As already observed, vector field realizations of Lie algebras have important applications in the analysis of systems of ordinary differential equations, such as the study of superposition formulae [12, 13]. A first-order system

$$\dot{x}^i = F^i\left(t, \mathbf{x}\right), \quad 1 \leq i \leq n \tag{12}$$

is said to have a fundamental system of solutions if the general solution can be written in terms of m independent particular solutions $\{\mathbf{y}_1, \cdots, \mathbf{y}_m\}$ and n constants $\{C_1, \cdots, C_n\}$:

$$x^i = \varphi^i\left(\mathbf{y}_1, \cdots, \mathbf{y}_m, C_1, \cdots, C_n\right). \tag{13}$$

An expression of the latter type is called a (nonlinear) superposition formula or principle. A classical criterion due to Lie [12] establishes that such a principle exists whenever the independent variable t and the dependent coordinates x^i can be separated by means of a realization $\Phi : \mathfrak{g} \to \mathfrak{X}(\mathbb{R}^n)$ such that the identity

$$\dot{x}^i = F^i\left(t, \mathbf{x}\right) = \varphi_1(t)\xi_1^i\left(\mathbf{x}\right) + \cdots + \varphi_r(t)\xi_r^i\left(\mathbf{x}\right) \tag{14}$$

holds for each index, where $\mathbf{X}_\alpha = \xi_\alpha^l\left(\mathbf{x}\right)\frac{\partial}{\partial x^l}$ with $1 \leq \alpha \leq r = \dim \mathfrak{g}$ form a basis of $\mathfrak{g}$. A system (12) can be thought of as a (nonautonomous) dynamical system in phase space (albeit not necessarily Hamiltonian), for which reason the notion of cyclic coordinates arises at once. If for some $i_0 \in \{1, \cdots, n\}$ the constraint $\xi_\alpha^{i_0} = 0$ is satisfied for $1 \leq \alpha \leq r$, then clearly $\dot{x}^{i_0} = 0$ and x^{i_0} can be seen as a cyclic coordinate in the classical sense, implying that the system can be reduced by one degree of freedom, a fact that translates to the realization of the Lie algebra:

$$\widetilde{\mathbf{X}}_\alpha = \sum_{i=1}^{i_0-1} \xi_\alpha^i\left(\mathbf{x}\right)\frac{\partial}{\partial x^i} + \sum_{i_0+1}^{r} \xi_\alpha^i\left(\mathbf{x}\right)\frac{\partial}{\partial x^i},\ 1 \leq \alpha \leq r. \tag{15}$$

If $\Phi : \mathfrak{g} \to \mathfrak{X}\left(\mathbb{R}^n\right)$ is a realization of $\mathfrak{g}$ of rank $r_\Phi\left(\mathfrak{g}\right)$ and J is an invariant, then it follows at once from (14) that

$$\dot{J} = \dot{x}^i\frac{\partial J}{\partial x^i} = \sum_{s=1}^{r} \varphi_s(t)\xi_s^i(\mathbf{x})\frac{\partial J}{\partial x^i} = 0, \tag{16}$$

a fact that enables us to consider a new reference $\left\{y^1, \cdots, y^n\right\}$ in which $y^n = J$ is a (generalized) cyclic coordinate, as well as to reduce the system by one degree of freedom. Clearly, the maximal number of such cyclic coordinates is given by χ_0.

Under specific circumstances, it may be useful to simplify a realization normalizing one of the vector fields, i.e., introducing a reference $\left\{y^1, \cdots, y^n\right\}$ in which $\widetilde{\mathbf{X}}_1 = \frac{\partial}{\partial y^1}$ holds. Clearly, the new coordinates $\left\{y^1, \cdots, y^n\right\}$ are obtained as solutions of the system

$$\mathbf{X}_1\left(y^k\right) = \xi_1^1\left(\widetilde{\mathbf{x}}\right)\frac{\partial y^k}{\partial x^1} + \cdots + \xi_1^n\left(\widetilde{\mathbf{x}}\right)\frac{\partial y^k}{\partial x^1} = \delta_1^k, \tag{17}$$

with $\widetilde{\mathbf{x}} = \left(x^1, \cdots, x^n\right)$ and $2 \leq k \leq n$. This implies that the coordinates $\left\{y^2, \cdots, y^n\right\}$ are invariant functions of the differential operator $\mathbf{X}_1$, while y^1 is obtained as a particular solution of an inhomogeneous equation.[2]

[2] Depending on the solutions chosen, standard forms can correspond to non-equivalent realizations of the Lie algebra $\mathfrak{g}$. This may alter the physical interpretation of the symmetry generators.

3. Linear representations and embeddings

In this section, we analyze some properties of realizations related to linear representations of Lie algebras and its consequences for branching rules of representations in embedding problems $\mathfrak{g}' \subset \mathfrak{g}$.

Realizations of a Lie algebra $\mathfrak{g}$ by vector fields having linear components

$$\mathbf{X}_\alpha = a_{jk}^{\alpha} x^j \frac{\partial}{\partial x^k},\ 1 \leq \alpha \leq r, \tag{18}$$

naturally correspond to a (faithful) representation F of $\mathfrak{g}$, the matrix elements of which are obtained by matrix transposition:

$$F\left(\mathbf{X}_\alpha\right) = A_{\mathbf{X}_\alpha};\ \left(A_{\mathbf{X}_\alpha}\right)_{jk} = a_{kj}^{\alpha},\ 1 \leq j,k \leq n. \tag{19}$$

In this context, we convene to call a realization $\Phi : \mathfrak{g} \rightarrow \mathfrak{X}\left(\mathbb{R}^n\right)$ nonlinear if it is not equivalent, by a change of coordinates, to a realization arising from a linear representation of $\mathfrak{g}$. It follows that if the Lie algebra $\mathfrak{g}$ does not have a faithful n-dimensional real representation, any realization in n coordinates $\Phi : \mathfrak{g} \rightarrow \mathfrak{X}\left(\mathbb{R}^n\right)$ is necessarily nonlinear.

Consider an embedding $f : \mathfrak{g}' \longrightarrow \mathfrak{g}$ of a (simple) Lie algebra $\mathfrak{g}'$ into a (simple) Lie algebra $\mathfrak{g}$. Any embedding of Lie algebras determines an integer factor j_f given by the relation

$$\left(f(X), f(X'\right) = j_f\left(X, X'\right), \tag{20}$$

where $(.,.)$ is the usual scalar product in $\mathfrak{g}$ obtained from the root system. The scalar j_f is generally called the embedding index of $\mathfrak{g}'$ in the Lie algebra $\mathfrak{g}$ [16, 17]. Given disjoint subalgebras $\mathfrak{g}'_j$ of $\mathfrak{g}$, the direct sum of the subalgebras defines an embedding $f = \sum f_j$, the index of which is simply the sum of the various indices j_{f_j}. Further, for reduction chains $\mathfrak{g} \supset \mathfrak{g}' \supset \mathfrak{g}''$, the index of the last algebra in $\mathfrak{g}$ is the product of the corresponding indices of the chain members [18]. Given an embedding $f : \mathfrak{g}' \rightarrow \mathfrak{g}$ and a linear representation F of $\mathfrak{g}$, the embedding index is determined by the formula:

$$j_f = \frac{l_{fF}}{l_F}, \tag{21}$$

where l_F and l_{fF} denote the index of F and the subduced representation fF on the subalgebra, respectively. We recall that the index l_F of any representation is obtained from the quadratic Casimir operator C_2 of $\mathfrak{g}$ by means of the formula

$$l_F = \frac{\dim F}{\dim \mathfrak{g}} C_2(F). \tag{22}$$

Embeddings (and hence the index) are characterized by the branching rule obtained from the lowest dimensional (irreducible) representation of $\mathfrak{g}$, and have been tabulated for all simple Lie algebras (see e.g. [17, 19]).

Let $\Phi : \mathfrak{g} \rightarrow \mathfrak{X}\left(\mathbb{R}^n\right)$ be a realization and $\mathfrak{g}' \subset \mathfrak{g}$ a subalgebra. The restriction $\Phi|_{\mathfrak{g}'} : \mathfrak{g}' \rightarrow \mathfrak{X}\left(\mathbb{R}^n\right)$ provides us with a realization of $\mathfrak{g}'$ such that $r_{\Phi|_{\mathfrak{g}'}}\left(\mathfrak{g}'\right) \leq r_\Phi\left(\mathfrak{g}\right)$. Now, as we are dealing with a subalgebra, the subduced realization may admit cyclic coordinates, arising from invariant functions specifically related to the embedding $\mathfrak{g}' \subset \mathfrak{g}$. In this context, we remark that the missing label problem corresponds to a special case of subduced realizations, associated to the coadjoint representation of Lie algebras [20, 21, 23].

Proposition 1 *Let $\mathfrak{g}' \subset \mathfrak{g}$ be a proper subalgebra and $\Phi(\mathfrak{g})$ the realization associated to the representation Γ of $\mathfrak{g}$. Let $\Gamma \downarrow \Gamma'_1 \oplus \cdots \oplus \Gamma'_l$ be the branching rule as sum of representations of $\mathfrak{g}'$. If the multiplicity $\mathrm{mult}_{\Gamma'_0}(\Gamma)$ of the trivial representation Γ'_0 of $\mathfrak{g}'$ in Γ is k, then the subduced realization $\Phi|_{\mathfrak{g}'} : \mathfrak{g}' \to \mathfrak{X}(\mathbb{R}^n)$ admits k linear invariants.*

Suppose that $\Gamma \downarrow \Gamma'_1 \oplus \cdots \oplus \Gamma'_l$ contains k copies of the trivial representation Γ'_0 of $\mathfrak{g}'$. Reordering the coordinates if necessary, for any generator $X \in \mathfrak{g}'$, the representation matrix $A_F(X)$ decomposes as a block matrix with respect to the branching rule of Γ:

$$A_\Gamma(X) = \begin{pmatrix} A_{\Gamma'_1}(X) & & & \\ & A_{\Gamma'_2}(X) & & \\ & & \ddots & \\ & & & A^k_{\Gamma'_0}(X) \end{pmatrix}, \tag{23}$$

and it follows from (18) that the vector field $\mathbf{X}$ associated to X is given by

$$\mathbf{X} = \mathbf{x}\, A_\Gamma(X) \frac{\partial}{\partial \mathbf{x}}. \tag{24}$$

Now take the basis of 1-forms $\{dx^1, \cdots, dx^n\}$ dual to the coordinates. It is immediate to verify that $dx^j(\mathbf{X}) = 0$ holds for $j = n-k+1, \cdots, n$, showing that $\{x^{n-k+1}, \cdots, x^n\}$ are invariants of the subduced realization, from which the assertion follows.

It is clear that any invariant of a realization of $\mathfrak{g}$ is also an invariant for the subduced realization associated to the embedding $\mathfrak{g}' \subset \mathfrak{g}$. Additionally it may happen, depending on the particular branching rule, that such an invariant splits up as a sum of the invariants of the subalgebra, hence leading to independent invariants in $\mathfrak{g}'$. In this frame, it is worthy to be mentioned that the computation of invariants of realizations associated to linear representations is deeply connected with the problem of determining the generalized Casimir invariants of inhomogeneous Lie algebras [24].

The converse of the preceding result also holds, and can be used as an indirect procedure to determine the embedding of a Lie algebra into another:

Proposition 2 *Let $\mathfrak{g}' \subset \mathfrak{g}$ be a proper subalgebra and Γ a representation of $\mathfrak{g}$. If the subduced realization of $\mathfrak{g}'$ by vector fields possesses k linear invariants, then the branching rule of Γ contains k copies of the trivial representation of $\mathfrak{g}'$.*

We indicate that the latter result, the proof of which follows at once reversing the argumentation in Proposition 1, can be used to distinguish non-conjugate subalgebras of a given Lie algebra, as well as to decide whether a Lie algebra $\mathfrak{g}' \subset \mathfrak{g}$ in a given realization results from the restriction of a realization of $\mathfrak{g}$ with the same number of coordinates.

To exemplify this situation, consider in $\mathfrak{X}(\mathbb{R}^3)$ the Lie algebra $\mathfrak{g}$ generated by the vector fields $\mathbf{X}_1 = x\frac{\partial}{\partial z}$, $\mathbf{X}_2 = y\frac{\partial}{\partial x}$ and $\mathbf{X}_3 = z\frac{\partial}{\partial y}$. A routine computation shows that $\mathfrak{g} \simeq \mathfrak{sl}(3,\mathbb{R})$ and that the realization arises from the fundamental three-dimensional representation $\Gamma = [1,0]$. Now consider the subalgebras generated by the vector fields

$$\mathbf{Y}_1 = x\frac{\partial}{\partial z},\ \mathbf{Y}_2 = z\frac{\partial}{\partial x},\ \mathbf{Y}_3 = x\frac{\partial}{\partial x} - z\frac{\partial}{\partial z};\ \mathbf{Z}_1 = x\frac{\partial}{\partial z} + z\frac{\partial}{\partial x},\ \mathbf{Z}_2 = y\frac{\partial}{\partial z} + z\frac{\partial}{\partial y},\ \mathbf{Z}_3 = y\frac{\partial}{\partial x} - x\frac{\partial}{\partial y}$$

The Lie algebras generated by the $\mathbf{Y}_i$ respectively $\mathbf{Z}_i$ are both isomorphic to $\mathfrak{sl}(2,\mathbb{R})$, as can be easily verified. It follows at once that the $\mathfrak{sl}(2,\mathbb{R})$-copy generated by the $\mathbf{Y}_i$ possesses the linear invariant y, indicating that the representation Γ branches as the direct sum of the irreducible 2-dimensional and the trivial representation of $\mathfrak{sl}(2,\mathbb{R})$. On the contrary, for the copy generated by the vector fields $\mathbf{Z}_i$ there is no linear invariant, and the subduced representation remains necessarily irreducible. As conjugate subalgebras must give rise to the same branching rules [18], we conclude that the two copies of $\mathfrak{sl}(2,\mathbb{R})$ are not conjugate, and in particular nonequivalent as $\mathfrak{sl}(2,\mathbb{R})$-subalgebras of vector fields in $\mathbb{R}^3$.

In a general frame, the detailed analysis of the branching rules of semisimple Lie algebras (see e.g. [17]) indicates additional patterns that can be potentially useful for the obtainment and characterization of invariants of (linear) realizations. Two of such properties, that still have to be justified for arbitrary representations of simple Lie algebras of rank $l \geq 8$ and their corresponding real forms of representations, can be enumerated as follows:

(i) If the Lie algebra $\mathfrak{g}'$ is irreducibly embedded into $\mathfrak{g}$ and $\Phi_\Gamma : \mathfrak{g} \rightarrow \mathfrak{X}(\mathbb{R}^n)$ is a realization associated to the lowest dimensional irreducible representation Γ of $\mathfrak{g}$ and $\dim\Gamma < \dim\mathfrak{g}'$, then $r_{\Gamma'}(\mathfrak{g}') = r_\Gamma(\mathfrak{g})$ and both realizations have the same invariants.

(ii) The rank of a realization of $\mathfrak{g}$ associated to a reducible representation $\Gamma \oplus \Gamma^*$ with $\dim\Gamma < \frac{1}{2}\dim\mathfrak{g}$ and Γ^* the dual representation satisfies the relation $r(\mathfrak{g}) \leq 2\, r_\Gamma(\mathfrak{g}) - 1$.

3.1. Subduced realizations of $\mathfrak{so}(7)$

In order to illustrate some of the preceding properties, we analyze the compact simple exceptional Lie algebra $G_{2(-14)}$ as a subalgebra of the compact orthogonal algebra $\mathfrak{so}(7)$ in some detail. Restricting the adjoint representation $[0,1,0]$ of $\mathfrak{so}(7)$ to G_2, we obtain the branching rule

$$[0,1,0] \downarrow [1,0] \oplus [0,1]\,, \tag{25}$$

where $[1,0]$ is the adjoint representation of G_2 and $[0,1]$ is defining 7-dimensional representation. Actually, this representation characterizes the embedding $G_{2(-14)} \subset \mathfrak{so}(7)$, that has index $j_f = 1$. From the realization of $\mathfrak{so}(7)$ associated to the fundamental 7-dimensional representation $[1,0,0]$, given by the vector fields $\mathbf{E}_{ij} = x^i\frac{\partial}{\partial x^j} - x^j\frac{\partial}{\partial x^i}$ $(i \neq j)$ the restriction to $G_{2(-14)}$ provides the following realization as vector fields in $\mathbb{R}^7$:

$$\begin{array}{lcl}
\mathbf{X}_1 & = & -x^7\frac{\partial}{\partial x^1} + x^4\frac{\partial}{\partial x^2} - x^2\frac{\partial}{\partial x^4} + x^1\frac{\partial}{\partial x^7}, \\
\mathbf{X}_2 & = & x^2\frac{\partial}{\partial x^1} - x^1\frac{\partial}{\partial x^2} - x^7\frac{\partial}{\partial x^4} + x^4\frac{\partial}{\partial x^7}, \\
\mathbf{X}_4 & = & x^6\frac{\partial}{\partial x^1} + x^4\frac{\partial}{\partial x^3} - x^3\frac{\partial}{\partial x^4} - x^1\frac{\partial}{\partial x^6}, \\
\mathbf{X}_{11} & = & -\sqrt{3}x^6\frac{\partial}{\partial x^1} - 2\sqrt{3}x^3\frac{\partial}{\partial x^2} + \sqrt{3}x^4\frac{\partial}{\partial x^3} - \sqrt{3}x^3\frac{\partial}{\partial x^4} + 2\sqrt{3}x^2\frac{\partial}{\partial x5} + \sqrt{3}x^1\frac{\partial}{\partial x^6}, \\
\mathbf{X}_{13} & = & -2\sqrt{3}x^5\frac{\partial}{\partial x^1} + \sqrt{3}x^6\frac{\partial}{\partial x^2} + \sqrt{3}x^7\frac{\partial}{\partial x^3} + 2\sqrt{3}x^1\frac{\partial}{\partial x5} - \sqrt{3}x^2\frac{\partial}{\partial x^6} - \sqrt{3}x^3\frac{\partial}{\partial x^7}.
\end{array} \tag{26}$$

The remaining vector fields corresponding to generators of $G_{2(-14)}$ follow from the commutators:

$$\begin{gathered}
[\mathbf{X}_1,\mathbf{X}_2] = \frac{1}{2}\mathbf{X}_3,\ [\mathbf{X}_1,\mathbf{X}_4] = \mathbf{X}_7,\ [\mathbf{X}_1,\mathbf{X}_{11}] = \mathbf{X}_{14},\ [\mathbf{X}_1,\mathbf{X}_{13}] = -\mathbf{X}_{12},\ [\mathbf{X}_2,\mathbf{X}_4] = \mathbf{X}_6, \\
[\mathbf{X}_4,\mathbf{X}_{13}] = \mathbf{X}_{10},\ [\mathbf{X}_1,\mathbf{X}_6] = \mathbf{X}_5,\ [\mathbf{X}_4,\mathbf{X}_{14}] = -\mathbf{X}_9,\ \ [\mathbf{X}_6,\mathbf{X}_7] - [\mathbf{X}_4,\mathbf{X}_5] = \frac{2}{\sqrt{3}}\mathbf{X}_8.
\end{gathered} \tag{27}$$

It hence suffices to analyze the properties of the symmetry generators $\mathbf{X}_1, \mathbf{X}_2, \mathbf{X}_4, \mathbf{X}_{11}$ and $\mathbf{X}_{13}$, as all the remaining elements of the basis are expressible as commutators of these five elements. As $\dim G_{2(-14)} > 7$, the fact that the reduction $[1,0,0] \downarrow [0,1]$ is irreducible implies that the rank of the realization matrix is preserved, according to the property (i) above. Clearly there is only one invariant function $F(\mathbf{x})$, given by the quadratic form $Q_2 = \sum_{k=1}^{7}\left(x^i\right)^2$. Introducing Q_2 as a new coordinate allows us to consider it as cyclic, and hence the realization can be reduced to a genuinely nonlinear one in $\mathbb{R}^6$, as $[0,1]$ is the lowest dimensional faithful representation of $G_{2(-14)}$. Further, considering the embedding $\mathfrak{su}(3) \subset G_{2(-14)}$, where the subalgebra is generated by $\{\mathbf{X}_1, \mathbf{X}_2, \mathbf{X}_4\}$ and their commutators, the branching rule is given by $[0,1] \downarrow (1,0) \oplus (0,1) \oplus (0,0)$, where the representation $(1,0) \oplus (0,1)$ is reducible as a complex representation, but irreducible as a real one. As a copy of the trivial representation of $\mathfrak{su}(3)$ appears, we have an additional coordinate that is cyclic. Hence the realization in $\mathbb{R}^7$ of $\mathfrak{su}(3)$ possesses two invariants, one being Q_2, as well as x^5 (observe that this provides an example of the property (ii) in the preceding paragraph). With respect to a new set of coordinates $\{y^1, \cdots, y^5\} = \{x^1, \cdots, x^4, x^6\}$, we write the realization of $\mathfrak{su}(3)$ in $\mathfrak{X}(\mathbb{R}^5)$ as

$$\begin{array}{lcl} \mathbf{X}_1 & = & -J_0 \frac{\partial}{\partial y^1} + y^4 \frac{\partial}{\partial y^2} - y^2 \frac{\partial}{\partial y^4} \\ \mathbf{X}_2 & = & y^2 \frac{\partial}{\partial y^1} - y^1 \frac{\partial}{\partial y^2} - J_0 \frac{\partial}{\partial y^4} \\ \mathbf{X}_4 & = & y^5 \frac{\partial}{\partial y^1} + y^4 \frac{\partial}{\partial y^3} - y^3 \frac{\partial}{\partial y^4} - y^1 \frac{\partial}{\partial y^5} \end{array}, \tag{28}$$

where $J_0 = \sqrt{\sum_{k=1}^{5} (y^k)^2 + \lambda}$ for some constant λ and the remaining vector fields are obtained from (27). Finally, if we extend the chain to $\mathfrak{so}(3) \subset \mathfrak{su}(3) \subset G_{2(-14)}$, the resulting real representation for $\mathfrak{so}(3)$ is $R^{II}_{1/2} \oplus \Gamma^3_0$, with invariants $J_1 = \sqrt{(y^1)^2 + (y^2)^2 + (y^4)^2 + \lambda}$ and three linear ones.

On the other hand let the realization of $\mathfrak{so}(3)$ as subalgebra of $\mathfrak{so}(7)$ be given by

$$\begin{array}{rl} \mathbf{X}_1 = & -\sqrt{\frac{3}{2}} x^4 \frac{\partial}{\partial x^1} + \sqrt{\frac{3}{2}} x^3 \frac{\partial}{\partial x^2} - \left(\sqrt{\frac{3}{2}} x^2 + \sqrt{\frac{5}{2}} x^6 \right) \frac{\partial}{\partial x^3} + \sqrt{\frac{3}{2}} x^1 \frac{\partial}{\partial x^4} \\ & + \sqrt{\frac{5}{2}} x^5 \frac{\partial}{\partial x^4} - \sqrt{\frac{5}{2}} x^4 \frac{\partial}{\partial x^5} + \sqrt{\frac{5}{2}} x^3 \frac{\partial}{\partial x^6} - \sqrt{6} x^7 \frac{\partial}{\partial x^6} + \sqrt{6} x^6 \frac{\partial}{\partial x^7}, \\ \mathbf{X}_2 = & -\sqrt{\frac{3}{2}} x^3 \frac{\partial}{\partial x^1} - \sqrt{\frac{3}{2}} x^4 \frac{\partial}{\partial x^2} + \left(+\sqrt{\frac{3}{2}} x^1 - \sqrt{\frac{5}{2}} x^5 \right) \frac{\partial}{\partial x^3} + \sqrt{\frac{3}{2}} x^2 \frac{\partial}{\partial x^4} \\ & - \sqrt{\frac{5}{2}} x^6 \frac{\partial}{\partial x^4} + \sqrt{\frac{5}{2}} x^3 \frac{\partial}{\partial x^5} + \sqrt{6} x^7 \frac{\partial}{\partial x^5} + \sqrt{\frac{5}{2}} x^4 \frac{\partial}{\partial x^6} - \sqrt{6} x^5 \frac{\partial}{\partial x^7}, \\ \mathbf{X}_2 = & 3x^2 \frac{\partial}{\partial x^1} - 3x^1 \frac{\partial}{\partial x^2} + 2x^4 \frac{\partial}{\partial x^3} - 2x^3 \frac{\partial}{\partial x^4} + x^6 \frac{\partial}{\partial x^5} - x^5 \frac{\partial}{\partial x^6}. \end{array} \tag{29}$$

It is clear that no linear invariants of the realization can exist as the rank of the realization matrix is three. By the preceding results, we conclude that the branching rule associated to $[1,0,0]$ does not contain a copy of the trivial representation of $\mathfrak{so}(3)$. As the latter is not a maximal subalgebra of $\mathfrak{so}(7)$, it follows that $\mathfrak{so}(3)$ must be contained in either $G_{2(-14)}$, $\mathfrak{su}(4)$ or $\mathfrak{su}(2)^3$, that correspond to the semisimple maximal subalgebras of $\mathfrak{so}(7)$. From the branching rules (see e.g. [17]) it follows that the only possibility to provide the irreducible representation associated to (29) is given by $G_{2(-14)}$, from which we deduce that $\mathfrak{so}(3)$ is irreducibly embedded into $G_{2(-14)}$ with embedding index $j_f = 28$.

4. The Lie algebra $\mathfrak{so}(3)$ as a principal subalgebra of $\mathfrak{so}(2J+1)$

As is well known, for any integer J the $\mathfrak{so}(3)$-representation R^I_J of dimension $2J+1$ and highest weight $2J$ is of the first class (that is, its complexification remains irreducible, see [25]). Therefore, $\mathfrak{so}(3)$ can be represented as a subalgebra of the compact Lie algebra $\mathfrak{so}(2J+1)$. Indeed, as follows from the structure theory, for $J \neq 3$ the Lie algebra $\mathfrak{so}(3)$ is irreducibly embedded into $\mathfrak{so}(2J+1)$ as a three-dimensional principal subalgebra, meaning in particular that the branching rule for the defining representation $\left[1, 0^{(J-1)}\right]$ of $\mathfrak{so}(2J+1)$ remains irreducible when restricted to $\mathfrak{so}(3)$.[3] The explicit construction of representation matrices for the real

[3] The exceptional case $J = 3$ corresponds to the chain $\mathfrak{so}(3) \subset G_{2(-14)} \subset \mathfrak{so}(7)$ that has been crucial in atomic spectroscopy [10].

irreducible representations of $\mathfrak{so}(3)$ was developed in [26], for which reason we merely recall the construction. For any $J \geq 1$ define the scalars

$$a_l = \sqrt{\frac{2l\,J - l\,(l-1)}{4}},\ 1 \leq l \leq J-1;\ \nu_0 = \sqrt{\frac{J\,(J+1)}{2}}. \tag{30}$$

Let $\{\mathbf{e}_1, \cdots, \mathbf{e}_{2J+1}\}$ denote a basis of the representation space of the real representation R_J^I. Further let $\left[\frac{n}{2}\right]$ denote the integer part of $\frac{n}{2}$. Then the matrix elements are given by

$$\begin{aligned}
\langle \mathbf{e}^k | R_J^I\,(X_1)\,|\mathbf{e}_l\rangle = {} & \left(\frac{1+(-1)^{k-1}}{2}\right)\left(\delta_{k+3}^l\, a_{([\frac{k+1}{2}])} + \delta_k^{l+1} a_{([\frac{k-1}{2}])}\right) - (a_J + \nu_0) \times \\
& \left(\delta_{2J+1}^l \delta_k^{2J} - \delta_{2J}^l \delta_k^{2J+1}\right) - \left(\frac{1+(-1)^k}{2}\right)\left(\delta_{k+1}^l a_{([\frac{k}{2}])} + \delta_k^{l+3} a_{([\frac{k-2}{2}])}\right), \\
\langle \mathbf{e}^k | R_J^I\,(X_2)\,|\mathbf{e}_l\rangle = {} & \delta_{k+2}^l\, a_{([\frac{k+1}{2}])} - \delta_k^{l+2} a_{([\frac{k-1}{2}])} - (a_J + \nu_0)\left(\delta_{2J+1}^l \delta_k^{2J-1} - \delta_{2J-1}^l \delta_k^{2J+1}\right), \\
\langle \mathbf{e}^k | R_J^I\,(X_3)\,|\mathbf{e}_l\rangle = {} & \frac{\left(1+(-1)^k\right)\delta_k^{l+1}\,(2J+2-k) + \left((-1)^k - 1\right)\delta_l^{k+1}\,(2J+1-k)}{4},
\end{aligned} \tag{31}$$

where $1 \leq k, l \leq 2J+1$. The matrices $R_J^I\,(X_k)$ constructed with these values are clearly skew-symmetric, and thus belong to $\mathfrak{so}\,(2J+1)$, showing that the linear map

$$\varphi_J : \mathfrak{so}\,(3) \rightarrow \mathfrak{so}\,(2J+1)\,; \quad X_k \mapsto R_J^I\,(X_k) \tag{32}$$

defines a Lie algebra homomorphism and an irreducible embedding. The case $J=1$ is somewhat degenerate, as it corresponds to the adjoint representation of $\mathfrak{so}(3)$, thus possessing the quadratic Casimir operator as only invariant. Denoting for $J \geq 2$

$$\langle \mathbf{e}^k | R_J^I\,(X_\alpha)\,|\mathbf{e}_l\rangle = \left(R_J^I\,(X_\alpha)\right)_{k,l},\ \alpha = 1,2,3, \tag{33}$$

it follows at once that the vector fields corresponding to each representation R_J^I are given by

$$\mathbf{X}_\alpha = x^k \left(R_J^I\,(X_\alpha)\right)_{k,l} \frac{\partial}{\partial x^l},\ \alpha = 1,2,3. \tag{34}$$

It can be easily verified that for any $j \in \{1, \cdots, 2J+1\}$ there is always an α such that the condition

$$dx^j\,(\mathbf{X}_\alpha) = \left(R_J^I\,(X_\alpha)\right)_{k,j} \neq 0, \tag{35}$$

is satisfied, showing that there are no linear invariants. On the other hand, due to the embedding $\mathfrak{so}(3) \subset \mathfrak{so}(2J+1)$, the quadratic form $Q = \sum_{p=1}^{2J+1} (x^p)^2$ is an invariant. In this case, for any $J \geq 2$, the rank of the coefficient matrix is always three, and thus the realization possesses exactly $2J-2$ fundamental invariants. The explicit obtainment of the invariants of these realizations, either in the semi-algebraic formulation developed in an earlier section, or solving the corresponding systems of PDEs, presents considerable difficulties, and is currently not explicitly solved. Without going into the computation details, that shall be given elsewhere, we remark that the invariants of realizations associated to the real representations R_j^I are divided into two cases, according to the parity of J:

(i) If $J = 2q \geq 2$ is even, then the fundamental invariants of (34) can be chosen to have degrees $d = 2, 3, \cdots, 2J-1$.

(ii) If $J = 2q + 1 \geq 1$ is odd, then any invariant of the realization (34) has even degree.

It may be observed that for half-integer values, the real irreducible representations $R^{II}_{J/2}$ of $\mathfrak{so}(3)$ corresponding to the real form of the representations $D_{J/2}$ with highest weight J are of the second class, meaning that their complexification is reducible as a complex representation [25]. Although these representations define an embedding $\mathfrak{so}(3) \subset \mathfrak{so}(4J+2)$, it is not embedded as a maximal subalgebra [18]. In this situation, the pertinent Lie algebras to be analyzed with respect to the embedding are the symplectic algebras [27, 28]. As for half-integer values $J/2$ the scalar J is necessarily odd, the corresponding invariants of the realization are all of even degree, in analogy to the integer case.

5. Symmetries of second-order dynamical (Lagrangian) systems

Invariants of realizations can also be used directly for the construction and symmetry analysis of second-order dynamical systems, either as Lie-point or Noether symmetries [4, 29]. Here, the representation theory can further be applied to study the symmetry-breaking phenomena of systems with a given (maximal) symmetry by means of additional potential terms, as well as to guarantee that a given realization of a Lie algebra corresponds to the maximal symmetry of a system.

Among the various methods to compute symmetries of differential equations, those of most common use are the direct prolongation method and the formulation in terms of differential operators [7]. Any system of n second-order ordinary differential equations

$$\ddot{\mathbf{x}} = \omega^{\alpha}(t, \mathbf{x}), \; 1 \leq \alpha \leq n, \tag{36}$$

can be rewritten in equivalent form as the partial differential equation

$$\mathbf{A}f = \left(\frac{\partial}{\partial t} + \dot{x}^i \frac{\partial}{\partial x^i} + \omega^i(t, \mathbf{x}) \frac{\partial}{\partial \dot{x}^i} \right) f = 0. \tag{37}$$

In this context, a vector field $X = \zeta(t, \mathbf{x}) \frac{\partial}{\partial t} + \eta_i(t, \mathbf{x}) \frac{\partial}{\partial x^i} \subset \mathfrak{X}\left(\mathbb{R}^{n+1}\right)$ is a Lie point symmetry of the system (36) whenever the prolonged vector field $\dot{X} = X + \dot{\eta}_i(t, \mathbf{x}, \dot{\mathbf{x}}) \frac{\partial}{\partial \dot{x}^i}$ satisfies the commutator

$$\left[\dot{X}, \mathbf{A}\right] = -\frac{d\zeta}{dt}\mathbf{A}, \tag{38}$$

where $\dot{\eta}_i = -\frac{d\zeta}{dt}\dot{x}_i + \frac{d\eta_i}{dt}$. The advantage of this reformulation is that the prolongation of the symmetry generator X is already contained in the commutator, the symmetry condition being given by the coefficients in $\frac{\partial}{\partial \dot{x}^i}$. Whenever the system arises from a variational principle, a point symmetry that satisfies the condition

$$\dot{\mathbf{X}}(L) + \mathbf{A}(\zeta) - \mathbf{A}(V) = 0 \tag{39}$$

is called a Noether symmetry, where $V(t, \mathbf{x})$ a function independent on $\dot{\mathbf{x}}$ and L is an admissible Lagrangian of the system. By some abuse of notation, we shall call a symmetry satisfying (39) with $\zeta = V = 0$ a variational symmetry. Noether symmetries can alternatively be deduced from a constant of the motion, i.e., a function $J(\dot{q}_j, q_j, t)$ such that

$$\frac{dJ}{dt} = \frac{\partial J}{\partial t} + \dot{x}^i \frac{\partial J}{\partial x^i} + \ddot{x}^i \frac{\partial J}{\partial \dot{x}^i} = \mathbf{A}(J) = 0. \tag{40}$$

Any first integral of (36) determines (up to a gauge term) a Noether symmetry

$$X_J = \zeta\left(t,\mathbf{x},\dot{\mathbf{x}}\right)\frac{\partial}{\partial t} + \eta^{\alpha}\left(t,\mathbf{x},\dot{\mathbf{x}}\right)\frac{\partial}{\partial x^{\alpha}} + \dot{\eta}^{\alpha}\left(t,\mathbf{x},\dot{\mathbf{x}}\right)\frac{\partial}{\partial \dot{q}^{\alpha}}, \tag{41}$$

the components of which are recovered using the relations

$$\frac{\partial^2 L}{\partial \dot{x}^{\alpha}\partial \dot{x}^{\beta}}\left(\eta_{\alpha} - \dot{x}^{\alpha}\zeta\right) + \frac{dJ}{d\dot{x}^{\beta}} = 0. \tag{42}$$

As vector field, X_J is always a point symmetry if the components do not depend on the $\dot{\mathbf{x}}$.[4]

Proposition 3 *Let $\Phi : \mathfrak{g} \rightarrow \mathfrak{X}(\mathbb{R}^n)$ be a realization of a Lie algebra $\mathfrak{g}$ associated to a linear skew-symmetric representation Γ and J an invariant of the realization. Then $\Phi(\mathfrak{g})$ can be realized as a Lie algebra of Lie point symmetries of the second-order dynamical system with equations of the motion*

$$\ddot{x}^i - \varphi(t)\frac{\partial J}{\partial x^i} = 0,\ 1 \leq i \leq n \tag{43}$$

arising from the regular Lagrangian $L = \frac{1}{2}\sum_{k=1}^{n}(\dot{x}^i)^2 + \varphi(t)J(\mathbf{x})$.

The equations of the motion, expressed as a vector field, are in this case

$$\mathbf{A} = \frac{\partial}{\partial t} + \dot{x}^i\frac{\partial}{\partial x^i} + \varphi\left(t\right)\frac{\partial J}{\partial x^i}\frac{\partial}{\partial \dot{x}^i},$$

while the first prolongation of $\mathbf{X}$ is given by

$$\dot{\mathbf{X}}_{\alpha} = \xi_{\alpha}\frac{\partial}{\partial x^j} + \dot{x}^i\frac{\partial \xi_{\alpha}^j}{\partial x^i}\frac{\partial}{\partial \dot{x}^j}.$$

Evaluating the commutator (38) gives

$$\left[\dot{\mathbf{X}}_{\alpha},\mathbf{A}\right] = \left(\dot{x}^i\frac{\partial \xi_{\alpha}^j}{\partial x^i} - \dot{\xi}_{\alpha}^j\right)\frac{\partial}{\partial x^j} + \left(\dot{x}^i\frac{\partial \dot{\xi}_{\alpha}^j}{\partial x^i} + \varphi\left(t\right)\frac{\partial J}{\partial x^i}\frac{\partial \dot{\xi}_{\alpha}^i}{\partial \dot{x}^j} - \varphi\left(t\right)\xi_{\alpha}^i\frac{\partial^2 J}{\partial x^i\partial x^j}\right)\frac{\partial}{\partial \dot{x}^j}. \tag{44}$$

The term in $\frac{\partial}{\partial x^j}$ is obviously zero by the definition of $\dot{\xi}_{\alpha}^j$, hence the symmetry condition reduces to

$$\dot{x}^i\frac{\partial \dot{\xi}_{\alpha}^j}{\partial x^i} + \varphi\left(t\right)\frac{\partial J}{\partial x^i}\frac{\partial \dot{\xi}_{\alpha}^i}{\partial \dot{x}^j} - \varphi\left(t\right)\xi_{\alpha}^i\frac{\partial^2 J}{\partial x^i\partial x^j} = 0,\ 1 \leq j \leq n \tag{45}$$

where

$$\frac{\partial \dot{\xi}_{\alpha}^j}{\partial x^i} = \dot{x}^l\frac{\partial^2 \xi_{\alpha}^j}{\partial x^i\partial x^l},\quad \frac{\partial \dot{\xi}_{\alpha}^j}{\partial \dot{x}^i} = \frac{\partial \xi_{\alpha}^j}{\partial x^i} = -\frac{\partial \xi_{\alpha}^i}{\partial x^j},\ 1 \leq i,j \leq n.$$

Inserting these expression in (45) gives the equation

$$\dot{x}^i\left(\dot{x}^l\frac{\partial^2 \xi_{\alpha}^j}{\partial x^i\partial x^l}\right) + \varphi\left(t\right)\left(\frac{\partial J}{\partial x^i}\frac{\partial \xi_{\alpha}^j}{\partial x^i} - \xi_{\alpha}^i\frac{\partial^2 J}{\partial x^i\partial x^j}\right) = \dot{x}^i\dot{x}^l\frac{\partial^2 \xi_{\alpha}^j}{\partial x^i\partial x^l} - \varphi\left(t\right)\frac{\partial}{\partial x^j}\left(\xi_{\alpha}^i\frac{\partial J}{\partial x^i}\right). \tag{46}$$

As J is an invariant of the realization, the last term is zero, while the first one vanishes whenever the component functions ξ_{α}^j are linear. The latter requirement is fulfilled because the realization

[4] Noether symmetries that are not point symmetries are sometimes called dynamical symmetries, like in [7], although there is no such restriction in Noether's original approach [30].

arises from a skew-symmetric linear representation of $\mathfrak{g}$, so that $\left[\dot{\mathbf{X}}_\alpha, \mathbf{A}\right] = 0$ holds and we conclude that $\mathbf{X}_\alpha$ is a point symmetry of the system.

As the result does not depend on the particular structure of the invariant, it allows us to construct dynamical systems with a given Lie algebra of point symmetries. We observe that the approach is valid for both conservative and non-conservative Lagrangian systems, as the function $\varphi(t)$ in the potential does not explicitly intervene in the commutator. The procedure can be easily extended to subalgebras. If $\mathfrak{g}' \subset \mathfrak{g}$ is a subalgebra and $\{J_1, \cdots, J_k\}$ are invariants of the subduced realization of $\mathfrak{g}'$, then for generic functions $\{\varphi_1(t), \cdots, \varphi_k(t)\}$ the system

$$\ddot{x}^i = \sum_{l=1}^{k} \varphi_l(t) \frac{\partial J_l}{\partial x^i} \; 1 \leq j \leq n \tag{47}$$

turns out to have exactly a Lie-point symmetry algebra isomorphic to $\mathfrak{g}'$. Starting from the system associated to an invariant of $\mathfrak{g}$, the restriction to subalgebras $\mathfrak{g}'$ and the addition of potential terms corresponding to invariants of the latter (and not of the former) can be seen as a recursive procedure of breaking the symmetry of a system. In connection with the branching rules, it also allows to determine whether a given subalgebra can be realized as a maximal symmetry algebra or not.

Let us illustrate this fact considering once more the compact Lie algebra $\mathfrak{so}(7)$ in the fundamental 7-dimensional representation $\Gamma = [100]$ and the quadratic invariant $J = \left(x^1\right)^2 + \cdots + \left(x^7\right)^2$ of the realization. As the rank of the realization remains irreducible when restricted to $G_{2(-14)}$, it turns out that a system of the type

$$\ddot{x}^i = \varphi(t) \frac{\partial F(J)}{\partial x^i}, \tag{48}$$

where $F(J)$ is an arbitrary function of J, is always $G_{2(-14)}$-invariant, but there is no possibility to break the maximal symmetry of $\mathfrak{so}(7)$ to the subalgebra, as both have the same invariant functions for the realization associated to the corresponding representations. A dimensional reduction of the symmetry algebra would require an additional term invariant by the generators of $G_{2(-14)}$, but not invariant by all of $\mathfrak{so}(7)$. Such a symmetry-breaking can be obtained, for example, taking the principal $\mathfrak{so}(3)$-subalgebra of index $j_f = 28$, as in this case, although the embedding is still irreducible, there are 3 invariants of the subduced realization. As commented before, the fundamental invariants can be chosen as J, as well as a fourth and sixth order invariants J_2 and J_3 respectively, with J_2 being explicitly given by:

$$\begin{aligned} J_2 = \; & \left(x^1\right)^4 + \left(x^2\right)^4 - \tfrac{3}{4}\left(\left(x^3\right)^4 + \left(x^4\right)^4\right) + \tfrac{3}{5}\left(\left(x^5\right)^4 + \left(x^6\right)^4 + \frac{\left(x^7\right)^4}{10}\right) + \\ & \left(4\left(\left(x^1\right)^2 + \left(x^2\right)^2\right) + 7\left(x^7\right)^2\right)\left(\left(x^5\right)^2 + \left(x^6\right)^2\right) + 2\left(x^1x^2\right)^2 - 3\left(x^3x^4\right)^2 \\ & + 2\left(\left(x^1\right)^2 + \left(x^2\right)^2\right)\left(\left(x^3\right)^2 + \left(x^4\right)^2\right) + 4\sqrt{\tfrac{3}{5}}\left(x^2\left(x^6\right)^3 - x^1\left(x^5\right)^3\right) \\ & + 2\sqrt{15}\left(x^1x^5 - x^2x^6\right)\left(\left(x^3\right)^2 - \left(x^4\right)^2\right) + 5\left(\left(x^3\right)^2 + \left(x^4\right)^2\right)\left(x^7\right)^2 \\ & + 4\sqrt{\tfrac{3}{5}}\left(\left(x^5\right)^2\left(x^3x^7 - 3x^2x^6\right) - \left(x^6\right)^2\left(x^3x^7 - 3x^1x^5\right)\right) + \tfrac{6}{5}\left(x^5\right)^2\left(x^6\right)^2 \\ & + \tfrac{1}{5}\left(x^7\right)^2\left(\left(x^5\right)^2 + \left(x^6\right)^2\right) + 4\sqrt{15}x^3x^4\left(x^1x^6 + x^2x^5\right) \\ & - 12x^2x^7\left(x^4x^5 + x^3x^6\right) - 12x^1x^7\left(x^3x^5 - x^4x^6\right) + 8\sqrt{\tfrac{3}{5}}x^4x^5x^6x^7. \end{aligned}$$

A routine computation shows that the equations of the motion

$$\ddot{x}^i = \varphi(t) \frac{\partial F(J)}{\partial x^i} + \psi(t) \frac{\partial G(J_2)}{\partial x^i} \tag{49}$$

are $\mathfrak{so}(3)$-invariant, and that no other generator of $G_{2(-14)}$ possesses J_2 as an invariant. For generic functions $\varphi(t)$ and $\psi(t)$, it is straightforward to verify that no other point symmetry, specifically with components in $\frac{\partial}{\partial t}$, can exist.

We finally comment how, for realizations related to orthogonal representations, the dynamical systems (48) having potentials associated to invariants of the realization are not only point symmetries, but also special Noether symmetries, namely variational symmetries.

Proposition 4 *Let $\Phi_\Gamma : \mathfrak{g} \to \mathfrak{X}(\mathbb{R}^n)$ be the subduced realization associated to the embedding $\mathfrak{g} \subset \mathfrak{so}(N)$. Let $\{J_1, \cdots, J_{\chi_0}\}$ be invariants of the realization. Then the generators $\mathbf{X}_\alpha$ of $\Phi(\mathfrak{g})$ are variational symmetries of the Lagrangian*

$$L = \frac{1}{2}\sum_{k=1}^{n}\left(\dot{x}^i\right)^2 - \sum_{k=1}^{\chi_0}\varphi_k(t)\,J_k. \tag{50}$$

In order to satisfy the Noether symmetry condition, the equation (39) must be satisfied identically. Evaluating it we obtain

$$\dot{\mathbf{X}}(L) - \mathbf{A}(\zeta) - \mathbf{A}(V) = -\varphi_k(t)\,\xi_\alpha^j(\mathbf{x})\frac{\partial J_k}{\partial x^j} + \dot{x}^l\frac{\partial \xi_\alpha^j}{\partial x^l}\dot{x}^j - \frac{\partial V}{\partial t} - \dot{x}^l\frac{\partial V}{\partial x^l}. \tag{51}$$

The first term automatically vanishes as J_k is an invariant of the realization for any k. The terms linear in $\dot{\mathbf{x}}$ imply that V is at most a time function, hence implying that V must be a constant in order to fulfill (51). Finally, the quadratic term in $\dot{\mathbf{x}}$ can be rewritten as

$$\sum_{k=1}^{n}\left(\dot{x}^k\right)^2\frac{\partial \xi_\alpha^k}{\partial x^k} + \sum_{l<j}\dot{x}^l\dot{x}^j\left(\frac{\partial \xi_\alpha^j}{\partial x^l} + \frac{\partial \xi_\alpha^l}{\partial x^j}\right) \tag{52}$$

As the realization arises from the restriction of a representation of $\mathfrak{so}(N)$, the representation matrices are skew-symmetric, and hence the vector fields satisfy the constraints

$$\frac{\partial \xi_\alpha^k}{\partial x^k} = 0,\ \frac{\partial \xi_\alpha^j}{\partial x^l} + \frac{\partial \xi_\alpha^l}{\partial x^j} = 0,\ 1 \leq \alpha \leq \dim\mathfrak{g}, \tag{53}$$

showing that (52) and hence (51) vanishes.

As an explicit example leading to systems of this type we may consider the vector fields in formula (29). For generic choices of the functions $\varphi(t)$ and $\psi(t)$ the Lie point and Noether symmetry algebras of the system determined by the equations of motion (49) coincide and are isomorphic to $\mathfrak{so}(3)$.

6. Concluding remarks

The approach to realizations by vector fields of Lie algebras by means of linear representations and their invariants has been shown to exhibit some interesting connections with the branching rules for the embedding problem of Lie algebras, such as the properties of irreducibly embedded subalgebras or the multiplicity of representations by subduction. This analysis allows to distinguish between nonconjugate subalgebras of a given Lie algebra by means of the invariants of a given realization, without knowing a priori the embedding index or the corresponding branching rule. Some properties that hold at least for low ranks have been recognized, and their validity in the arbitrary case could be of potential use in the context of a further systematization of computational criteria for branching rules, as well as for the Clebsch-Gordan problem. On the other hand, the study of the orders and structure of the invariants of a realization suggests, in some sense, a natural extension of the analysis of generalized Casimir invariants in Lie algebras and the associated missing label problem, a relevant problem in applications, where the exact implications with the invariant theory of inhomogeneous Lie algebras has still to be studied in detail.

Some applications to the construction of second-order dynamical systems with prescribed symmetry, as well as a procedure to break the symmetry to that of a proper subalgebra, have been indicated. An interesting question that arises in this context is whether similar statements concerning second-order dynamical systems and potentials determined by the invariants of a nonlinear realization can be established, and to what extent these criteria are further related with the representation theory, e.g. by means of projections. Within this frame, a problem that is worthy to be mentioned is the possibility of extending the constructions to symmetries explicitly depending on the velocities, generalizing the method to Noether symmetries that are no more point symmetries of a system. Even if such an ansatz only succeeds partially, any progress in this direction would be of interest for the structural analysis of the constants of the motion of dynamical systems.

Acknowledgments
During the preparation of this work, the author was financially supported by the research project MTM2016-79422-P of the AEI/FEDER (EU).

References

[1] Cartan E 1945 *Systèmes différentiels extérieurs et leurs applications géométriques* (Paris: Hermann)
[2] Flanders H 1963 *Differential Forms with Applications to the Physical Sciences* (New York: Academic Press)
[3] Fuks D B 1984 *Cohomology of Infinite-Dimensional Lie Algebras* (Moscow: Nauka) (Russian)
[4] Olver P J 1986 *Applications of Lie Groups to Differential Equations* (New York: Springer)
[5] Petrov A Z 1960 *Einstein Spaces* (Moscow: Fizmatlit) (Russian)
[6] Dickson L E 1924 *Annals of Math.* **4** 287
[7] Stephani H 1993 *Differentialgleichungen. Symmetrien und Lösungsmethoden* (Heidelberg: Spektrum Akademischer Verlag)
[8] Ovsyannikov L V, Ibragimov N Kh 2013 *Lectures on the Theory of Group Properties of Differential Equations* (Singapore: World Scientific)
[9] Hamermesh M 1962 *Group Theory and its Applications to Physical Problems* (Reading: Addison-Wesley)
[10] Racah G 1951 *Group Theory and Spectroscopy* (New Jersey: Princeton Univ. Press)
[11] Iachello F 2006 *Lie Algebras and Applications* (Berlin: Springer Verlag)
[12] Lie S 1883 *C. R. Acad. Sci.* **116** 1233
[13] Bountis T C, Papageorgiou V, Winternitz P 1986 *J. Math. Phys.* **27** 1215
[14] Cariñena J F, de Lucas J 2011 *Dissertationes Math.* **479** 1
[15] Kamke E 1962 *Differentialgleichungen. Lösungsmethoden und Lösungen. Band II* (Leipzig, Akademische Verlagsgesellschaft)
[16] Tits J 1967 *Tabellen zu den einfachen Lie Gruppen und ihren Darstellungen* (Berlin: Springer Verlag)
[17] Patera J, Sankoff D 1973 *Tables of Branching Rules for Representations of Simple Lie Algebras* (Montréal: Presses de l'Université de Montréal)
[18] Dynkin E B 1952 *Mat. Sb.* **30** 349
[19] Lorente M, Gruber B 1972 *J. Math. Phys.* **13** 1639
[20] Sharp R T and Pieper S C 1968 *J. Math. Phys.* **9** 663
[21] Ališauskas S J 1974 *Lit. Fiz. Sb.* **14** 709
[22] Rowe D J 1995 *J. Math. Phys.* **36** 1520
[23] Campoamor-Stursberg R 2012 *J. Phys. Conf. Ser.* **343** 012021
[24] Campoamor-Stursberg R 2006 *J. Phys. A: Math. Gen.* **39** 2325
[25] Iwahori N 1959 *Nagoya Math. J.* **14** 59
[26] Campoamor-Stursberg R 2015 *Symmetry* **7** 1655
[27] Zhelobenko D P 1962 *Uspekhi Mat. Nauk* **17** 27

[28] Campoamor-Stursberg 2012 *Lith. J. Phys.* **53** 71
[29] Udrişte C, Nicola I R 2007 *J. Dyn. Syst. Geom. Theor.* **5** 85
[30] Logan J D 1973 *J. Math. Anal. Appl.* **42** 191

IOP Conf. Series: Journal of Physics: Conf. Series **1071** (2018) 012006 doi:10.1088/1742-6596/1071/1/012006

Universal critical behaviour of 3-level atoms interacting dipolarly with radiation

O. Castaños, S. Cordero, E. Nahmad-Achar, R. López-Peña

Instituto de Ciencias Nucleares, Universidad Nacional Autónoma de México, Apdo. Postal 70-543, 04510, Cd.Mx., Mexico

E-mail: ocasta@nucleares.unam.mx

Abstract. A system of 3-level atoms interacting with 2-modes of electromagnetic radiation is studied by means of a variational procedure. This allows us to establish the corresponding quantum phase diagram and demonstrate that the super-radiant region is divided in monochromatic sub-regions where only one coupling strength μ_{ij} and one radiation field mode rule the behaviour of the system. In this contribution we prove that the variational coherent states, the symmetry-adapted states, and the exact quantum solution exhibit a universal parametric curve associated essentially with the two-level population and the number of photons exciting these levels.

1. Introduction

Studies of quantum phase transitions have attracted the attention of physicists of the nuclear, molecular, quantum optics, and condense matter communities. The concept of quantum phase transitions was introduced in the studies of quantum critical phenomena in spin lattice systems at zero temperature [1]. They result from competing ground state phases, and revisions of the field in condensed matter are given in [2, 3]. In nuclear physics the phase transitions were studied using an algorithm introduced at the end of the seventies by Gilmore [4]. Applications of the procedure to describe the shape transitions in nuclei can be found in [5, 6].

The Gilmore algorithm has been applied to many body systems in quantum mechanics: As examples we can mention the Lipkin-Meshkov-Glick (LMG) collective model to describe N identical nucleons which can occupy two possible states separated by a given energy difference; and the Dicke model, considered a key piece to study the field-matter interactions, which describes the interaction between a low-density gas of two-level atoms with one mode of an electromagnetic field in the long wave approximation [7].

A study of the critical behaviour of the nucleus within the LMG model or, equivalently, to the infinitely coordinated ferromagnetic Ising model, was carried out at the beginning of the eighties [8, 9]. These systems have been used to illustrate the theory of phase transitions, where the case with $T = 0$ is included [10].

The quantum phase transition in the Dicke model was determined since the early seventies by Hepp and Lieb [11] and the free energy of the system in the thermodynamic limit was calculated analytically by Wang and Hioe [12]. Experimentally, this phase transition has been realised in an open system formed by a Bose-Einstein condensate coupled to an optical cavity, observing a superradiant phase and a quantitative description is obtained using the Dicke model [13].

Generalisations of the Dicke model to 3-level atomic systems were introduced in the middle of the eighties and a review of the dynamics of an atom interacting with radiation can be found in [14] and references therein.

The superradiant transition has been thoroughly studied and a recent review is found in [15]. Lately, it has been established that symmetry-adapted coherent states can describe very well the superradiant region [16]. A nice characterisation of the quantum phase transition can be obtained by means of the Husimi function, Rényi-Wehrl entropies, and second moment of the Husimi function [17, 18]. The critical points in the Dicke model, with a finite number of particles, has been determined by means of semiclassical and exact numerical solutions [16]. In the semiclassical approach we used a variational method with coherent and symmetry adapted coherent states, whereas for the exact quantum case we use the concept of fidelity introduced in quantum information theory. For the coherent states (CS), symmetry-adapted coherent states (SAS), and the quantum exact solutions, we have shown the existence of a universal parametric curve for the expectation values of the number of photons and the relative population as function of the matter-field coupling strength [19].

Generalisations of the quantum phase diagram to 3-level atoms interacting dipolarly with 2-modes of an electromagnetic field has also been established in [20], where the phase diagram is divided into two regions: (i) the normal region where all the atoms are in their lowest energy level and the state has zero photons, and (ii) the superradiant region, where there is an enhancement in the emission rates of the excited atoms. The latter region divides itself into subregions where only one coupling strength μ_{ij} and one radiation field mode Ω_{ij} rule the behaviour of the system. This behaviour has been explicitly studied for the λ and N atomic configurations of 4-level atoms. As an example, for the λ case we determine three superradiant subregions associated to two electromagnetic modes denoted by S_Λ, S_{23} and S_{34} [21].

These studies allow us to generalise the result for n-level atoms interacting dipolarly with ℓ-modes of an electromagnetic field in a cavity [22]. The procedure to determine its quantum phase diagram is the following: A test function is constructed from the tensorial product of coherent states associated to the matter and field components. Then the expectation value of the Hamiltonian with respect to the variational state is calculated defining the so-called energy surface. The critical points of this energy surface then determine, in the parameter space of the dipolar matter-field couplings, the locus where the ground state of the system suffers quantum phase transitions.

In this contribution we show that the CS, SAS, and the exact quantum solutions for the 3-level atomic V-configuration follow the Universal curve established for the Dicke model in the different superradiant subregions of the generalised Dicke model. Thus, for the region S_{12} the expectation values of the number of photons in the electromagnetic mode Ω_{12} and the relative population between the atomic levels with energies proportional to ω_2 and ω_1, as function of μ_{12} fall on a Universal parametric curve. A similar result is obtained for the region S_{13}, where the relevant parameters are Ω_{13}, for the photon field, and ω_3 and ω_1 for the matter part with small values of μ_{12}.

This contribution is organised as follows. In section 2 the generalised Dicke model for 3-level atoms interacting with electromagnetic radiation is presented. The energy surface is calculated with respect to the tensorial product of coherent states of the matter and field sectors. The stability properties of the energy surface are determined together with the variational ground state energies and eigenfunctions in section 3. These results allow us to calculate the quantum phase diagram of the atomic V-configuration and contour plots of the energy surface are shown to illustrate its behaviour when a quantum phase transition is crossed. In section 4 two independent constants of motion of the Hamiltonian for the V quantum system, in the rotating wave approximation (RWA), are obtained, which allow to divide the Hilbert space into four orthogonal subspaces. The corresponding quantum phase diagram together with the symmetries of the Hamiltonian suggest a truncation of the Hilbert space into the basis states of two subgroups, viz., the $SU_{12}(2)$ and $SU_{13}(2)$ subgroups of the three dimensional unitary group of the 3-level system, $U(3)$. We present the calculation of the Universal parametric curve for the Dicke model in section 5 together with the variational coherent state solution and the exact quantum calculation for a small number of particles. Additionally, the corresponding quantum phase diagram for a finite number of atoms is calculated. In section 6 a summary of the results is presented together with additional remarks.

2. Generalised Dicke model (GDM)

We are considering a low density gas of 3-level atoms interacting dipolarly with 2-modes of electromagnetic radiation in the long wave approximation, that is, neglecting the position dependence of the atoms [20].

The Hamiltonian of the 3-level atoms is given by

$$\boldsymbol{H}_M = \hbar\,\omega_1\,\boldsymbol{A}_{11} + \hbar\,\omega_2\,\boldsymbol{A}_{22} + \hbar\,\omega_3\,\boldsymbol{A}_{33}\,, \tag{1}$$

with $\omega_1 \leq \omega_2 \leq \omega_3$ and $\boldsymbol{A}_{jj}$ indicating the number operator of particles in level j.

The 2-mode electromagnetic field is described by the Hamiltonian

$$\boldsymbol{H}_F = \hbar\,\Omega_{12}\,\boldsymbol{a}^\dagger_{12}\boldsymbol{a}_{12} + \hbar\,\Omega_{13}\,\boldsymbol{a}^\dagger_{13}\boldsymbol{a}_{13}\,, \tag{2}$$

with Ω_{jk} denoting the frequency of the photon creation and annihilation operators $(\boldsymbol{a}^\dagger_{jk}, \boldsymbol{a}_{jk})$. These are boson operators satisfying the commutation relations $[\boldsymbol{a}_{jk}, \boldsymbol{a}^\dagger_{mn}] = \delta_{jm}\,\delta_{kn}$ and $[\boldsymbol{a}_{jk}, \boldsymbol{a}_{mn}] = [\boldsymbol{a}^\dagger_{jk}, \boldsymbol{a}^\dagger_{mn}] = 0$. The energy separation between the (i,j) levels satisfies the expression $\omega_j - \omega_i = \Omega_{ji} + \Delta_{ij}$, with Δ_{ij} defining the corresponding detuning parameter.

In this model there is no direct interaction between the atoms, the atoms only interact through the quantised electromagnetic field. The coupling interaction between the matter and field is constituted by two terms called rotating $\boldsymbol{O}_{\rm R}$ and counterrotating terms $\boldsymbol{O}_{\rm CR}$, given by

$$\begin{aligned}\boldsymbol{H}_{FM} &= \boldsymbol{O}_{\rm R} + \boldsymbol{O}_{\rm CR}\\ &= -\frac{\hbar}{\sqrt{N_a}}\Big\{\mu_{12}(\boldsymbol{A}_{12}\,\boldsymbol{a}^\dagger_{12} + \boldsymbol{A}_{21}\,\boldsymbol{a}_{12}) + \mu_{13}(\boldsymbol{A}_{13}\,\boldsymbol{a}^\dagger_{13} + \boldsymbol{A}_{31}\,\boldsymbol{a}_{13})\Big\}\\ &\quad -\frac{\hbar}{\sqrt{N_a}}\Big\{\mu_{12}(\boldsymbol{A}_{21}\,\boldsymbol{a}^\dagger_{12} + \boldsymbol{A}_{12}\,\boldsymbol{a}_{12}) + \mu_{13}(\boldsymbol{A}_{31}\,\boldsymbol{a}^\dagger_{13} + \boldsymbol{A}_{13}\,\boldsymbol{a}_{13})\Big\}\,,\end{aligned} \tag{3}$$

with $\mu_{jk} = (\omega_j - \omega_k) d_{jk} \sqrt{\frac{2\,\pi\,\rho_m}{\hbar\,\Omega_{jk}}}$ denoting the strength of the interaction proportional to the dipolar interaction d_{ij} between the atomic levels i and j. Here, N_a is the total number of atoms and ρ_m is the density of atoms in the quantisation volume.

The collective (many-body) operators $\boldsymbol{A}_{jk} = \sum_{s=1}^{N_a} |j\,s\rangle\langle k\,s| \equiv \boldsymbol{b}^\dagger_j\,\boldsymbol{b}_k$ satisfy the commutation relations of a unitary algebra in three dimensions $u(3)$,

$$[\boldsymbol{A}_{jk}, \boldsymbol{A}_{mn}] = \boldsymbol{A}_{jn}\,\delta_{mk} - \boldsymbol{A}_{mk}\,\delta_{jn}. \tag{4}$$

Next we use as variational state a condensate formed by the tensorial product of coherent states for the matter and field parts because they survey properly the available Hilbert space of the system.

IOP Conf. Series: Journal of Physics: Conf. Series **1071** (2018) 012006 doi:10.1088/1742-6596/1071/1/012006

3. Energy surface of the GDM

The coherent states have played a central importance in quantum mechanics and quantum optics [23]. In what follows we define the coherent states of the matter and field components of the condensate.

The coherent states were introduced in quantum mechanics by Schrödinger to avoid the unwanted features of the spreading of wave packets. For the description of quantised electromagnetic fields (field oscillators) they were investigated by R. Glauber [24], E. Sudarshan [25], and J. Klauder [26]. The two-mode coherent states are written in terms of two complex parameters α_{12} and α_{13}, as

$$\begin{aligned} |\alpha_{12},\alpha_{13}\rangle &= |\alpha_{12}\rangle\otimes|\alpha_{13}\rangle \\ &= e^{-\frac{|\alpha_{12}|^2+|\alpha_{13}|^2}{2}}\sum_{k_1,k_2}\frac{(\alpha_{12})^{k_1}\,(\alpha_{13})^{k_2}}{\sqrt{k_1!\,k_2!}}|k_1,k_2\rangle\,, \end{aligned}$$

where $|k_1,\,k_2\rangle$ define eigenstates of the photon number operators, $\boldsymbol{\nu}_{12}$ and $\boldsymbol{\nu}_{13}$. These states minimise the uncertainty relation between the quadratures components of the 2-mode electromagnetic field.

For the matter sector we consider the coherent states associated to the totally symmetric representation of $u(3)$ algebra [27]. These coherent states have been used in nuclear physics to describe the shapes within the interacting boson model [5]. In our case we have two complex parameters γ_2 and γ_3 describing the matter condensate,

$$|\gamma_2,\gamma_3\rangle = \frac{1}{\sqrt{N_a!}}\boldsymbol{\Gamma}^{\dagger\,N_a}|0\rangle\,, \tag{5}$$

where

$$\boldsymbol{\Gamma}^\dagger = \frac{1}{\sqrt{1+|\gamma_2|^2+|\gamma_3|^2}}\left(\boldsymbol{b}_1^\dagger+\gamma_2\,\boldsymbol{b}_2^\dagger+\gamma_3\,\boldsymbol{b}_3^\dagger\right). \tag{6}$$

with $\boldsymbol{b}_k|0\rangle=0$, $k=1,2,3$. Notice that $\boldsymbol{\Gamma}$ and $\boldsymbol{\Gamma}^\dagger$ satisfy the boson commutation relations.

The tensorial product of the two coherent states allows us to establish the variational function: $|\alpha_{12}\,\alpha_{13}\rangle\otimes|\gamma_2\,\gamma_3\rangle$, which has eight real parameters.

The energy surface of the system is calculated taking the expectation value of the Hamiltonian with respect to the test function and dividing the result by the number of atoms to have an intensive quantity. Then, by means of the expectation values: (i) For the matter part,

$$\begin{aligned} \langle\gamma_2\,\gamma_3|\boldsymbol{A}_{1\,k}|\gamma_2\,\gamma_3\rangle &= \frac{N_a\gamma_k}{1+|\gamma_2|^2+|\gamma_3|^2}\,, \quad \langle\gamma_2\,\gamma_3|\boldsymbol{A}_{k\,k}|\gamma_2\,\gamma_3\rangle = \frac{N_a|\gamma_k|^2}{1+|\gamma_2|^2+|\gamma_3|^2}\,, \\ \langle\gamma_2\,\gamma_3|\boldsymbol{A}_{2\,3}|\gamma_2\,\gamma_3\rangle &= \frac{N_a\,\gamma_2^*\,\gamma_3}{1+|\gamma_2|^2+|\gamma_3|^2}\,, \quad \langle\gamma_2\,\gamma_3|\boldsymbol{A}_{1\,1}|\gamma_2\,\gamma_3\rangle = \frac{N_a}{1+|\gamma_2|^2+|\gamma_3|^2}. \end{aligned} \tag{7}$$

with $k=2,3$ and (ii) for the electromagnetic field,

$$\langle\alpha_{12}\,\alpha_{13}|\boldsymbol{a}_{1\,2}|\alpha_{12}\,\alpha_{13}\rangle = \sqrt{N_a}\,r_{12}\,e^{i\,\theta_{12}}\,,\quad \langle\alpha_{12}\,\alpha_{13}|\boldsymbol{a}_{1\,3}|\alpha_{12}\,\alpha_{13}\rangle = \sqrt{N_a}\,r_{13}\,e^{i\,\theta_{13}}\,. \tag{8}$$

where $\alpha=\sqrt{N_a}\,r\,e^{i\,\theta}$, with $r\in\Re$ for each mode field, one arrives to the energy surface of the generalised Dicke model (GDM),

$$\begin{aligned} E &= \Omega_{12}\,r_{12}^2+\Omega_{13}\,r_{13}^2+\frac{\omega_1\,+\omega_2\,\rho_2^2+\omega_3\,\rho_3^2}{1+\rho_2^2+\rho_3^2} \\ &- \frac{4\Big(\mu_{12}\,r_{12}\,\rho_2\,\cos\theta_{12}\cos\phi_2+\mu_{13}\,r_{13}\,\rho_3\,\cos\theta_{13}\cos\phi_3\Big)}{1+\rho_2^2+\rho_3^2} \end{aligned} \tag{9}$$

IOP Conf. Series: Journal of Physics: Conf. Series **1071** (2018) 012006 doi:10.1088/1742-6596/1071/1/012006

with $\gamma_2 = \rho_2 \, e^{i\,\phi_2}$ and $\gamma_3 = \rho_3 \, e^{i\,\phi_3}$. Notice that $|\alpha_{12}|^2$ and $|\alpha_{13}|^2$ measure the intensity of the electromagnetic modes (equivalently to the number of photons) while $|\gamma_2|^2 = (n_2/n_1)$ and $|\gamma_3|^2 = (n_3/n_1)$ measure the ratio of the populations n_k, with $k = 1, 2, 3$, of the atomic levels and E is an scalar intensive quantity.

3.1. Critical points

The critical points of the variational energy surface are obtained by solving the set of coupled equations

$$\left(\frac{\partial E}{\partial \theta_{1k}}, \frac{\partial E}{\partial \phi_k}, \frac{\partial E}{\partial r_{1k}}, \frac{\partial E}{\partial \rho_k}\right) = 0 \,, \tag{10}$$

with $k = 2, 3$.

From expression (10), one determines the critical points for the angle variables $\theta^c_{12} = 0, \pi$, $\theta^c_{13} = 0$, π, $\phi^c_2 = 0, \pi$, and $\phi^c_3 = 0$, π. These critical points provide a minimum value of the energy surface when $\mu_{jk} \cos\theta^c_{jk} \cos\phi^c_j \geq 0$, which under the assumption that the coupling interactions $\mu_{jk} \geq 0$, implies $\theta^c_{jk} = \phi^c_j$.

The critical points r^c_{12} and r^c_{13} that satisfy the expression (10) are given by

$$r^c_{12} = \frac{2\,\mu_{12}\,\rho_2}{\Omega_{12}(1+\rho_2^2+\rho_3^2)}\,, \qquad r^c_{13} = \frac{2\,\mu_{13}\,\rho_3}{\Omega_{13}(1+\rho_2^2+\rho_3^2)}\,. \tag{11}$$

They determine the expectation value of the number of photons in each mode of the electromagnetic field.

The variables ρ^c_2 and ρ^c_2 are also obtained from expression (10), it is straightforward to show that $\rho^c_2 = \rho^c_3 = 0$ is a solution and the corresponding energy surface is $E_N = 0$, where we take without loss of generality $\omega_1 = 0$. This solution defines the normal region, which has all the atoms in their lowest level and zero photons. Therefore, the variational solution is given by the state

$$|\psi\rangle_N = |0,0\rangle_F \otimes |N_a, 0, 0\rangle_M\,, \quad \text{with} \;\; E_N = 0\,. \tag{12}$$

Considering $\rho^c_3 = 0$ one has solutions for ρ^c_2 and the corresponding energy

$$\rho^c_2 = \sqrt{\frac{4\,\mu_{12}^2 - \Omega_{12}\,\omega_2}{4\,\mu_{12}^2 + \Omega_{12}\,\omega_2}}\,, \qquad E_{12} = -\frac{\left(4\mu_{12}^2 - \Omega_{12}\,\omega_2\right)^2}{16\,\mu_{12}^2\,\Omega_{12}}\,, \tag{13}$$

which define the superradiant subregion S_{12}. In this region one has $\mu_{12} \geq \sqrt{\Omega_{12}\,\omega_2}/2$, and the expectation value of the photons of the mode Ω_{13} is zero. In consequence the test function takes the form

$$|\psi\rangle_{S_{12}} \;=\; \frac{e^{-\frac{N_a\, {r^c_{12}}^2}{2}}}{(1+{\rho^c_2}^2)^{\frac{N_a}{2}}} \sum_{n\,\nu} \left(\begin{array}{c} N_a \\ n \end{array}\right)^{1/2} \frac{\left(\sqrt{N_a}\, r^c_{12}\right)^{\nu} \, {\rho^c_2}^{\,n}}{\sqrt{\nu!}} \, |\nu, 0\rangle_F \otimes |N_a - n, n, 0\rangle_M\,. \tag{14}$$

Observe that the the expectation value of the number of atoms in the lowest energy level is bounded by $1/2 \leq \langle \boldsymbol{A}_{11}\rangle/N_a \leq 1$, indicating also that $0 \leq \langle \boldsymbol{A}_{22}\rangle/N_a \leq 1/2$ and there are no atoms in the energy level ω_3.

Now we consider the opposite case, that is, $\rho^c_2 = 0$ which implies the solutions for ρ^c_3 and its associated energy as follows

$$\rho^c_3 = \sqrt{\frac{4\,\mu_{13}^2 - \Omega_{13}\,\omega_3}{4\,\mu_{13}^2 + \Omega_{13}\,\omega_3}}\,, \qquad E_{13} = -\frac{\left(4\mu_{13}^2 - \Omega_{13}\,\omega_3\right)^2}{16\,\mu_{13}^2\,\Omega_{13}}\,; \tag{15}$$

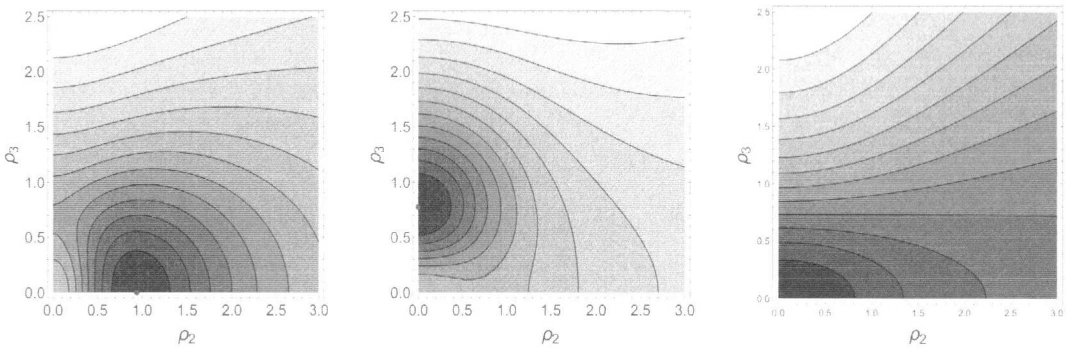

Figure 1. Contour plots for the energy surface of the atomic V-configuration. The plots are for ($\mu_{12} = 1/2$, $\mu_{13} = 4/5$) at left in the region S_{12}, ($\mu_{12} = 1/5, \mu_{13} = 1$) at the middle in the region S_{13}, and ($\mu_{12} = 1/10, \mu_{13} = 1/3$) at the right in the normal region. The red dots indicate the corresponding critical minimum values. We use the parameters $\Omega_{13} = 1$, $\Omega_{12} = 0.8$, $\omega_2 = 0.8$ and $\omega_3 = 1$.

this solution determines the other superradiant subregion S_{13} of the atomic V-configuration. This region exists for coupling interactions satisfying the condition $\mu_{13} \geq \sqrt{\Omega_{13}\,\omega_3}/2$, the expectation value of $\boldsymbol{\nu}_{12}$ is zero, and the expectation value of the number of atoms occupying ω_1 is again bounded by $1/2 \leq \langle \boldsymbol{A}_{11} \rangle / N_a \leq 1$, and thus $0 \leq \langle \boldsymbol{A}_{33} \rangle / N_a \leq 1/2$ and there are no atoms in energy level with ω_2. In this region the variational solution takes the form

$$|\psi\rangle_{S_{13}} = \frac{e^{-\frac{N_a\, r_{23}^{c\,2}}{2}}}{(1+\rho_3^{c\,2})^{N_a/2}} \sum_{n\,\nu} \begin{pmatrix} N_a \\ n \end{pmatrix}^{1/2} \frac{\left(\sqrt{N_a}\, r_{13}^c\right)^{\nu} \rho_3^{c\; n}}{\sqrt{\nu!}} \, |0,\nu\rangle_F \otimes |N_a - n, 0, n\rangle_M \, . \quad (16)$$

Next we establish explicitly the quantum phase diagram for the atomic V-configuration, where the previous results play a fundamental role. Contour plots for the energy surface of the system in terms of the variables ρ_2 and ρ_3 are shown in Fig. 1, we consider dipolar interactions of the normal, S_{12} and S_{13} regions.

3.2. Quantum phase diagram

We have found that the atomic V-configuration exhibits a quantum phase transition from the no photons present to monochromatic superradiant subregions at two critical atom-field coupling interactions $\mu_{12}^c = \sqrt{\Omega_{12}\,\omega_2}/2$ and $\mu_{13}^c = \sqrt{\Omega_{13}\,\omega_3}/2$. This is equivalent to consider the equalities of the energies,

$$E_N = E_{12} \Rightarrow 4\,\mu_{12}^2 = \Omega_{12}\,\omega_2 \, , \quad E_N = E_{13} \Rightarrow 4\,\mu_{13}^2 = \Omega_{13}\,\omega_3 \, . \quad (17)$$

At these values the critical points of the superradiant subregions coalesce to $(\rho_2^c, \rho_3^c) = (0,0)$. Then, from the catastrophe formalism point of view, the borders of the normal to the monochromatic regions are both bifurcations and Maxwell sets [28].

Taking $E_{12} = E_{13}$, one finds

$$\mu_{12}^2 = \frac{\Omega_{12}\left(\mu_{13}^2 - \frac{\Omega_{13}\,\omega_3}{4}\right)^2}{2\,\Omega_{13}\,\mu_{13}^2} \left(1 + \frac{\mu_{13}^2\,\Omega_{13}\,\omega_3}{2\,(\mu_{13}^2 - \frac{\Omega_{13}\,\omega_3}{4})^2} + \sqrt{1 + \frac{\mu_{13}^2\,\Omega_{13}\,\omega_2}{(\mu_{13}^2 - \frac{\Omega_{13}\,\omega_3}{4})^2}}\right) , \quad (18)$$

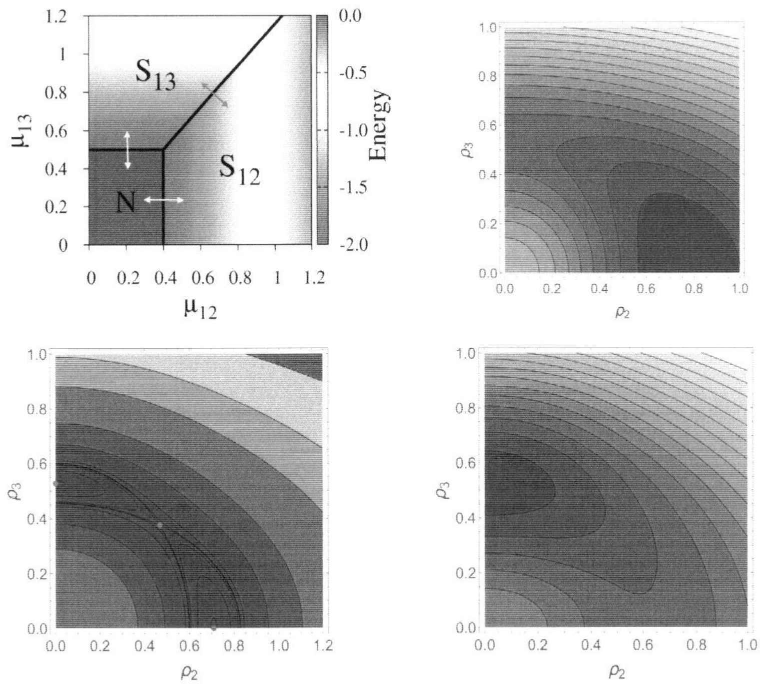

Figure 2. Energy and phase diagram plots for the atomic V-configuration are shown at the top left. The yellow arrows indicates second order phase transitions while the red arrow specify first order phase transitions. Contour plots of the energy surface are displayed at top right for the region S_{12}, at the left side of bottom left on the separatrix of the superradiant subregions, and at the bottom right for the region S_{13}.

which must satisfy the conditions associated to the existence of the critical points $4\,\mu_{12}^2 \geq \Omega_{12}\,\omega_2$, and $4\,\mu_{13}^2 \geq \Omega_3\,\omega_{31}$. This curve separates the subregion S_{12}, where the photons of the mode Ω_{12} dominate, from the sector S_{13}, where the photons of the mode Ω_{13} dominate.

For the parameters $\Omega_{13} = 1$, $\Omega_{12} = 0.8$, $\omega_2 = 0.8$, and $\omega_3 = 1$ we display in Fig. 2 the minimum energy as function of the matter-field dipolar intensities μ_{12} and μ_{13}. The energy scale is shown at the right side of the figure. The normal region is indicated by N (black region) while the superradiant subregions by S_{12} and S_{13}. The quantum phase diagram of the 3 level system is also shown (see white lines) together with the order of the transitions. These are established by means of the Ehrenfest classification [28].

The behaviour of the energy surface, when there is a crossing between the superradiant subregions S_{12} and S_{13}, as function of the variables ρ_2 and ρ_3 is also exhibited in Fig. 2. Notice that the crossing corresponds to a saddle point, i.e., a critical point unstable under a small perturbation, going to the minima in the variable ρ_2 or ρ_3. The parameters used for the different contour plots of the energy surface at sector S_{12}, at the separatrix, and at S_{13}, are $(\mu_{12}, \mu_{13}) = \{(0.18763, 2/3), (0.21763, 2/3), (0.24763, 2/3)\}$, respectively.

4. Symmetries of the atomic V-configuration Hamiltonian

The symmetries of the GDM can be established by considering first the rotating wave approximation (RWA). Then we propose as constant of the motion the operator

$$\boldsymbol{K} = \eta_1\,\boldsymbol{\nu}_{12} + \eta_2\,\boldsymbol{\nu}_{13} + \lambda_1\,\boldsymbol{A}_{11} + \lambda_2\,\boldsymbol{A}_{22} + \lambda_3\,\boldsymbol{A}_{33}\,,$$

where the set of real parameters η_k and λ_j, with $k = 1, 2$ and $j = 1, 2, 3$, will be determined by asking that the commutation relation $[\boldsymbol{K}, \boldsymbol{O}_\mathrm{R}] = 0$ holds. One finds

$$\boldsymbol{K} = \eta_1(\boldsymbol{\nu}_{12} + \boldsymbol{A}_{22}) + \eta_2(\boldsymbol{\nu}_{13} + \boldsymbol{A}_{33}) + \lambda_1(\boldsymbol{A}_{11} + \boldsymbol{A}_{22} + \boldsymbol{A}_{33}) \,,$$

which implies three independent constants of motion,

$$\boldsymbol{K}_1 = \boldsymbol{A}_{11} - \boldsymbol{\nu}_{12} - \boldsymbol{\nu}_{13} \,, \quad \boldsymbol{K}_2 = \boldsymbol{A}_{22} + \boldsymbol{\nu}_{12} \,, \quad \boldsymbol{K}_3 = \boldsymbol{A}_{33} + \boldsymbol{\nu}_{13} \,. \tag{19}$$

Observe that the number of atoms is also a constant of the motion, which can be obtained by summing the previous invariants.

For the counter-rotating term one finds that the parity of the constants of motion is an invariant, that is, $e^{i\pi \boldsymbol{K}_j}\, \boldsymbol{O}_\mathrm{CR}\, e^{-i\pi \boldsymbol{K}_j} = \boldsymbol{O}_\mathrm{CR}$ because

$$e^{i\pi \boldsymbol{K}_\ell}\, \boldsymbol{a}_{jk}\, e^{-i\pi \boldsymbol{K}_\ell} = e^{-i\pi}\, \boldsymbol{a}_{jk} \,, \quad e^{i\pi \boldsymbol{K}_\ell}\, \boldsymbol{A}_{jk}\, e^{-i\pi \boldsymbol{K}_\ell} = e^{-i\pi}\, \boldsymbol{A}_{jk} \,,$$

for all possible values of j, k and ℓ. The basis for the atomic V-configuration Hamiltonian can be defined in terms of the infinite dimensional set of Fock states: $|\nu_{12}, \nu_{13}, n_1, n_2, n_3\rangle$, two of them describing the photons and three related with the populations of the atomic levels.

The quantum labels for the field part can be replaced by the eigenvalues of the constants of motion $\boldsymbol{K}_2$ and $\boldsymbol{K}_3$ while the total number of atoms in the lowest level in terms of number of atoms of the system, $N_a = n_1 + n_2 + n_3$. Thus, the infinite dimensional Hilbert space is divided into four orthogonal subspaces related with the parities of the eigenvalues K_2 and K_3,

- $\mathcal{H}_{e\,e} : \{|\nu_{12}, \nu_{13}; n_1, n_2, n_3\rangle\}$ with $\nu_{12} + n_2$ and $\nu_{13} + n_3$ even values.
- $\mathcal{H}_{e\,o} : \{|\nu_{12}, \nu_{13}; n_1, n_2, n_3\rangle\}$ with $\nu_{12} + n_2$ even and $\nu_{13} + n_3$ odd values.
- $\mathcal{H}_{o\,e} : \{|\nu_{12}, \nu_{13}; n_1, n_2, n_3\rangle\}$ with $\nu_{12} + n_2$ odd and $\nu_{13} + n_3$ even values.
- $\mathcal{H}_{o\,o} : \{|\nu_{12}, \nu_{13}; n_1, n_2, n_3\rangle\}$ with $\nu_{12} + n_2$ and $\nu_{13} + n_3$ odd values.

Therefore, from the variational solution we are able to get four independent energy bands of the system, associated to the four Hilbert subspaces. We have found that the ground state is always determined within the Hilbert space $\mathcal{H}_{e\,e}$.

The calculation of the eigensystem of the Hamiltonian requires a truncation of the infinite dimensional Hilbert space. This is done by establishing the maxima values of $K_{2\,\mathrm{Max}}$ and $K_{3\,\mathrm{Max}}$ for a given interval of the dipolar interactions μ_{12} and μ_{13} (details will be published elsewhere). Then, to construct the matrix hamiltonian we propose the finite dimensional set of states,

$$\mathcal{H}_{e\,e} : \cup_{q=0}^{N_2} \cup_{s=0}^{N_3} \cup_{n_2,n_3=0}^{N_a} |2\,q - n_2, 2\,s - n_3; N_a - n_2 - n_3, n_2, n_3\rangle \tag{20}$$

with $N_2 \equiv \lfloor K_{2\mathrm{Max}}/2 \rfloor$, $N_3 \equiv \lfloor K_{3\mathrm{Max}}/2 \rfloor$ together with the conditions $2\,q - n_2 \geq 0$ and $2\,s - n_3 \geq 0$.

For $\mu_{13} = 0$, the Hilbert space $\mathcal{H}_{e\,e}^{(12)}$ can be generated by (20) with $s = n_3 = 0$, which is equivalent to the Hilbert space of the Dicke model. For $\mu_{12} = 0$, we have an analogous result, i.e., the Hilbert space $\mathcal{H}_{e\,e}^{(13)}$ is obtained from (20) with $q = n_2 = 0$.

For $(\mu_{13}, \mu_{12}) \neq 0$, the symmetries of the Hamiltonian and the variational ground state result lead us to propose as zero order approximation a Hilbert space of the form

$$\mathcal{H}_{e\,e}(0) = \mathcal{H}_{e\,e}^{(12)} \oplus \mathcal{H}_{e\,e}^{(13)} \,.$$

The approximation can be improved by considering the union of the Hilbert space $\mathcal{H}_{e\,e}$ with $s = 0, 1, \cdots k$, which we denote by $\mathcal{H}_{e\,e}^{(12)}(k)$ and $\mathcal{H}_{e\,e}$ with $q = 0, 1, \cdots k$, which we denote by $\mathcal{H}_{e\,e}^{(13)}(k)$. Therefore a truncated basis of kth-order can be written as follows

$$\mathcal{H}_{e\,e}(k) = \mathcal{H}_{e\,e}^{(12)}(k) \oplus \mathcal{H}_{e\,e}^{(13)}(k) \,. \tag{21}$$

5. Universal curves

We will extend extend the result obtained for the Dicke model in relation with the determination of the Universal parametric curve for the expectation value of the number of photons of the electromagnetic field and the relative atomic population of the levels as implicit function of the dipolar interaction [16].

5.1. Universal parametric curve for the Dicke model

We give here a brief review of the determination of the variational ground state for the Dicke model. Considering the same notation as for the 3-level case, the Dicke Hamiltonian takes the form

$$\boldsymbol{H} = \hbar\omega_1\,\boldsymbol{A}_{11} + \hbar\omega_2\,\boldsymbol{A}_{22} + \hbar\,\Omega\,\boldsymbol{a}^\dagger\boldsymbol{a} - \frac{\hbar\,\mu}{\sqrt{N_a}}(\boldsymbol{A}_{12}+\boldsymbol{A}_{21})(\boldsymbol{a}^\dagger+\boldsymbol{a})\,. \tag{22}$$

The expectation value of the Hamiltonian with respect to the corresponding tensorial product of the matter and field coherent states $|\alpha\,\gamma\rangle$ determines the energy surface,

$$\begin{aligned} E &= \langle\alpha\gamma|H|\alpha\gamma\rangle/\hbar\,N_a \\ &= \Omega\,r^2 + \frac{\omega_1\; + \omega_2\,\rho^2}{1+\rho^2} - \frac{4\mu\,r\,\rho\,\cos\theta_\rho\cos\phi_\gamma}{1+\rho^2}\,, \end{aligned} \tag{23}$$

with $\alpha = \sqrt{N_a}\,r\,e^{i\theta_\rho}$, and $\gamma = \rho\,e^{i\phi_\gamma}$. Notice that in this two-level problem the coherent states of matter are equivalent to those introduced by Radcliffe related with the so-called spin coherent states [29]

The critical points of the energy surface are obtained by solving the set of coupled equations

$$\left(\frac{\partial E}{\partial\theta_\rho}, \frac{\partial E}{\partial\phi_\gamma}, \frac{\partial E}{\partial r}, \frac{\partial E}{\partial\rho}\right) = 0\,.$$

The first three equations are solved immediately

$$\theta^c_\rho = 0,\,\pi\,,\quad \phi^c_\gamma = 0,\,\pi\,,\quad r^c = \frac{2\,\mu\,\rho}{\Omega(1+\rho^2)}\,, \tag{24}$$

which provide a minimum value for the energy surface when $\mu\,\cos\theta^c_\rho\,\cos\phi^c_\gamma \geq 0$. Then, by assuming $\mu \geq 0$, the above condition implies $\theta^c_\rho = \phi^c_\gamma$.

From $\frac{\partial E}{\partial\rho} = 0$, one finds the following solutions: (i) $\rho^c = 0$, which yields $E_N = 0$, where we take $\omega_1 = 0$ without lose of generality. (ii) Considering $\rho^c \neq 0$, one gets

$$\rho^c = \sqrt{\frac{4\,\mu^2 - \Omega\,\omega_2}{4\,\mu^2 + \Omega\,\omega_2}}\quad,\quad E = -\frac{\left(4\mu^2 - \Omega\,\omega_2\right)^2}{16\,\mu^2\,\Omega}. \tag{25}$$

Notice that this minimum critical point exists for $\mu \geq \sqrt{\Omega\,\omega_2}/2$. Therefore there is a quantum phase transition between the zero field with all the atoms in the lowest energy level to the superradiant region when the dipolar coupling parameter is larger than

$$\mu^c = \sqrt{\Omega\,\omega_2}/2 \tag{26}$$

in agreement with the result for the superradiant subregions of the phase diagram of the 3-level system.

The symmetry preserved by the Hamiltonian (22) is the parity of the constant of motion in the RWA, i.e., $\boldsymbol{M} = \boldsymbol{A}_{22} + \boldsymbol{a}^\dagger\,\boldsymbol{a}$, because we have

$$e^{i\pi\boldsymbol{M}}\,\boldsymbol{a}\,e^{-i\pi\boldsymbol{M}} = e^{-i\,\pi}\,\boldsymbol{a}\,,\qquad e^{i\pi\boldsymbol{M}}\,\boldsymbol{A}_{12}\,e^{-i\pi\boldsymbol{M}} = e^{-i\,\pi}\,\boldsymbol{A}_{12}\,.$$

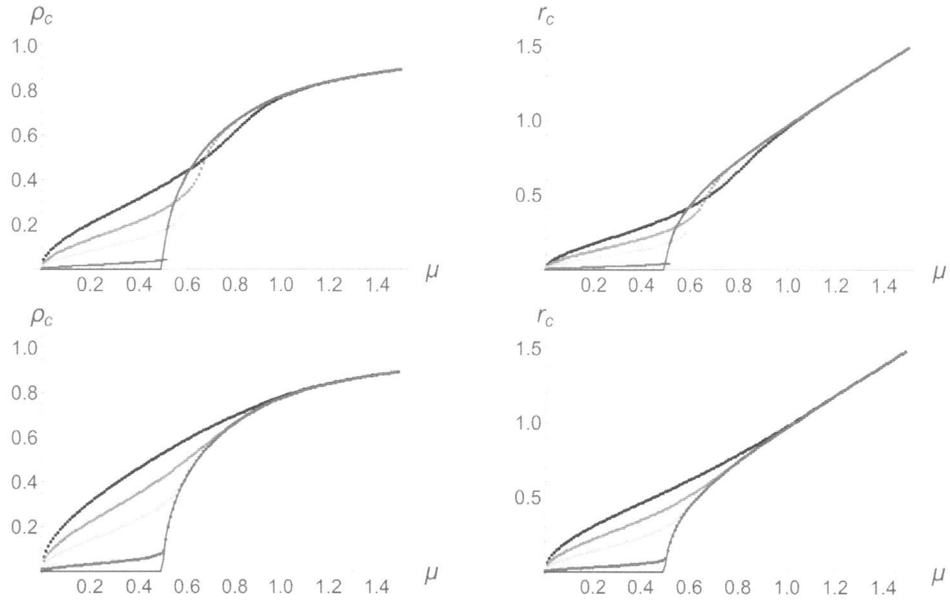

Figure 3. The minima critical points for the even and odd variational coherent states for the following cases: 2 particles (blue); 4 particles (orange); 10 particles (green); 200 particles (gray), and for the coherent states (red). We use $\Omega = 1$, $\omega_2 = 1$, and $\omega_1 = 0$.

The Hamiltonian symmetry is restored on the variational states by means of the projectors $\mathcal{P}_\pm = \frac{1}{2}\left(1 \pm e^{i\,\pi\,\boldsymbol{M}}\right)$. Then the symmetry adapted coherent states (SAS) are given by

$$|\alpha\,\gamma\rangle_\pm = \mathcal{N}_\pm\left(|\alpha\rangle \otimes |\gamma\rangle \pm |-\alpha\rangle \otimes |-\gamma\rangle\right)\,,$$

where the normalisation constant is given by

$$\mathcal{N}_\pm = \frac{1}{\sqrt{2}}\left(1 \pm e^{-2\,N_a\,r^2}\left(\frac{1-\rho^2}{1+\rho^2}\right)^{N_a}\right)^{-\frac{1}{2}}\,.$$

The expectation value of the Hamiltonian with respect to the SAS can be calculated and thus the corresponding energy surface takes the form

$$E_\pm = \Omega\,r^2\left(\frac{\mp 1 + \left(\frac{1-\rho^2}{1+\rho^2}\right)^{-N_a} e^{2\,N_a\,r^2}}{1 \pm \left(\frac{1-\rho^2}{1+\rho^2}\right)^{-N_a} e^{2\,N_a\,r^2}}\right) - \frac{\omega}{2}\left(\left(\frac{1-\rho^2}{1+\rho^2}\right)^{\pm 1} \pm \frac{\frac{4\,\rho^2}{1-\rho^4}}{1 \pm \left(\frac{1-\rho^2}{1+\rho^2}\right)^{-N_a} e^{2\,N_a\,r^2}}\right)$$
$$-\left(\frac{4\,\mu\,r\,\rho}{1+\rho^2}\right)\frac{\left(\frac{1-\rho^2}{1+\rho^2}\right)^{-N_a} e^{2\,N_a\,r^2}}{\left(\frac{1-\rho^2}{1+\rho^2}\right)^{-N_a} e^{2\,N_a\,r^2} \pm 1}\,. \tag{27}$$

Firstly, one finds, in the thermodynamic limit,

$$\lim_{N_a\to\infty} E_\pm = \Omega\,r^2 - (\omega_2 - \omega_1)\frac{1-\rho^2}{1+\rho^2} - \frac{4\,r\,\mu\,\rho}{1+\rho^2}\,, \tag{28}$$

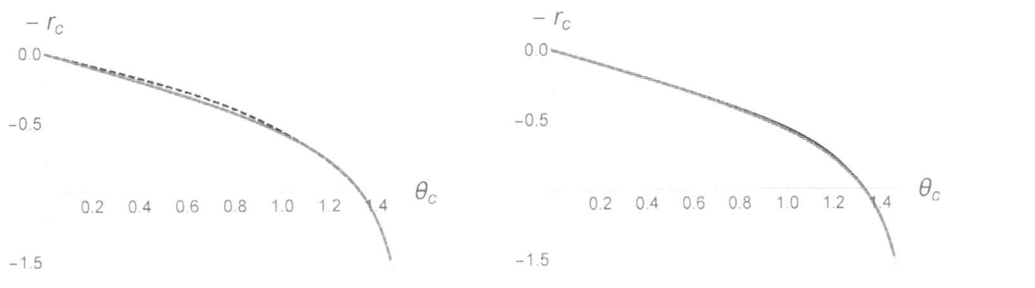

Figure 4. Even (left) and odd (right) results for: 2 particles (blue); 4 particles (orange); 10 particles (green); 200 particles (gray), and for the coherent states (red). We use $\Omega = 1$, $\omega_2 = 1$, and $\omega_1 = 0$.

which corresponds exactly to the energy surface obtained for the variational coherent state. Next we show that the minima critical points of the energy surface given in (27), coincide with those of the expression (28) in the thermodynamic limit.

In Fig. 3 the minima critical points of $E_\pm$ are displayed as functions of the dipolar coupling interaction for the following number of particles: $N_a = 2, 4, 10, 200$. One can observe that the critical points approximate the results obtained with the coherent state variational function when the number of atoms is large. Notice that the normal region shows more differences than the superradiant one, except in the vicinity of the quantum phase transition.

The variables r^c and θ^c, in the superradiant region, can be written as functions of the dipolar coupling interaction μ,

$$r^c = \frac{\mu}{\Omega}\sqrt{1-\left(\frac{\mu^c}{\mu}\right)^4}\,, \qquad \theta^c = \mathrm{ArcCos}(\mu^c/\mu)^2\,, \tag{29}$$

where θ^c defines the critical polar angle on the Bloch sphere [16]. In the normal region they are given by $r^c = 0$ and $\theta^c = 0$. The critical value of the coupling constant is given in expression (26). Then one can make a parametric plot of $-r^c$ vs. θ^c as functions of μ. The whole of the normal region is located at the origin and the expression (29) is valid for $\mu > \mu^c$ and for $N_a \geq 2$. The result is indicated by the dashed red line at the left and right sides of Fig. 4.

The same parametric plot can be done for the results obtained with the SAS states by means of a numerical determination of the minima critical points of the energy surface (27), after the substitution of $r^c = \frac{2\,\mu\,\rho}{\Omega(1+\rho^2)}$.

We have displayed in Fig. 4 the parametric plot of $-r_c$ vs. $\theta_c = \mathrm{ArcCos}\frac{1-\rho_c^2}{1+\rho_c^2}$ for the even and odd coherent states, with different number of particles. Notice that they are almost indistinguishable under the change in the number of particles, besides of having the same result for the even and odd cases.

The quantum calculation is done through the following identification

$$-r_c \to -\sqrt{\frac{\langle \boldsymbol{a}^\dagger\, \boldsymbol{a}\rangle}{N_a}}\,, \qquad \theta_c \to \mathrm{ArcCos}\left\{-\frac{\langle \boldsymbol{A}_{22} - \boldsymbol{A}_{11}\rangle}{N_a}\right\}\,. \tag{30}$$

The calculation of the expectation values in numerical form satisfies the mentioned Universal parametric plot as function of μ, and it is valid for any number of particles $N_a \geq 2$.

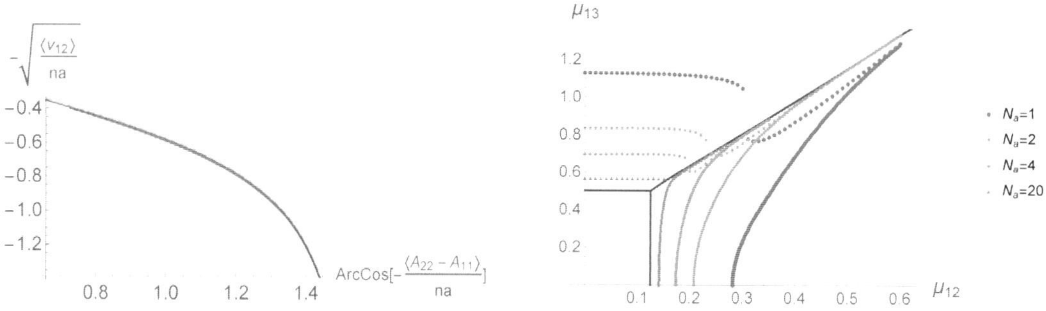

Figure 5. At left we display the universal parametric plot for $N_a = 12$. We use $\mu_{13} = 0$ (gray), $\mu_{13} = 0.1$ (red), $\mu_{13} = 0.3$ (green), and $\mu_{13} = 0.5$ (blue), and the 2-level case (red-dashed line). At right the quantum phase diagram for a finite number of atoms as function of the dipolar interactions is plotted. The continuous blue line defines the case when $N_a \to \infty$. In both panels we use the parameters $\Omega_{13} = 1$, $\Omega_{12} = 0.25$, $\omega_2 = 0.25$ and $\omega_3 = 1$.

5.2. Universal parametric curve for the GDM

We study the existence of a Universal curve of the type discussed in the previous subsection. This is motivated by the quantum phase diagram that we have obtained for the 3-level atomic V-configuration, i.e., the superradiant region is divided into sectors where there is dominance of only one photon mode of the electromagnetic field. Therefore, essentially they behave as 2-level systems. This can be seen in the expressions for the energy and corresponding eigenstates of the ground state of the system. For the normal region they are given in (12) while for the superradiant sectors S_{12} in (13) and (14), and for S_{13} in (15) and (16).

The numerical quantum calculations are done diagonalising the Hamiltonian matrix with the zero order basis $\mathcal{H}_{ee}(0)$. The universal parametric plot as a function of the dipolar strength μ_{12} was calculated for several values of the dipolar interaction μ_{13}, which are indicated in the left side of Fig. 5. All the calculated points curves fall approximately along the same curve together with the universal curve for the Dicke model indicated by a dashed red line. We have found that this agreement improves when the number of atoms is large; in the plot we consider $N_a = 12$.

The determination of the finite phase diagram is done by calculating the minima of the fidelity, a concept introduced in quantum information theory [30, 31]. The fidelity is a measure of the similarity of two probability distributions, which for pure states $|\xi\rangle$ and $|\phi\rangle$ reduces to scalar product, i.e., $\mathcal{F} = |\langle\xi|\phi\rangle|^2$. Here we use the fidelity to determine when the ground state exhibits a sudden change as function of the control parameters μ_{12} and μ_{13}. We keep μ_{13} fixed and get the minimum value of F when μ_{12} is suffering infinitesimal changes (see the dotted lines of the right side of Fig. 5) and vice versa (see the continuous lines at the right side of the figure). In this way, in Fig. 5 we have displayed the calculated phase diagrams for the atomic V-configuration and different number of atoms $N_a = 1, 2, 4, 20$ as indicated in the plot.

6. Conclusions

For the GDM the energy surface is determined together with its stability properties, which allows us to establish the quantum phase diagram. We have explicitly studied the quantum phase diagram for the 3-level atomic V-configuration, where there are two regions: the normal one, characterised by a zero-field and all the atoms in the lowest energy level, and the superradiant, which it is divided into two sectors with monochromatic dominance of one-mode photons.

Additionally, the contour plots of the energy surface are exhibited where there is a crossing between the two superradiant sectors S_{12} and S_{13}.

For the full Hamiltonian, without the RWA approximation, the constants of motion (parities) divide the Hilbert space into 4 orthogonal subspaces. Here the quantum numerical calculation is done for the even-even Hilbert space, that is, for even-even eigenvalues of the parity operators of $\boldsymbol{K}_2$ and $\boldsymbol{K}_3$. A truncated basis suggested by the symmetries of the Hamiltonian together with the quantum phase diagram allows us to determine the finite quantum crossovers of the GDM. The analogous quantum crossovers were also determined for a finite number of atoms, and they approach the quantum phase diagram for a large number of particles. This was done by choosing properly the basis of the Hilbert space because the dimension grows very rapidly with the number of particles.

The behaviour of the critical points of the energy surface for the even and odd variational coherent states of the Dicke model were compared with the critical points of the energy surface calculated with the coherent states, as functions of the dipolar strength.

We have reviewed the determination of the universal parametric curve of the Dicke model in terms of the number of photons and the relative atomic population between the levels. This universal curve exists also for the 3-level atomic V-configuration in the superradiant sectors, although it is perturbed for large values of the dipolar strength, i.e., for the subregion S_{12} troubled by μ_{13} while for S_{13} bothered by μ_{12}.

The variational state for the Dicke and GDM models does not exhibit entanglement and does not possess the symmetries of the Hamiltonian. This is, however, not so for the SAS states. We have demonstrated that the phase diagram obtained in the thermodynamic limit does not depend on the variational state chosen. As the ground state function must satisfy the symmetry properties of the Hamiltonian, we may conclude implying that there is entanglement between the matter and field components.

From the results presented here, we propose the existence of the universal parametric curve for the atomic Ξ- and Λ-configurations of 3-level systems, as well as for matter-field systems of n-levels and ℓ-modes of the electromagnetic field under dipolar interactions.

6.1. Acknowledgments

This work was supported in part by CONACyT (under project No. 238494), and DGAPA-UNAM (under Project No. IN101217).

References

[1] Hertz J A 1976 Phys. Rev. B **14** 1165.
[2] Sachdev S 2011 *Quantum Phase Transitions* Cambridge University Press 2nd. ed.
[3] Vojta M, 2003 Rep. Prog. Phys. **66** 2069.
[4] Gilmore R 1979 J. Math. Phys. **20** 891.
[5] Iachello F, and Arima A 1987 *The interacting boson model* Cambridge University Press.
[6] Cejnar P, Jolie J, and Casten R F 2010 Rev. Mod. Phys. **82** 2155.
[7] Dicke R H 1954 Phys. Rev. **93** 99.
[8] Botet R, Jullien R, and Pfeuty P 1982 Phys. Rev. Lett. **49** 478.
[9] Botet R and Jullien R 1983 Phys. Rev. B **28** 3995.
[10] Kittel C and Shore H 1965 Phys. Rev. **138** 1165.
[11] Hepp K and Lieb E H 1973 Ann. of Phys. **76** 360.
[12] Wang Y K and Hioe F T 1973 Phys. Rev A **7** 831.
[13] Baumann K, Guerlin C, Brennecke F, and Esslinger T 2010 Nature (London) **464**, 1301.
[14] Yoo H I and Eberly J H 1985 Phys. Rep. **118** 239.
[15] B. M. Garraway 2011 Philos. Trans. R. Soc. A **369** 1137.
[16] Castaños O, Nahmad-Achar E, López-Peña, and Hirsch J G 2011 Phys. Rev A **84** 013819.
[17] Romera E, del Real R, and Calixto M 2012 Phys. Rev. A **85** 053831.
[18] Romera E, Calixto M, and Nagy A 2012 Europhys. Lett. **97** 20011.
[19] Castaños O, Nahmad-Achar E, López-Peña R, and Hirsch J G 2011 Phys. Rev A **86** 023814.
[20] Cordero S et al 2015 Phys. Rev. A **92** 053843.
[21] Cordero S et al 2016 Phys. Rev. A **94** 013802.

IOP Conf. Series: Journal of Physics: Conf. Series **1071** (2018) 012006 doi:10.1088/1742-6596/1071/1/012006

[22] Cordero S, Nahmad-Achar E, Castaños O, and López-Peña R, 2017 Phys. Scr. **92** 044004.
[23] Schleich W P 2001 *Quantum Optics in Phase Space* Wiley-VCH.
[24] Glauber R J 1963 Phys. Rev. Lett. **10** 84.
[25] Sudarshan E C G 1963 Phys. Rev. Lett. **10** 277.
[26] Klauder J R 1963 J. Math. Phys. **4** 1055.
[27] Perelomov A 1986 *Generalised coherent states and their applications* Springer-Verlag.
[28] Gilmore R 1993 *Catastrophe theory for scientists and engineers* Dover Publications.
[29] Radcliffe J M 1971 J. Phys. A **4** 313.
[30] Uhlmann A 1976 Rep. Math. Phys. **9** 273.
[31] Zanardi P and Paunkovic N 2006 Phys. Rev E **74** 031123.

IOP Conf. Series: Journal of Physics: Conf. Series **1071** (2018) 012007 doi:10.1088/1742-6596/1071/1/012007

Weak measurement and weak values — New insights and effects in reflectivity and scattering processes

C Aris Chatzidimitriou-Dreismann

Institute of Chemistry, Sekr. C2, Faculty II, Technical University of Berlin, D-10623 Berlin, Germany

E-mail: dreismann@chem.tu-berlin.de

Abstract. Recently, the notions of Weak Measurement (WM), Weak Value (WV) and Two-State-Vector Formalism (TSVF), firstly introduced by Aharonov and collaborators, have extended the theoretical frame of standard quantum mechanics, thus providing a quantum-theoretical formalism for extracting new information from a system in the limit of small disturbance to its state. Here we provide an application to the case of two-body scattering with one body weakly interacting with its environment — e.g. a neutron being scattered from a H_2 molecule physisorbed in a carbon nanotube. In particular, we make contact with the field of incoherent inelastic neutron scattering from condensed systems. We provide a physically compelling prediction of a new quantum effect — a momentum transfer deficit; or equivalently, an enhanced energy transfer; or an apparent reduction of the mass of the struck particle. E.g., when a neutron collides with a H_2 molecule in a C-nanotube and excites its translational motion along the nanotube, it apparently exchanges energy and momentum with a fictitious particle with mass of 0.64 atomic mass units. Experimental results are shown and discussed in the new theoretical frame. The effect under consideration has no conventional interpretation, thus also supporting the novelty of the quantum theoretical framework of WV and TVSF. Some speculative remarks about possible applications being of technological interest (fuel cells and hydrogen storage; Li^+ batteries; information and communication technology) are shortly mentioned.

1. Introduction

The fundamental t-inversion symmetry of the Schrödinger equation plays a crucial role in the novel theory of weak measurement (WM), weak values (WV) and the two-state vector formalism (TSVF) of Aharonov and collaborators [1, 2, 3, 4, 5, 6, 7]. Based on this theory, new experiments were suggested and several novel quantum effects were discovered; see the cited references.

Very recently, Aharonov et al. [8] provided a remarkably simple and clear example demonstrating the predictive power of the theory, revealing an "anomalous" momentum exchange between photons (or particles) passing through a Mach-Zehnder interferometer (MZI) and one of its mirrors. Here the measured photon's (particle's) final state, being post-selected in a specific output of the MZI, plays a crucial and succinct role, together with the basic t-inversion symmetry of quantum mechanics, as captured by the Aharonov-Bergmann-Lebowitz rule [9]. In essence, the revealed effect (which has no conventional interpretation) is as follows: Although the photons (particles) collide with the considered mirror only from the inside of the MZI, they do not push the mirror outwards, but rather they somehow succeed to pull it in [8]; see Sec. 2.

Inspired by this remarkable theoretical result, we recently proposed an application of the theory under consideration, in the context of incoherent inelastic neutron scattering off atoms and/or molecules in condensed matter [10]. Related experimental results obtained by incoherent inelastic neutron scattering (IINS, or INS) and deep-inelastic neutron scattering (DINS) — also called neutron Compton scattering (NCS) — were presented and discussed [10].

In this paper, after a short introduction to the new theory and the aforementioned "anomalous" momentum transfer in a MZI, we present the specification of the theoretical frame to real scattering experiments. In particular, we discuss a counter-intuitive experimental result of neutron scattering from H_2 in carbon nanotubes and related materials, as obtained with conventional INS as well as modern 2-*dimensional* neutron spectrometers, e.g. ARCS [11]. Additionally, an "anomalous" DINS result from H of a solid polymer is presented. In short, our findings correspond to a striking strong *mass deficit* of the scattering objects. These observations have no known conventional interpretation.

The experimental findings and their theoretical interpretation support the view that the quantum theory of WM, WV and TSVF sheds new light on interpretational issues concerning fundamental quantum theory. Moreover, this t-symmetric theoretical formalism offers a new guide for our intuition to design new experiments and discover novel quantum effects.

2. Motivation—A recent theoretical result of quantum optics

The experimental context of this paper concerns
(a) the measurements of *momentum* and *energy* transfer in real scattering experiments,
(b) the predictions of conventional (classical or quantum) theory and
(c) comparison of the experimental results with a new prediction based on the formalism of WM and TSVF.
The latter point may be best motivated by referring directly to some of the intriguing results presented in a recent theoretical paper by Aharonor et al. [8].

Let us first refer to a surprising theoretical prediction derived by Aharonov et al. in Ref. [8]. Here we consider one specific result of this work only, which concerns "anomalous" momentum transfer between two quantum objects (i.e. a photon and a mirror) and thus appears to be in intimate connection with the neutron-atom collision (or scattering) experiments considered in the following sections.

The schematic representation of the experimental setup is shown in Fig. 1.

A photon (or particle) beam enters a device similar to a usual Mach-Zehnder interferometer (MZI), with the exception that one reflecting mirror is sufficiently small (say, a meso- or nanoscopic object M) in order that its momentum distribution may be detectable by a suitable non-demolition measurement [12]. In Ref. [8] the authors show the following astonishing result. Although the post-selected photons (as all photons do, of course) collide with the mirror M only from the *inside* of the MZI, they do not push M outwards, but rather they somehow succeed to *pull it in*. It is obvious that this result cannot have any conventional theoretical interpretation.

Let us now consider a straightforward derivation of this effect, following the presentation of some derivations (relevant to our purposes) of Ref. [8] closely.

The two identical beam splitters of MZI have nonequal reflectivity r and transmissivity t (both real, with $r^2 + t^2 = 1$), say $r > t$. Now we are interested in the momentum kicks given to the mirror M due to the photon-M collisions inside the interferometer, but especially for photons emerging toward detector D_2 in Fig. 1). This corresponds to a post-selection condition. (The effect of the photons emerging toward the other detector, D_1, is of less relevance here.)

Using the standard convention, an incoming state $|\text{in}\rangle$ impinging on the beam splitter will emerge as a superposition of a reflected state $|R\rangle$ and a transmitted state $|T\rangle$; $|\text{in}\rangle \to ir|R\rangle + t|T\rangle$. Hence when a single photon impinges from the left on BS_1, as illustrated in Fig. 1, the effect of

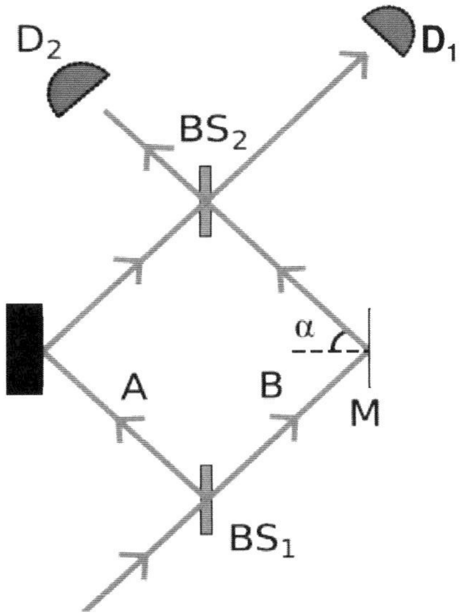

Figure 1. Mach-Zehnder interferometer and path of a light beam (or a particle, etc.). The mirror M in the rhs is a mesoscopic *quantum* object. Of particular interest are the photons (particles) emerging toward detector D_2. (Adapted from Fig. 2 of [8].)

the beam splitter is to produce inside the interferometer the state

$$|\Psi\rangle = ir|A\rangle + t|B\rangle \tag{1}$$

where $|A\rangle$ and $|B\rangle$ denote the photon propagating along the A and B arms of the interferometer, respectively.

The second beamsplitter, BS_2, is identical to the first. As one can readily check, a photon in the quantum state $|\Phi_1\rangle = t|A\rangle - ir|B\rangle$ impinging on the second beamsplitter, emerges towards detector D_1 while a photon in the orthogonal state $|\Phi_2\rangle = -ir|A\rangle + t|B\rangle$ emerges towards detector D_2.

Thus, when a single photon (particle) enters the interferometer by impinging on the left side of the beamsplitter BS_1, the probabilities to be found in the arms A and B are r^2 and t^2 respectively. The probability of emerging towards detector D_1 is $|\langle\Phi_1|\Psi\rangle|^2 = 4r^2t^2$ while the probability of emerging towards D_2 is $|\langle\Phi_2|\Psi\rangle|^2 = (r^2 - t^2)^2 = 1 - 4r^2t^2$.

Firstly, we send a *classical* light (particle) beam of intensity I towards this interferometer. The light (particle) intensity in arm A is then $I_A = r^2I$ while the intensity in arm B is $I_B = t^2I$. The intensities of the output beams, toward the tow detectors, are $I_{D_1} = 4r^2t^2I$ and $I_{D_2} = (1-4r^2t^2)I$. The momentum given to the mirror M by the beam inside the interferometer is $2I_B \cos\alpha = 2t^2I\cos\alpha$. Clearly, this pushes the mirror M outward.

Secondly, let us now send a *quantum* beam of photons (or particles) into the MZI. Each photon incident on M gives it a momentum kick δ. Note that each individual momentum kick must be much smaller than the spread Δ of momentum of the mirror M. This is a general property of any interferometer. It has to be so in order to maintain the coherence of the beam in the interferometer; otherwise the photons (particles) will become entangled with the mirror

M. To show this, we note that a photon when going through arm A will produce no kick to M while when going through arm B it will. Accordingly, if $\phi(p)$ is the initial quantum state of M and by $|\Psi\rangle$ the quantum state of the photon after the input beam splitter BS_1, but before reaching the mirror, the reflection on the mirror results in

$$|\Psi\rangle\phi(p) = (ir|A\rangle + t|B\rangle)\phi(p) \rightarrow ir|A\rangle\phi(p) + t|B\rangle\phi(p-\delta) \tag{2}$$

If $\phi(p)$ is orthogonal to $\phi(p-\delta)$ where δ is the kick given by the photon, then the photon ends up entangled with the mirror and coherence is lost. Another way of looking at this is to note that the mirror has to be localized within a distance smaller than the wavelength of light, otherwise there will be phase fluctuations larger than 2π and interference is lost.

[In fact the spread Δ in the momentum of the mirror has to be many times bigger — of order $\sqrt{\overline{n}}$ times - than that of an individual kick to ensure coherence when a beam with an average of $\overline{n}$ photons and a spread $\sqrt{\overline{n}}$ goes through the interferometer. At the same time Δ, being of order $\sqrt{\overline{n}}\delta$, is small enough so that the average kick, which is of order $\overline{n}\delta$ is detectable.]

For simplicity we take the state of the mirror to be (up to normalisation) $\phi(p) = \exp(-\frac{p^2}{2\Delta^2})$. Consider now a single photon propagating through the interferometer. Given that $\delta \ll \Delta$, we can approximate the state (2) of the photon and mirror just before the photon reaches the output beamsplitter by

$$\begin{aligned} |\Psi\rangle\phi(p) &\approx ir|A\rangle\phi(p) + t|B\rangle\Big(\phi(p) - \frac{d\phi(p)}{dp}\delta\Big) \\ &= |\Psi\rangle\phi(p) - t|B\rangle\frac{d\phi(p)}{dp}\delta \end{aligned} \tag{3}$$

Suppose now that the photon emerges in the beam directed towards detector D_2. The state of the mirror M is then given (up to normalization) by projecting the joint state onto the state of the photon corresponding to this beam, i.e.

$$\begin{aligned} & \langle\Phi_2|\Big(|\Psi\rangle\phi(p) - t|B\rangle\frac{d\phi(p)}{dp}\delta\Big) \\ = \; & \langle\Phi_2|\Psi\rangle\Big(\phi(p) - \frac{t\langle\Phi_2|B\rangle}{\langle\Phi_2|\Psi\rangle}\frac{d\phi(p)}{dp}\delta\Big) \\ = \; & \langle\Phi_2|\Psi\rangle\Big(\phi(p) - \frac{\langle\Phi_2|\mathbf{P}_B|\Psi\rangle}{\langle\Phi_2|\Psi\rangle}\frac{d\phi(p)}{dp}\delta\Big) \\ = \; & \langle\Phi_2|\Psi\rangle\phi(p - P_B^w\delta) \end{aligned} \tag{4}$$

Here $\mathbf{P}_B = |B\rangle\langle B|$ is the *projection operator* on state $|B\rangle$ and

$$P_B^w = \frac{\langle\Phi_2|\mathbf{P}_B|\Psi\rangle}{\langle\Phi_2|\Psi\rangle}$$

is the so called *weak value* (WV) of $\mathbf{P}_B$ between the initial state $|\Psi\rangle$ and the final state $|\Phi_2\rangle$ [1, 2, 3, 4, 5, 6, 7]. The value of P_B^w is readily found to be

$$\begin{aligned} P_B^w &= \frac{\langle\Phi_2|\mathbf{P}_B|\Psi\rangle}{\langle\Phi_2|\Psi\rangle} = \frac{(ir\langle A| + t\langle B|)\mathbf{P}_B(ir|A\rangle + t|B\rangle)}{(ir\langle A| + t\langle B|)(ir|A\rangle + t|B\rangle)} \\ &= -\frac{t^2}{r^2 - t^2} \end{aligned} \tag{5}$$

Hence the momentum kick received by the mirror due to a photon emerging towards D_2 is

$$\delta p_M = P_B^w \, \delta = -\frac{t^2}{r^2 - t^2}\, \delta \tag{6}$$

The appearance in the above expressions of the WV of the projector $\mathbf{P}_B$ is not accidental. Indeed, we can view the mirror as a measuring device measuring whether or not the photon is in arm B or not. The momentum of the mirror acts as a "pointer" (no kick — the photon is in arm A; kick — the photon is in arm B). However, since the photon can only change the position of the pointer (i.e. the momentum of the mirror) by far less than its spread, we are in the so called "weak measurement" [1, 2, 3, 4, 5, 6, 7] regime.

Finally, the total momentum given to the mirror by all the photons emerging towards D_2 is given by the momentum due to each photon times the number of photons in the beam. Using the fact that the probability of a photon to end in this beam is $(r^2 - t^2)^2$ we obtain

$$\delta p_M = \overline{n}(r^2 - t^2)^2 \frac{-t^2}{r^2 - t^2}\delta = -t^2 \overline{n}(r^2 - t^2)\,\delta < 0 \tag{7}$$

Since we investigated the case with $r > t$, the sign of the momentum received by the mirror is negative, hence the mirror is pushed towards the inside of the MZI. This momentum change is a result of the mirror receiving a *superposition* between a kick δ and no kick at all, corresponding to the photon (particle) propagating through the two MZI-arms.

To summarize: The physical insight obtained from quantum mechanics is dramatically different from the classical one. In fact, according to quantum mechanics, the photons that end up in the D_2 beam give negative momentum kick on M. Astonishingly, although they collide with the mirror only from the inside of the MZI, they do not push the mirror outwards; rather they somehow succeed to pull it in! This is realized by a superposition of giving the mirror zero momentum and positive momentum — the superposition results in the mirror gaining negative momentum.

Concluding the above short discussions, one may say that the new insights and/or predictions made possible within the theoretical frame of WV and TSVF are not limited to interpretational issues only. The revised intuitions can lead one to find novel quantum effects that can be measured in real experiments.

3. On post-selection, weak measurement, weak values and two-state-vector formalism

Weak quantum measurement (WM) is unique in measuring noncommuting operators and revealing new counter-intuitive effects predicted by the two-state-vector-formalism (TSVF) [3, 6]. The main aim of this article is to point out certain new (and experimentally observable) features of elementary scattering processes predicted within the theoretical frame of WM and TSVF, and which contradict every conventional expectation; cf. [10]. Concretely, we have in mind incoherent-inelastic scattering of single (massive) particles, (e.g. neutrons or electrons) from nuclei and/or atoms.

As the starting point of the new theory under consideration one usually considers the paper [2] by Aharonov, Albert and Vaidman, and the earlier paper [9] by Aharonov, Bergmann and Lebowitz. Here some short remarks may be helpful.

A standard von Neumann (also called "strong") measurement yields the eigenvalues of the measured observable, but at the same time disturbs the measured system. According to standard theory [13], the final state of the system after the measurement becomes an eigenstate of the measured observable. This usually changes the initial state of the system. On the other hand, by coupling a measuring device to a system sufficiently weakly, it may be possible to read out certain information while limiting the disturbance induced by the measurement to the system. As Aharonov and collaborators originally proposed [2, 9], one may achieve new physical insights when one furthermore post-selects on a particular outcome of the experiment. In this case the eigenvalues of the measured observable are no longer the relevant quantities; rather the measuring device consistently indicates the *weak value* (WV) [2, 9] given by

$$A_w \equiv \frac{\langle\psi_f|\hat{A}|\psi_i\rangle}{\langle\psi_f|\psi_i\rangle} \tag{8}$$

where $\hat{A}$ is the operator whose value is being ascertained, $|\psi_i\rangle$ is the initial state of the system, and $|\psi_f\rangle$ is the state that is post-selected (e.g. by performing a specific measurement). Note that the number A_w may be complex.

The significance of this formula is as follows. Let us couple a measuring device whose pointer has position coordinate q to the system $\mathcal{S}$, and subsequently measure its conjugated momentum p. The coupling interaction is taken to be the standard von Neumann measurement interaction [13]

$$\hat{H} = -g\,\hat{q} \otimes \hat{A} \tag{9}$$

The coupling constant g is assumed to be appropriately small and the interaction time sufficiently short. Then the mean value $\langle p\rangle$ of the pointer momentum is given by [2]

$$\langle p\rangle = g\ Re[A_w] \tag{10}$$

where Re denotes the real part. This formula requires the initial pointer momentum wavefunction to be real and centered at $\langle p\rangle = 0$ before measurement, but these assumptions can easily be relaxed; see cited references.

The formula (8) implies that, if the initial state $|\psi_i\rangle$ is an eigenstate of a measurement operator A, then the weak value post-conditioned on that eigenstate is the same as the classical (strong) measurement result. When there is a definite outcome, therefore, strong and weak measurements agree. Interestingly, a WM can yield values outside the range of measurement results predicted by conventional theory [2].

For some experiments, but not in those considered in this paper, a WV can also be complex, with its imaginary part affecting the pointer position. I.e., the mean of the pointer position after measurement is given by

$$\langle q\rangle = 2gv\ Im[A_w] \tag{11}$$

where Im denotes the imaginary part and v is the variance in the initial pointer spatial position [2], with $\langle q\rangle = 0$ before measurement.

The situation we shall consider is where a system $\mathcal{S}$ evolves unitarily from an initial state $|\psi_i\rangle$ to a final post-selected measurement outcome $\langle\psi_f|$. At various time points inbetween, observables may be measured weakly. Here we consider the scenario where there is a single copy of the system, with the measuring device weakly coupled to it.

In the simplest case where there is just one observable A, we assume the evolution from $|\psi_i\rangle$ to the point where A is measured is given by U, and from this point to the post-selection the evolution is given by V. Then we can rewrite (8) as:

$$A_w = \frac{\langle\psi_f|\hat{V}\hat{A}\hat{U}|\psi_i\rangle}{\langle\psi_f|\hat{V}\hat{U}|\psi_i\rangle} \tag{12}$$

and the mean of the pointer is given by (10) as before.

The fact that one only sufficiently "weakly" disturbs the system in making WMs means that one can in principle measure different variables in succession. This theoretical observation has led to a great number of experimental applications and discovery of several new effects; see e.g. [1, 2, 3, 4, 5, 6, 7, 10] and references cited therein.

4. Elementary scattering in view of WV and TSVF

4.1. Momentum transfer in impulsive two-body collisions

In this theoretical section, we present the basic result of WV and TSVF as applied to scattering processes, especially non-relativistic neutron scattering, firstly obtained in Ref. [10].

Here we mainly follow the presentation of that reference.

The position and momentum of the neutron (probe particle) are denoted as (q, p). Similarly, the position and momentum of the scatterer (atom, nucleus) are denoted as (Q, P). $\hat{X}$ may represent the corresponding operator of a physical quantity X. For simplicity, let us consider here a one-dimensional quantum model for momentum exchange in a two-particle impulsive collision.

In the usually considered, but extremely limiting case in which both particles occupy states with well defined momenta (i.e. plane waves), the initial state of the two-body system is standardly assumed as a product (uncorrelated) state

$$\Psi_{in} = \phi_n(p) \otimes \Xi_A(P) \tag{13}$$

(indices n and A refer to neutron and atom, respectively). In a first attempt, an impulsive scattering process may be formally approached by the interaction Hamiltonian

$$\hat{V} = F(t)\,(\hat{q} - \hat{Q}) \tag{14}$$

where the function $F(t)$ represents a non-zero force during a short time interval τ, the duration of the collision. We may assume that $F(t)$ is proportional to a delta function. Furthermore, the integral

$$\int_0^\tau F(t)\,dt = \hbar K \tag{15}$$

provides the momentum transfer $\hbar K$ caused by the collision.

Neutron and atom observables commute, so $[\hat{q}, \hat{Q}] = 0$, and the associated unitary evolution operator is

$$\hat{U}(\tau) = e^{-(\imath/\hbar)\int \hat{V} dt} = e^{-(\imath/\hbar)\,\hbar K\,(\hat{q}-\hat{Q})} \equiv e^{-\imath K \hat{q}} e^{+\imath K \hat{Q}} \tag{16}$$

The *shift operator* $e^{\imath \hbar K\, Q/\hbar}$ describes the momentum shift of an atomic momentum eigenket as $e^{\imath \hbar K\hat{Q}/\hbar}|P\rangle = |P + \hbar K\rangle$ while the operator $e^{-\imath\hbar K\,\hat{q}/\hbar}$shifts neutron's momentum eigenket as $e^{-\imath\hbar K\hat{q}/\hbar}|p\rangle = |p - \hbar K\rangle$ of the impinging particle (neutron). Immediately after the momentum exchange, the state of the two-particle system in the momentum representation is

$$\begin{aligned} \Psi_f &= \hat{U}(\tau)\,\phi_n(p) \otimes \Xi_A(P) \\ &= e^{-i\hbar K\,\hat{q}/\hbar}\phi_n(p) \otimes e^{i\hbar K\,\hat{Q}/\hbar}\Xi_A(P) \\ &= \phi_n(p + \hbar K) \otimes \Xi_A(P - \hbar K) \end{aligned} \tag{17}$$

which is a not entangled final state, due to the trivial form of $\hat{V}$. These introductory considerations may motivate the search for a more appropiate (i.e. more realistic) two-body impulsive interaction Hamiltonian, which is presented in the following section.

4.2. WM, weak interaction Hamiltonian and momentum transfer

Let us now assume that the struck atom is initially at rest. For simplicity of notation, it is sufficient to consider the atomic momentum component parallel to **K**, which from now on we shall denote simply by P—and the associated operator simply by $\hat{P}$. In other words, we consider a one-dimensional problem.

IOP Conf. Series: Journal of Physics: Conf. Series **1071** (2018) 012007 doi:10.1088/1742-6596/1071/1/012007

For illustration and motivation of the derivations presented below, we first provide a heuristic derivation of a von Neumann-type [13] interaction Hamiltonian for momentum transfer. Let us start with a formal one-body model Hamiltonian describing momentum transfer $-\hbar K \equiv +\hbar K_n$ *on the neutron* due to the collision with the atom:

$$\hat{V}_n(t) = \delta(t)\, \hbar K\, \hat{q} \tag{18}$$

The associated evolution operator acting on the space of the neutron

$$\hat{U}_n(\tau) = \exp(-\frac{i}{\hbar}\hbar K \hat{q}) \tag{19}$$

shifts a momentum eigenstate of the impinging particle (neutron) as $e^{-i\hbar K\hat{q}/\hbar}|p\rangle = |p - \hbar K\rangle$.

Due to momentum conservation in the two-body collision, one has

$$-\hbar K_n = \hbar K_A \equiv \hbar K \tag{20}$$

where $\hbar K_A$ is the momentum transfer *on the atom* due to the collision. (We choose K_A as positive, following standard notation; see e.g. the textbook [14] or the review article [15].)

Let the scattering atom be at rest before collision, $\langle \hat{P} \rangle_i = 0$. Hence, after the collision, it holds

$$\hbar K_A = +\hbar K = \langle \hat{P} \rangle_f = \langle \hat{P} \rangle_f - \langle \hat{P} \rangle_i \tag{21}$$

and correspondingly for the neutron momentum

$$\hbar K_n = -\hbar K = \langle \hat{p} \rangle_f - \langle \hat{p} \rangle_i \tag{22}$$

Hence, the aforementioned operator $\hat{U}_n(\tau)$ of the neutron, Eq. (19), may be written as

$$\hat{U}_n(\tau) = \exp(-\frac{i}{\hbar}\langle \hat{P} \rangle_f\, \hat{q}) \tag{23}$$

To apply the theory of weak measurement (WM) and two-state-vector formalism (TSVF), a *von Neumann* two-body interaction Hamiltonian is needed; see the associated references cited in the Introduction. Thus one is guided to search of a two-body generalization of the one-body evolution operator $\hat{U}_n(\tau)$ of the form

$$\hat{U}(\tau) = \exp(-\frac{i}{\hbar}\hat{q}\, \hat{P}) \tag{24}$$

However, this heuristically derived expression still has no direct context to the *realistic* experimental situations under consideration (see Sec. 6 below). To achieve this, we now may proceed as follows.

To begin with, let us refer to the so-called *impulse approximation* (IA) of standard theory [15, 22, 21] and Eq. (38) regarding energy conservation (see Sec. 5):

$$E = \frac{(\hbar K)^2}{2M} + \frac{\hbar \mathbf{K}\, \mathbf{P}}{M}$$

Looking at this equation, and having the WV and TSVF in mind, one sees that the *larger* recoil term $\frac{(\hbar K)^2}{2M}$ may be viewed to result from a *strong* impulsive interaction (associated with momentum transfer $+\hbar K$ on the atom). The theoretical treatment of this part of the interaction can be found in standard textbooks (e.g. [14]) but is not within the scope of the present paper. Since in the impulse approximation (IA) holds $|K| \gg |P|$, the *smaller* Doppler term $\frac{\hbar \mathbf{K}\cdot\mathbf{P}}{M}$

may correspond to a *weaker* interaction, in which the atomic momentum $\hat{P}$ couples with an appropriate dynamical variable of the neutron, say $\hat{\mathcal{O}}_n$. Looking at the preceding formulas (23,24) for the model operator effectuating momentum transfer, it appears that this dynamical variable should be $\hat{q}$, that is $\hat{\mathcal{O}}_n = \hat{q}$.

In view of the theory of WV and TSVF, the weak interaction is expected to cause weak *deviations* from the conventionally expected *large* momentum transfer $\hbar K$. This can be formally captured by replacing $\hat{P}$ with the *small* momentum difference $\hat{P} - \hbar K\hat{I}_A$, and also including a positive *smallness* factor

$$0 < \lambda \ll 1$$

in the model interaction Hamiltonian. Summarizing these considerations, let us assume the model interaction Hamiltonian

$$\hat{H}_{int}(t) = +\lambda\,\delta(t)\,\hat{q} \otimes (\hat{P} - \hbar K\hat{I}_A) \tag{25}$$

It should be pointed out that the *plus sign* in front of this expression is *not* arbitrary, since it is consistent with the aforementioned definitions (20) of momentum transfer. This point plays a decisive role in the context of the new quantum effect of momentum transfer deficit.

For further physical motivation of the two parts of the model Hamiltonian of Eq. (25), it may be helpful to compare the above reasoning with an example by Aharonov et al. [16], p. 3:

"Consider, for example, an ensemble of electrons hitting a nucleus in a particle collider. [...] Their initial states are known, and a specific post-selection is done after the interaction. The main interaction is purely electromagnetic, but there is also a relativistic and spin-orbit correction in higher orders which can be manifested now in the form of a weak interaction."

4.3. WV of atomic momentum operator $\hat{P}$ and the effect of "anomalous" momentum transfer

Here, the atom represents the system of the general formalism. Since the WV of the identity operator is $(\hat{I}_A)_w = 1$, for the WV of the atomic coupling operator $\hat{P} - \hbar K\hat{I}_A$ in the above interaction Hamiltonian holds:

$$(\hat{P} - \hbar K\hat{I}_A)_w = P_w - \hbar K \tag{26}$$

To proceed, we first calculate the WV P_w of $\hat{P}$ for some characteristic (and experimentally relevant) final state in momentum space. The derivation is rather straightforward and reveals a striking deviation — more precisely, a deficit — from the conventionally expected momentum transfer to the neutron. The latter represents the pointer of general theory, and the pointer momentum variable $\hat{p}$ is conjugated to $\hat{q}$ contained in Eq. (25).

Moreover, for the calculation of the WV, it is natural to use the momentum space representation, as scattering experiments usually measure momenta (rather than the positions of the scatterers in real space).

Let the atom initially be in a spatially confined state and at rest; e.g. in a potential representing physisorption on a surface; cf. examples in experimental section below. Usually, the initial atomic wave function $\Xi(P)_i$ is approximated by a Gaussian G_A centered at zero momentum,

$$\Xi(P)_i \approx G_A(P).$$

At sufficiently deep temperature the atom will be in its ground state, and the width of $\Xi(P)_i$ is determined by the quantum uncertainty.

It follows that the struck atom moves in the direction of momentum transfer $\hbar K_A = \hbar K$; therefore, to simplify notations, in the following calculations P represents the atomic momentum along the momentum transfer direction.

To be specific, as well as to facilitate the derivations, let us make the following simplifying assumption concerning the final atomic state:

- The final atomic state has the same width in momentum space as the initial state.

However, it may be noted that this assumption is very common in molecular (optical) spectroscopy. It captures the viewpoint that the impulsive transition is very fast and so the atomic environment didn't have sufficient time to change configuation and adapt to the "disturbance" due to the atomic final state.

In other words, let the final atomic state have the same shape as the initial state, but its center should be shifted from zero to the transferred momentum, i.e.

$$\Xi(P)_f = \Xi(P - \hbar K_A)_i. \tag{27}$$

The WV of the atomic momentum operator $\hat{P}$ is calculated as follows:

$$\begin{aligned} P_w &= \frac{\langle \Xi_f | \hat{P} | \Xi_i \rangle}{\langle \Xi_f | \Xi_i \rangle} \\ &= \frac{\int dP\, \Xi(P - \hbar K_A)_i\, P \Xi(P)_i}{\int dP\, \Xi(P - \hbar K_A)_i\, \Xi(P)_i} \\ &= +\frac{\hbar K_A}{2} \\ &= +\frac{\hbar K}{2} \end{aligned} \tag{28}$$

The last equality follows immediately from (a) the two symmetrically distributed Ξ functions around their central position $\bar{P} = \hbar K_A/2$ and (b) the linear term P in the integral of the nominator.

It should be noted that this result does not depend on the width of Ξ, as long as the two Ξ functions are not orthogonal to each other.

Thus we arrived at the following striking conclusion: The momentum transfer comes out to be only half of the conventionally expected value $\hbar K$.

This is a quite interesting result because it represents a momentum-transfer deficit of 50%; i.e. the scattered neutron measures a momentum kick being only half of the conventionally expected value. In more detail, it holds

$$(\hat{P} - \hbar K \hat{I}_A)_w = +\frac{\hbar K_A}{2} - \hbar K = -\frac{\hbar K}{2} \tag{29}$$

To proceed, one should note the change of sign between the usually applied interaction Hamiltonian of Eq. (9), as e.g. treated in [2], and the presently used specific form (25). Therefore, there is a corresponding change of sign in Eq. (10),

$$\langle \hat{p} \rangle_f - \langle \hat{p} \rangle_i = -g\, \mathrm{Re}[A_w] \tag{30}$$

Thus one obtains for the *correction* to the shift of the meter pointer variable:

$$\langle \hat{p} \rangle_f - \langle \hat{p} \rangle_i = -\lambda\, (\hat{P} - \hbar K\, \hat{I}_A)_w = +\lambda\, \frac{\hbar K}{2} \tag{31}$$

Moreover, for the *total* momentum transfer shown by the pointer of the measuring device (here: of neutrons) we have

$$\begin{aligned} [\langle \hat{p} \rangle_f - \langle \hat{p} \rangle_i]_{\text{total}} &= [\langle \hat{p} \rangle_f - \langle \hat{p} \rangle_i]_{\text{conventional}} \\ &\quad + [\langle \hat{p} \rangle_f - \langle \hat{p} \rangle_i]_{\text{correction}} \\ &= -\hbar K + \lambda\, \frac{\hbar K_A}{2} \end{aligned} \tag{32}$$

IOP Conf. Series: Journal of Physics: Conf. Series **1071** (2018) 012007 doi:10.1088/1742-6596/1071/1/012007

This expression represents the new quantum effect of *momentum-transfer deficit*: the absolute value of momentum transfer on the neutron predicted by the new theory is *smaller* than that predicted by conventional theory:

$$|-\hbar K+\lambda\hbar K_A/2| \le |-\hbar K| \tag{33}$$

Recall that by definition holds $\hbar K_A = \hbar K$; see Eq. (20).

4.3.1. Plane waves — The limiting case of conventional momentum transfer. Moreover, it is interesting to point out the succinct reason of this large "anomaly" by the following calculation. Let us now make the usual assumption of conventional theory (see e.g. [14, 15, 22]) which is:

- The final state should be a plane wave (i.e. it has vanishing width in momentum space) — as assumed in general conventional theory and the IA.

Here we will show straightforwardly that the result of conventional theory is reproduced.

Namely, now the final state is a delta function (in momentum space), that is, the momentum wave function is a delta function δ_A centered at the assumed transferred momentum $\hbar K_A$,

$$\Xi(P)_f = \delta_A(P-\hbar K_A). \tag{34}$$

The weak value of $\hat{P}$ follows straightforward:

$$\begin{aligned} P_w &= \frac{\langle \Xi_f|\hat{P}|\Xi_i\rangle}{\langle \Xi_f|\Xi_i\rangle} \\ &= \frac{\int dP\,\delta_A(P-\hbar K_A)\,P\Xi(P)_i}{\int dP\,\delta_A(P-\hbar K_A)\,\Xi(P)_i} \\ &= \frac{\hbar K_A\,\Xi(\hbar K_A)_i}{\Xi(\hbar K_A)_i} \\ &= +\hbar K_A \equiv +\hbar K \end{aligned} \tag{35}$$

(Recall the notations of Eq. (20).) Hence, the WV of the system coupling operator $(\hat{P}-\hbar K\hat{I}_A)$ is just zero,

$$(\hat{P}-\hbar K\hat{I}_A)_w = P_w - \hbar K = 0 \tag{36}$$

[According to standard quantum scattering theory, the scattered wave may acquire an additional phase factor, say $e^{i\chi}$, which does not affect the preceding result because this factor cancels out in the fractions of Eqs. (35).]

This physically means that, under the usual assumption of "plane waves", the new theory yields *no correction* to the conventionally expected value of momentum transfer:

$$\langle\hat{p}\rangle_f - \langle\hat{p}\rangle_i = -\lambda\,(\hat{P}-\hbar K\hat{I}_A)_w = 0 \tag{37}$$

And hence the pointer of the measuring device will show the conventionally expected value $-\hbar K$. Thus, the result of Eq. (36) is consistent with conventional theory of the impulse approximation (IA, see below) and, more generally, of incoherent neutron scattering. Clearly, as the standard result of conventional theory is now reproduced, this is a very satisfactory result.

Comparison of the aforementioned two derivations shows immediately that the degree of the momentum transfer "anomaly" under considerations depends on the "deformation" of the shape of the final atomic state. E.g., if the final state is, say, "nearly" a delta function (plane wave), then the anomalous momentum transfer deficit will be "small" — and perhaps remain undetectable.

The rather *general* scheme of the above derivations provides evidence that the new effect under consideration is of a general character and thus may apply to any field of scattering physics, (perhaps even to relativistic scattering in high-energy physics). The immediate implication is that the general theory of WV and TSVF may be relevant for a broad range of modern scattering experiments.

5. Two-body collision — Momentum- and energy-transfer

In a typical scattering experiment, the involved amounts of energy and momentum transfer are determined. (Angular momentum or spin measurements play no role in this paper.) However, this determination involves not only directly measured quantities, but also several theoretically motivated "beliefs" and/or insights based on some "commonly applied" theory. Let us have a short look at these issues.

Modern neutron spectrometers are time-of-flight (TOF) instruments; cf. Fig. (2). Here, a short pulse of neutrons produced at a spallation source (e.g. SNS or ISIS) reaches the first monitor of the spectrometer which triggers the measurement of TOF. A neutron scatters from the sample and may reache the detector, which stops the TOF measurement.

Note that instrumental details play a crucial role in the theoretical framework under consideration, because they concern pre-selection and, in particular, post-selection being essential for the TSVF. Therefore the following facts should be pointed out.

(*a*) From the measured TOF-values, but without using the actual value of scattering angle θ, one determines the value of $k_1 = |\mathbf{k}_1|$, and consequently of energy transfer $E = \hbar\omega$; see Eq. 38.

(*b*) Momentum transfer $\hbar\mathbf{K}$, as defined in Eq. (39), is determined from both the scattering angle θ and the TOF value.

(*c*) Furthermore, from each TOF-value measured with the detector at θ, the associated transfers of momentum ($\hbar K$) and energy ($E = \hbar\omega$) of the neutron to the struck particle are uniquely determined; see e.g. [10]. Hence a specific detector measures only one specific trajectory in the whole K–E plane. Obviously, this fact is related to the *post-selection* of the WV and TSVF theory under consideration

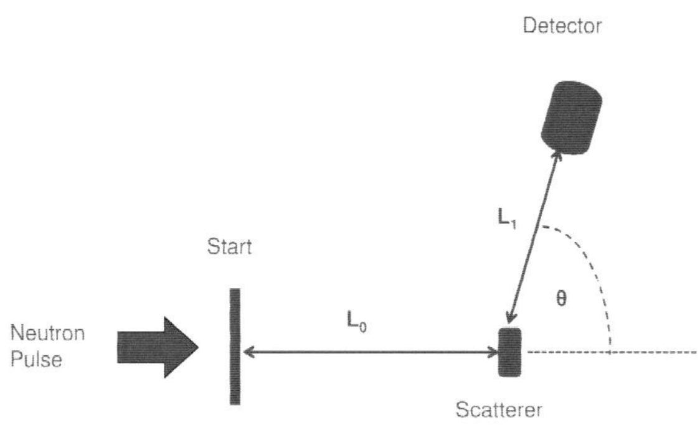

Figure 2. Schematic presentation of a time-of-flight scattering experiment.

In the following sections of this paper, the preceding theoretical considerations and results are applied to concrete neutron scattering experiments, especially to *incoherent* scattering. In simple terms, "incoherent" means that the impinging neutron (more generally: photon, electron, atom, etc.) collides with, and scatters from, a *single* particle (nucleus, atom, molecule, etc.) As concerns neutron scattering from protons (commonly referred to also as H-atoms), the scattering is mainly incoherent due to the spin-flip mechanism of neutron-proton collision; see e.g. [14]. A clear first-principles explanation of *coherent* versus *incoherent* scattering may be found in Section 3-3 of the well-known Feynman Lectures [17].

Due to scattering, the impinging neutron causes a momentum transfer $\hbar\mathbf{K}$ and an energy transfer E to the atom. For simplicity, in the following we consider isotropic scattering only, which is overwhelmingly dominant in non-relativistic neutron scattering [14]. These quantities are experimentally determined by standard methods; see e.g. Refs. [10, 14, 15, 20, 21].

Due to momentum conservation, it follows that the neutron receives the opposite momentum $-\hbar\mathbf{K}$. The elastic collision of a neutron and a (free) atom with mass M and initial momentum $\mathbf{P}$ results in the neutron's lost energy $E \equiv \hbar\omega$ being transferred to the struck atom:

$$E = E_i - E_f = \hbar\omega = \frac{(\hbar\mathbf{K} + \mathbf{P})^2}{2M} - \frac{P^2}{2M} = \frac{(\hbar K)^2}{2M} + \frac{\hbar\mathbf{K}\,\mathbf{P}}{M} \tag{38}$$

E_i and E_f are the neutron's initial and final kinetic energy, respectively. This equation represents energy conservation. Furthermore,

$$\hbar\mathbf{K} = \hbar\mathbf{k}_i - \hbar\mathbf{k}_f \tag{39}$$

where $\mathbf{k}_i$ and $\mathbf{k}_f$ are the neutron's initial and final wavevectors, respectively. The absolute value of $\mathbf{K}$ is given by

$$|\mathbf{K}| = K = \sqrt{k_i^2 + k_f^2 - 2k_i k_f \cos\theta} \tag{40}$$

The first term in the right-hand-side (rhs) of Eq. (38) is the so-called recoil energy,

$$E_{rec} = \hbar\omega_{rec} = \frac{(\hbar K)^2}{2M} \tag{41}$$

and represents the kinetic energy of a recoiling atom being initially at rest, i.e. $\langle P\rangle = 0$. Hence one may write

$$\langle E\rangle = \frac{\hbar^2\langle K\rangle^2}{2M} \equiv \bar{E}_{rec} \tag{42}$$

which holds for the E and $\mathbf{K}$ of the peak center. Thus incoherent scattering from (say, a gaseous sample of) such atoms leads to a experimental recoil peak centered at energy transfer $\bar{E}_{rec}$, and exhibiting a width being caused by the term $\hbar\mathbf{K}\cdot\mathbf{P}/M$, which represents the well known effect of *Doppler broadening* — being the second term in the rhs of Eq. (38).

$$E_{Doppler} = \frac{\hbar\mathbf{K}\,\mathbf{P}}{M} \equiv \frac{\hbar K P_{\|}}{M} \tag{43}$$

where $P_{\|}$ denotes the atomic-momentum component parallel to $\mathbf{K}$. For *isotropic* systems (gases, liquids, amorphous solids etc., as those considered below) the specific direction determined by $\mathbf{K}$ becomes immaterial, and thus $P_{\|}$ may represent the projection along any direction.

The above simple formulas of this section capture the features of the so-called *Impulse Approximation* (IA) [15, 21, 20, 22].

Relation (41), also called recoil parabola, is nicely illustrated in Fig. 3, which shows experimental data of incoherent inelastic (Compton, DINS) scattering from ^{4}He atoms in the liquid phase; see [18] for details.

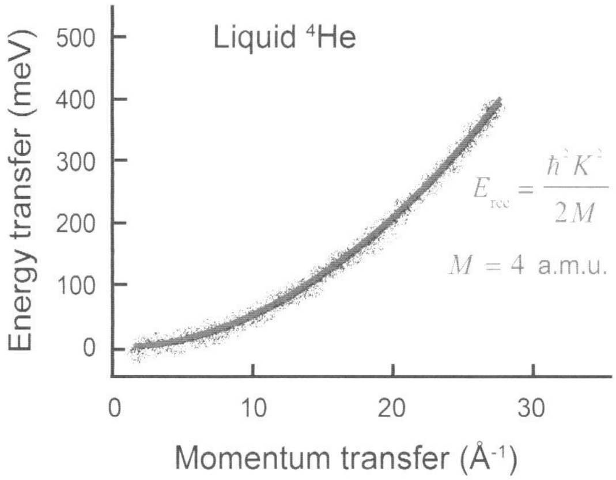

Figure 3. Schematic representation (blue points) of measured dynamic structure factor $S(K, E)$ of liquid helium [18]. The red line is the calculated recoil parabola, according to Eq. (41), for the mass of ^{4}He, shown as a guide to the eye. The white-blue ribbon around the recoil parabola represent data points measured with the time-of-flight spectrometer ARCS [11] (Figure taken from Ref. [10] with permission from Quanta.)

5.1. Effective mass as measured in the scattering experiment

In real experiments, deviations from the impulse approximation are well known; see e.g. [15, 20, 21]. Within conventional theory such deviations are understood as follows.

The energy conservation relation (38) for a two-body collision holds in the so-called impulse approximation, which holds exactly for infinite momentum transfer and, consequently, for quasi-free scattering particles. But it is not completely fulfilled at finite momentum and energy transfers of actual experiments, and thus so-called *final-state effects* (FSE) may become apparent [15, 20, 21]. (In fact, this term commonly includes both *initial and final* state effects). FSE are caused by environmental interactions with the struck particle, which affect both initial and final states of it. Here, we shall shortly discuss this effect in the frame of conventional theory.

Firstly, as Feynman puts it: "We use the term *mass* as a quantitative measure of *inertia* ..." ([19], section 9-1). Therefore, when a scattering particle with mass M is not completely free but *partially bound* to other adjacent particles, then the impinging (or probe, e.g. neutron) particle scatters on an object with *higher* measure of inertia — or higher mass — than M. This is because the adjacent massive particles excert forces on the scattering particle. These forces may be due to some conventional binding mechanisms (such as ionic or van der Waals forces; chemi- or physisorption). However these forces can only hinder the motion of the scattering particle, and thus can never cause an increase of the particle's mobility. In other terms, the particle is *dressed* by certain environmental degrees of freedom, and this dressing increases its inertia, or equivalently, its *effective* mass M_{eff}

$$M_{eff} \geq M \equiv M_{free}. \tag{44}$$

This reasoning corresponds to a well understood effect, often observed in scattering from condensed systems; cf. [15, 20, 21]. The last relation holds under the conditions of NCS (DINS) and INS too.

This effect can be also shown by referring to the aforementioned energy conservation relation, see (42), here including a term $E_{int} > 0$ describing the atom-environment interaction:

$$\bar{E} = \frac{\hbar^2 \bar{K}^2}{2M} + E_{int} \tag{45}$$

where $\bar{E}$ and $\bar{K}$ refer to the center of a peak (measured by a specific detector). Thus there will be a reduced amount of energy, $\bar{E} - E_{int}$, available as kinetic energy to the recoiling particle. As pointed out above, $\bar{E}$ and $\bar{K}$ are determined from the kinematics of the neutron, in contrast to E_{int}, which is a quantity of the scattering system.

Let us now try to fulfill this equation with a pair (E_{IA}, K_{IA}) being determined by the conventional theory in the impulse approximation (IA), for which holds

$$E_{IA} = \frac{\hbar^2 K_{IA}^2}{2M_{eff}} \tag{46}$$

Obviously, from the last two equations follows $M_{eff} > M$.

This effect may be demonstrated with the aid of a DINS result by Sokol et al. [23], obtained from H atoms produced by chemisorbed, dissociated H_2 in the graphite intercalation compound (GIC) C_8K. The experiment was done with a so-called *inverse geometry* TOF spectrometer (of Argone Nat. Lab., USA) with energy selection in the *final* flight path and chosen momentum trasfer with $K = 39$ Å^{-1}, which is high enough for the IA to be valid. Additionally, an accompanying DINS experiment from the ionic solid KH was done with the same instrumental setup. The similarity between the H recoil peaks in KH and the GIC, see Fig. 4, led to the conclusion that the state of H is similar in both samples [23]. The measured position of the H peaks corresponds to a very high effective mass, i.e. $M_{eff} = 1.2$ a.m.u.

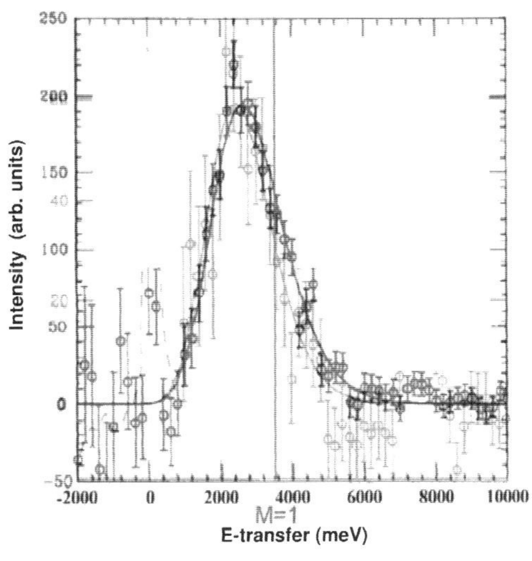

Figure 4. The conventional effect $M_{eff} > M$ — DINS from chemisorbed H in C_8K, adapted from figues of [23]. The maxima of the H peaks are about 2620 meV The vertical red line indicates the position of a $M = 1$ a.m.u. recoil peak (at 3540 meV), assuming that the conventional theory (here: the IA) is obeyed. — Black points and fitted line: KH, Grey points and fitted line: $C_8K_{0.9}H_{0.13}$. The spectra were recorded with a selected final neutron energy E_f =4280 meV. The peak maxima correspond to $M_{eff} = 1.2$ a.m.u.

IOP Conf. Series: Journal of Physics: Conf. Series **1071** (2018) 012007 doi:10.1088/1742-6596/1071/1/012007

Thus one obtains the following experimentally testable predictions of conventional theory:

In specifically designed scattering experiments with fixed K-transfer — the so-called *constant-K* measurements — the measured recoil (i.e. kinetic) energy of an atom with mass M will be *smaller* than that predicted by the IA. This is tantamount to a two-body collision of a probe particle (neutron) with a fictitious free atom of mass $M_{eff} > M$.

For more references about FSE and associated discussions of "effective mass" the interested reader may consult Ref. [10].

6. Experiments on the new scattering effect

In this section, the obtained theoretical results are compared with some actually performed experiments. The derivations of the preceding section should apply to all neutron scattering subfields of interest (i.e. INS, NCS, DINS) as the derivations do not contain any specific assumption being valid in one subfield only. The presented experimental results may be considered as examples of the WV-TSVF-theoretical analysis of Section 5.

The revised phyiscal intuitions offered by the theoretical result [8] outlined in Sec. 2 may be understood as the reason that led us to the analysis of the experiments considered here (and several others; see discussion below). In short: The measured scattering signal by a detector is due to a quantum superposition of the neutron

(1) colliding with the atom (and thus giving a "positive" momentum transfer $\hbar K$ to the atom) and

(2) being transmitted without collision, i.e. giving a "zero" momentum transfer,

with the superposition of both causing the atom to receive an additional "negative" component $-\Delta(\hbar K)$ to its total momentum.

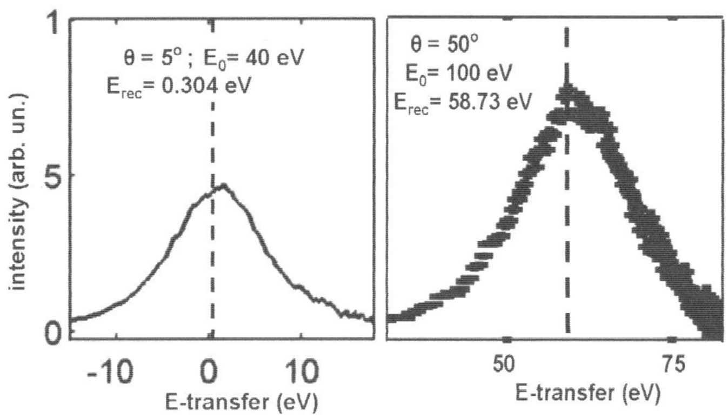

Figure 5. Two examples of DINS spectra from a solid polymer (polyethylene, $[-CH_2-]_n$) measured with the TOF spectrometer MARI of ISIS [24]. Peak-shifts to *higher* energy transfers than the conventionally expected recoil energy E_{rec} (vertical lines) are clearly visible. This effect corresponds to a *lower* effective mass $M_{eff}(\mathrm{H})$ of the recoiling H: $M_{eff}(\mathrm{H}) \approx 0.91 - 0.96$ a.m.u. See [24] for experimental details and more examples.

6.1. NCS/DINS from H atoms of a solid polymer—MARI experiment

An experimental demonstration of the new quantum effect under consideration can be found in the data of [24]. The NCS (or DINS) experiments were carried out by Cowley and collaborators with the TOF spectrometer MARI [25] of the neutron spallation source ISIS (Rutherford Appleton Laboratory, UK). The sample (a foil of a solid polymer, low-density polyethylene) was at room temperature.

Fig. 5 shows two examples of the extensive measurements reported in [24]. The depicted recoil peaks are mainly due to scattering from protons (H atoms), due to the high incoherent scattering cross-section of H [14, 15]. The vertical lines show the E-transfer values according to conventional theory, Eq. (41). The centroids of the measured recoil peaks are markedly shifted to *higher* energy transfer than conventionally expected. As discussed in subsection 5.1, this shift is equivalent to a *smaller* effective mass of the recoiling H atom.

One might object that the shown data contain an additional small contribution from the C recoil. However this is located at smaller E-transfers than that of H, due to their mass difference [15]. Therefore the preceding qualitative conclusion remains unaffected.

It may be noted that the shown NCS peaks are very broad and asymmetric, which is due to the (very) low resolution of the employed modified setup of MARI [24]. This makes a quantitative analysis to determine the peak-position impossible. Nevertheless, visual inspection of the data shows that the centroids of the peaks are *shifted to higher energy* roughly by 5-10 % of the recoil energy, which equivalently means that the effective mass M_{eff} of the recoiling H atoms is *smaller* than $M_H = 1.0079$ a.m.u. of a free H by the same percentage

$$M_{eff}(\mathrm{H}) \approx 0.91 - 0.96 \quad \text{a.m.u.} \tag{47}$$

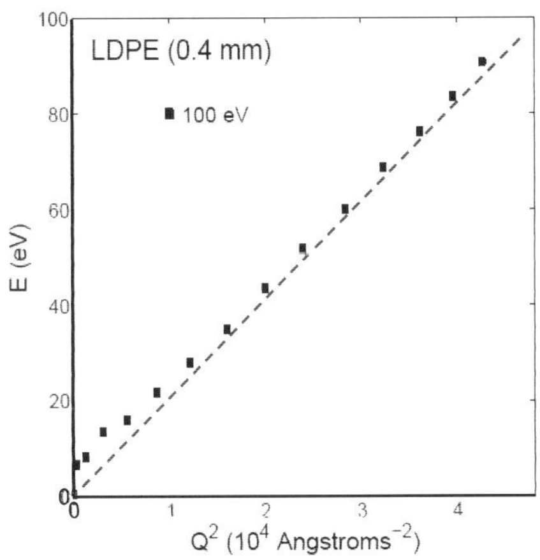

Figure 6. Data from NCS (DINS) from a solid polymer (low-density polyethylene, LDPE), $[-CH_2-]_n$) measured with the TOF spectrometer MARI of ISIS with excitation energy $E_0 = 100$ eV. [24]. The straight line corresponds to the recoil parabola of conventional theory. All data points (center-of-gravity of recoil peaks) show the effect under consideration. i.e. an apparent reduced effective mass of the scattering H-atom. For experimental details and more examples, see [24]. (Data taken from Fig. 12 of [24].)

Throughout this section we presumed that the *calibration* of MARI as employed in the experiment [24] was correct, i.e. that the applied TOF method (see Sec. 5) used correct instrumental parameters.

Moreover, it seems important to look at the additional (about 50) DINS-results being reported in Fig. 12 of Ref. [24], which were obtained at three different excitation energies ($E_i = 20, 40, 100$ eV). Remarkably, *all* those H-recoil peaks — i.e. with no exception! — appeared to show *positive* E-transfers; cf. Fig. 6 which shows data obtained with neutron's initial energy $E_i = 100$ eV. This observation underlines considerably the reliability of the considered "anomalous" E-transfer shift. However, the main aim of that investigation [24] was motivated differently — i.e., the focus was the measurment of the cross-section of H, and to show a weakness of the "inverse geometry" TOF-spectrometer VESUVIO of ISIS — this striking experimental finding remained fully unnoticed and/or uncommented in the discussions of Ref. [24].

6.2. INS from single H_2 molecules in carbon nanotubes

Another surprising result from incoherent inelastic neutron scattering was observed by Olsen et al. [26] in the quantum excitation spectrum of H_2 adsorbed in multi-walled nanoporous carbon (with pore diameter about 8–20 Å).

The INS (or IINS) experiments were carried out at the new generation TOF spectrometer ARCS of Spallation Neutron Source SNS (Oak Ridge Nat. Lab., USA) [11]. In this experiment, the temperature was $T = 23$ K, and the incident neutron energy E_i was 90 meV. The latter implies that the energy transfer cannot excite molecular vibrations (and thus cannot break the molecular bond), but only excite rotation and translation (also called recoil) of H_2 which interacts only weakly with the substrate:

$$E = E_{rot} + E_{trans} \tag{48}$$

The experimental two-dimensional incoherent inelastic neutron scattering intensity map $S(K, E)$ of H (after background subtraction) is shown in Fig. 7, which is taken from the original paper [26]. The following features are clearly visible. First, the intensive peak centered at $E_{rot} \approx 14.7$ meV is due to the well known first rotational excitation $J = 0 \to 1$ of the H_2 molecule [27]. Furthermore, the wave vector transfer of this peak is $K_{rot} \approx 2.7$ Å^{-1}. Thus the peak position in the K–E plane shows that the experimentally determined mass of H that fulfills the relation $E_{rot} = (\hbar K_{rot})^2/2M_H$ is (within experimental error) the mass M_H of the free H atom,

$$\text{rotation:} \quad M_H = 1.0079 \ \text{ a.m.u.} \tag{49}$$

namely, $M_{eff}(H) = M_H$. (a.m.u.: atomic mass units.) In other words, the location of this rotational excitation in the K–E plane agrees with conventional theoretical expectations for IINS, according to which each neutron scatters from a single H [27]. Recall that an agreement with conventional theory was also observed in the case of scattering from ^{4}He [18]; see Fig. 3.

Moreover, the authors provide a detailed analysis of the roto-recoil data from incoherent inelastic neutron scattering, as shown in Fig. 7, and extracted a strongly reduced effective mass of the whole recoiling H_2 molecule (left parabola, green line); see Eq. (41):

$$\text{translation (recoil):} \quad M_{eff}(\mathrm{H}_2) \approx 0.64 \pm 0.07 \ \text{ a.m.u.} \tag{50}$$

This is in blatant contrast to the conventionally expected value $M(\mathrm{H}_2) = 2.01$ a.m.u. for a freely recoiling H_2 molecule (right parabola, red line). (Recall that the neutron-molecule collision does not break the molecular H-H bond.)

An extensive numerical analysis of the data is presented in [26], being based on time-of-flight data analysis (cf. Section 4) and the analysis of the measured data within conventional theory [14, 27].

This strong reduction of effective mass, which is far beyond any conceivable experimental error, corresponds to a strong reduction of momentum transfer by the factor 0.566. Namely, the

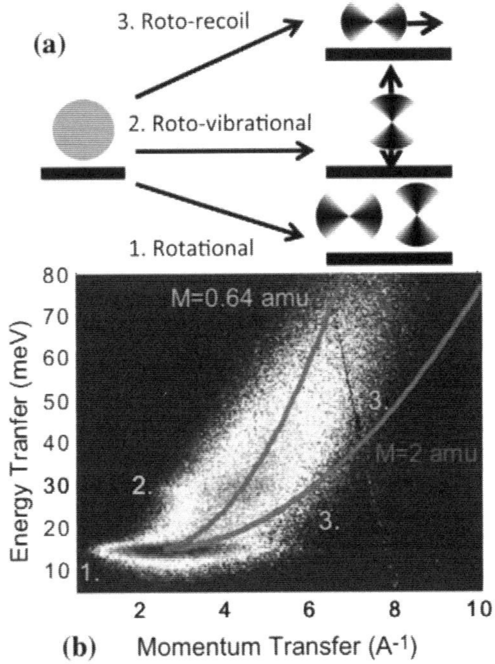

Figure 7. (*a*) Sketch depicting different types of transitions of the hydrogen molecule predicted to be observed by incoherent inelastic neutron scattering (IINS). (*b*) IINS (or INS) intensity map of hydrogen adsorbed in multi-walled carbon nanotubes after background subtraction, with intensity plotted on a log scale. Incident neutron energy is 90 meV and temperature is 23 K. The types of transitions are identified, and a fit to the roto recoil tail is shown (green curve); Fig. 1 of Ref. [26]. — The translation motion of the recoiling H_2 molecules causes the observed continuum of intensity, usually called *roto-recoil* (yellow-orange ribbon), starting at the well visible first rotational excitation of H_2 being centered at $E \approx 14.7$ meV and $K \approx 2.7$ Å^{-1} (red ellipsoid). The K–E position of the latter agrees with conventional theory. In contrast, a detailed fit (green parabola; full line) to the roto-recoil data reveals a strong reduction of the effective mass of recoiling H_2, which appears to be only 0.64 a.m.u. The red curve (on the right) represents the theoretical parabola with the conventionally expected effective mass 2 a.m.u. of a recoiling H_2 molecule. For full details about the data analysis see [26]. (Reproduced from Ref. [26], Fig. 1, with permission from Elsevier.)

observed momentum transfer deficit is about -43% of the conventionally expected momentum transfer. This provides first experimental evidence of the new *anomalous* effect of momentum-transfer deficit in an elementary neutron collision with a recoiling molecule.

Recall that, as explained above (see Subsec. 5.1), every H_2-substrate conventional binding must increase the molecule's effective mass. Thus these findings from IINS are in clear contrast to every conventional (classical or quantum) theoretical expectation. However, they have a natural (albeit qualitative, at present) interpretation in the frame of modern theory of WV and TSVF.

Incidentally, it may be noted that the same *calibration* of ARCS was used in both experiments [18] and [26].

Let us now make a comparison with a related 1-*dimensional* experiment, which may shed more light into the observations under consideration.

The above experimental results also show that the two-dimensional spectroscopic technique, as offered by the advanced TOF-spectrometer ARCS, represents a powerful method that provides novel insights into quantum dynamics of molecules and condensed matter. Clearly, this is due to the fact that K and E transfers can be measured over a broad region of the K–E plane. This advantage makes these new instruments superior to the common one-dimensional ones (like TOSCA at ISIS spallation source, UK), in which the detectors can only measure along a single specific trajectory in the K–E plane. (TOSCA measures along two such trajectories [29].)

As an example, consider the results of [28] from molecular H_2 adsorbed in single-wall carbon nanotubes (which is similar to the material of [26]) at $T \approx 20$ K, as investigated with TOSCA. Also this paper reports the measurement of the roto-recoil spectrum, but as a function of E only (due to the aforementioned single trajectory in the K–E plane being instrumentally accessible). Therefore for the theoretical analysis of the data the mass of H_2 was *fixed* to its conventionally expected value of 2 a.m.u., and thus the strong anomalous effect, Eq. (50), remained unnoticed; see [28], p. 903.

7. Discussion

Due to the novelty of the theory of WV and TSVF, several counter-intuitive features therein and in its applications may appear "strange" or even "wrong". Aiming to clarify some of them, here we provide additional remarks and explanations.

(A) *Instrumental calibration* — The results presented above have considerable consequences for the calibration of the associated TOF spectrometers. In particular, in NCS (DINS) instrumentation, it is a common practice to use the recoil peaks of light atoms (typically H, D or He) to achieve an "optimal" calibration of the spectrometer. In other words, the measured positions of a peak in the K–E plane — together with the standard free-atom recoil expression of Eq. (41), sometimes also improved with FSE — are used in order to "fine tune" the numerical values of the instrument's parameters determining the measured TOF values. Obviously, such a "optimization" of the calibration leads automatically to an artificial agreement of the data with conventional theory.

(B) Note that in in our approach, the two operators $\hat{q}$ and $\hat{P}$ occurring in the von Neumann-type interaction Hamiltonian of Eq. (25) refer to two *different* quantum systems. Thus, according to Vaidman [30], the concept of WV arises here due to the interference of a quantum *entangled* wave and therefore it has no analog in classical wave interference. This further supports the conclusion that WV is a genuinely quantum concept and not some kind of "approximation".

In contrast, conventional neutron scattering theory treats the neutron as a classical mass point. Namely, consider the basic formulas of the theory, e.g. the expression for the partial differential cross-section

$$\frac{d^2\sigma_A}{d\Omega d\omega} = \frac{k_f}{k_i} b_A^2 S(\mathbf{K}, \omega) \tag{51}$$

($E = \hbar\omega$, $d\Omega$: solid angle measured/covered by the detector.) $S(\mathbf{K}, \omega)$ is the dynamic structure factor of the scattering system, which contains degrees-of-freedom of the scattering particles only. This equation contains *no dynamical* variable of the neutron at all — the neutron is "degraded" to a classical object and only its scattering length b_A — a c-number capturing its scattering properties from atom A — appears here. This fact is a specific feature of the first Born approximation and/or first-order perturbation theory which the conventional theory is based on [14, 15].

(C) The momentum-transfer deficit, and the associated effective-mass reduction of the scattering particle A, may appear to someone as violating the energy and momentum conservation laws of basic physics. However, this is not the case, because the scatterer A is *not* an isolated, but an *open* quantum system. Thus we may say that the quantum dynamics of the "environment" of A, which participates to the neutron-A scattering, is indispensable for the new WV-TSVF effect under consideration. Thus it may be helpful to write down the "conservation" relations

$$E_{H+env} = -E_n \quad \text{and} \quad \hbar K_{H+env} = -\hbar K_n \tag{52}$$

which express energy and momentum conservation for the case that the *environment* (indicated with the subscript "*env*") of the scattering H is not neglected. The theoretically derived momentum transfer deficit, when interpreted (or better: misinterpreted) in terms of conventional theory, is equivalent to scattering of a neutron from a ficticious particle representing the *whole* H+environment system, which has a mass M_{eff} being smaller than the mass of a free H-atom — or, in other terms, the environment of H seems to have a negative mass!

(D) The INS results from H_2 in C-nanotubes [26] discussed in subsection 6.2 appear contradictory — in the frame of conventional theory — because of the following two main points:

(i) The observed $J = 0 \to 1$ rotational excitation of the H_2 molecule by the collision with a neutron exhibits a $M_{eff} \approx 1$ a.m.u., see Eq. (49), as conventionally expected. Namely, the scattering is incoherent, i.e. the neutron exchanges energy and momentum with one H only.

(ii) In contrast, in the same experiment, the M_{eff} of the observed roto-recoil response of the whole H_2 molecule is not 2 a.m.u. as it should (because the whole H_2 undergoes a translational motion), but only $M_{eff} \approx 0.64$ a.m.u., see Eq. (50).

However, this "contradiction" just disappears in the light of the new theory, because it simply implies that the quantum *environment* of H in case (i) must be *different* from that in case (ii) — the locally rotating H_2 is much less influenced by its environmental interactions than the translating H_2, which necessarily interacts with a greater environmental part during its translational motion caused by its collision with the neutron.

(E) In the scientific literature one often meets the criticism that *post-selection* just means "throwing out some data". In the experimental context at issue, however, post-selection certainly means "performing a concrete measurement on the system, using a well defined detector, and analyzing the measured data only". In this context it may be helpful to mention the paper by Wu [31], in which he offers a slight extension of the WV and TSVF that does not discard any data. Furthermore, it is shown how the matrix elements of any generalized state (pure or mixed) can be directly read from appropriate weak measurements.

(F) Concerning the applicability of WV (and WM) to the neutron scattering process, one may object that the neutron-nucleus potential (i.e., the conventional Fermi pseudopotential [14]) may be not "weak" in the specific sense of the new theory. Therefore it is helpful to mention the recent generalization of the considered WM and WV theory by Oreshkov and Brun [32]. The authors showed that WM are universal, in the sense that any generalized measurement can be decomposed into a sequence of WMs. This important theoretical result is further supported by the work of Qin et al. [33], who showed that the main WM results can be extended to the realm of *arbitrary* measurement strength.

(G) Very recently, some qualitatively new theoretical results of WV-TSVF have been obtained, a few of which should be mentioned here:

(a) The quantum Cheshire-Cat effect [34], which has also been observed experimentally [35].
(b) The predicted quantum violation of the Pigeonhole principle [36].
These new results point once more to a very interesting structure of quantum mechanics that was hitherto unnoticed. Moreover, they also shed new light on the very notions of separability, quantum correlations and quantum entanglement [37].

(H) The exploration of the "anomalous" momentum exchange taking place on a mirror of a MZI [8], as discussed in Sec. 2, indicates that this effect could also be relevant in the context of the widely applied methods of small-angle neutron scattering (SANS), small-angle X-ray scattering (SAXS) and reflectometry on surfaces and/or thin multilayer structures of physico-chemical and biological materials.

(I) Recently, quantum interference and entanglement [37] effects have been recognized to be crucial for quantum computing and quantum information theory; cf. [38]. Hence it seems interesting to explore the potential applicability of the new quantum features connected with WV, TSVF and the momentum transfer deficit (or energy transfer surplus) to recently emerging fields of quantum computational complexity theory [39, 40], and/or ICT (information and communication technology).

(J) In view of the experimentally detected effects of Sec. 6, and in particular of the striking reduction of effective mass due to quantum interference, it seems appropriate to mention here some speculative ideas concerning a possible *practical* and/or *technological* importance of the new theory. Namely:
(1) a more mobile (i.e. with smaller effective mass) H atom (or H^+ ion) in a properly designed fuel cell material would facilitate proton transfer and thus also increase the efficiency of the fuel cell;
(2) a properly designed solid environment (or nanostructure) of Li^+ ions in a Li-ion-battery would result in a lower Li^+ effective mass, thus allowing a faster ionic mobility and a *faster recharging* process — with obvious advantages for current technological efforts.

Concluding, the present author believes that the theoretical formalism of WM, WV, and TSVF not only sheds new light on interpretational issues concerning fundamental quantum theory (like e.g. the time-inversion invariance of the basic physical laws, the meaning of quantum entanglement and correlations, etc.) but it also offers a fascinating new guide for our intuition to predict new effects, and also helps to plan and carry out new experiments and reveal novel quantum phenomena.

Acknowledgments
I wish to thank Erik B. Karlsson (Uppsala) and Ingmari Tietje (CERN) for helpful discussions, and COST Action MP1403 – Nanoscale Quantum Optics – for partial financial support.

[1] Aharonov Y and Rohrlich D 2005 *Quantum Paradoxes: Quantum Theory for the Perplexed* (Weinheim: Wiley-VCH)
[2] Aharonov Y, Albert D Z and Vaidman L 1988 *Phys. Rev. Lett.* **60** 1351
[3] Aharonov Y and Vaidman L 1990 *Phys. Rev. A* **41** 11
[4] Kofman A G, Ashhab S and Nori F 2012 *Phys. Rep.* **520** 43
[5] Tamir B and Cohen E 2013 *Quanta* 2013; **2** 7
[6] Aharonov Y, Cohen E and Elitzur A C 2014 *Phys. Rev. A* **89** 052105
[7] Dressel J, Malik M, Miatto F M, Jordan A N and Boyd R W 2014 *Rev. Mod. Phys.* **86** 307
[8] Aharonov Y, Botero A, Nussinov S, Popescu S, Tollaksen J and Vaidman L 2013 *New J. Phys.* **15** 093006
[9] Aharonov Y, Bergmann P G and Lebowitz J L 1964 *Phys. Rev.* **134**, B1410
[10] Chatzidimitriou-Dreismann C A 2016 *Quanta* **5** 61-84

[11] https://neutrons.ornl.gov/ARCS
[12] Scully M O and Zubairy M S 1997 *Quantum Optics* (Cambridge: Cambridge University Press)
[13] von Neumann J 1932 *Mathematische Grundlagen der Quantenmechanik* (Berlin: Springer Verlag) [English translation: 1955 *Mathematical Foundations of Quantum Mechanics* (Princeton: Princeton University Press]
[14] Squires G L 1996 *Introduction to the Theory of Thermal Neutron Scattering* (Mineola: Dover)
[15] Watson G I 1996 *J. Phys.: Condens. Matter* **8** 5955
[16] Aharonov Y, Cohen E and Ben-Moshe S 2014 *EPJ Web of Conferences* **70**, 00053
[17] Feynman R P, Leighton R B and Sands M 1965 *The Feynman Lectures on Physics, Vol. III, Quantum Mechanics* (Reading: Addison-Wesley Publ.) [Also available at: http://www.feynmanlectures.caltech.edu/]
[18] Diallo S O, Azuah R T, Abernathy D L, Rota R, Boronat J and Glyde H R 2012 *Phys. Rev. B* **85** 140505
[19] Feynman R P, Leighton R B and Sands M 1965 *The Feynman Lectures on Physics, Vol. I* (Reading: Addison-Wesley Publ.)
[20] Tietje I C 2012 *J. Phys.: Conf. Ser.* **380** 012014
[21] Sears V F 1984 *Phys. Rev. B* **30** 44
[22] Hohenberg P C and Platzman P M 1966 *Phys. Rev.* **152** 198
[23] Schirato B S, Herwig K W, Sokol P E and White J W 1990 *Chem. Phys. Lett.* **165** 453
[24] Stock C, Cowley R A, Taylor J W amd Bennington S M 2010 *Phys. Rev. B* **81** 024303
[25] Science and Technology Facilities Council, ISIS facility, Oxfordshire, UK, spectrometer MARI. (http://www.isis.stfc.ac.uk/instruments/MARI/)
[26] Olsen R J, Beckner M, Stone M B, Pfeifer P, Wexler C and Taub H 2013 *Carbon* **58** 46
[27] Mitchell P C H, Parker S F, Ramirez-Cuesta A J and Tomkinson J 2005 *Vibrational Spectroscopy with Neutrons.* (New Jersey: World Scientific)
[28] Georgiev P A, Ross D K, De Monte A, Montaretto-Marullo U, Edwards R A H, Ramirez-Cuesta A J, Adams M A and Colognesi D 2005 *Carbon* **43** 895
[29] http://www.isis.stfc.ac.uk/instruments/TOSCA/
[30] Vaidman L 2014 *Comment on "How the result of a single coin toss can turn out to be 100 heads"* arXiv:1409.5386 [quant-ph].
[31] Wu S 2013 *Sci. Rep.* **3** 1193
[32] Oreshkov O and Brun T A 2005 *Phys. Rev. Lett.* **95** 110409
[33] Qin L, Feng W and Li X-Q 2016 *Sci. Rep.* **6** 20286
[34] Aharonov Y, Popescu S and Skrzypczyk P 2013 *New J. Phys.* **15** 113015
[35] Denkmayr T, Geppert H, Sponar S, Lemmel H, Matzkin A, Tollaksen J and Hasegawa Y 2014 *Nat. Commun.* **5** 4492
[36] Aharono Y, Colombo F, Popescu S, Sabadini I, Struppa D C and Tollaksen J 2016 *Proc. Nat. Acad. Sci. USA* **113** 532
[37] Horodecki R, Horodecki P, Horodecki M and Horodecki K 2009 *Rev. Mod. Phys.* **81** 865
[38] Nielsen M A and Chuang I L 2000 *Quantum Computation and Quantum Information* (Cambridge: Cambridge University Press)
[39] Arora S and Barak B 2009 *Computational Complexity – A Modern Approach* (Cambridge: Cambridge University Press)
[40] Fortnow L 2009 *Communications of the ACM* **52** 78

IOP Conf. Series: Journal of Physics: Conf. Series **1071** (2018) 012008 doi:10.1088/1742-6596/1071/1/012008

Quantum suprematism picture of Triada of Malevich's squares for spin states and the parametric oscillator evolution in the probability representation of quantum mechanics

V. N. Chernega[1], O. V. Man'ko[1,2], V. I. Man'ko[1,3,4]

[1] - Lebedev Physical Institute, Russian Academy of Sciences, Leninskii Prospect 53, Moscow 119991, Russia

[2] - Bauman Moscow State Technical University, The 2nd Baumanskaya Str. 5, Moscow 105005, Russia

[3] - Moscow Institute of Physics and Technology (State University), Institutskii per. 9, Dolgoprudnyi, Moscow Region 141700, Russia

[4] - Tomsk State University, Department of Physics, Lenin Avenue 36, Tomsk 634050, Russia, Lebedev Physical Institute, Leninskii Prospect 53, Moscow 119991, Russia

E-mail: omanko@sci.lebedev.ru

Abstract. Review of tomographic probability representation of quantum states is presented both for oscillator systems with continious variables and spin–systems with discrete variables. New entropy–information inequalities are obtained for Franck–Condon factors. Density matrices of qudit states are expressed in terms of probabilities of artificial qubits as well as the quantum suprematism approach to geometry of these states using the triadas of Malevich squares is developed. Examples of qubits, qutrits and ququarts are considered.

Keywords: qubit, qudit, tomographic probability representation, entropy–information inequalities, Franck–Condon factors, quantum suprematism approach.

1. Introduction

In quantum mechanics the system states are associated either with vectors $|\psi\rangle$ in Hilbert spaces [1] as well as with wave functions (of position) $\psi(q)$ [2] (for pure states) or with density matrices [3,4] (for mixed states). The physical observables are described by Hermitian operators acting in the Hilbert space.

This formalism is radically different from the formalism of classical mechanics where the states of a particle are associated with its position q and momentum p (if there is no fluctuations of these classical observables) or with probability distributions $f(q,p)$ (if there present fluctuations of these observables). We wish to create equivalent quantum formalism which is closer to classical formalism yielding the result obtained by Wigner [5] who introduced the Wigner function $W(q,p)$. This function is similar to classical probability distribution $f(q,p)$ but the function can take negative values and in view of this it is not probability distribution. Analogous functions on phase–space are Glauber–Sudarshan P-function [6–8], Husimi–Kano Q–function [9,10] connected with the Wigner function by integral transform. These functions are called quasidistributions. For electron spin the pure state is associated either with vector $|\psi\rangle$ in two–dimensional Hilbert space (spinor) or for mixed state with 2×2 - density matrix (density operator $\hat{\rho}$) acting in this Hilbert space. In the work of Stratonovich [11] the formalism of analog of the quasidistributions for spin-1/2 particle was introduced. But in classical mechanics or classical statistical mechanics namely probability distributions are used to describe the system states. The probability representation for states of quantum systems with continious variables was introduced in [12,13] and discussed in [14]. This representation called tomographic probability representation is based on the known tool to reconstruct the state Wigner function in quantum–optical experiments [15] using relation [16,17] of this function with measurable optical tomogram $w(X|\theta)$, which is fair conditional probability distribution [18] of variable X called in quantum optics homodyne quadrature, depending on local oscillator phase θ, by means of integral Radon transform [19]. It turned out that the spin states also can be associated with fair probability distributions called spin–tomograms [20–25]. Since the quantum states in the probability representation are identified with probability distributions all the instruments and results of the classical probability theory were applied to get new results in quantum mechanics, quantum optics and quantum information (see e.g. reviews [26–28]). The aim of this work is to review the probability representation for quantum states on example of parametric oscillator and spin–states. Using the known physical meaning of Frank–Condon factors in molecular spectroscopy as probability distributions we describe some new entropy–information inequalities studied in quantum mechanics [29] for these factors. These inequalities can be useful in study of vibronic spectra of polyatomic molecules discussed e.g. in [30–33]. Also for spin–states we present new formulas for the density matrices expressed in terms only of probabilities of spin projections [34–37] and provide the quantum suprematism picture of the spin–states where triangle geometry of the states is associated with triadas of Malevich's squares [35–37]. The review of suprematism approach in art is given in [38].

The paper is organised as follows.

In Sec. 2 we review the optical and symplectic tomographic probability distributions on examples of oscillator states including parametric oscillator states. In Sec. 3 we discuss new entropy–information inequalities for Franc–Condon factors of diatomic and polyatomic molecules. In Sec. 4 we review the spin–tomography for qudit states. In Sec. 5 we present explicit expressions of spin–state density matrices for qubit and qutrit states in terms of probabilities of spin–1/2 projections on perpendicular directions and discuss the expressions of such density matrices for states of arbitrary spins. In Sec. 6 we review the quantum suprematism representation of qudit states. The conclusions and prospectives are given in Sec. 7.

2. Tomographic probability distributions of states with continious variables

We consider the systems like oscillator with wave function $\psi(x), -\infty \leq x \leq \infty$. The Wigner function of the system is given by relation ($\hbar = m = 1$)

$$W(q,p) = \int \psi(q + u/2)\psi^*(q - u/2)e^{-ipu}du. \tag{1}$$

The Wigner function is real function and for normalized states, i.e. $\int |\psi(x)|^2 dx = 1$, it is normalized

$$\frac{1}{2\pi}\int W(q,p)dqdp = 1. \tag{2}$$

The density matrix of the state $\rho_\psi(x,x') = \psi(x)\psi^*(x')$ is given by the transform

$$\rho_\psi(x,x') = \frac{1}{2\pi}\int W(\frac{x+x'}{2},p)e^{ip(x-x')}dp. \tag{3}$$

The tomographic probability distribution called symplectic tomogram $w(X|\mu,\nu)$ where $-\infty \leq X,\mu,\nu \leq \infty$ is defined by the Radon transform of the Wigner function [39]

$$w(X|\mu,\nu) = \int W(q,p)\delta(X-\mu q-\nu p)\frac{dqdq}{2\pi}. \tag{4}$$

The function is nonnegative and normalized for arbitrary parameters μ and ν, i.e.

$$\int w(X|\mu,\nu)dX = 1. \tag{5}$$

It can be easily checked using (1) that the symplectic tomogram of pure state with wave function $\psi(y)$ is given by the fractional Fourier transform of the wave function [40]

$$w(X|\mu,\nu) = \frac{1}{2\pi|\nu|}|\int \psi(y)exp\left(\frac{i\mu y^2}{2\nu}-\frac{iXy}{\nu}\right)dy|^2. \tag{6}$$

For the parameters $\mu = \cos\theta$, $\nu = \sin\theta$ the symplectic tomogram coincides with probability distribution called optical tomogram $w(X|\theta)$ which reads

$$w(X|\theta) = \int W(q,p)\delta(X-q\cos\theta-p\sin\theta)\frac{dqdp}{2\pi}. \tag{7}$$

For pure state the optical tomogram has the following form

$$w(X|\theta) = |\int \psi(y)\frac{\exp\left(\frac{i\cot\theta}{2}(X^2+y^2)-\frac{iXy}{\sin\theta}\right)}{\sqrt{2\pi i\sin\theta}}dy|^2. \tag{8}$$

The symplectic tomogram determines the Wigner function

$$W(q,p) = \frac{1}{2\pi}\int w(X|\mu,\nu)\exp(i(X-\mu q-\nu p))dXd\mu d\nu \tag{9}$$

and also the density operator $\hat{\rho}$ of the state with the Wigner function, namely [13]

$$\hat{\rho} = \frac{1}{2\pi}\int w(X|\mu,\nu)\exp(i(X\hat{1}-\mu\hat{q}-\nu\hat{p}))dXd\mu d\nu. \tag{10}$$

Here $\hat{q}$ and $\hat{p}$ are position and momentum operators.

3. Frank-Condon factors on example of parametric oscillator

For oscillator with varying frequency $\omega(t)$ and the Hamiltonian

$$\hat{H} = \frac{\hat{p}^2}{2}+\frac{\omega^2(t)\hat{q}^2}{2}, \quad \omega(0)=1 \tag{11}$$

the solutions of Schrödinger equation can be obtained using the integrals of motion $\hat{A}(t)$, $\hat{A}^\dagger(t)$ found in [41], namely

$$\hat{A}(t) = \frac{i}{\sqrt{2}}\left(\epsilon(t)\hat{p}-\dot{\epsilon}(t)\hat{q}\right), \tag{12}$$

where

$$\ddot{\epsilon}(t)+\omega^2(t)\epsilon(t)=0 \tag{13}$$

and $\epsilon(0) = 1$, $\dot{\epsilon}(0) = i$, $\hat{A}(0) = \frac{\hat{q}+i\hat{p}}{\sqrt{2}}$. The product of the invariants $\hat{A}^\dagger(t)\hat{A}(t)$ equals to Ermakov invariant [42]. Commutation relations of the integrals of motion (12) are $[\hat{A}(t), \hat{A}^\dagger(t)] = 1$. The ground–like state $\psi_0(x,t)$ of the parametric oscillator satisfying the condition $\hat{A}(t)\psi_0(x,t) = 0$ reads

$$\psi_0(x,t) = \pi^{-1/4}(\epsilon(t))^{-1/2} \exp\left[\frac{i\dot{\epsilon}(t)x^2}{2\epsilon(t)}\right]. \tag{14}$$

Fock states $\psi_n(x,t)$, $n = 0, 1, 2, \ldots$, solutions of Schrödinger equation are expressed in terms of the integrals of motion and ground–like state as

$$\psi_n(x,t) = \frac{(\hat{A}^\dagger(t))^n}{\sqrt{n!}}\psi_0(x,t), \quad n = 0, 1, 2, \ldots \tag{15}$$

They are expressed in terms of Hermite polynomials (see,e.g. [43]). The Frank-Condon factors are probabilities given by the scalar–product

$$P_m(n,t) = |\langle m, 0|n, t\rangle|^2 \tag{16}$$

where $|m, 0\rangle$ is the state at time $t = 0$. The Frank–Condon factor is related to vibronic structure of electronic line of two–atomic molecules [30, 31, 44–46].

The tomographic probability distribution (6) of the ground-like state (14) has the form of normal distribution

$$w_0(X|\mu,\nu) = \frac{1}{\sqrt{2\pi\sigma^2}} \exp\left(-\frac{1}{2\sigma^2}(X - \bar{X})^2\right), \tag{17}$$

where $\bar{X} = 0$ and

$$\sigma^2 = \frac{|\epsilon|^2}{2}\mu^2 + \frac{|\dot{\epsilon}|^2}{2}\nu^2 + \mu\nu(|\dot{\epsilon}\epsilon^*|^2 - 1)^{1/2}.$$

One can obtain a new result characterising the properties of the Franck–Condon factors (16). These factors provide the probability distribution. In view of this the general properties of the probability distributions are present in Franck–Condon factors. For example, Shannon entropy associated with the probability distribution and the inequalities known for this entropy are valid for the factors. If the probability distribution is a joint probability distribution of two random variables $P(a,b)$ there exists inequality for Shannon entropy of the system $H(1,2) = -\sum_a \sum_b P(a,b) \ln P(a,b)$ and two entropies corresponding to marginal distributions $H(1) = -\sum_b (\sum_a P(a,b)) \ln (\sum_a P(a,b))$ and $H(2) = -\sum_a (\sum_b P(a,b)) \ln (\sum_b P(a,b))$ which means the nonnegativity of the Shannon information

$$I = H(1) + H(2) - H(1,2) \geq 0.$$

If the random variables are not correlated, i.e. $P(a,b) = \Pi(a)\mathcal{P}(b)$ the information $I = 0$. The stronger are correlations the larger is the Shannon information. Using the partition tool [48, 49] to present the probability distribution (16) as the joint probability distribution we can apply the properties of Shannon information to the Franck–Condon factors.

We present new entropic inequality for the Franck–Condon factors (16) of the form

$$-\sum_{n=0}^{\infty} |\langle m, 0|n, t\rangle|^2 \ln |\langle m, 0|n, t\rangle|^2 \leq -\sum_{k=0}^{\infty}\left[\left(\sum_{j=0}^{1} |\langle m, 0|2k+j, t\rangle|^2\right) \ln \left(\sum_{j=0}^{1} |\langle m, 0|2k+j, t\rangle|^2\right)\right]$$

$$-\sum_{j=0}^{1}\left[\left(\sum_{k=0}^{\infty} |\langle m, 0|2k+j, t\rangle|^2\right) \ln \left(\sum_{k=0}^{\infty} |\langle m, 0|2k+j, t\rangle|^2\right)\right]. \tag{18}$$

The difference of the right–hand side and left–hand side (18) is Shannon [47] information. It is nonnegative. The entropic inequality is characterizing the correlations in the vibronic structure of the electronic line in diatomic molecules. This inequality can be checked experimentally. It is worthy to note that the relation and analogies of the molecular spectroscopy method and quantum information technique were studied in [44–46]. One can obtain other entropy–information inequalities for the Franck–Condon factors of diatomic and polyatomic molecules using the tool of mapping the integers $s = 1, 2, 3, \ldots, N = \prod_{k=1}^{M} n_k$ onto sets of of integers $j_1, j_2, \ldots, j_M$ where $j_1 = 1, 2, , \ldots, n_1$, $j_2 = 1, 2, \ldots, n_2, \ldots, j_n = 1, 2, \ldots, n_M$ [48, 49]. This tool provides possibility to consider Franck-Condon factors as joint probability distributions. These distributions can be mapped onto the joint probability distributions of system with M artificial subsystems. Then the known in probability theory entropy–information inequalities are obviously valid for Franck–Condon factors. The correlations of Franck–Condon factors reflect the correlations of line intensities in vibronic structure of the electronic lines in the molecular spectra. The Franck–Condon factors can be also related to tomographic–probability distribution of quantum states of molecules [33]. In this case the known probability properties of the quantum tomograms provide possibility to obtain new relations for the Franck–Condon factors. It is worthy to note that oscillator with time–dependent parameters describing the dissipation in framework of using Ermakov quadratic invariant [42, 53] was studied in [54–57].

4. Spin–tomography

In [20–22] the tomographic probability distribution $w(m|\tilde{n})$ was introduced for spin j-states and expressed in terms of density matrix ρ and its unitary transform u as diagonal element of the product of three matrices, i.e.

$$w(m|\vec{n}) = \Big(u(\vec{n})\rho u^\dagger(\vec{n})\Big)_{mm}. \tag{19}$$

Here $m = -j, -j+1, \ldots, j-1, j$ is spin–projection on the direction determined by unit vector $\vec{n} = (\sin\theta\cos\phi, \sin\theta\sin\phi, \cos\theta)$. The unitary matrix $u(\vec{n})$ is the matrix of irreducible representation of rotation group corresponding to spin j. It depends of three Euler angles (ψ, θ, ϕ) where $\psi = 0$. The probability distribution $w(m|\vec{n})$ called spin–tomogram determines the density matrix ρ. It is normalized

$$\sum_{m=-j}^{j} w(m|\vec{n}) = 1 \tag{20}$$

for arbitrary direction $\vec{n}$. The density matrix of spin state $\rho = \rho^\dagger$, $\mathrm{Tr}\rho = 1$, $\rho \geq 0$ depends on $(2j+1)^2 - 1$ real parameters. In view of this the minimal number of different unit vectors $\vec{n}$ which is sufficient to express the density matrix in terms of the tomographic probabilities equals also to $(2j+1)^2 - 1$. For $j = 1/2$ it is sufficient to use three such vectors, e.g. to take vectors coinciding with $\vec{x}$, $\vec{y}$, $\vec{z}$ axes, respectively. Let us introduce the notation $w(m = +\frac{1}{2}|\vec{x}) = p_1$, $w(m = +\frac{1}{2}|\vec{y}) = p_2$, $w(m = +\frac{1}{2}|\vec{z}) = p_3$. As it was pointed out [34–37] the density 2×2-matrix ρ is expressed in terms of these three probabilities p_1, p_2, p_3 to have positive spin projection $+\frac{1}{2}$ on the axes $\vec{x}, \vec{y}, \vec{z}$, respectively in the form

$$\rho = \begin{pmatrix} p_3 & p_1 - ip_2 - \frac{1-i}{2} \\ p_1 + ip_2 - \frac{1+i}{2} & 1 - p_3 \end{pmatrix}. \tag{21}$$

The tomogram (19) of the spin-1/2 state $w(+\frac{1}{2}|\vec{n})$ is expressed in terms of the probabilities p_1, p_2, p_3 described by the vector $\vec{p} = (p_1, p_2, p_3)$ and vector $\vec{n}_0 = (1/2, 1/2, 1/2)$ as the scalar product

$$w(m|\vec{n}) = \vec{n}(\vec{p} - \vec{p}_0) + 1/2. \tag{22}$$

IOP Conf. Series: Journal of Physics: Conf. Series **1071** (2018) 012008 doi:10.1088/1742-6596/1071/1/012008

Thus the state of the spin-1/2 is identified with three probability distributions determined by three probability vectors $\vec{\mathcal{P}}_1 = (p_1, 1-p_1)$, $\vec{\mathcal{P}}_2 = (p_2, 1-p_2)$ and $\vec{\mathcal{P}}_3 = (p_3, 1-p_3)$, $1-p_k$, $k = 1, 2, 3$ are the probabilities to have spin projection $m = -1/2$ on the axes $\vec{x}$, $\vec{y}$, $\vec{z}$, respectively. The probabilities must satisfy the quantum relation corresponding to nonnegativity of the density matrix (21) [34]

$$\sum_{k=1}^{3}(p_k - \frac{1}{2})^2 \leq \frac{1}{4}. \tag{23}$$

One can measure in any spin–1/2 state the probabilities of spin–projection $m = +1/2$ onto three perpendicular directions. These probabilities must satisfy the above inequality. It was pointed out in [34] that this inequality can be checked in experiments with superconducting circuits where analogs of spin states are realized in Josephson junction devices. Different aspects of statistical properties of Josephson junction were discussed in [50–52].

5. Probability representation of arbitrary qudit states

In [36] the matrix elements of density matrix of qutrit state (spin–1 system) were expressed in terms of probabilities of positive spin–1/2 projections for three artificial qubits. We rederive this result in the form which can be used to generalize the expression for the density matrix of qutrit and obtain the expression of density matrix of arbitrary qudit state in terms of probabilities associated with the artificial qubits (spin–1/2 projections). Let us consider the 3×3-density matrix of a qutrit state and construct two density 4×4 matrices of the qudit states with specific zero columns and rows. The tool we use is to start from qutrit density matrix ρ

$$\rho = \begin{pmatrix} \rho_{11} & \rho_{12} & \rho_{13} \\ \rho_{21} & \rho_{22} & \rho_{23} \\ \rho_{31} & \rho_{32} & \rho_{33} \end{pmatrix}. \tag{24}$$

and construct the 4×4-matrices

$$\rho(1) = \begin{pmatrix} \rho_{11} & \rho_{12} & \rho_{13} & 0 \\ \rho_{21} & \rho_{22} & \rho_{23} & 0 \\ \rho_{31} & \rho_{32} & \rho_{33} & 0 \\ 0 & 0 & 0 & 0 \end{pmatrix}, \quad \rho(2) = \begin{pmatrix} 0 & 0 & 0 & 0 \\ 0 & \rho_{11} & \rho_{12} & \rho_{13} \\ 0 & \rho_{21} & \rho_{22} & \rho_{23} \\ 0 & \rho_{31} & \rho_{32} & \rho_{33} \end{pmatrix}. \tag{25}$$

Interpretating these matrices as the density matrices of two qubit states we construct using the partial tracing procedure four density matrices of the artificial qubit states related to matrix $\rho(1)$, i.e.

$$R(1) = \begin{pmatrix} \rho_{11} + \rho_{22} & \rho_{13} \\ \rho_{31} & \rho_{33} \end{pmatrix}, \quad R(2) = \begin{pmatrix} \rho_{11} + \rho_{33} & \rho_{12} \\ \rho_{21} & \rho_{22} \end{pmatrix}, \tag{26}$$

and matrix $\rho(2)$

$$R(3) = \begin{pmatrix} \rho_{11} & \rho_{13} \\ \rho_{31} & \rho_{33} + \rho_{22} \end{pmatrix}, \quad R(4) = \begin{pmatrix} \rho_{22} & \rho_{23} \\ \rho_{32} & \rho_{11} + \rho_{33} \end{pmatrix}, \tag{27}$$

Since these matrices are the density matrices of qubit states one can present these matrices using the probabilities of positive spin–1/2 projections, namely

$$R(k) = \begin{pmatrix} p_3^{(k)} & (p_1^{(k)} - \frac{1}{2}) - i(p_2^{(k)} - \frac{1}{2}) \\ (p_1^{(k)} - \frac{1}{2}) + i(p_2^{(k)} - \frac{1}{2}) & 1 - p_3^{(k)} \end{pmatrix}, \quad k = 1, 2, 3, 4 \tag{28}$$

where $p_1^{(k)}$, $p_2^{(k)}$, $p_3^{(k)}$ are probabilities of spin-1/2 projections equal to $+1/2$ on axes $\vec{x}$, $\vec{y}$, $\vec{z}$, respectively for all four artificial qubits. Comparing matrix elements of matrices (26) and (27) with matrix elements of the same matrices given in the form (28) we obtain the following relations for diagonal matrix elements

$$\begin{aligned} \rho_{33} &= 1 - p_3^{(1)}, \\ \rho_{22} &= 1 - p_3^{(2)} = p_3^{(4)}, \\ \rho_{11} &= p_3^{(1)} + p_3^{(2)} - 1 \end{aligned} \tag{29}$$

and for offdiagonal matrix elements

$$\begin{aligned} \rho_{21} &= (p_1^{(2)} - \frac{1}{2}) + i(p_2^{(2)} - \frac{1}{2}), \\ \rho_{31} &= (p_1^{(1)} - \frac{1}{2}) + i(p_2^{(1)} - \frac{1}{2}), \\ \rho_{32} &= (p_1^{(4)} - \frac{1}{2}) + i(p_2^{(4)} - \frac{1}{2}). \end{aligned} \tag{30}$$

One has the equalities

$$p_1^{(1)} = p_1^{(3)},\, p_2^{(1)} = p_2^{(3)},\, p_3^{(4)} = 1 - p_3^{(2)},\, p_3^{(1)} = p_3^{(3)} + p_3^{(4)}. \tag{31}$$

For four qubits we have 12 parameters but the given equalities mean that 8 probabilities determine the state (24). Thus the density 3×3-matrix ρ (24) can be written in the form

$$\rho = \begin{pmatrix} p_3^{(1)} + p_3^{(2)} - 1 & (p_1^{(2)} - \frac{1}{2}) - i(p_2^{(2)} - \frac{1}{2}) & (p_1^{(1)} - \frac{1}{2}) - i(p_2^{(1)} - \frac{1}{2}) \\ (p_1^{(2)} - \frac{1}{2}) + i(p_2^{(2)} - \frac{1}{2}) & 1 - p_3^{(2)} & (p_1^{(4)} - \frac{1}{2}) - i(p_2^{(4)} - \frac{1}{2}) \\ (p_1^{(1)} - \frac{1}{2}) + i(p_2^{(1)} - \frac{1}{2}) & (p_1^{(4)} - \frac{1}{2}) + i(p_2^{(4)} - \frac{1}{2}) & 1 - p_3^{(1)} \end{pmatrix} \tag{32}$$

Here the probabilities $p_j^{(k)}$, $j = 1, 2, 3$, $k = 1, 2, 3, 4$ satisfy the inequalities (23). Thus the inequality for one qubit state density matrix (23) is valid for all artificial qubit states density matrices associated with qutrit state (32). Also the inequality $\det \rho \geq 0$ provides the relation where cubic polynomial constructed from the probabilities $p_j^{(k)}$ must take nonnegative values.

In order to generalise the result obtained for the qutrit state we introduce the following notations

$$\begin{aligned} &p_3^{(1)} = p_3^{(33)}, \quad p_3^{(2)} = p_3^{(22)}, \quad p_1^{(1)} = p_1^{(31)}, \quad p_2^{(1)} = p_2^{(31)}, \\ &p_1^{(2)} = p_1^{(21)}, \quad p_2^{(2)} = p_2^{(21)}, \quad p_1^{(4)} = p_1^{(32)}, \quad p_2^{(4)} = p_2^{(32)}. \end{aligned}$$

Using the introduced notations we can rewrite the matrix elements of density matrix (32) in the form

$$\begin{aligned} \rho_{jk} &= (p_1^{(jk)} - \frac{1}{2}) + i(p_2^{(jk)} - \frac{1}{2}), \quad j > k \\ \rho_{jj} &= 1 - p_3^{(jj)}, \quad j \geq 2, \\ \rho_{11} &= 1 - \sum_{j=2}^{3} \rho_{jj}. \end{aligned} \tag{33}$$

This form is preserved if we consider arbitrary qudit density $N\times N$-matrix ρ_{jk}, $j,k=1,2,\ldots,N$. In this case we have the artificial qubit probabilities $p_{1,2}^{(jk)}$, $j>k$; $p_3^{(jj)}$, $j\geq 2$. For example, qutrit density matrix reads

$$\rho=\begin{pmatrix} p_3^{(33)}+p_3^{(22)}-1 & (p_1^{(21)}-\frac{1}{2})-i(p_2^{(21)}-\frac{1}{2}) & (p_1^{(31)}-\frac{1}{2})-i(p_2^{(31)}-\frac{1}{2}) \\ (p_1^{(21)}-\frac{1}{2})+i(p_2^{(21)}-\frac{1}{2}) & 1-p_3^{(22)} & (p_1^{(32)}-\frac{1}{2})-i(p_2^{(32)}-\frac{1}{2}) \\ (p_1^{(31)}-\frac{1}{2})+i(p_2^{(31)}-\frac{1}{2}) & (p_1^{(32)}-\frac{1}{2})+i(p_2^{(32)}-\frac{1}{2}) & 1-p_3^{(33)} \end{pmatrix} \tag{34}$$

It is worthy to note that in [35] the qutrit density matrix was expressed in terms of probabilities but there was used different artificial qubit state.

The proof of the expression of the arbitrary density $N\times N$ - matrix in terms of probabilities $p_{1,2}^{(jk)}$, $p_3^{(j,j)}$ where $j>k$ describing spin-1/2 projections on axes $\vec{x}$, $\vec{y}$, $\vec{z}$, respectively can be done using the induction method. We illustrate this statement considering the case of $N=4$. The density matrix ρ_{jk}, $j.k=1,2,3,4$ can be embedded into 6×6-density matrices,

$$\rho(1)=\begin{pmatrix} \rho & 0 \\ 0 & 0 \end{pmatrix}, \quad \rho(2)=\begin{pmatrix} 0 & 0 \\ 0 & \rho \end{pmatrix}. \tag{35}$$

Considering these matrices as density matrices of qutrit–qubit systems and applying the partial tracing procedure we get four density matrices $R(1)$, $R(2)$, $R(3)$, $R(4)$ of the form

$$R(1)=\begin{pmatrix} \rho_{11}+\rho_{22}+\rho_{33} & \rho_{14} \\ \rho_{41} & \rho_{44} \end{pmatrix}, \quad R(2)=\begin{pmatrix} \rho_{11}+\rho_{44} & \rho_{12} & \rho_{13} \\ \rho_{21} & \rho_{22} & \rho_{23} \\ \rho_{31} & \rho_{32} & \rho_{34} \end{pmatrix},$$

$$R(3)=\begin{pmatrix} \rho_{11} & \rho_{14} \\ \rho_{41} & \rho_{22}+\rho_{33}+\rho_{44} \end{pmatrix}, \quad R(4)=\begin{pmatrix} \rho_{22} & \rho_{23} & \rho_{24} \\ \rho_{32} & \rho_{33} & \rho_{34} \\ \rho_{42} & \rho_{43} & \rho_{44}+\rho_{11} \end{pmatrix}. \tag{36}$$

Since we know the expressions for the qubit states $R(1)$, $R(3)$ and qutrit states $R(2)$, $R(4)$ in terms of probabilities we can express the density matrix ρ_{jk}, $j,k=1,2,3,4$ in terms of these probabilities. We have shown that the transition from qubit density matrix (21) to qutrit density matrix (24) gives the expression of the qutrit density matrix in the form (34). The analogous procedure is used to consider the 4×4-matrix and it repeats all the steps of the procedure to consider the 3×3-matrix and only adds extra matrix elements. Thus we have the density matrix of the ququart state in the form

$$\rho=$$

$$\begin{pmatrix} p_3^{(44)}+p_3^{(22)}+p_3^{(33)}-2 & p_1^{(21)}-\frac{1}{2}-i(p_2^{(21)}-\frac{1}{2}) & p_1^{(31)}-\frac{1}{2}-i(p_2^{(31)}-\frac{1}{2}) & p_1^{(41)}-\frac{1}{2}-i(p_2^{(41)}-\frac{1}{2}) \\ p_1^{(21)}-\frac{1}{2}+i(p_2^{(21)}-\frac{1}{2}) & 1-p_3^{(22)} & p_1^{(32)}-\frac{1}{2}-i(p_2^{(32)}-\frac{1}{2}) & p_1^{(42)}-\frac{1}{2}-i(p_2^{(42)}-\frac{1}{2}) \\ p_1^{(31)}-\frac{1}{2}+i(p_2^{(31)}-\frac{1}{2}) & p_1^{(32)}-\frac{1}{2}+i(p_2^{(32)}-\frac{1}{2}) & 1-p_3^{(33)} & p_1^{(43)}-\frac{1}{2}-i(p_2^{(43)}-\frac{1}{2}) \\ p_1^{(41)}-\frac{1}{2}+i(p_2^{(41)}-\frac{1}{2}) & p_1^{(42)}-\frac{1}{2}+i(p_2^{(42)}-\frac{1}{2}) & p_1^{(43)}-\frac{1}{2}+i(p_2^{(43)}-\frac{1}{2}) & 1-p_3^{(44)} \end{pmatrix}. \tag{37}$$

Since the numbers $p_{1,2,3}^{(jk)}$ are probabilities the inequalities for all density matrices hold

$$\mathrm{Re}\,\rho_{jk}+\frac{1}{2}\geq 0, \quad \mathrm{Im}\,\rho_{jk}\leq\frac{1}{2}. \tag{38}$$

Also for arbitrary qudit density $N \times N$-matrix one has new entropic inequality

$$\left(\frac{1}{2} - \mathrm{Im}\,\rho_{jk}\right) \ln \left[\frac{\left(\frac{1}{2} - \mathrm{Im}\,\rho_{jk}\right)}{\left(\frac{1}{2} - \mathrm{Im}\,\rho_{j'k'}\right)}\right] + \left(\frac{1}{2} + \mathrm{Im}\,\rho_{jk}\right) \ln \left[\frac{\left(\frac{1}{2} + \mathrm{Im}\,\rho_{jk}\right)}{\left(\frac{1}{2} + \mathrm{Im}\,\rho_{j'k'}\right)}\right] \geq 0 \qquad (39)$$

Here $j \neq k$, $j' \neq k'$, $i, k, j', k' = 1, 2, \ldots, N$. Another new entropic inequality for arbitrary qudit state reads

$$\rho_{jj} \ln \left[\frac{\rho_{jj}}{\left(\frac{1}{2} \mp \mathrm{Im}\,\rho_{j'k}\right)}\right] + (1 - \rho_{jj}) \ln \left[\frac{(1 - \rho_{jj})}{\left(\frac{1}{2} \pm \mathrm{Im}\,\rho_{j'k}\right)}\right] \geq 0. \qquad (40)$$

The entropic inequality of the form

$$\ln 2 \geq -\left(\frac{1}{2} \mp \mathrm{Im}\rho_{jk}\right) \ln \left(\frac{1}{2} \mp \mathrm{Im}\rho_{jk}\right) - \left(\frac{1}{2} \pm \mathrm{Im}\rho_{jk}\right) \ln \left(\frac{1}{2} \pm \mathrm{Im}\rho_{jk}\right) \geq 0 \qquad (41)$$

holds for arbitrary density matrix of qudit state.

6. Triada of Malevich's squares and quantum suprematism picture of qudit states

The discussed probability representation of qudit states can be pictorially illustrated [35–37] by using Malevich's squares known in suprematism direction in art [38]. It means that for arbitrary probabilities $p_1^{(jk)}$, $p_2^{(jk)}$, $p_3^{(jk)}$ determined by matrix elements ρ_{jk} of density $N \times N$ - matrix ρ one can introduce the triangles with three sides

$$l_{jk}^{(s)} = (2 + 2(p_s^{(jk)})^2 - 4p_s^{(jk)} - 2p_{s+1}^{(jk)} + 2(p_{s+1}^{(jk)})^2 + 2p_s^{(jk)}p_{s+1}^{(jk)})^{1/2}, \quad s = 1, 2, 3.$$

Three squares (red, black and white) called "triada of Malevich's squares" have the areas

$$S_{jk}^{(s)} = \left[l_{jk}^{(s)}\right]^2, \quad s = 1, 2, 3.$$

There is the bijective map of the density matrix of the qudit states with the matrix elements ρ_{jk} and the set of the triadas of Malevich's squares. Due to nonnegativity condition for density matrix ρ_{jk} the areas of the Malevich's squares satisfy the inequalities described by the inequalities for polynomials expressed in terms of the probabilities $p_s^{(jk)}$, $s = 1, 2, 3$. The number of parameters determining the density matrix ρ_{jk}, $j, k = 1, 2, \ldots, N$ equals $N^2 - 1$. This number provides the number of triadas of Malevich's squares describing the qudit states. For example $N = 3$ (qutrit) considered in [37] one has three triadas of Malevich's squares with extra constraints for the sides of the squares. In Fig.1 we give a pictorial description of qudit state with N=4 (spin 3/2 state) using a mosaic constructed from five triadas of Malevich's squares. The ququart density matrix is mapped onto this "quantum suprematism" picture on the plane. Each square in five triadas has the area determined by the matrix elements of the spin 3/2 density matrix (37) expressed in terms of the probabilities $p_s^{(jk)}$. Thus five triadas of Malevich's squares illustrate a new parametrization of the density matrix by the 15 independent probabilities. There are quantum correlations in the geometric picture of the squares corresponding to nonnegativity condition of the density matrix. These conditions are expressed as nonnegativity conditions for eigenvalues of the matrix (37).

One can also map the probabilities $p_s^{(jk)}$ onto the parameters $X_{jk}^{(s)}$ of Bloch ball describing the artificial qubits in view of the relation $2p_s^{(jk)} - 1 = X_{jk}^{(s)}$. These parameters satisfy the inequalities $\sum_{s=1}^{3} \left(X_{jk}^{(s)}\right)^2 \leq 1$ as well as other quantum constraints associated with nonnegativity of the

IOP Conf. Series: Journal of Physics: Conf. Series **1071** (2018) 012008 doi:10.1088/1742-6596/1071/1/012008

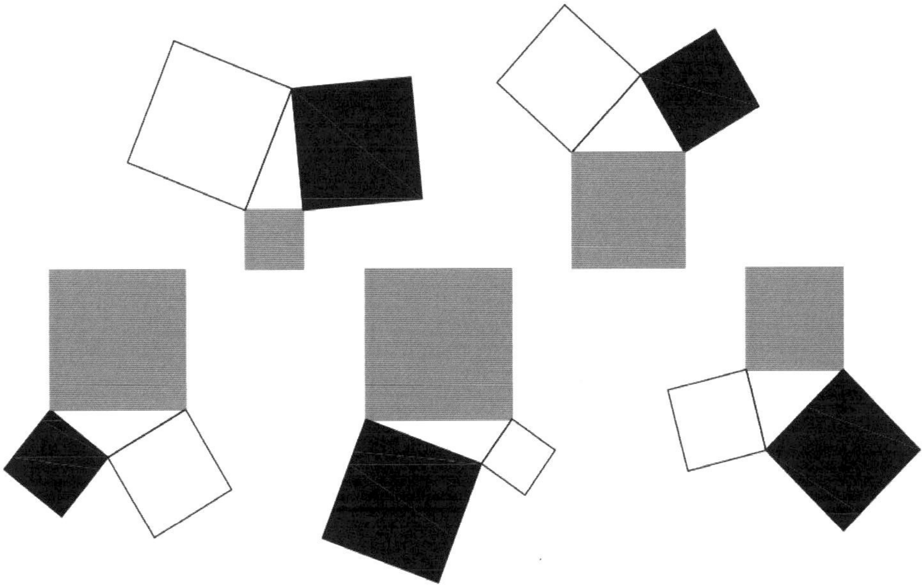

Figure 1. "Quantum suprematism" picture on the plane: five triadas of Malevich's squares associated with density matrix of ququart state.

density matrix. One can introduce the entropic characteristic of the qudit state as the sum of entropies of small artificial qubits. For example, for qubit (N=2) the entropy reads

$$H = -p_1 \ln p_1 - (1-p_1)\ln(1-p_1) - p_2 \ln p_2 - (1-p_2)\ln(1-p_2) - p_3 \ln p_3 - (1-p_3)\ln(1-p_3).$$

The relation of this entropy with von–Neumann entropy S can be clarified if one takes into account the expression of entropy S with the probabilities p_1, p_2, p_3, i.e.

$$\begin{aligned} S = \quad & -\left(\frac{1}{2} + \sqrt{\sum_{k=1}^{3}(p_k - 1/2)^2}\right) \ln\left(\frac{1}{2} + \sqrt{\sum_{k=1}^{3}(p_k - 1/2)^2}\right) \\ & -\left(\frac{1}{2} - \sqrt{\sum_{k=1}^{3}(p_k - 1/2)^2}\right) \ln\left(\frac{1}{2} - \sqrt{\sum_{k=1}^{3}(p_k - 1/2)^2}\right). \end{aligned}$$

Another entropic characteristics of the qudit state is connected with areas of Malevich's squares. The sum Σ of the areas of all the Malevich's squares determining the qudit state provides the probability characteristic of the state

$$\mathcal{P}_{jk}^{(s)} = \frac{\Sigma_{jk}^{(s)}}{\Sigma},$$

where $\sum_{jk}^{(s)}$ is the area of one square. The Shannon entropy

$$H_\Sigma = -\sum_{jk}\sum_{(s)}\mathcal{P}_{jk}^{(s)}\ln\mathcal{P}_{jk}^{(s)}$$

gives the characteristics associated with the properties of the qudit state.

7. Conclusion

To resume we point out the main results of our work. We show that an arbitrary state of the qudit (spin-j state, N-level atom state) can be described by N^2-1 probability distributions of artificial classical variables. These variables can be identified with variables associated with classical coins. It means that arbitrary $N \times N$ - density matrix of N-level system contains offdiagonal matrix elements of the form $p_1^{(jk)} - \frac{1}{2} - i\left(p_2^{(jk)} - \frac{1}{2}\right)$, $j < k$, where the nonnegative numbers $p_1^{(jk)}$, $1 - p_1^{(jk)}$ and $p_2^{(jk)}$, $1 - p_2^{(jk)}$ are probability distributions describing the classical coin position "up" and "down". These probabilities correspond to $j\,k$-th spin-projection $+1/2$ on direction x and direction y, respectively. The diagonal matrix elements ρ_{jj}, where $1 < j \leq N$ have the form $1-p_3^{(jj)}$. Thus the density matrix is mapped (see Eq.(33)) onto N^2-1 probabilities which can be interpreted as probabilities from distributions associated with random positions of N^2-1 classical coins. It is the main result of our work. The density matrix can be demonstrated on the plane pictorially as the set of triadas of Malevich's squares which have been introduced in the framework of quantum suprematism picture of the qubit and qutrit states in [35]- [37]. New entropy–information inequalities are obtained for matrix elements of the density matrix using the relation of the matrix elements with probability distributions. Another result of our work is considering the parametric oscillator in probability representation. We obtain the new entropic relations for Frank–Condon factors. The vibronic structure of the electronic lines in polyatomic molecules is measured experimentally. Since this structure is determined by the Franck-Condon factors the correlations in the factors correspond to the correlations in the line intensities. Consequently the values of Shannon information given, e.g. by Eq.(18) reflect the correlations in vibronic structure of electronic lines in polyatomic spectra. The deeper clarification of the relation of the dependence of Shannon information value to the correlations in the polyatomic molecule spectra will be considered in future publication. It is worthy to note that the results known in molecular spectroscopy can be related to properties of systems used in quantum optics and information (see, e.g. [44, 46]). These relations (inequalities) can be checked in experiments with superconducting circuits. In [36] it was found for qubit state that the quantum observables (arbitrary Hermitian matrices) are mapped onto sets of classical random variables. This map can be constructed for arbitrary qudit observables.

Thus the quantum statistics formalism for arbitrary system can be mapped onto classical statistics formalism of the sets of classical coins and corresponding random variables. This problem will be considered in future publication.

Acknowledgments
V. I. M. thanks Professor D. Schuch and Professor M. Ramek for invitation to participate in Symmetries in Science Symposium, Bregenz, 2017, and for the hospitality. We thank P. I. Lysikhin, artist for help with mosaic presentation of Malevich's squares.

References

[1] Dirac P 1930 *The Principles of Quantum Mechanics*, Oxford University Press
[2] Schrödinger E 1926 *Ann. Phys.* **79**, 361; **81**, 109
[3] Landau L D 1927 *Z. Phys.* **45** 430

[4] von Neumann J 1927 *Göttingenische Nachrichten* **11** 245
[5] Wigner E 1932 *Phys. Rev.* **40** 749
[6] Glauber R J 1963 *Phys. Rev. Lett.* **10** 84
[7] Sudarshan E C C 1963 *Phys. Rev. Lett* **10** 277
[8] Cahill K E and Glauber R J 1969 *Phys. Rev.* **177** 1882
[9] Husimi K 1940 *Proc. Phys. Math. Soc. Jpn* **23** 264
[10] Kano Y 1965 *J. Math. Phys.* **6** 1913
[11] Stratonovich R L 1957 *J. Exp. Theor. Phys.* **5** 1206
[12] Mancini S Man'ko V I and Tombesi P 1996 *Phys. Lett. A* **213** 1
[13] Mancini S Man'ko V I and Tombesi P 1997 *Found. Phys.* **27** 801
[14] Man'ko O V and Man'ko V I 1997 *J. Russ. Las. Res.* **18** 407
[15] Smithey D T, Beck M, Raymer M G and Faridani A 1993 *Phys. Rev. Lett.* **70** 1244
[16] Bertrand J and Bertrand P 1989 *Found. Phys.* **17** 397
[17] Vogel K and Risken H 1989 *Phys. Rev. A* **40**, 2847
[18] Man'ko M A and Man'ko V I 2012 "Tomographic entropic inequalities in the probability representation of quantum mechanics," in: R. Bijker, (Ed.), *Beauty in Physics: Theory and Experiment: in Honor of Francesco Iachello on the Occasion of His 70th Birthday, Hacienda Cocoyoc, Mexico, 14-18 May 2012*, AIP Conference Proceedings, Vol. 1488, pp. 110–121
[19] Radon J 1917 *Berichte Srächsische Akademie der Wissenschaften* **29**, 262
[20] Dodonov V V and Man'ko V I 1997 *Phys. Lett. A* **229** 335
[21] Man'ko V I and Man'ko O V 1997 *J. Exp. Theor. Phys.* **85** 430
[22] Man'ko O V 1998 "Tomography of spin states and classical formulation of quantum mechanics," in: B. Gruber and M. Ramek (Eds.), *Proceedings of International Conference "Symmetries in Science X" (Bregenz, Austria, 1997)*, Plenum Press, New York , p. 207.
[23] Weigert S 2000 *Phys. Rev. Lett.* **84** 802
[24] Amiet J P and Weigert S 1999 *J. Opt. B: Quantum Semiclass. Opt.* **1** L5
[25] D'Ariano G M Maccone L and Paini M 2003 *J. Opt. B: Quantum Semiclass. Opt.* **5** 77
[26] Ibort A Man'ko V I Marmo G Simoni A and Ventriglia F 2009 *Phys. Scr.* **79** 065013
[27] Asorey M Ibort A Marmo G and Ventriglia F 2015 *Phys. Scr.* **90** 074031
[28] Man'ko M A Man'ko V I and Mendes R V 2006 *J. Russ. Las. Res.* **27** 507
[29] Man'ko M A Man'ko V I 2018 *J. Russ. Las. Res.* **39** 1
[30] Doktorov E V Malkin I A and Man'ko V I 1975 *J. Mol. Spectroscop.* **56** 1
[31] Doktorov E V Malkin I A and Man'ko V I 1977 *J. Mol. Spectroscop.* **64** 302
[32] Huh J and Berger R 2012 *J. Phys.: Conf. Ser.* **380** 1
[33] Zhebrak E D 2016 *J. Rus. Las. Res.* **37** 123
[34] Man'ko V I Marmo G Ventriglia F and Vitale P 2017 *J. Phys. A: Math. and Theor.* **50** 335302
[35] Chernega V N Man'ko O V and Man'ko V I 2017 *J. Rus. Las. Res.* **38** 1
[36] Chernega V N Man'ko O V and Man'ko V I 2017 *J. Rus. Las. Res.* **38** 324
[37] Chernega V N Man'ko O V and Man'ko V I 2017 *J. Russ. Las. Res.* **38** 141
[38] Aleksandra Shatskikh, *Black Square: Malevich and the Origin of Suprematism*, Yale University Press, New Haven (2012).
[39] Mancini S Man'ko V I and Tombesi P. 1995 *J. Opt. B: Quantum Semiclass. Opt.* **7** 615
[40] Man'ko M A Man'ko V I and Mendes R V 2001 *J. Phys. A: Math. and Gen.* **34** 8321
[41] Malkin I A Man'ko V I and Trifonov D A 1969 *Phys. Let. A* **30** 414
[42] Ermakov V P 1880 *Universitetskie Izvestiya*, Kiev, **No. 9** 1 [in Russian]
[43] Dodonov V V and Man'ko V I 1987 in: M. A. Markov (Ed.), *"Invariants and the Evolution of Nonstationary Quantum Systems", Proceedings of Lebedev Physics Institute*, Vol. 183 Nauka, Moscow, p. 182
[44] Shen Y Y Huh J J Lu Y Y Zhang J J Zhang K K Zhang S S and Kim K K 2018 *Chem. Sci.* **9** 836
[45] Huh J Guerreschi G G Peropadre B McClean J R and Aspuru-Guzik A 2015 *Nature Photon.* **9** 615
[46] Huh J and Yung M H 2017 *Sci. Rep* **7** 7462
[47] Shannon C E 1948 *Bell. Syst. Tech. J.* **27** 379; 623
[48] Man'ko V I and Seilov Z 2017 *J. Russ. Las. Res.* **38** 50
[49] Man'ko V I and Seilov Z 2016 *J. Russ. Las. Res.* **37** 591
[50] Man'ko O V 2012 "Quantum tomography of current and voltage states in nanoelectri circuits," in: M. D.'Ariano, S.-M. Feri, E. Haven, B. Hiesmayer, G. Jaeger, A. Khrennikov and J.-A. Larsson, (Ed.), *Foundation of Probabilility and Physics-6, Vaxjo, Sweden, 14-16 June 2011*, AIP Conference Proceedings, Vol. 1424, pp. 221–233
[51] Man'ko O V 2013 *Phys. Scr.* **T153** 014046
[52] Fedorov A K Kiktenko E O Man'ko O V and Man'ko V I 2015 *Phys. Rev. A* **91** 042312

[53] Lewis H R and Riesenfeld W B 1969 *J. Math. Phys.* **10** 1458
[54] Cruz H Schuch D Castanos O and Rosas-Ortiz O 2015 *Annals of Physics* **360** 44
[55] Schuch D 2014 *J. Phys.: Conf. Ser.* **538** 012019
[56] Schuch D Guerrero J Lopez-Ruiz F F and Aldaya V 2015 *Phys. Scr.* **90** 045209
[57] Rosas-Ortiz O Castanos O and Schuch D 2015 *J. Phys. A: Math. Theor.* **48** 445302

IOP Conf. Series: Journal of Physics: Conf. Series **1071** (2018) 012009 doi:10.1088/1742-6596/1071/1/012009

Quantum Computing for Molecular Vibronic Spectra and Gaussian Boson Sampling

Seungbeom Chin and Joonsuk Huh
Department of Chemistry, Sungkyunkwan University, Suwon 16419, Korea
E-mail: joonsukhuh@skku.edu

Abstract. Boson sampling (BS) is a multimode linear optical problem that is expected to be intractable on classical computers. It was recently suggested that molecular vibronic spectroscopy (MVS) is computationally as complex as BS. In this review, we discuss the correspondence relation between BS and MVS and briefly introduce the experimental demonstrations of the molecular spectroscopic process using quantum devices. The similarity of the two theories results in another BS setup, which is called "vibronic BS". The hierarchical structure of vibronic BS, which includes the original BS and other Gaussian BS, is also explained.

1 Introduction

A multimode quantum optical system called "Boson Sampling (BS)" [1] is considered to demonstrate quantum speedup with fewer implements than the universal quantum computer. The original BS is a strong candidate for a counterexample to the extended Church-Turing (ECT) thesis, which states that everything efficiently computable with actual devices are efficiently solvable by a Turing machine. In the BS system, the input state is a set of non-interacting identical single photons, which are sent through a linear optical network and result in another photon distribution. This process is not classically accessible at a sufficiently large scale.

Instead of non-interacting single photon sources, Gaussian states recently have been suggested as input states for BS [2, 3]. An interesting application of Gaussian BS is that it can simulate molecular spectroscopy [4]. In the process of molecular vibronic spectroscopy (MVS), we consider the vibrational transitions between two electronic states of a molecule. A molecule undergoes a structural deformation through vibronic transition. An experiment was designed in [4] to simulate MVS with BS, since the Frank-Condon profiles (FCPs, defined as the probability distribution regarding a given transition) correspond to the transition amplitudes of the Fock state BS. Actual experiments were performed recently with trapped ions [5] and photonics [6].

On the other hand, this intimate relation between BS and MVS yielded another version of BS, which is named "vibronic BS" [7]. With the most general Gaussian states as an input and twice as many modes as the original setup, one can find a hierarchical structure that embraces every instance of the known Gaussian BS.

In this review, we discuss the correspondence between BS and MVS, and briefly introduce the experimental demonstrations of the molecular spectroscopic process using the correspondence. Then the concept of vibronic BS and its hierarchical structure [7] is explained.

2 Boson sampling and vibronic transition

We first summarize the theoretical background for the whole discussion, focusing on the definition of Gaussian BS and molecular vibronic transition.

2.1 *Boson Sampling and the Gaussian version*

The general form of $P_{\hat{\Pi}}$, the probability of a post-selected measurement $\hat{\Pi}$ for sampling Bosons, is given by

$$P_{\hat{\Pi}} = \mathrm{tr}\left[\hat{O}\hat{\rho}_{\mathrm{in}}\hat{O}^{\dagger}\hat{\Pi}\right], \tag{1}$$

where $\hat{O}$ is a transition operator that represents the given quantum optical network, and $\hat{\rho}_{\mathrm{in}}$ is an input state. There exist various boson sampling setups that are determined by $(\hat{\rho}_{\mathrm{in}}, \hat{O}, \hat{\Pi})$, e.g.,

- The original boson sampling [1]: $\hat{\Pi} = |\mathbf{m}\rangle\langle\mathbf{m}|$, $\hat{\rho}_{\mathrm{in}} = |\mathbf{n}\rangle\langle\mathbf{n}|$ (here $\mathbf{m}$ and $\mathbf{n}$ are photon distribution vectors of N photons in M ($\gg N$) modes, and the components of $(\mathbf{m}, \mathbf{n})$ are not larger than 1, therefore $\hat{\prod}$ is a photon number projector), and $\hat{O} = \hat{R}_U$, a beam splitting operator with a unitary matrix U. The exact definition of $\hat{R}_U$ is given in Appendix Appendix A, which shows that $\hat{O}$ does not change the total photon number N.
- Gaussian boson sampling [3]: $\hat{\Pi} = |\mathbf{m}\rangle\langle\mathbf{m}|$, $\hat{\rho}_{\mathrm{in}}$ are products of Gaussian modes (*e.g.*, squeezed vacuum states or thermal states), and $\hat{O} = \hat{R}_U$.

The most general expression for a Gaussian state $\hat{\rho}_{\mathrm{G}}$ is obtained from a thermal state $\hat{\rho}_{\mathrm{th}}$ and a Gaussian operator $\hat{O}_{\mathrm{G}}$ [8, 9], i.e. $\hat{\rho}_{\mathrm{G}} = \hat{O}_{\mathrm{G}}\hat{\rho}_{\mathrm{th}}\hat{O}_{\mathrm{G}}^{\dagger}$. The action of $\hat{O}_{\mathrm{G}}$ on the Boson creation operator, $\hat{\mathbf{a}}'^{\dagger} = \hat{O}_{\mathrm{G}}^{\dagger}\hat{\mathbf{a}}^{\dagger}\hat{O}_{\mathrm{G}}$ is generally expressed as

$$\hat{\mathbf{a}}'^{\dagger} = \hat{O}_{\mathrm{G}}^{\dagger}\hat{\mathbf{a}}^{\dagger}\hat{O}_{\mathrm{G}} = X\hat{\mathbf{a}} + Y\hat{\mathbf{a}}^{\dagger} + \mathbf{z}, \tag{2}$$

where X and Y ($M \times M$ matrices) satisfy $XX^{\dagger} - YY^{\dagger} = I$ and $XY^{\mathrm{t}} = YX^{\mathrm{t}}$, and $\hat{\mathbf{a}}'^{\dagger}$ and $\hat{\mathbf{a}}^{\dagger}$ are the M-dimensional boson creation operator column vector of output and input states, respectively (see Appendix Appendix A). Since $\hat{O}_{\mathrm{G}}$ can be decomposed into quantum optical operators as

$$\hat{O}_{\mathrm{G}} = \hat{D}_{\mathbf{z}^*}\hat{R}_{U_{\mathrm{L}}}\hat{S}_{\Sigma}\hat{R}_{U_{\mathrm{R}}^{\dagger}}, \tag{3}$$

($\hat{D}_{\mathbf{z}^*}$ is the displacement operator, $\hat{R}_U$ is the rotation operator, and $\hat{S}_{\Sigma}$ is the squeezing operator. For the detailed definitions and relations for these operators, see Appendix Appendix A), X and Y in Eq. 2 are identified by singular value decomposition with two unitary matrices U_{L} and U_{R} and a real diagonal matrix Σ as follows: $X = U_{\mathrm{L}}\sinh(\Sigma)U_{\mathrm{R}}^{\mathrm{t}}$ and $Y = U_{\mathrm{L}}\cosh(\Sigma)U_{\mathrm{R}}^{\dagger}$.

2.2 *Vibronic transition and Franck-Condon Profile*

It was first pointed out in [4] that a system of N photons in M modes is isomorphic to that of N phonons in M vibrational modes. To understand the relation clearly, we review the concept of vibrational transition of molecules and Franck-Condon Profile (FCP).

The total molecular wavefunction of the nuclei at the position vector $\hat{\mathbf{R}} = (\hat{R}_1, \hat{R}_2, ...)$ and electrons at $\hat{\mathbf{r}} = (\hat{r}_1, \hat{r}_2, ...)$ is simplified as $\psi(\hat{\mathbf{r}}, \hat{\mathbf{R}})\phi(\hat{\mathbf{R}})$ under the Born-Oppenheimer approximation (note that the electronic wavefunction $\psi(\hat{\mathbf{r}}, \hat{\mathbf{R}})$ depends on the nuclear coordinates $\hat{\mathbf{R}}$). Therefore, for transitions between two electronic levels $|g\rangle$ (the ground state) and $|e\rangle$ (the excited state), the molecular Hamiltonian $\mathcal{H}_{\mathrm{mol}}$ is approximately given by

$$\begin{aligned}\mathcal{H}_{\mathrm{mol}} &= |g\rangle\langle g| \otimes \left(\textstyle\sum_k \hbar\omega_k \hat{a}_k^{\dagger}\hat{a}_k\right) + |e\rangle\langle e| \otimes \left(\textstyle\sum_k \hbar\omega_k' \hat{a}_k'^{\dagger}\hat{a}_k' + \hbar\omega_{\mathrm{ad}}\right) \\ &\equiv |g\rangle\langle g| \otimes H_g + |e\rangle\langle e| \otimes H_e,\end{aligned} \tag{4}$$

where H_g is the nuclear Hamiltonian for the ground state $|g\rangle$ and H_e for the excited state $|e\rangle$ (ω_k and ω'_k are the angular frequencies for k-th mode of the initial and final electronic states, respectively.). The electronic adiabatic transition frequency is denoted as ω_{ad}.

The normal coordinate of initial ($\mathbf{Q}$) and final ($\mathbf{Q}'$) states have the Duschinsky relation [10], $\mathbf{Q}' = \mathbf{U}\mathbf{Q} + \mathbf{d}$. $\mathbf{U}$ and $\mathbf{d}$ are the Duschinsky rotation matrix and displacement vector, respectively. The corresponding Bogoliubov transformation for the vibronic transition is given by the Doktorov operator $\hat{U}_{\mathrm{D}}$ [11], which is expressed as

$$\hat{\mathbf{a}}'^{\dagger} = \hat{U}_{\mathrm{D}}^{\dagger}\hat{\mathbf{a}}^{\dagger}\hat{U}_{\mathrm{D}} = \frac{1}{2}\left(J - (J^{\mathrm{t}})^{-1}\right)\hat{\mathbf{a}} + \frac{1}{2}\left(J + (J^{\mathrm{t}})^{-1}\right)\hat{\mathbf{a}}^{\dagger} + \frac{1}{\sqrt{2}}\delta, \tag{5}$$

where J and δ are defined as

$$J_{ij} = \Omega'_i U_{ij} \Omega_j, \qquad \delta_i = \hbar^{-\frac{1}{2}}\Omega'_i d_i, \tag{6}$$

where $\mathbf{\Omega}'$ and $\mathbf{\Omega}$ are the vectors for square roots of angular frequencies of the final and initial states, i.e., $\mathbf{\Omega}' = (\sqrt{\omega'_1}, \ldots, \sqrt{\omega'_N})$ and $\mathbf{\Omega} = (\sqrt{\omega_1}^{-1}, \ldots, \sqrt{\omega_N}^{-1})$. It is straightforward to see that Eq. (5) is a special case of Eq. (2) with

$$\hat{O}_{\mathrm{G}} = \hat{U}_{\mathrm{D}}, \quad X = \frac{1}{2}\left(J - (J^{\mathrm{t}})^{-1}\right), \quad Y = \frac{1}{2}\left(J + (J^{\mathrm{t}})^{-1}\right), \quad \mathbf{z} = \frac{1}{\sqrt{2}}\delta. \tag{7}$$

The Doktorov operator $\hat{U}_{\mathrm{D}}$ is expressed with optical operators as [11]

$$\hat{U}_{\mathrm{D}} = \hat{D}_{\delta/\sqrt{2}}\hat{S}^{\dagger}_{\Omega'}\hat{R}_U\hat{S}_{\Omega}, \tag{8}$$

where $\hat{D}$, $\hat{R}$, and $\hat{S}$ are the displacement, rotation, squeezing operators (see Appendix Appendix A). FCP is an essential quantity that is proportional to the molecular absorption lineshape,

$$\mathrm{FCP}\left(\omega_{\mathrm{V}}\right) = \sum_{\mathbf{m}} |\langle \mathbf{m}|\hat{U}_D|\mathbf{n}\rangle|^2 P_{\mathrm{in}}(\mathbf{n})\delta_{\omega_v}(\Delta). \tag{9}$$

Here $|\langle \mathbf{m}|\hat{U}_{\mathrm{D}}|\mathbf{n}\rangle|^2$ is the Franck-Condon factor, $\delta_{\omega_{\mathrm{v}}}(\Delta)$ with $\Delta = \omega_{\mathrm{v}} - \mathbf{m}\cdot\omega' + \mathbf{n}\cdot\omega$ imposes the energy conservation, and $P_{\mathrm{in}}(\mathbf{n})$ denotes the initial distribution of the phonon mode.

The relation of operators given by Eq. (8) renders a quantum optical network with multi-modes to simulate the FCP efficiently, as will be explained in more detail in the next section.

3 Simulating MVS with BS

Since the vibronic transition is reexpressed as a special case of the Gaussian state operation as can be seen in Eq. (8), we can set up a BS system so that it can simulate MVS. By taking the optical operators Eq. (8) subsequently to the vacuum state and measuring the final state in the Fock basis as in Eq. (9), one can simulate the FCP with quantum devices. The brief summary of the simulation protocol, with trapped-ion as a concrete example [5], is as follows: (i) initialize the ion in the ground state, (ii) apply the Doktorove operation (Eq. (8)) to the ion, (iii) detect the vibronic spectrum of final state with the projection measurements.

In this section we introduce such BS simulations with various devices. Even though the standard BS requires less optical operations than the Gaussian BS, the difficulties in preparing the initial states render the former more challenging in practical optical systems. Indeed, it is hard to create single photon states (the original boson sampling) and squeezed coherent states (the molecular simulation) in realistic laboratories. The candidates suggested for scalable BS are trapped ions [12, 13] and superconducting circuits [14]. The photoelectron spectra of SO_2 and SO_2^- were demonstrated with a trapped ion device by Yangchao et al. [5]. Recently, Clements et al. [6] attempted to simulate the absorption spectrum of tropolone ($C_7H_6O_2$) with a photonic setup.

3.1 Simulation with trapped ions
The first experimental demonstration of the MVS of SO_2 with trapped-ions was reported in [5]. For a reliable Gaussian BS, the experimental technology for phonons to be operated with displacement, squeezing, and rotation is developed. The operators that can simulate the optical systems are installed through Raman laser beams, resulting in a Bogoliubov transformation. As a result, the photoelectron spectroscopies of SO_2 and SO_2^- is reproduced with the trapped-ion simulator.

3.2 Simulation with optics
It was demonstrated in [6] that an approximated pattern of the molecular vibronic spectra can be simulated with imperfect quantum simulators. The partial vibronic spectrum of tropolone is experimentally simulated to prove the approximation claim. The result shows that there exists the efficient error bound on the estimations of the spectra and such simulations can surpass a classicality criterion.

3.3 Simulation with superconducting circuits
Spectroscopy is an important technique for interpreting quantum systems with electromagnetic radiation. However, technical limitations restrict the practical application of molecular spectroscopy. It was demonstrated in [15] that a superconducting quantum simulator can perform a simple one-dimensional computation.

4 Vibronic boson sampling
In this section, we introduce a hierarchical structure [7] that encompasses all the known types of BS, including scattershot BS [2], and Gaussian BS [3]. We also show that it is possible to incorporate the initial thermal Gaussian correlation with the Gaussian BS with no correlation. It is achieved by doubling the number of modes of Scattershot BS and the size of the photon network.

We first present the definition of vibronic BS, which has an initial state that has the purification form of mixed thermal states with ancillar modes i.e., $|\mathbf{0}(\beta)\rangle = \sum_{\mathbf{n}=\mathbf{0}}^{\infty} \sqrt{\langle\mathbf{n}|\,\hat{\rho}_{\text{th}}\,|\mathbf{n}\rangle}\,|\mathbf{n}\rangle \otimes |\mathbf{n}\rangle_{\text{B}}$, where $\hat{\rho}_{\text{th}}$ is the thermal state and $|\mathbf{n}\rangle_{\text{B}}$ is the Fock state of the ancillary system with M modes. Note that $\text{tr}_{\text{B}}[|\mathbf{0}(\beta)\rangle\langle\mathbf{0}(\beta)|] = \hat{\rho}_{\text{th}}$ (tr_{B} is the partial trace over the ancillary system). The pure state $|\mathbf{0}(\beta)\rangle$ can be rewritten as

$$|\mathbf{0}(\beta)\rangle = \hat{V}(\beta)\,|\mathbf{0}\rangle \otimes |\mathbf{0}\rangle_{\text{B}}\,, \tag{10}$$

where the operator $\hat{V}(\beta)$ is a tensor product of two-mode squeezing operators, i.e., $\hat{V}(\beta) = \bigotimes_{k=1}^{M} \exp(\theta_k(\hat{a}_k^\dagger \hat{b}_k^\dagger - \hat{a}_k \hat{b}_k)/2)$, where $\tanh(\theta_k/2) = \mathrm{e}^{-\beta_k \hbar\omega_k/2} = \sqrt{\bar{n}_k/(\bar{n}_k+1)}$, $\hat{b}_k$ ($\hat{b}_k^\dagger$) is the annihilation (creation) operator for the ancillary system, and $\bar{n}_k = 1/(\mathrm{e}^{\beta_k\hbar\omega_k} - 1)$ is the average number of photons in the k-th mode. The linear optical operation $\hat{O}_{\text{G}}$ acts on both the original and ancillary Hilbert space. As a result, the BS with initial sampling of the Fock state is translated to a problem only with the post-selection measurement, i.e.,

$$P(\mathbf{m},\mathbf{n}) = \text{tr}\Big[\hat{O}_{\text{G}}\;\;|\mathbf{0}(\beta)\rangle\langle\mathbf{0}(\beta)|\,\hat{O}_{\text{G}}^\dagger\,|\mathbf{m}\rangle\langle\mathbf{m}| \otimes |\mathbf{n}\rangle_{\text{B}}\,{}_{\text{B}}\langle\mathbf{n}|\Big], \tag{11}$$

where the trace is on the overall Hilbert space.

When the Gaussian operator is the rotation operator, i.e. $\hat{O}_{\text{G}} = \hat{R}_U$, Eq. (11) is reduced to the scattershot BS [2]. Therefore, we can state that the scattershot BS is a special case of vibronic BS.

In the following, we show that a Bogoliubov transformation can eliminate the mode correlation from the two-mode squeezing operator $\hat{V}$ [7]. With the definition

$$\hat{\boldsymbol{U}}(\boldsymbol{\beta}) \equiv \hat{\boldsymbol{O}}_{\mathrm{G}}\ \hat{\boldsymbol{V}}(\boldsymbol{\beta}), \tag{12}$$

the Bogoliubov transformation by $\hat{U}(\beta)$ on the collective creation operator vector $\mathbf{c}^{\dagger}$ ($\equiv (\mathbf{a}^{\dagger}, \mathbf{b}^{\dagger})^{\mathrm{t}}$) of the extended space is given by

$$\hat{\mathbf{c}}'^{\dagger} = \hat{\boldsymbol{U}}(\boldsymbol{\beta})^{\dagger}\hat{\mathbf{c}}^{\dagger}\hat{\boldsymbol{U}}(\boldsymbol{\beta}) = \boldsymbol{\mathcal{X}}\hat{\mathbf{c}} + \boldsymbol{\mathcal{Y}}\hat{\mathbf{c}}^{\dagger} + \boldsymbol{\gamma}, \tag{13}$$

which is an extension of Eq. (2). The parameters $\mathcal{X}$, $\mathcal{Y}$, and γ are written as

$$\mathcal{X} = \begin{pmatrix} XF & YG \\ G & \mathrm{diag}(\mathbf{0}) \end{pmatrix}, \quad \mathcal{Y} = \begin{pmatrix} YF & XG \\ \mathrm{diag}(\mathbf{0}) & F \end{pmatrix}, \quad \gamma = \begin{pmatrix} \mathbf{z} \\ \mathbf{0} \end{pmatrix} \tag{14}$$

where $F \equiv \mathrm{diag}(\sqrt{\bar{n}_1+1}, \ldots, \sqrt{\bar{n}_M+1})$ and $G \equiv \mathrm{diag}(\sqrt{\bar{n}_1}, \ldots, \sqrt{\bar{n}_M})$.

Using Eq. (13), one can convert vibronic BS into Gaussian BS with squeezed coherence states ($\gamma \neq \mathbf{0}$) or squeezed vacuum states ($\gamma = \mathbf{0}$) as inputs. It is achieved by the singular valued decompostion of the matrices $\mathcal{X} = \mathcal{C}_{\mathrm{L}} \sinh(\mathcal{S})\mathcal{C}_{\mathrm{R}}^{\mathrm{t}}$ and $\mathcal{Y} = \mathcal{C}_{\mathrm{L}} \cosh(\mathcal{S})\mathcal{C}_{\mathrm{R}}^{\dagger}$, where $\mathcal{C}_{\mathrm{L}}$ and $\mathcal{C}_{\mathrm{R}}$ are the unitary matrices and $\mathcal{S} = \mathrm{diag}(s_1, \ldots, s_{2M})$ is a real diagonal matrix that includes the squeezing parameters.

The extended unitary operator is decomposed as

$$\hat{U}(\beta) = \hat{D}_{\gamma^*}\hat{R}_{\mathcal{C}_{\mathrm{L}}}\hat{S}_{\mathcal{S}}\hat{R}_{\mathcal{C}_{\mathrm{R}}^{\dagger}} = \hat{R}_{\mathcal{C}_{\mathrm{L}}}\hat{S}_{\mathcal{S}}\hat{R}_{\mathcal{C}_{\mathrm{R}}^{\dagger}}\hat{D}_{\gamma'}, \tag{15}$$

where the displacement parameter vector $\gamma' = \gamma_{\mathrm{R}}' + \mathrm{i}\gamma_{\mathrm{I}}'$ is given by,

$$\begin{pmatrix} \gamma_{\mathrm{R}}' \\ \gamma_{\mathrm{I}}' \end{pmatrix} = \begin{pmatrix} \mathcal{X}_{\mathrm{R}} + \mathcal{Y}_{\mathrm{R}} & -\mathcal{X}_{\mathrm{I}} + \mathcal{Y}_{\mathrm{I}} \\ \mathcal{X}_{\mathrm{I}} + \mathcal{Y}_{\mathrm{I}} & \mathcal{X}_{\mathrm{R}} - \mathcal{Y}_{\mathrm{R}} \end{pmatrix}^{-1} \begin{pmatrix} \gamma_{\mathrm{R}} \\ \gamma_{\mathrm{I}} \end{pmatrix}, \tag{16}$$

with $\gamma = \gamma_{\mathrm{R}} + \mathrm{i}\gamma_{\mathrm{I}}$, $\mathcal{X} = \mathcal{X}_{\mathrm{R}} + \mathrm{i}\mathcal{X}_{\mathrm{I}}$ and $\mathcal{Y} = \mathcal{Y}_{\mathrm{R}} + \mathrm{i}\mathcal{Y}_{\mathrm{I}}$. One can implement the operator decomposed as the second equality of Eq. (15) by preparing the $2M$-mode squeezed coherent states

$$\hat{\boldsymbol{S}}_{\boldsymbol{\mathcal{S}}}\hat{\boldsymbol{R}}_{\boldsymbol{\mathcal{C}}_{\mathrm{R}}^{\dagger}}\hat{\boldsymbol{D}}_{\boldsymbol{\gamma}'}\,|\mathbf{0}\rangle \equiv \hat{\boldsymbol{S}}_{\boldsymbol{\mathcal{S}}}|\boldsymbol{\gamma}''\rangle = \bigotimes_{k=1}^{2M} \hat{\boldsymbol{S}}_{s_k}|\gamma_k''\rangle\ , \tag{17}$$

which are injected to the quantum optical device, and then taking the linear optical operation ($\hat{R}_{\mathcal{C}_{\mathrm{L}}}$). The vector γ'' is obtained by a rotation of γ', i.e. $\gamma'' = \mathcal{C}_{\mathrm{R}}^{\mathrm{t}}\gamma'$. Hence, we see that the $2M$-mode squeezed coherent states replace the correlated squeezed thermal state $\hat{S}\hat{R}\rho_{\mathrm{th}}\hat{R}^{\dagger}\hat{S}^{\dagger}$.

The hierarchical structure of Gaussican BS is described as follows:

Gaussian BS (M modes, T=0) = vibronic BS (M modes, T=0)
$\subseteq$ Gaussian BS (M modes, T$\neq$0) $\subseteq$ vibronic BS (M modes, T$\neq$0)
$\subseteq$ Gaussian BS (2M modes, T=0) $\subseteq$ vibronic BS (M modes, T=0)

The hierarchy of Gaussian BS is arrayed along the number of modes #(mode) and temperature T. When #(mode) and T are the same, vibronic BS is always a more general case of Gaussian BS. The only exception is when T=0, for which the two types of BS are equivalent [4]. The Gaussian BS with #(mode)=M and T$\neq$ 0 is the case analyzed in [3]. In addtion, the hierarchical relation "original BS $\subseteq$ scattershot BS $\subseteq$ vibronic BS" holds in the extended Hilbert space, which is confirmed from the fact that the scattershot BS with post-selected measurement becomes the original BS [2].

5 Conclusions

In this review, we explained how the isomorphism between BS and MVS presents a distinctive viewpoint on some aspects of quantum computation problems. The mathematical relation of Gaussian and vibronic BS is exploited in experiments to simulate MVS with non-optical BS systems, and to construct the hierarchical structure of vibronic BS that includes other known BS setups. We expect that the complexity analysis of vibronic BS would provide a new insight to understand the extended Church-Turing thesis with BS. Finally, we comment on the possible extension of vibronic BS to go beyond the harmonic picture of the oscillators and the Condon approximation for the transition dipole moment. For the former, we would be able to incorporate the simple anharmonicity by a Morse oscillator [16, 17]. For the latter, one can approximate the non-Condon operator [18] as a unitary operator so that the current vibronic boson sampling setup can be simply used. Currently, we are working in these directions for the quantum simulation of vibronic problems.

Acknowledgements

This work was supported by Basic Science Research Program through the National Research Foundation of Korea (NRF) funded by the Ministry of Education, Science and Technology (NRF-2015R1A6A3A04059773 and 2017R1A4A1015770).

Appendix A Quantum optical operators

We denote the tranformation of boson operator vector $\hat{\mathbf{x}}$ as $\hat{A}\ \hat{\mathbf{x}}\ \hat{B} \equiv (\hat{A}\hat{x}_1\hat{B}, \ldots, \hat{A}\hat{x}_M\hat{B})^{\mathrm{t}}$. The optical unitary operators are defined as

$\hat{A}\ \hat{\mathbf{x}}\ \hat{B} \equiv (\hat{A}\hat{x}_1\hat{B}, \ldots, \hat{A}\hat{x}_M\hat{B})^{\mathrm{t}}$. The optical unitary operators are defined as

$$\begin{aligned} \hat{R}_U &= \exp((\hat{\mathbf{a}}^\dagger)^{\mathrm{t}} \ln U^* \hat{\mathbf{a}}) \qquad \text{(rotation)}, \\ \hat{D}_\alpha &= \exp(\alpha^{\mathrm{t}}\hat{\mathbf{a}}^\dagger - \alpha^\dagger \hat{\mathbf{a}}) \qquad \text{(displacement)}, \\ \hat{S}_\Sigma &= \exp(((\hat{\mathbf{a}}^\dagger)^{\mathrm{t}}\Sigma\hat{\mathbf{a}}^\dagger - \hat{\mathbf{a}}^{\mathrm{t}}\Sigma\hat{\mathbf{a}})/2) \qquad \text{(squeezing)}, \end{aligned} \tag{A.1}$$

where α is a coherent state phase vector in $d = M$, Σ is a diagonal matrix, and U is an $M \times M$ unitary matrix. The optical operators act on $\hat{\mathbf{a}}^\dagger$ as

$$\begin{aligned} \hat{R}_U^\dagger \hat{\mathbf{a}}^\dagger \hat{R}_U &= U\hat{\mathbf{a}}^\dagger, \\ \hat{D}_\alpha^\dagger \hat{\mathbf{a}}^\dagger \hat{D}_\alpha &= \hat{\mathbf{a}}^\dagger + \alpha^*, \\ \hat{S}_\Sigma^\dagger \hat{\mathbf{a}}^\dagger \hat{S}_\Sigma &= \sinh(\Sigma)\hat{\mathbf{a}} + \cosh(\Sigma)\hat{\mathbf{a}}^\dagger, \end{aligned} \tag{A.2}$$

and the squeezing operator $\hat{V}(\beta) = \bigotimes_{k=1}^{M} \exp(\theta_k(\hat{a}_k^\dagger \hat{b}_k^\dagger - \hat{a}_k\hat{b}_k)/2)$ acts as

$$\begin{aligned} \hat{V}(\beta)^\dagger \hat{a}_k \hat{V}(\beta) &= \cosh(\theta_k/2)\ \hat{a}_k + \sinh(\theta_k/2)\ \hat{b}_k^\dagger, \\ \hat{V}(\beta)^\dagger \hat{b}_k \hat{V}(\beta) &= \sinh(\theta_k/2)\ \hat{a}_k^\dagger + \cosh(\theta_k/2)\ \hat{b}_k. \end{aligned} \tag{A.3}$$

References

[1] Aaronson S and Arkhipov A 2011 *Proceedings of the forty-third annual ACM symposium on Theory of computing* (ACM) pp 333–342
[2] Lund A, Laing A, Rahimi-Keshari S, Rudolph T, OBrien J L and Ralph T 2014 *Physical review letters* **113** 100502
[3] Rahimi-Keshari S, Lund A P and Ralph T C 2015 *Physical review letters* **114** 060501
[4] Huh J, Guerreschi G G, Peropadre B, McClean J R and Aspuru-Guzik A 2015 *Nature Photonics* **9** 615–620
[5] Shen Y, Lu Y, Zhang K, Zhang J, Zhang S, Huh J and Kim K 2017 *Chemical Science* **9** 836–840
[6] Clements W R, Renema J J, Eckstein A, Valido A A, Lita A, Gerrits T, Nam S W, Kolthammer W S, Huh J and Walmsley I A 2017 *arXiv preprint arXiv:1710.08655*

IOP Conf. Series: Journal of Physics: Conf. Series **1071** (2018) 012009 doi:10.1088/1742-6596/1071/1/012009

[7] Huh J and Yung M H 2017 *Scientific Reports* **7** 7462
[8] Weedbrook C, Pirandola S, García-Patrón R, Cerf N J, Ralph T C, Shapiro J H and Lloyd S 2012 *Reviews of Modern Physics* **84** 621
[9] Adesso G, Ragy S and Lee A R 2014 *Open Systems & Information Dynamics* **21** 1440001
[10] Duschinsky F 1937 *Acta Physicochim. URSS* **7** 551–566
[11] Doktorov E, Malkin I and Man'ko V 1976 *Journal of Physics B: Atomic and Molecular Physics* **9** 507
[12] Lau H K and James D F 2012 *Physical Review A* **85** 062329
[13] Shen C, Zhang Z and Duan L M 2014 *Physical review letters* **112** 050504
[14] Peropadre B, Guerreschi G G, Huh J and Aspuru-Guzik A 2016 *Physical review letters* **117** 140505
[15] Hu L, Ma Y, Xu Y, Wang W, Ma Y, Liu K, Yung M H and Sun L 2017 *arXiv preprint arXiv:1703.03300*
[16] Iachello F and Ibrahim M 1998 *J. Phys. Chem. A* **102** 9427
[17] Santiago R D, Arias J M, Gómez-Camacho J and Lemus R 2017 *Mol. Phys.* **115** 3206–3223
[18] Huh J and Berger R 2012 *J. Phys. Conf. Ser.* **380** 012019

IOP Conf. Series: Journal of Physics: Conf. Series **1071** (2018) 012010 doi:10.1088/1742-6596/1071/1/012010

Contact Hamiltonian mechanics. An extension of symplectic Hamiltonian mechanics

Hans Cruz

Institut für Theoretische Physik, Goethe Universität

E-mail: prado@th.physik.uni-frankfurt.de

Abstract. This contribution gives a possible solution of the major question whether it is possible to construct a classical mechanical theory which not only contains all the advantages of the Hamiltonian formalism, but also takes into account the effects of the environment on the system.

1. Introduction

The Hamiltonian formulation of classical mechanics consists of an even-dimensional phase space with symplectic structure, the dynamics is determined by the Hamiltonian vector field associated with the differential of the Hamiltonian. In addition, we can mention that the Hamiltonian structure is directly connected with other areas of physics such as optics, statistical physics and quantum mechanics. Despite all these advantages, this formulation exclusively describes isolated systems with reversible dynamics, while real systems are constantly in interaction with an environment that introduces the phenomena of dissipation and irreversibility.

Recently, in order to include the dissipation phenomenon in a Hamiltonian theory, the contact Hamiltonian formalism has been proposed in [1]. This formalism consists of enlarging the symplectic phase space of classical mechanics by adding an additional dimension, the new phase space is an odd-dimensional manifold but now with a contact structure. In analogy to the standard theory, the dynamics is determined by the contact vector field induced by the differential of the so called contact Hamiltonian function. There are two major advantage of this formalism, the first is the fact that the symplectic phase space maybe embedded in it and the second is that the phenomenon of irreversible dissipation can be defined straightforward.

This work gives a concise and short survey of the concept of contact Hamiltonian dynamics and its application in classical dissipative system, enhancing the fact that this theory includes all the important properties of the symplectic one. Besides, along this discussion, we provide examples which enable us to show the usefulness of our formalism. Finally, we conclude this contribution introducing a possible quantization route of the contact formalism leading to a nonlinear Schödinger equation (NLSE) for the description of quantum dissipative systems, which provides a theoretical justification for the introduction of the nonlinear phenomenological Schrödinger equation described in the literature [2, 3, 4]. In particular, we show the solution of the NLSE for the parametric oscillator systems and its corresponding Wigner function.

2. Contact Hamiltonian mechanics

The phase space of a conservative system is the cotangent bundle of the configuration manifold, that is an $2n$-dimensional manifold whit local system of coordinates (p_a, q^a), for $a = 1, \ldots, n$. Then any point in the phase space locally is given by a point in the configuration manifold and one in the corresponding fiber in the cotangent bundle. Here we are interested in adding an additional dimension to the phase space to model dissipative phenomena, then we are dealing with a $(2n+1)$-dimensional manifold. However an odd-dimensional manifold cannot admit a symplectic structure. The analogue to the symplectic structure for odd-dimensional manifolds is the contact structure [5].

The contact phase space is define to be a *contact manifold* $\mathcal{T}$, a $(2n+1)$-dimensional manifold endowed with a global differential *contact form* 1-form η, that satisfies the condition

$$\eta \wedge (\mathrm{d}\eta)^n \neq 0\,, \tag{1}$$

everywhere on $\mathcal{T}$, where $\wedge$ denotes the exterior product and the exponent denotes the nth exterior power. The last property is known as the condition of *maximal nonintegrability* [5, 6], note that a contact manifold in this sense is orientable. The left hand side in (1) provides a *volume form* on $\mathcal{T}$. Associated with the definition of the contact 1-form on a contact manifold, there is another fundamental object called the *Reeb vector field* ξ, which is defined intrinsically by the conditions

$$i_\xi \eta = 1, \qquad i_\xi \mathrm{d}\eta = 0\,, \tag{2}$$

and then

$$\pounds_\xi \eta = i_\xi \mathrm{d}\eta + \mathrm{d} i_\xi \eta = 0\,. \tag{3}$$

The Reeb vector field is unique and defines at every point a 'vertical' direction with respect to $2n$-dimensional horizontal distribution $\ker(\eta)$. Therefore, the contact structure of the new phase space is completely determine by the triad $(\mathcal{T}, \eta, \xi)$.

Now, assuming that the phase space has a contact structure we can define a dynamics on it. Given the 1-form η, one can associate to every differentiable function $\mathscr{H} : \mathcal{T} \to \mathbb{R}$, a vector field $X_{\mathscr{H}}$, called the *contact Hamiltonian vector field generated by* $\mathscr{H}$, defined through the two (intrinsic) relations

$$\pounds_{X_{\mathscr{H}}} \eta = f_{\mathscr{H}}\, \eta \qquad \text{and} \qquad \mathscr{H} = -i_{X_{\mathscr{H}}} \eta\,, \tag{4}$$

where $\pounds_{X_{\mathscr{H}}}$ is the Lie derivative along a vector field $X_{\mathscr{H}}$ and $f_{\mathscr{H}} \in C^\infty(\mathcal{T})$ is a function depending on $\mathscr{H}$ defined by the condition $f_{\mathscr{H}} = -\xi(\mathscr{H})$. The function $\mathscr{H}$ is called the *contact Hamiltonian* [6, 7]. Using the Cartan's identity

$$\begin{aligned} f_{\mathscr{H}} \eta = \pounds_{X_{\mathscr{H}}} \eta &= i_{X_{\mathscr{H}}} \mathrm{d}\eta + \mathrm{d} i_{X_{\mathscr{H}}} \eta & (5)\\ &= i_{X_{\mathscr{H}}} \mathrm{d}\eta - \mathrm{d}\mathscr{H}\,, & (6)\end{aligned}$$

and the conditions in (4), it follows that

$$i_{X_{\mathscr{H}}} \mathrm{d}\eta = \mathrm{d}\mathscr{H} - (\pounds_\xi \mathscr{H})\, \eta\,, \tag{7}$$

this equation determines uniquely $X_{\mathscr{H}}$ up to a vector field proportional to ξ. This additional term is fixed by the requirement $\mathscr{H} = -i_{X_{\mathscr{H}}} \eta$. In addition, from Eq. (7) it is direct that the evolution of the contact Hamiltonian function along its flow is

$$\pounds_{X_{\mathscr{H}}} \mathscr{H} = f_{\mathscr{H}} \mathscr{H}\,. \tag{8}$$

Therefore, at difference to the symplectic theory, for the contact mechanics the contact Hamiltonian is not preserved along the evolution.

IOP Conf. Series: Journal of Physics: Conf. Series **1071** (2018) 012010 doi:10.1088/1742-6596/1071/1/012010

It is important to observe that from Eq. (7), if we consider $\pounds_\xi \mathscr{H} = 0$, we recover the usual symplectic result. Therefore this term is responsible to introduce dissipation into the dynamics, because this term introduce a dissipative force in the direction of the Reeb vector field.

Assigning a set of local (Darboux) coordinates (q^a, p_a, S) for $\mathcal{T}$ [5], referred to as *contact coordinates*, the 1-form η and the Reeb vector field ξ can be written as

$$\eta = \mathrm{d}S - p_a \mathrm{d}q^a, \qquad \xi = \frac{\partial}{\partial S}, \tag{9}$$

where here and in the following Einstein's summation convention over repeated indices is assumed. The second expression in (9) directly implies that in these coordinates

$$f_{\mathscr{H}} = -\frac{\partial \mathscr{H}}{\partial S}. \tag{10}$$

Therefore, the contact Hamiltonian vector field $X_{\mathscr{H}}$ takes the form

$$X_{\mathscr{H}} = \left(p_a \frac{\partial \mathscr{H}}{\partial p_a} - \mathscr{H}\right)\frac{\partial}{\partial S} - \left(p_a \frac{\partial \mathscr{H}}{\partial S} + \frac{\partial \mathscr{H}}{\partial q^a}\right)\frac{\partial}{\partial p_a} + \left(\frac{\partial \mathscr{H}}{\partial p_a}\right)\frac{\partial}{\partial q^a}, \tag{11}$$

where it is direct that the flow of $X_{\mathscr{H}}$ obtains in contact coordinates by solving

$$\dot{q}^a = \frac{\partial \mathscr{H}}{\partial p_a}, \tag{12}$$

$$\dot{p}_a = -\frac{\partial \mathscr{H}}{\partial q^a} - p_a \frac{\partial \mathscr{H}}{\partial S}, \tag{13}$$

$$\dot{S} = p_a \frac{\partial \mathscr{H}}{\partial p_a} - \mathscr{H}. \tag{14}$$

The similarity of Eqs. (12)–(14) with Hamilton's equations of symplectic mechanics is evident. In fact, this set of equations is a generalization of Hamilton's equations, because, when $\mathscr{H}$ does not depend on the S-variable, Eqs. (12) and (13) give exactly Hamilton's equations in the symplectic phase space and the remaining Eq. (14) in this case is the usual definition of Hamilton's principal function. Finally, given an arbitrary function in the contact phase space $\mathscr{F} \in C^\infty(\mathcal{T})$, we can define its evolution according to Eqs. (12)–(14) by $X_{\mathscr{H}}[\mathscr{F}]$, i.e.

$$\begin{aligned}\frac{\mathrm{d}\mathscr{F}}{\mathrm{d}t} &= -\mathscr{H}\frac{\partial \mathscr{F}}{\partial S} + p_a\left[\frac{\partial \mathscr{F}}{\partial S}\frac{\partial \mathscr{H}}{\partial p_a} - \frac{\partial \mathscr{F}}{\partial p_a}\frac{\partial \mathscr{H}}{\partial S}\right] + \frac{\partial \mathscr{F}}{\partial q^a}\frac{\partial \mathscr{H}}{\partial p_a} - \frac{\partial \mathscr{F}}{\partial p_a}\frac{\partial \mathscr{H}}{\partial q^a} \\ &= -\mathscr{H}\frac{\partial \mathscr{F}}{\partial S} + p_a\{\mathscr{F}, \mathscr{H}\}_{(S,p_a)} + \{\mathscr{F}, \mathscr{H}\}_{(q^a,p_a)},\end{aligned} \tag{15}$$

where $\{\ ,\ \}_{(q^a,p_a)}$ is the standard Poisson bracket and the remaining terms are contact corrections. Therefore, the contact Hamilton's theory can include a much larger class of models than the symplectic formalism.

As in the symplectic case, the contact phase space is a manifold equipped with a standard volume form: $\mathrm{V} = \eta \wedge (\mathrm{d}\eta)^n$. However, the Lie derivative of this volume form with respect to the contact flow $X_{\mathscr{H}}$ is different to zero, in fact it contracts (or expands) along the flow according to

$$\pounds_{X_{\mathscr{H}}} \mathrm{V} = -(n+1)\frac{\partial \mathscr{H}}{\partial S}\mathrm{V}, \tag{16}$$

which means that the contact flow has a non-zero divergence, namely

$$\mathrm{div}(X_{\mathscr{H}}) = -(n+1)\frac{\partial \mathscr{H}}{\partial S} \tag{17}$$

and therefore Liouville's theorem does not hold. However, It is important mention that the non-vanishing divergence of the contact vector field depends on the choice of the volume form. For instance, let us consider the volume form $\tilde{\mathrm{V}} = \mathscr{H}^{-(n+1)}\mathrm{V}$ proposed in [8]. Then for such volume form we have that

$$\pounds_{X_{\mathscr{H}}}\tilde{\mathrm{V}} = \left(\pounds_{X_{\mathscr{H}}}\mathscr{H}^{-(n+1)}\right)\mathrm{V} - \mathscr{H}^{-(n+1)}\pounds_{X_{\mathscr{H}}}\mathrm{V} \tag{18}$$

$$= \left(X_{\mathscr{H}}[\mathscr{H}^{-(n+1)}] - (n+1)\mathscr{H}^{-(n+1)}\frac{\partial \mathscr{H}}{\partial S}\right)\mathrm{V}\,, \tag{19}$$

where the last equality follows from (16). Finally, applying the contact Hamiltonian vector field to $\mathscr{H}^{-(n+1)}$, namely

$$X_{\mathscr{H}}[\mathscr{H}^{-(n+1)}] = (n+1)\mathscr{H}^{-(n+1)}\frac{\partial \mathscr{H}}{\partial S}\,, \tag{20}$$

one finds that

$$\pounds_{X_{\mathscr{H}}}\tilde{\mathrm{V}} = 0\,. \tag{21}$$

Therefore, for this choice of the volume form the contact flow has zero divergence.

3. Contact mechanics for dissipative systems

To model dissipative system by means of contact mechanics, we can use the fact that this formalism includes a new variable in the dynamics, the S-variable, which now affects the evolution of the system. Then, it is possible to connect this variable with the degree of freedom of loss of mechanical energy between the system and the environment. Let us consider the case, in which the contact Hamiltonian is describe by

$$\mathscr{H} = H_{\mathrm{mec}}(q^a, p_a) + h(S)\,, \tag{22}$$

where $H_{\mathrm{mec}}(q^a, p_a)$ is the mechanical energy of the system and $h(S)$ characterizes effectively the interaction with the environment. Then, from (15), the evolution of the mechanical energy is

$$\frac{\mathrm{d}H_{\mathrm{mec}}}{\mathrm{d}t} = -p_a\frac{\partial H_{\mathrm{mec}}}{\partial p_a}\,h'(S)\,, \tag{23}$$

from which it is clear that $h(S)$ behaves like a potential that generates dissipative forces. For example, we can consider the one-dimensional contact Hamiltonian system

$$\mathscr{H}_{\mathrm{S}} = \frac{p^2}{2m} + V(q) + \gamma\,S\,, \tag{24}$$

where $V(q)$ is the mechanical potential and γ is a constant parameter associated with the friction coefficient, also known in the literature as *Ohmic damping*. Then for this case, the rate of dissipation of the mechanical energy is

$$\frac{\mathrm{d}H_{\mathrm{mec}}}{\mathrm{d}t} = -\frac{\gamma}{m}\,p^2\,, \tag{25}$$

which agrees with standard results based on Rayleigh's dissipation function [9]. Then, the equations of motion (12)-(14) for this system are given by

$$\dot{q} = \frac{p}{m}\,, \tag{26}$$

$$\dot{p} = -\frac{\partial V(q)}{\partial q} - \gamma\,p\,, \tag{27}$$

$$\dot{S} = \frac{p^2}{2m} - V(q) - \gamma\,S\,. \tag{28}$$

From Eqs. (26) and (27) one can derive the damped Newtonian equation

$$\ddot{q} + \gamma \dot{q} + \frac{1}{m}\frac{\partial V(q)}{\partial q} = 0\,, \tag{29}$$

which describes all one-dimensional systems with a friction force that depends linearly on the velocity. Therefore, the contact phase space is a natural space of evolution for position and momentum quantities for dissipative systems, because of the Eq. (27) cannot be derived from a Hamiltonian function with the canonical symplectic structure of the cotangent bundle.

Finally, as we mention in the previous section the contact Hamiltonian vectorial field with respect to the standard volume form has a non-zero divergence. This property of the field is usually interpreted as a sign of dissipation [10, 11]. However, in this formalism this property does not classify the systems as non-dissipative or dissipative, because, as we already proved, we can always find a volume form such as the vector field has a zero divergence. Therefore, we can argue that being dissipative is not an intrinsic property of the vector field.

4. Time-dependent contact Hamiltonian systems

To include time-dependence in the system we extend the contact phase space by adding the time variable to it. Then we have an extended phase space $\mathcal{T}^{\mathrm{E}} = \mathcal{T} \times \mathbb{R}$ with coordinates as (q^a, p_a, S, t) and extended 1-form

$$\eta^{\mathrm{E}} = \mathrm{d}S - p_a \mathrm{d}q^a + \mathscr{H}\mathrm{d}t\,, \tag{30}$$

where $\mathscr{H}$ is the contact Hamiltonian. To define the dynamics on $\mathcal{T}^{\mathrm{E}}$ under the assumption that $\mathrm{d}\eta^{\mathrm{E}}$ is not degenerate, we set the two (intrinsic) simultaneous conditions

$$i_{X^{\mathrm{E}}\mathrm{d}\eta^{\mathrm{E}}} + \mathrm{d}\eta^{\mathrm{E}} i_{X^{\mathrm{E}}\eta^{\mathrm{E}}} = g_{\mathscr{H}}\,\eta^{\mathrm{E}} \qquad \text{and} \qquad i_{X^{\mathrm{E}}_{\mathscr{H}}}\eta^{\mathrm{E}} = 0\,, \tag{31}$$

with $g_{\mathscr{H}} \in C^\infty(\mathcal{T}^{\mathrm{E}})$ a function depending on $\mathscr{H}$, where this factor in coordinates corresponds to

$$g_{\mathscr{H}} = -\frac{\partial \mathscr{H}}{\partial S}\,. \tag{32}$$

Notice that (31) is the natural extension of (4) to $\mathcal{T}^{\mathrm{E}}$. These two conditions define a vector field $X^{\mathrm{E}}_{\mathscr{H}}$ on $\mathcal{T}^{\mathrm{E}}$ which is completely equivalent to the contact Hamiltonian flow (11) with general form

$$X^{\mathrm{E}}_{\mathscr{H}} = X_{\mathscr{H}} + \frac{\partial}{\partial t}\,, \tag{33}$$

with $X_{\mathscr{H}}$ given by (11), the proof of this result can be found in [1]. From this, it can be recognised immediately that the equations of motion given by such a field on $\mathcal{T}^{\mathrm{E}}$ are the same as those of the contact Hamiltonian vector field (11), with the addition of the trivial equation $\dot{t} = 1$. We call a system defined by a contact Hamiltonian $\mathscr{H}(q^a, p_a, S, t)$ and by the vector field $X^{\mathrm{E}}_{\mathscr{H}}$ of the form (33) a *time-dependent contact Hamiltonian system*. Besides, from (33) and (15) it follows that the evolution of any function $\mathscr{F} \in C^\infty(\mathcal{T}^{\mathrm{E}})$ under the dynamics given by a time-dependent contact Hamiltonian system is

$$\frac{\mathrm{d}\mathscr{F}}{\mathrm{d}t} = -\mathscr{H}\frac{\partial \mathscr{F}}{\partial S} + p_a\{\mathscr{F}, \mathscr{H}\}_{(S,p_a)} + \{\mathscr{F}, \mathscr{H}\}_{(q^a,p_a)} + \frac{\partial \mathscr{F}}{\partial t}\,. \tag{34}$$

Therefore, in general, we say that a function $\mathscr{F} \in C^\infty(\mathcal{T}^{\mathrm{E}})$ is a *first integral* (or *invariant*) of the contact dynamics given by $X^{\mathrm{E}}_{\mathscr{H}}$ if $\mathscr{F}$ is constant along the flow of $X^{\mathrm{E}}_{\mathscr{H}}$, that is if $X^{\mathrm{E}}_{\mathscr{H}}[\mathscr{F}] = 0$.

5. Hamilton-Jacobi formulation

In the symplectic theory, the Hamilton–Jacobi formulation is a powerful tool which enables to re-express Hamilton's equations in terms of a single partial differential equation whose solution, a function of the configuration space, has all the necessary information to obtain the trajectories of the mechanical system. The same formulation can be obtained from the contact theory. Therefore, in this section we introduce a Hamilton–Jacobi formulation of contact Hamiltonian systems, which establishes a connection with the configuration space, where the phenomenological equations are defined.

As we mentioned before, the Hamilton–Jacobi equation is a re-formulation of the dynamical equations in terms of a single partial differential equation (PDE) for the function $S(q^a, t)$. Thus, we are looking for a PDE of the form

$$\mathbb{F}\left(q^a, \frac{\partial S}{\partial q^a}, S, t, \frac{\partial S}{\partial t}\right) = 0\,, \tag{35}$$

and whose characteristic curves are equivalent to the contact Hamiltonian dynamics (12)–(14). To construct such PDE, let us define the function

$$\mathbb{F}\,(q^a, p_a, S, t, E) \equiv E - \mathscr{H}(q^a, p_a, S, t)\,. \tag{36}$$

It turns out that the solution of the Eq. $\mathbb{F} = 0$ is defined by

$$\eta^{\mathrm{E}} = \mathrm{d}S - p_a \mathrm{d}q^a + \mathscr{H}\mathrm{d}t = 0\,, \tag{37}$$

that is by the two conditions

$$p_a = \frac{\partial S}{\partial q^a} \quad \text{and} \quad \mathscr{H}\left(q^a, \frac{\partial S}{\partial q^a}, S, t\right) = -\frac{\partial S}{\partial t} \tag{38}$$

gives exactly the contact Hamiltonian Eqs. (12)–(14). Therefore we call the second equation in (38) the *contact Hamilton-Jacobi equation.* Additionally, we fount that the S-variable is the generalization of Hamilton's principal function which satisfies the contact version of the Hamilton–Jacobi equation.

6. Damped parametric oscillator

For its wide range of applications, the study of the dynamics of the parametric oscillator, i.e. an oscillator with time-dependent frequency, has been the subject of major interest, leading to the search of exact invariants in the Hamiltonian theory [12, 13]. However, the study of a parametric oscillator that includes the effects of the environment on the Hamiltonian theory has not been entirely satisfactory. The most traditional route to describe dissipative systems in the standard theory is to consider systems with explicit time dependence. The idea is to introduce a convenient time-dependence into the Hamiltonian such that it reproduces the phenomenological equations of motion with energy dissipation. As an example, we have the Caldirola [14] and Kanai [15] approach. But the major problem in such models is that the canonical and physical momenta and positions do not coincide [16, 17]. Here we study the damped parametric oscillator in the contact formalism by means of invariants in the contact phase space, which enables us to show the usefulness of our formalism and also show that the contact formalism does not have the ambiguity existing in the traditional approaches to dissipative systems.

The contact Hamiltonian for the system under consideration is given by

$$\mathscr{H} = \frac{p^2}{2m} + \frac{1}{2}m\omega^2(t)q^2 + \gamma\,S\,, \tag{39}$$

and the equations of motion can be easily deduced by the contact Hamiltonian Eqs. (26)–(28). Now, by direct substitution into result (34) it has been proved in [1] that the damped parametric oscillator possesses the quadratic invariant

$$\mathscr{I}(q,p,t) = \frac{m\,e^{\gamma t}}{2}\left[\left(\alpha(t)\frac{p}{m} - \left[\dot{\alpha}(t) - \frac{\gamma}{2}\alpha(t)\right]q\right)^2 + \left(\frac{q}{\alpha(t)}\right)^2\right], \tag{40}$$

where the time-dependent function $\alpha(t)$ satisfies the *Ermakov equation* [18]

$$\ddot{\alpha} + \left(\omega^2(t) - \frac{\gamma^2}{4}\right)\alpha = \frac{1}{\alpha^3}, \tag{41}$$

and the S-dependent invariant

$$\mathscr{G}(q,p,S,t) = e^{\gamma t}\left[S - \frac{q\,p}{2}\right]. \tag{42}$$

The invariant $\mathscr{I}(q,p,t)$ is a generalization of the canonical invariant found by H. R. Lewis Jr. for the parametric oscillator [12], which is recovered when $\gamma \to 0$. Now, by means of these invariants it is possible to find the general solution of the system, by means of the time-dependent contact transformation, see Appendix A,

$$\tilde{Q} = \arctan\left(\alpha\left[\dot{\alpha} - \frac{\gamma}{2}\alpha\right] - \alpha^2\frac{p}{m\,q}\right) \tag{43}$$

$$\tilde{P} = \mathscr{I}(q,p,t) \tag{44}$$

$$\tilde{S} = \mathscr{G}(q,p,S,t)\,. \tag{45}$$

with conformal factor $f = e^{\gamma t}$. The contact Hamiltonian expressed in this new variables takes the simple form

$$\mathscr{K} = \frac{\mathscr{I}}{\alpha^2}. \tag{46}$$

Therefore, as $\mathscr{K}$ does not depends on the variables $\tilde{Q}$ and $\tilde{S}$, the new contact Hamiltonian equations have the trivial form

$$\dot{\tilde{Q}}^a = \frac{1}{\alpha^2}, \quad \dot{\tilde{P}}_a = 0, \quad \dot{\tilde{S}} = 0, \tag{47}$$

with solutions

$$\tilde{Q}(t) = \int^t \frac{\mathrm{d}\tau}{\alpha^2(\tau)}, \quad \tilde{P}(t) = \mathscr{I} \quad \text{and} \quad \tilde{S}(t) = \mathscr{G}\,. \tag{48}$$

Finally, inverting the transformation (43)–(45), one obtains the solutions in the original (physical) coordinates, i.e.

$$q(t) = \sqrt{\frac{2\mathscr{I}}{m}}e^{\gamma t}\alpha(t)\cos\phi(t)\,, \tag{49}$$

$$p(t) = \sqrt{2\,m\mathscr{I}}\,e^{\gamma t}\left[\left(\dot{\alpha} - \frac{\gamma}{2}\alpha\right)\cos\phi(t) - \frac{1}{\alpha}\sin\phi(t)\right], \tag{50}$$

$$S(t) = \mathrm{e}^{-\gamma t}\mathscr{G} + \frac{q(t)p(t)}{2}, \tag{51}$$

where $\phi(t) = \tilde{Q}(t)$ and the values of the constants $\mathscr{I}$ and $\mathscr{G}$ are determined by the initial conditions. Therefore, we have derived here the solutions of the equations of motion of the damped parametric oscillator using the invariants of the contact Hamiltonian system, where all the dynamics of the system is encoded in the Ermakov equation (41).

7. Schrödinger quantization of contact Hamiltonian systems

In this section we sketch briefly a possible quantization of the contact formalism. Thus using the fact that the additional contact variable is a generalization of Hamilton's principal function, and that the canonical momenta and positions in our formalism coincide with the physical ones, then we suggest based on the Schödinger quantization introduced in the early days of the quantum mechanics [19], a canonical quantization of the contact Hamiltonian based on the standard rules of canonical quantization in the position representation, namely

$$p_a \to \frac{\hbar}{\mathrm{i}}\frac{\partial}{\partial q^a}\,, \quad q^a \to \hat{q}^a\,, \quad S(q^a,t) \to \frac{\hbar}{\mathrm{i}} \ln \Psi(q^a,t)\,, \tag{52}$$

where the last relation is Schrödinger's original definition of the wave function in terms of the Hamilton's principal function S [19]. Using such rules to quantize the contact Hamiltonian $\mathscr{H}$ and obtain the operator $\hat{\mathscr{H}}$, one can define the *contact Schrödinger equation*

$$\mathrm{i}\hbar\frac{\partial\Psi}{\partial t} = \hat{\mathscr{H}}\Psi\,. \tag{53}$$

This equation has a fundamental property: in the case in which the contact Hamiltonian reduces to a symplectic Hamiltonian (i.e. when $\mathscr{H}$ does not depend on S explicitly) the dynamics reduces to the standard quantum dynamics. However, this equation has the disadvantage that in general it does not conserve the norm of the wave function. But for systems with contact Hamiltonian of the form $\mathscr{H} = H_{\rm mec} + h(S)$, normalization is achieved following the procedure of Gisin [20], which consists in subtracting the mean value of $\hat{h}$, that is $\hat{h} \to \hat{h} - \langle\hat{h}\rangle$ where

$$\langle\hat{h}\rangle = \int dq^a\, \Psi^* \hat{h}(S)\, \Psi\,. \tag{54}$$

This leads to the nonlinear Schrödinger equation

$$\mathrm{i}\hbar\frac{\partial\Psi}{\partial t} = \hat{H}_{\rm mec}\Psi + (\hat{h} - \langle\hat{h}\rangle)\Psi\,. \tag{55}$$

We can mention that at variance with Gisin's proposal [20] we have specified the choice of $\hat{h}$ associated with the dependence of the contact Hamiltonian on the S-variable that is responsible for dissipation.

As an example we can apply this nonlinear Schrödinger equation (NLSE) to the contact Hamiltonian with arbitrary potential and linear dependence on S given in Eq. (24), leading to the nonlinear evolution equation

$$i\hbar\frac{\partial}{\partial t}\Psi = \left[-\frac{\hbar^2}{2m}\frac{\partial^2}{\partial q^2} + V(q) + \gamma\frac{\hbar}{i}\left[\ln\Psi - \langle\ln\Psi\rangle\right]\right]\Psi\,, \tag{56}$$

which is exactly the phenomenological NLSE introduced in [2, 3, 4] for the description of dissipative systems and has been a matter of recent research [21, 17]. Therefore, we have provided a theoretical justification for the introduction of the nonlinear phenomenological Schrödinger equation that displays consistency between the classical and quantum descriptions in terms of the contact formalism.

This derivation of the NLSE has important implications. It is well-known, in the standard theory, that the classical canonical evolution is associated in quantum mechanics with a unitary evolution. Then, because the NLSE has been obtained from the contact formalism, whose evolution is governed by contact transformations the evolution of the wave function Ψ is no longer unitary. On the other hand, from Eq. (56) it is possible to deduce the Smoluchowski equation

$$\frac{\partial \rho}{\partial t} + \frac{\partial}{\partial q} j(q,t) - \mathrm{D}(t) \frac{\partial^2 \rho}{\partial q^2} = 0 \tag{57}$$

for the probability density $\rho(q,t) = \Psi(q,t)\Psi^*(q,t)$, where the current of probability has the well-known form

$$j(q,t) = \rho(q,t) v(q,t) = \frac{\hbar}{2im} \left(\Psi^* \frac{\partial \Psi}{\partial q} - \Psi \frac{\partial \Psi^*}{\partial q} \right) \tag{58}$$

and the diffusion term is defined by the relation

$$-\mathrm{D}(t) \frac{1}{\rho} \frac{\partial^2 \rho}{\partial q} = \gamma \left[\ln \rho - \langle \ln \rho \rangle \right] . \tag{59}$$

Therefore, the continuity equation for this case is substituted by a Smoluchowski equation with an extra (possible time-dependent) diffusion coefficient $\mathrm{D}(t)$.

In particular, we can consider the one dimensional contact Hamiltonian for the parametric oscillator (39). The NLSE associated with this contact Hamiltonian possesses analytic wave packet solution of the form [16]

$$\Psi_{\mathrm{NL}}(q,t) = \left[\frac{m}{\pi \hbar \, \alpha(t)^2} \right]^{\frac{1}{4}} \exp\left\{ \frac{im}{2\hbar} \left[\frac{\dot{\alpha}(t)}{\alpha(t)} - \frac{\gamma}{2} + \frac{i}{\alpha^2(t)} \right] (q - \langle q \rangle)^2 + \frac{i}{\hbar} \langle p \rangle (q - \langle q \rangle) + \frac{i}{2\hbar} \langle q \rangle \langle p \rangle - i \phi_{\mathrm{NL}}(t) \right\} \tag{60}$$

where the time-dependent functions $(\langle q \rangle(t), \langle p \rangle(t))$ satisfy the contact Hamilton's equation:

$$\frac{\mathrm{d}\langle q \rangle}{\mathrm{d}t} = \frac{\langle p \rangle}{m}, \tag{61}$$

$$\frac{\mathrm{d}\langle p \rangle}{\mathrm{d}t} = -\omega^2(t) \langle q \rangle - \gamma \langle p \rangle , \tag{62}$$

the function $\alpha(t)$ obeys the nonlinear Ermakov equation (41) and the purely time-dependent phase

$$\phi_{\mathrm{NL}}(t) = \frac{1}{2} \int_0^t dt' \left(\frac{1}{\alpha^2(t')} - \frac{\gamma}{2} \alpha(t') \left[\dot{\alpha}(t') - \frac{\gamma}{2} \alpha(t') \right] \right) . \tag{63}$$

Then, from the form of the wave packet we can see that the maximum of the density distribution follows the classical trajectories according to Ehrenfest's theorem. Additionally, it is possible to prove that the position and the momentum uncertainties as well as their correlation obey the relations

$$\sigma_x^2(t) = \frac{\hbar}{2m} \alpha^2(t) , \tag{64}$$

$$\sigma_p^2(t) = \frac{m\hbar}{2} \left[\left(\dot{\alpha}(t) - \frac{\gamma}{2} \alpha(t) \right)^2 + \frac{1}{\alpha^2(t)} \right] , \tag{65}$$

$$\sigma_{xp}(t) = \frac{\hbar}{2} \alpha(t) \left[\dot{\alpha}(t) - \frac{\gamma}{2} \alpha(t) \right] . \tag{66}$$

where one can deduce that the diffusion term in this case correspond to $\mathrm{D}(t) = \frac{\gamma}{2} \sigma_q^2(t)$.

Because the Wigner phase space representation of quantum mechanics provides physical insight of the system as well as a natural comparison between quantum and classical dynamics, in the sense of correspondence and non-correspondence concepts [22, 23], we want to conclude this section presenting the Wigner function associated with the Gaussian wave packet (60).

IOP Conf. Series: Journal of Physics: Conf. Series **1071** (2018) 012010 doi:10.1088/1742-6596/1071/1/012010

Then, by means of the Wigner transformation [22], the Wigner function corresponding to the wave packet (60) corresponds to

$$W(x,p;t) = \frac{1}{\pi\hbar}\exp\left\{-\frac{2}{\hbar}\mathscr{I}(q-\langle q\rangle, p-\langle q\rangle, t)\right\}. \tag{67}$$

where $\mathscr{I}(q,p,t)$ is identical to the contact invariant (40), only replacing

$$q \to q - \langle q\rangle(t) \quad \text{and} \quad p \to p - \langle p\rangle(t)\,. \tag{68}$$

Therefore, the Wigner function for the damped parametric oscillator is completely determined by a contact invariant of the contact phase space.

8. Conclusion

Throughout this work we have introduced the contact structure on the phase space for the study of classical system, by means of the contact Hamiltonian vectorial field generated by a contact Hamiltonian. This geometrical description not only contains all the properties of the standard symplectic theory (to describe isolated systems), but also allows to include classical dissipative systems. Then, taking advantage of the fact that we have adding an extra dimension in the contact formalism, we connect this new variable with the degree of freedom for the exchange of energy between the system and the environment. This connection allowed us to include the dynamics of basic dissipative systems.

This formalism can also include time-dependent and time-independent systems. Additionally, in both cases the contact transformations, see Appendix A, which are transformation on the contact manifold that preserve the contact structure: $(\mathcal{T}, \eta, \xi)$, and also leave the equations of motion invariant. The usefulness of such transformation, as well as all the concepts developed in the work, were shown in detail for a specific important example, the damped parametric oscillator, for which we have solved the dynamics using the invariants of the system.

To close the study and the comparison of the contact formalism with the symplectic one, the contact Hamilton–Jacobi formalism of contact theory has been introduced, which encloses all the dynamics of the system in a single partial differential equation, called contact Hamilton–Jacobi equation. In addition, this formulation allowed us to show that the additional contact variable is the generalization of the standard Hamilton's principal function.

Due to the importance of the symplectic Hamiltonian theory in quantum mechanics, we have given a possible quantum version of the contact formalism. So based on the Schödinger quantization, introduced in the early days of the quantum mechanics, this quantization leads to a nonlinear Schrödinger equation that has been already proposed in the literature for the description of quantum dissipative system. There are important consequences in the NLSE derived from the contact theory compared to the standard one, like the fact that the evolution of the wave packet solution is no longer a unitary one, besides the density probability obeys a Smoluchowski equation instead of the standard continuity equation. In particular, we showed an analytical solution of the NLSE for the damped parametric oscillator, where the maximum of the probability distribution follows the classical equation of motion according to the Ehrenfest's theorem. In addition, given the Gaussian wave packet solution we presented the Wigner function associated with it and we enhance that it is completely determined by a contact invariant, which displays an intriguing consistency between the classical and quantum descriptions in the contact formalism.

Finally, it is important mention that if we are not able to find a contact structure we cannot proceed with the analysis carried on in the work and we cannot describe dissipation. To be able to introduce the Hamilton-Jacobi equation we need to identify a "configuration space". This would not be possible on a compact contact manifold, for instance like the three-dimensional

IOP Conf. Series: Journal of Physics: Conf. Series **1071** (2018) 012010 doi:10.1088/1742-6596/1071/1/012010

sphere. Therefore, the contact manifolds which are being considered in the work are of a very special kind. In this sense, in physical terms we can say that the contact formalism describes a subset of all dissipative systems.

Acknowledgments
The author is grateful for the scholarship provided by CONACyT México, with reference number 379177, and thanks Prof. Giuseppe Marmo and Prof. Dieter Schuch for their comments and the critical reading of the manuscript.

Appendix A. Contact transformations

A *contact transformations* is a transformation that leaves the contact form invariant up to multiplication by conformal factor f [24, 25], namely $\tilde{\eta} = f\,\eta$. In coordinates the transformation from (q^a, p_a, S) to $(\tilde{Q}^a, \tilde{P}_a, \tilde{S})$ is

$$\mathrm{d}\tilde{S} - \tilde{P}_a \mathrm{d}\tilde{Q}^a = f(\mathrm{d}S - p_a \mathrm{d}q^a)\,. \tag{A.1}$$

which is equivalent to

$$f = \frac{\partial \tilde{S}}{\partial S} - \tilde{P}_a \frac{\partial \tilde{Q}^a}{\partial S} \tag{A.2}$$

$$-f p_i = \frac{\partial \tilde{S}}{\partial q^i} - \tilde{P}_a \frac{\partial \tilde{Q}^a}{\partial q^i} \tag{A.3}$$

$$0 = \frac{\partial \tilde{S}}{\partial p_i} - \tilde{P}_a \frac{\partial \tilde{Q}^a}{\partial p_i}\,. \tag{A.4}$$

In addition, we can obtain the generating function of a contact transformation. Assuming that the coordinates $(q^a, \tilde{Q}^a, S)$ are independent, we compute the differential of the generating function $\tilde{S}(q^a, \tilde{Q}^a, S)$, namely

$$\mathrm{d}\tilde{S} = \frac{\partial \tilde{S}}{\partial S}\mathrm{d}S + \frac{\partial \tilde{S}}{\partial q^a}\mathrm{d}q^a + \frac{\partial \tilde{S}}{\partial \tilde{Q}^a}\mathrm{d}\tilde{Q}^a\,. \tag{A.5}$$

Substituting (A.5) into (A.1) we obtain the following conditions for $\tilde{S}$

$$f = \frac{\partial \tilde{S}}{\partial S}\,, \quad f\,p_a = -\frac{\partial \tilde{S}}{\partial q^a}\,, \quad \tilde{P}_a = \frac{\partial \tilde{S}}{\partial \tilde{Q}^a}\,. \tag{A.6}$$

In particular, for contact transformations with $f = 1$ the conditions in (A.6) imply that the generating function has the form

$$\tilde{S} = S - F_1(q^a, \tilde{Q}^a)\,, \tag{A.7}$$

where $F_1(q^a, \tilde{Q}^a)$ is the generating function of a symplectic canonical transformation. This result implies that all canonical transformations are a special case of contact transformations corresponding to $f = 1$.

For the time-dependent case *time-dependent contact transformations* are transformation of coordinates

$$(q^a, p_a, S, t) \to (\tilde{Q}^a, \tilde{P}_a, \tilde{S}, t)\,, \tag{A.8}$$

that preserve the contact structure of the extended phase space (30) up to multiplication by a non-zero conformal function f, that is

$$f\,(\mathrm{d}S - p_a \mathrm{d}q^a + \mathscr{H}\mathrm{d}t) = \mathrm{d}\tilde{S} - \tilde{P}_a \mathrm{d}\tilde{Q}^a + \mathscr{K}\,\mathrm{d}t\,, \tag{A.9}$$

where $\mathscr{K}$ is a function on $\mathcal{T}^{\mathrm{E}}$ which is going to be the new contact Hamiltonian in the transformed coordinates. From this conditions one obtains the standard conditions (A.2)–(A.4) for a time-independent contact transformation, together with the following rule for the transformation of the Hamiltonians

$$f\,\mathscr{H} = \frac{\partial \tilde{S}}{\partial t} - \tilde{P}_a \frac{\partial \tilde{Q}^a}{\partial t} + \mathscr{K}\,. \tag{A.10}$$

that relates the Hamiltonians.

References

[1] Bravetti A, Cruz H and Tapias D 2017 *Ann. Phys.* **376** 17–39
[2] Schuch D, Chung K M and Hartmann H 1983 *J. Math. Phys.* **24** 1652–1660
[3] Schuch D, Chung K M and Hartmann H 1984 *J. Math. Phys.* **25** 3086–3092
[4] Schuch D and Chung K M 1986 *Int. J. Quan. Chem.* **29** 1561–1573
[5] Arnold V I 1989 *Mathematical methods of classical mechanics* vol 60 (Springer)
[6] Arnold V I and Novikov S P 2001 *Dynamical Systems IV: Symplectic Geometry and Its Applications* Dinamicheskie sistemy (Springer)
[7] Boyer C P 2011 *SIGMA* **7** 22–58
[8] Bravetti and Tapias D 2015 *J. Phys. A* **48** 245001
[9] Goldstein H P C and J S 2002 *Classical mechanics*
[10] Daems D and Nicolis G 1999 *Phys. Rev. E* **59** 4000
[11] Gallavotti G and Cohen E 2004 *Phys. Rev. E* **69** 035104
[12] Lewis Jr H R 1968 *J. Math. Phys.* **9** 1976–1986
[13] Lewis Jr H R and Riesenfeld W 1969 *J. Math. Phys.* **10** 1458–1473
[14] Caldirola P 1941 *Nuovo Cim. (1924-1942)* **18** 393–400
[15] Kanai E 1948 *Prog. Th. Phys.* **3** 440–442
[16] Schuch D 2011 Dynamical invariants in systems with and without broken time-reversal symmetry *LATIN-AMERICAN SCHOOL OF PHYSICSL ELAF: Symmetries in Physics* vol 1334 (AIP Publishing) pp 291–340
[17] Cruz H, Schuch D, Castanos O and Rosas-Ortiz O 2016 *Ann. Phys.* **373** 609–630
[18] Leach P and Andriopoulos K 2008 *App. An. and Dis. Math.* 146–157
[19] Schrödinger E 1926 *Ann. Phys.* **385** 437–490
[20] Gisin N 1983 *J. Math. Phys.* 24 1779–1782
[21] Cruz H, Schuch D, Casta nos O and Rosas-Ortiz O 2015 *Ann. Phys.* **360** 44–60
[22] Wigner E 1932 *Phys. Rev.* **40** 749
[23] Hillery M, O Connell R F, Scully M O and Wigner E P 1997 Distribution functions in physics: fundamentals *Part I: Physical Chemistry. Part II: Solid State Physics* (Springer) pp 273–317
[24] Rajeev S 2008 *Ann. Phys.* **323** 768–782
[25] Rajeev S 2008 *Ann. Phys.* **323** 2265–22

IOP Conf. Series: Journal of Physics: Conf. Series **1071** (2018) 012011 doi:10.1088/1742-6596/1071/1/012011

Fundamental quantum structures — Conclusions with respect to cosmology and interactions

T Görnitz

Fachbereich Physik
J. W. Goethe-Universität Frankfurt/Main, Germany

E-mail: goernitz@em.uni-frankfurt.de

Abstract. An explanation can be seen as the process of reducing complicated structures to simple structures. For more than two millennia, the guiding idea was to reduce phenomena to "small particles", to some types of "atoms". In other words: the smaller, the simpler.
Up to the atoms of the chemistry this concept was extremely successful. Structures smaller than atomic nuclei, however, require ever higher energies for their production and their analysis. That these ever higher energies lead to ever simpler structures is implausible.
The simplest structures that are mathematically possible are abstract bits of quantum information - AQIs. Taken as the ultimate substance, referred to as protyposis, the AQI concept affords a unifying explanation of the structure of space and time and the evolution of the cosmos and its inventory.
Some thirty years ago, the model of a closed cosmos expanding at the speed of light was developed deriving from basic quantum-theoretical arguments. In this model, the equation of state reads $\rho+3p=0$ (ρ and p denoting cosmological energy density and pressure). At the core of the model, there are the quantum bits of absolute quantum information (AQI), the simplest of all possible quantum structures.
Recent astrophysical investigations declare that the "$\rho+3p=0$"-model is in better agreement with the observation data than the present standard flat-ΛCDM model. Ad hoc assumptions such as Dark Energy and Inflation are dispensable.
The protyposis theory has been advanced to account for the formation of relativistic quantum particles from AQI bits. Moreover, a rationalization has been given of the General Theory of Relativity and of the three non-gravitational forces.
Ultimately, the protyposis concept will allow us to understand the emergence of both matter and consciousness from the AQIs

1. Introduction: The idea of smallest "elementary particles"

The theories of scientific explanations are a huge field. (see e.g. [39]) In physics, however, one is less interested in the philosophical subtleties. Since the end of the 19th century, atomism has become the guiding principle of scientific explanations in physics. This idea is more than two millennia old. The idea was to reduce phenomena to "small particles", to some types of "indivisibilities", in Greek "atoms".

> “Atomism, any doctrine that explains complex phenomena in terms of aggregates of fixed particles or units. This philosophy has found its most successful application in natural science: according to the atomistic view, the material universe is composed of minute particles, which are considered to be relatively simple and immutable and too small to be visible. The multiplicity of visible forms in nature,

IOP Conf. Series: Journal of Physics: Conf. Series **1071** (2018) 012011 doi:10.1088/1742-6596/1071/1/012011

then, is based upon differences in these particles and in their configurations; hence, any observable changes must be reduced to changes in these configurations." [1]

The principal explanatory motive in contemporary physics can be formulated in brief as:

"The smaller, the simpler."

Up to the atoms of the chemistry this concept was extremely successful.

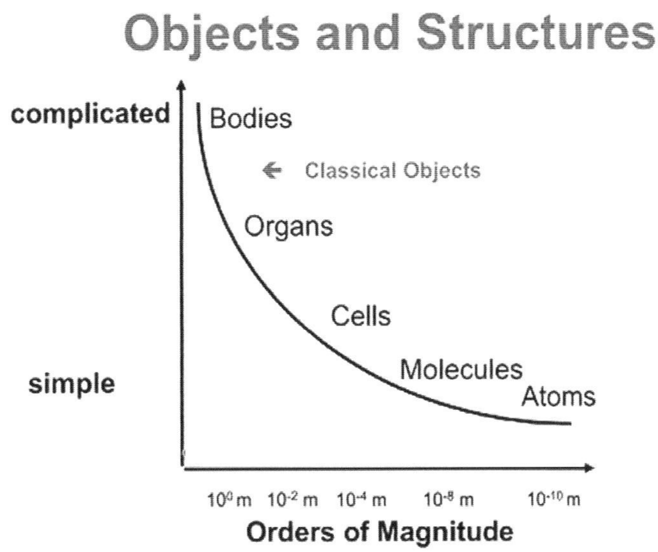

Fig. 1: Smaller becomes simpler

This idea is deeply rooted in the subconscious of the natural sciences. Nowadays, when the call for "a new physics" gets loud, it means that even smaller particles are intended.

It is trivial that a living being is more complicated than its organs. A cell, in turn, is simpler than an organ. A molecule is simpler than a cell, and an atom is simpler than a molecule. The idea that decomposing leads to something simpler is a basic principle of classical physics. In classical physics, the state space of a system is the direct sum of the state spaces of its subsystems. Another characteristic of classical physics is the identification of possibilities with missing knowledge. In classical physics it is therefore denied that possibilities can produce real effects. Nor does classical physics refer to the fact that reality is intrinsically relational.

1.1. Remarks on quantum theory

Quantum theory is based on two principles of common sense:

Relations are fundamental, even in inanimate nature, therefore:

Mostly a whole is more than the sum of its parts.

Mathematically, this universal property of nature is taken into account by constructing the state space of a system by the tensor product of the state spaces of its subsystems.

The second principle takes into account that even in inanimate nature,

mere possibilities can have a real impact on processes.

Such possibilities are "virtual facts" and not mere ignorance.

Measurement results and other facts are known to be indicated by real numbers. Possibilities that influence processes are described in quantum theory by imaginary numbers. Therefore, quantum theory is represented by the body of complex numbers.

IOP Conf. Series: Journal of Physics: Conf. Series **1071** (2018) 012011 doi:10.1088/1742-6596/1071/1/012011

The second principle takes into account that even in inanimate nature,

mere possibilities can have a real impact on processes.

Such possibilities are "virtual facts" and not mere ignorance.

Measurement results and other facts are known to be indicated by real numbers. Possibilities that influence processes are described in quantum theory by imaginary numbers. Therefore, quantum theory is represented by the body of complex numbers.

1.2. The dead-end-road of "elementary particles"

We pointed out above that in the atomistic concept the basic idea is:

The smaller the simpler.

The basis of quantum theory can be found in Planck's and Einstein's formula

(1.1) $$m c^2 = E = h \nu = h c / \lambda$$

In quantum theory, the finding since Planck is:

The more (matter or energy), the smaller.

Both theses should now be combined.

"the more, the smaller" and "the smaller the simpler".

That would imply the statement:

"The more, the simpler"

But that is obviously implausible.

Max Planck had called the quantum of action as a "threatening explosive device" in the realm of physics. This threat is directed in depth against the idea that one should seek the simple on a small scale. Planck's words already show a paradigm shift that will prove fundamental. The history of physics shows that before the introduction of such a new paradigm, the previous paradigm was artfully extended with elaborate mathematical refinements. An illustrative example is the cosmological model of Tycho Brahe. It was a very clever invention to explain the observation data without dropping the basic idea of cosmic circles. (see e.g. [40]) Before Kepler, it was an incontrovertible belief in astronomy that all celestial bodies move on circular paths, possibly with additional epicycles.

Looking at domains that are spatially smaller than the atomic nuclei, it turns out that the structures and theories are becoming increasingly complicated - and of course mathematically very interesting.

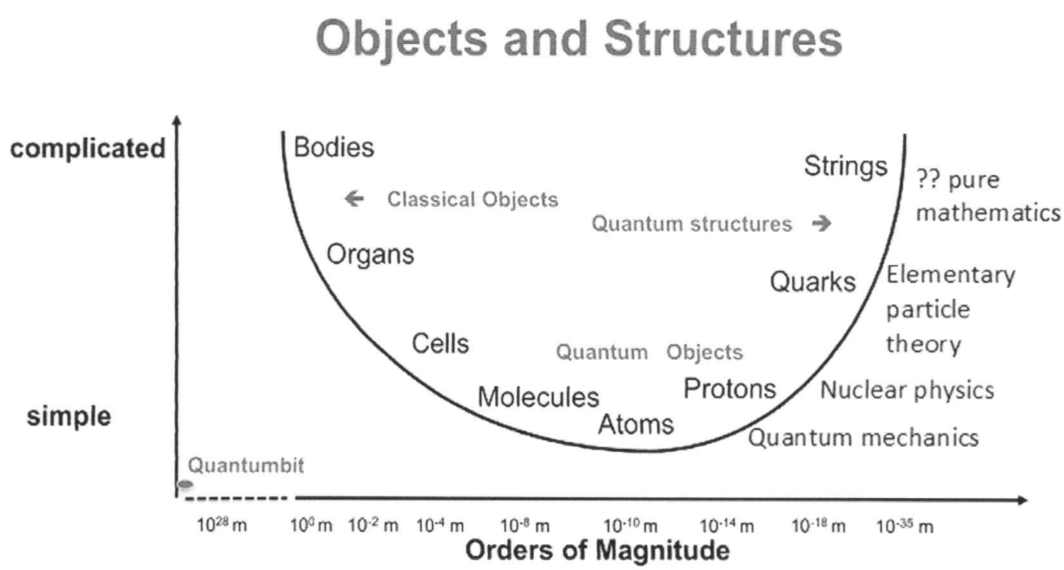

Fig. 2: The smaller, the more complicated.

IOP Conf. Series: Journal of Physics: Conf. Series **1071** (2018) 012011 doi:10.1088/1742-6596/1071/1/012011

Nuclear physics is much more complicated than quantum mechanics, which is responsible for the behavior of the electrons in the atomic shell. Quantum mechanics can presuppose the electrons and atomic nuclei and need not consider changes in their particle number. Nuclear physics must consider the formation and destruction of particles because of the much stronger forces. Nuclear physics must apply methods of quantum field theory.

The Standard Model of Elementary Particle Theory is the most complicated mathematical structure still related to the experiments. However, the standard model needs 18 free - therefore unexplained - parameters. The essential structures of the theory, the quarks and gluons, must be understood as virtual objects. They can not appear as real particles in a vacuum. Therefore, at the latest there, the two and a half millennia old pictures of "fundamental small elementary particles" reach their limits.

The strings are touted as "even much smaller". String theory is a very interesting area of the theory of several complex variables. However, a relationship to reality is largely lost. Complicated mathematical spaces are declared to be the "actual space-time structure," and the real space in which humanity is located would be a bad approximation.

That with more and more energy ever smaller and smaller structures can be studied has been accepted in physics for over a century. The enormous experimental success of this concept has helped to displace the absurdity of this fact. Nevertheless, ever smaller structures and substructures are courageously propagated.

However, the basic structure of quantum theory becomes quite clear when it is not about energy, but about information. It is obvious that with more and more information one can limit a space area more and more narrowly.

1.3. Abstract Quantum bits of Information (AQIs)

An AQI is a "quantum pre-structure" with a two-dimensional complex valued state space. It is the simplest and at the same time the most abstract construction that can be introduced and mathematically formulated in terms of quantum theory.

The reduction to such abstract quantum bits, however, would remain merely a philosophical concept if a connection to established physics were not possible.

The tensor product structure of quantum theory allows the following: Many extended entities can converge into a strongly localized object.

A simple example is the sine wave. When the sine, a widely extended function, is often multiplied by itself - as a quantum theoretical combination requires - we obtain a sharply-localized graph. If we combine different values of the exponent, then a single sharp localization can be obtained.

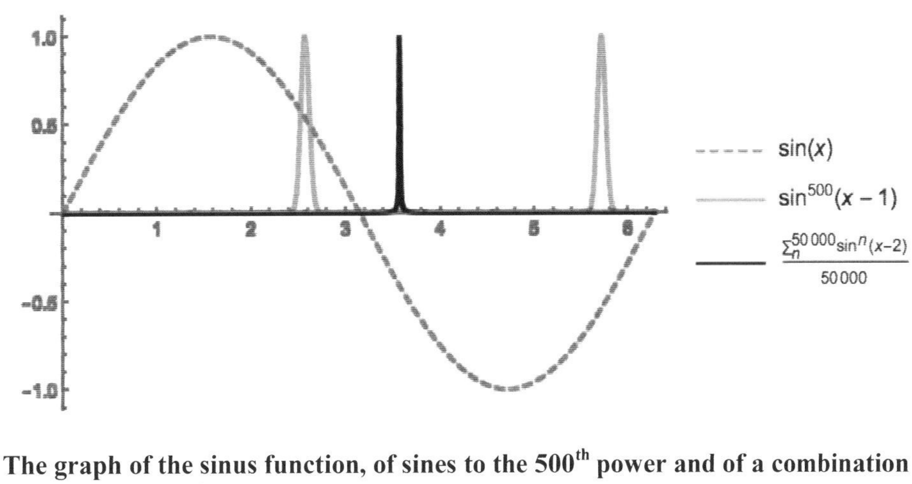

Fig. 3: The graph of the sinus function, of sines to the 500th power and of a combination of different powers of the sine.

IOP Conf. Series: Journal of Physics: Conf. Series **1071** (2018) 012011 doi:10.1088/1742-6596/1071/1/012011

2. Cosmological considerations

Recent examinations of extensive astronomic data sets [1 – 7] have shown that the best agreement with the data is afforded by a model referred to as $R_h=ct$ universe by the authors of these studies. It is thus a universe expanding at constant speed, the speed of light.

In the $R_h=ct$ universe, problems arising in the previous flat-ΛCDM standard model are absent. As the authors López-Corredoira et al. [2] put it:

"Cosmological models with a geometry different from that in the current standard model have fallen out of favour and are rarely considered in ongoing tests using the latest high-precision measurements. However, even within the framework of the standard model, not all the data fit together tension free. At least some controversy still surrounds the interpretation of various measurements, and other competing models often fit at least some of these observations better than the concordance model does. It is therefore useful to re-examine how these alternative scenarios fare compared to ΛCDM when new, improved data become available."

"But whereas this optimization of parameters in ΛCDM/wCDM creates some tension with their concordance values, the Rh=ct universe has the advantage of fitting the QSO and AP data without any free parameters."

This is corroborated by [4]:

"In contrast to the perception based on Type Ia SNe that ΛCDM can best account for the observed expansion of the Universe, the conclusion from these other studies is that the cosmic dynamics is better described by a cosmology we refer to as the $R_h = ct$ Universe."

Another quote reads:

"As we shall show here, the use of this diagnostic, [...], disfavors the current concordance (ΛCDM) model at 2.3σ. Within the context of expanding Friedmann-Robertson-Walker (FRW) cosmologies, these data instead favor the zero active mass equation-of-state, $\rho + 3p = 0$, where ρ and p are, respectively, the total density and pressure of the cosmic fluid, the basis for the $R_h = ct$ universe." [3]

See also the conclusions in the work of [5, 6]], confirmed by [7]:

"The 'standard' model of cosmology is founded on the basis that the expansion rate of the universe is accelerating at present — as was inferred originally from the Hubble diagram of Type Ia supernovae. There exists now a much bigger database of supernovae so we can perform rigorous statistical tests to check whether these 'standardisable candles' indeed indicate cosmic acceleration. Taking account of the empirical procedure by which corrections are made to their absolute magnitudes to allow for the varying shape of the light curve and extinction by dust, we find, rather surprisingly, that the data are still quite consistent with a constant rate of expansion."

However, a constant expansion is not sufficient yet. Essential for the interpretation of the data is the equation of state $\rho+3p=0$. The Authors call this as "the zero-active mass condition". This condition is critical to the whole picture. There are various ways of attaining a constant expansion rate in the Universe including the famous Milne cosmology (i.e., an empty Universe). However, only this one with the equation of state $\rho+3p=0$ produces measurable quantities, such as the Hubble expansion rate and luminosity distance, that are consistent with the observations. All the other constant-expansion rate scenarios have been ruled out at a very high statistical significance. [8]

This better model than the present flat-ΛCDM model was derived using fundamental quantum-theoretical arguments 30 years ago. [9-14]

Today, it is referred to as the "protyposis model". Its basic entities are the mathematically simplest quantum structures, the abstract quantum bits of absolute (free-of-meaning) information – AQI bits.

This model was opposing the mainstream then, because at that time almost all of the experts believed that the cosmos would re-collapse – at least if its volume is not infinitely large. Presently the model is at odds with the view of an accelerated cosmic expansion, a view still held by many experts.

The quantum-theoretically founded protyposis model has the great epistemological advantage that it dispenses with the concepts needed and conceived in the standard model, such as "Dark Energy" and "inflation" together with the required, freely adjustable parameters.

The protyposis model eliminates a large part of the all but intractable "cosmological problems". The hitherto mysterious "Dark Energy" can readily be explained. In the case of "Dark Matter", various fictitious particles were postulated to explain the apparent gravitational effect, which have yet to be

IOP Conf. Series: Journal of Physics: Conf. Series **1071** (2018) 012011 doi:10.1088/1742-6596/1071/1/012011

found. In the protyposis model, there is no need for such particles. The gravitation effect due to non-luminescent entities derives from the model. The AQI bits entail dark matter-type local gravitational effects, without the necessity of appearing as massive particles or massless photons.

2.1. The quantum theoretical foundation of cosmology

Regarding the connections between quantum theory and cosmology, there have been misconceptions in the historical development of physics, which, however, can be cleared in light of the present knowledge.

Historically, quantum theory was the very theory that, for the first time, allowed one to really understand microphysical phenomena, that is, atomic and intra-atomic processes. Also for historical reasons, quantum theory was seen as carrying forward the thousands of years old concept of "indivisibilities", of "atoms".

It has long been believed that the path would lead to ever smaller "elemental structures", e.g., preons and strings. Here "smallness" was always seen as a spatial aspect. That this is still the predominant view is exemplified in the "holographic universe". There the "smallest structures" are supposedly areas of a Planck-length diameter.

Einstein's General Theory of Relativity (GRT) allowed physicists to understand, for the first time, space and time as dynamic entities and describe adequately the interaction of space and time with the material and energetical content of space.

The applications of GRT, from the perihelia rotation of the Mercury to the pairs of neutron stars and the detection of gravity waves, shows that the theory describes the gravitational effects extremely well. These successful applications are normally based on approximations to the GRT, and as approximations they can be quantized, such as in the spin-2 quantization of gravitational waves. However, it is seen a major theoretical problem that one has not yet succeeded in the quantization of the full GRT.

Unlimited exact solutions of the GRT always describe a whole universe. Of these infinitely many solutions at most a single one can apply to the actual cosmos. Thus, the GRT describes much more than exists in the reality.

2.2. A new look at the foundations

From Planck's and Einstein's fundamental formulas

$$E = mc^2 = h\nu - hc/\lambda \qquad (2.1)$$

it is apparent that a smaller wavelength or Compton wavelength comes with higher energy or mass, respectively.

Taking off the blinders of the antique atom conceptions, would not these formulas make it implausible that the path into the small will lead to simpler structures?

Yet the answer is not entirely simple, as the great historical success of the concept of atoms shows. In fact, down to the atoms of chemistry the disassembling into smaller parts, e.g., of a molecule into its atoms, affords a gain in simplicity.

"The higher the energy, the simpler the system": this becomes more and more an entirely incomprehensible thesis the deeper one proceeds into the interior of the atoms. The required increase of energy causes a concomitant increase of the probability for the occurrence of ever more virtual and also short-lived real quanta.

By contrast, what is plausible is that the simplest quantum structures possess the smallest energies and thereby the largest extensions. This insight makes it possible to release quantum theory out of the "ghetto of the spatially small". As ever more sophisticated experiments show, quantum systems extending over large distances can nowadays be handled. [15]

With regard to cosmology, a relevant aspect of quantum theory is the composition of subsystems via tensor products of their state spaces. From a mathematical point of view, this structure allows for holistic entities lacking parts while being extended.

What are the simplest structures?

For good reasons, quantum field theories are at the core of modern theoretical studies. They involve the most complex mathematical structures with which one can operate in physics. The simplest way to understand quantum fields is to see them as representing an infinite number of field quanta, that is, quantum particles. This of course implies that quantum particles are considerably simpler structures than quantum fields.

In analogy to the field-particle concept, it has been shown that relativistic quantum particles can be understood as an infinite number of quantum bits. [16, 17]

While quantum fields have a state space of uncountable infinite dimensions and the state space of quantum particles is countably infinite, that of the quantum bits is just two-dimensional.

By mathematical reasons the quantum bits are the simplest possible quantum structures.

One bit is the smallest possible amount of information. Obviously, a quantum bit must not be conceived as something "spatially small". The more accurately a position in space is to be determined, the more information is required to this end. This fact is however often hidden as a bit can be "attached" to a "carrier", like the spin of an electron; by constructing ever-smaller carriers, ever more bits of the special intended information can be stored in a given volume.

In accepting the quantum bits as the basic constituents of physics, one has to dispense with the familiar connotation "information = meaning". "Meaning" always has a large subjective contribution and is therefore not suitable for defining objectifiable physical quantities.

In descending the stages from the complex to the simple, "quantum fields -> quantum particles -> quantum bits", the physical structure at the lowest stage is formed by bits of **a**bstract and **a**bsolute (free-of-meaning) quantum information (AQI) referred to as "protyposis" (Greek: pre-formation). (see [18 – 20]) The introduction of a new designation was motivated by the need to distinguish from the commonplace understanding of "information". The abstractness and absoluteness of the AQI bits also implies the absence of any reference to the notions of a transmitter and a receiver.

2.3. The construction of the physical space from the elementary structures of quantum theory

A basic fact of human experience is that the reality is three-dimensional. An actual foundation of this finding can be derived from the simplest quantum structures, the AQI bits.

More than half a century ago, Carl Friedrich v. Weizsäcker [21 – 23] had shown that the physical space is three-dimensional as a consequence of the simplest quantum-theoretical entities which we now call AQI bits. Weizsäcker had termed these quantum objects "Ure" (singular: Ur) or "Ur-Alternativen".

To relate to the familiar physics, Weizsäcker's concept of information, still committed to "meaning" and "knowledge", had to be carried further and made radically abstract, which is reflected in the notational change from the "Ur" to the AQI bit.

The AQI bits are the simplest possible structures – already for mathematical reasons. Therefrom space and time and their dynamics originate, as well as the material and energetic structures encountered.

Weizsäcker's summarizing considerations are to be found in his book "Aufbau der Physik" [22, 23]. A brief sketch is as follows: The symmetry groups of the two-dimensional complex state space of the quantum bit are the $SU(2)$ and $U(1)$ groups. These groups span a real 4-dimensional manifold. The $U(1)$ group was linked to the empirical time. From the group-theoretical point of view, the physical space was supposed to be a homogeneous space of the $SU(2)$, that is, the group itself, being a S^3, the tree-dimensional boundary of a four-dimensional sphere.

Missing in Weizsäcker's design was a metric. [22, p. 399] Using group-theoretical arguments a metric could be established. [9 – 11]

2.4. Establishing the metric of space-time via the elementary structures of quantum theory

The regular representation of a semi-simple compact topological group such as the $SU(2)$ operates in the Hilbert space of L^2-functions on the maximal homogeneous space of the group, that is the group as

S^3. The regular representation comprises all irreducible representations of the group. An irreducible representation stands for an indivisible quantum system, a whole without parts.

The states of an AQI bit generate a two-dimensional representation of the *SU(2)*. In physics, for obvious reasons, this is often referred to as "spin-½ representation". The representation is formed by functions on the S^3, which divide the S^3 into two halves. For an analogy, consider the sine

Let ${}^{2k+1}\boldsymbol{D}_k$ denote an irreducible representation of the *SU(2)*. Here, the subscript indicates a "spin"-value, *k=0, ½, 1, 3/2, ...,* while the superscript *2k+1* specifies the respective dimension of the representation.

A tensor product of two-dimensional representations can be decomposed into a direct sum of irreducible representations. From a physical point of view, this means that a manifold of qubits will appear, with a certain probability, as an indivisible quantum system.

Let N be the number of AQI bits in the cosmos. The state space of an AQI bit is the two-dimensional space of a spin-½ representation. Thus, the state space of the tensor product of N of the AQI bits has the dimension 2^N. [10]

The decomposition of this tensor product into irreducible representations can be written as

(2.2) $$\left({}^{2}D_{1/2}\right)^{\otimes N} = \oplus_{j=0}^{\{N/2\}} f(N,j)^{2\{N/2\}-2j+1}\mathrm{D}_{\{N/2\}-j}$$

where *{N/2}* denotes the integer *N/2* or *(N-1)/2* for even or odd N, respectively. *f(N, j)* is an integer factor given by

(2.3) $$f(N,j) = \frac{N!(N+1-2j)}{(N+1-j)!j!}$$

The quantum-physical meaning of this decomposition is that, with a specific probability, the system will be in a state belonging a representation ${}^{2k+1}\boldsymbol{D}_k$. That probability is proportional to the factor *f(N, j),* being the frequency with which the respective representation occurs in the decomposition of the tensor product.

As to be expected, the weighted sum of the dimensions becomes

(2.4) $$\oplus_{j=0}^{\{N/2\}} f(N,j)(2\{N/2\} - 2j + 1) = 2^N$$

To obtain probabilities for the occurrence of a representation, the frequency factors have to be divided by the total number Z of all representations:

(2.5)
$$Z = \sum_{j=0}^{2k} \frac{(2k)!\,(2k+1-2j)}{(2k+1-j)!\,j!} = \frac{(2k)!}{(k!)^2} \quad \text{for } N = 2k$$

(2.6)
$$Z = \sum_{j=0}^{2k} \frac{(2k+1)!\,(2k+2-2j)}{(2k+2-j)!\,j!} = \frac{(2k+1)!}{k!\,(k+1)!} \approx 2\frac{(2k)!}{(k!)^2} \quad \text{for } N = 2k+1$$

The maximum of the frequency factor *f(N, j)* is assumed for

(2.7) $$j_{max} = (1/2)[N-\sqrt{(N+2)}]$$

which for $N \gg 1$ can be approximated as

(2.8) $$j_{max} \approx (1/2)[N-\sqrt{(N)}]$$

Using the approximation formula for the factorial,

(2.9) $$\ln(n!) \approx [n+(1/2)]\ln n - n + (1/2)\ln(2\pi)$$

one obtains the following orders of magnitudes

(2.10) $$f(N, \{N/2\}) \approx O(2^N N^{-3/2})$$

(2.11) $$f(N, j_{max}) \approx f(N, (1/2)[N-\sqrt{N}]) \approx O(2^N N^{-1})$$

for the frequencies of the one-(or two-)dimensional representation ($j=\{N/2\}$) and the $2^{\sqrt{N}}$ dimensional representation ($j=j_{max}$), respectively.

Below j_{max} the frequencies decrease exponentially towards the value 1 for $j = 0$. A numerical example can give an illustration (see Fig.4).

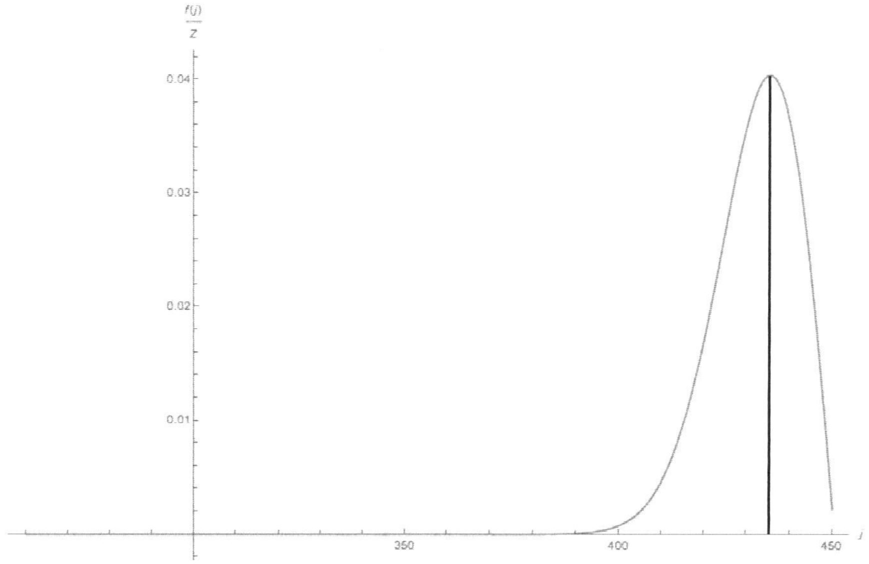

Fig. 4: Relative frequencies of irreducible representations in the decomposition of the N-fold tensor product of two-dimensional representations of the *SU(2)* group.

The chosen example is: $N=900$, $N/2 = 450$, $(N - \sqrt{N})/2 = 435$, $N/2-\sqrt{(2N)} = 407.5$. The largest wavelength (or smallest dimension) belongs to the representations $j=N/2$ (here 450), the most abundant ones to $j=(N-\sqrt{N})/2$ (here 435), and the smallest wavelengths (largest dimensions) with non-negligible frequency to $j=(N/2-\sqrt{2N})$ (here 407.5).

A numerical evaluation of the area below the curve gives $0.999284 \approx 1$. The integral from $j= 407.5$ to $j=450$ amounts to 0.984347. As this graphical example shows, the frequencies of representations with $j < N/2 - \sqrt{(2N)}$, that is, representations of the dimension between $\sqrt{(2N)}$ und N, can be safely neglected. In these representations, the corresponding wave-functions are of a wavelength that is considerably shorter than that associated with the representation of maximal frequency at $j=(N-\sqrt{N})/2$.

Let R denote the radius of the S^3, of the cosmic space. An *SU(2)* representation of dimension n implies wave functions that allow for a partitioning of the space down to a size of the order of R/n. (See Appendix1)

In the quantum-theoretical context of our argumentation "facts" are not given *a priori*, but have to be defined in an appropriate way. It is reasonable to relate facts to large probabilities.

Accordingly, an S^3-space accommodating N of the AQI bits will have a smallest "factual" length of an extension of the order $R/\sqrt{(2N)}$, which we shall identify with the Planck length.

Smaller lengths are to be seen as being merely "virtually" existent. This insight is relevant for our perception of the reality.

As a consequence of Planck‘s formula $E=hc/\lambda$ together with Einstein's formula $E=mc^2$, the characteristic extension of a quantum system, that is, its wavelength or Compton wavelength for massless or massive quanta, respectively, decreases with increasing energy or mass. Corresponding to a smallest "factually possible" length, there is a largest factually possible energy or mass, and thus also a largest number of AQI bits for which a system can be regarded a factually indivisible quantum particle.

For systems with energy or mass larger than the Planck mass to be considered as "indivisible quantum systems", the "particle"-model is no longer adequate. Here the Black Hole model may apply.

Objects with a mass exceeding the Planck mass, other than Black Holes, cannot be treated as factually indivisible quantum objects.

This is of relevance to the issue of the transition from a quantum to the classical description. That transition does not relate to the spatial extension but rather to the energies or masses here involved.

The border is the Planck mass. More massive systems (other than Black Holes) are at most "virtually indivisible quantum systems".

2.5. Plenty of extended becomes "small"

A single AQI bit was represented by a state function spreading out over S^3, the entire cosmic space. It seems to be at odds with common sense that it should be possible to construct from many of those AQI bits a strongly localized wave function. While our experience is "plenty" implies "big", it is postulated here that "plenty" results in "small". Seen from the information perspective, however, it is apparent that plenty of information affords a better localization than little information.

What comes into effect with the AQI bits is an important quantum-physical feature. In quantum theory there is a multiplicative composition of a system from parts – as opposed by the additive composition in classical physics.

3. Establishing a metric

As a science founding on empiricism, physics cannot be solely based on *a priori* arguments, but needs the connection to observations and experimental results.

As already recognized by Max Planck in 1899, the three fundamental natural constants – the velocity of light, the gravitational constant, and the action quantum or Planck's constant – allow one to derive units of length, time, and mass. These units are called Planck units nowadays. The Planck length and Planck time mark the smallest length and the shortest time that can be conceived as being factually realizable. The Planck mass characterizes the most massive quantum particle and also the border to the smallest conceivable black hole.

- Backed by this empirical knowledge we will identify the smallest realizable length following from our group-theoretical arguments with the Planck length l_{Pl}. Thus, the relation between the radius R of the S^3 space and the number of the AQI bits reads

(3.1) $$R = l_{Pl}\sqrt{2N}$$

This allows us to define distances in the S^3 space, that is, the physical space according to the derivation given here.

Assuming an invariable Planck length l_{Pl}, the radius R of the S^3 space will increase with a growing number N of the AQI bits. (While an expanding cosmos better conforms to common sense, one might equally well postulate $R=constant$ and assume a correspondingly shrinking of the Planck length l_{Pl}.)

When we speak of a "growing number" of AQI bits, a definition of time is required in which such a process shall be measured.

Having a definition of a distance via the Planck length, time (or duration) follows if a velocity is given.

- There is a distinguished velocity in nature, namely the velocity c of light in vacuum.

For the present cosmological model, this suggests to postulate a definition of the cosmic time t such that the cosmic radius R changes at the rate of the distinguished velocity c:

(3.2) $$R = ct$$

Using Planck units, $c=1$, this simply reads $R = t$.

A basic formula of quantum theory is Planck's relation between energy and the characteristic extension.

(3.3) $$E = h\nu = hc/\lambda$$

The "wavelength" λ of an AQI bit is of the order of the radius R of the S^3 space. Corresponding to that largest possible length, there is a smallest energy which can be understood as a factual quantity, that is, the energy E_{AQI} of a qubit $E_{AQI} \sim 1/R$

Since $R=\sqrt{(2N)}$ or $N=R^2/2$, the total energy of N AQI bits in the S^3 cosmos relates to R according to

(3.5) $$U \sim N/R \sim R/2$$

We consider a multitude of AQI bits in an expanding volume. Accordingly, it is necessary to resort also to statistical arguments and observe thermodynamic relations. The first law of thermodynamics addresses the behavior of the energy U of a system with varying volume, that is, for $dV \neq 0$.

Here a few explanatory remarks are in order, addressing also aspects of the GTR, even though that theory will not be used in deriving our arguments. Originally, the first law of thermodynamics allowed one to rationalize the performance of heat engines. An inflow of an amount of heat δQ can result in an increase of the internal energy or in an expansion of the volume dV towards an environment at pressure p: $\delta Q = dU + pdV$. In a thermally isolated system $\delta Q = 0$. Obviously, the cosmos is thermally isolated so that $\delta Q = 0$ is a reasonable assumption.

It is, however, important to note that the AQI bits are no particles. Arguments deriving from the thermodynamics of gases cannot simply be transferred to the present model.

For the "normal" materials considered in thermodynamics, almost always a positive pressure applies. It takes a positive pressure to increase the volume against external forces. However, such arguments apply only as long as the GTR is left out of consideration.

In the framework of the GTR all components of the energy-momentum tensor and thus also a positive pressure contribute to the gravitational effect and thereby to a contraction. There are two different aspects of the pressure. For a system in a volume the pressure means a force against the boundary. Beside this the pressure has the dimension of an energy density and any energy density reacts on the structure of space and time. Therefore, a sufficiently large positive pressure will cause an appreciable contracting effect by its reaction on space. In a sufficiently large neutron star an increasing positive internal pressure contributes to the gravitational collapse rather than preventing it. In contrast to our everyday conceptions, where a positive pressure inflates a tire, a negative pressure in the GTR counteracts a collapse.

Due to the quantum-theoretical relations for the AQI bits (3.5) the internal energy U of the cosmos and its volume grow in the same way. Accordingly, in the present model the inner pressure assumes a negative value as will be shown below.

The negative pressure associated with the fictitious Dark Energy is of such a magnitude that – if real – it would effect an accelerated cosmic expansion. Without resorting to the GTR, the value of the negative pressure deriving from the protyposis model entails a constant expansion.

- For our cosmological model featuring a closed yet varying volume the first law of thermodynamics applies,

(3.6) $$dU + p\, dV = 0$$

Since $dU \sim dR/2$ we obtain

(3.7) $$dR/2 + 2\pi^2 p\, 3R^2\, dR = 0$$

As the cosmic radius is changing, $dR \neq 0$, yielding the relation

(3.8) $$p = -1/4\pi^2 3R^2$$

Defining an energy density ρ according to $U = \rho V = \rho\, 2\pi^2 R^3$, ρ can be written as

(3.9) $$U = R/2 = \rho\, 2\pi^2 R^3 \quad \text{or} \quad \rho = 1/4\pi^2 R^2$$

The result is an equation of state for the AQI bits of the protyposis of the form

(3.10) $$p = -\rho/3 \quad \text{or} \quad \rho + 3p = 0$$

The authors of the studies on the $R_h = ct$ universe refer to this equation of state as the "zero active mass condition".

The equation of the quantum-cosmological protyposis model suggests to understand the cosmos as a splitting of the AQI vacuum into two equally large parts: a positive energy part comprising also matter and light, and a negative pressure part to be associated with a gravitational effect, invoking here an interpretation according to Newton's theory. In that interpretation gravitation is assigned, so to say, to a "negative energy".

Even though both parts become ever bigger due to the increase of the AQI bits and the thereby effected expansion of the cosmos, the sum of the parts stays always zero. This calls to mind the antique and medieval philosophers who pondered on a "creatio ex nihilo".

The protyposis concept establishes a real and moreover understandable foundation of the "substance of the cosmos" and its evolution.

The arbitrary fabrications such as "inflation" or "Dark Energy" become no longer necessary. The problems to be tackled by these hypotheses are solved in a natural way via the negative pressure entailed by the protyposis concept.

4. The cosmological term and Dark Matter

The cosmological constant Λ presents an interesting aspect in the history of scientific cosmology. Albert Einstein firmly believed that the universe should not have a temporal beginning. He had to realize, however, that his equations did not allow for a corresponding solution. Hence, he changed them by introducing an additional free constant Λ. This constant, having a vacuum-type energy-momentum tensor, enables a cosmos with matter but without a temporal onset. That "Einstein cosmos" is an S^3 space with a constant radius. The negative pressure introduced by Λ can, suitably adapted, balance a collapse of the space to be expected otherwise as an effect of the gravitating matter.

Shortly after, the redshift of the galaxies was detected and thereby the expansion of the cosmic space – implying a beginning in time, and Λ was dropped.

4.1. The cosmic substrate as an "ideal fluid"

If both the energy density and the pressure are uniform in the entire volume, then the physical model of an "ideal fluid" applies. In this model the energy-momentum tensor T_{ik} assumes the form of a diagonal matrix with the elements ρ (energy density) and p (pressure) depending only on time but not of the position in space or of the orientation.

The vacuum is a special case. Here the energy-momentum tensor is the unit matrix multiplied by a factor.

$$(4.1) \qquad T^i_k = \begin{bmatrix} \rho & & & \\ & p & & \\ & & p & \\ & & & p \end{bmatrix} \qquad {}_{Vakuum}T^i_k = \Lambda \begin{bmatrix} 1 & & & \\ & 1 & & \\ & & 1 & \\ & & & 1 \end{bmatrix}$$

Due to the indefinite metric of the Minkowski space, a positive pressure appears with a minus sign.

When in the 1980-ties the observation data improved, it was realized that the solutions of Einstein's equations seen as valid then, differed too much from the data. Therefore, the Λ concept was revived, allowing one to adapt a free parameter to the new data, without having to resort to new ideas.

The problem with Λ is that it has to be extremely small, though larger than zero.

The protyposis concept affords a solution to that problem. In cosmology, it is necessary to distinguish matter and light. Trying to partition the AQI substance accordingly into forms appearing as matter and as light, one finds that another part is needed that has to be of the form of an energy-momentum tensor of a "vacuum". As noted, the energy-momentum tensor associated with Λ is of this form. Accordingly, we denote that part by Λ_{eff}. As will be seen, the theoretically derived value for Λ_{eff} leads to good agreement with the empirical findings. However, Λ_{eff} cannot be a time-independent constant. Hence, we refer to Λ_{eff} , being temporally variable, as an "effective cosmological constant" or, preferably, as a "cosmological term".

Moreover, as mentioned before, Λ_{eff} solves the issue of the "Dark Matter" sought for since long. Trying to associate those particular forms of the AQI bits with a known quantum field, such a field would have to display the quantum characteristics of a vacuum. The Higgs field with charge and spin zero might be a possible candidate. However, there are arguments against that. The quanta of the Higgs field are of extreme mass and therefore decay utmost rapidly. While this would not yet be a compelling case against that assumption, it is more likely that the Dark Matter will not appear in the form of any kind of particles.

The protyposis energy-momentum tensor ${}_{AQI}T_{ik}$, in which the negative pressure enters with a positive sign, can be partitioned into contributions traditionally considered in cosmology. There the astronomers distinguish pressure-less dust, comprising stars, planets, and Black Holes, from light.

The AQI bits of the protyposis divide into matter, that is, ${}_{dust}T_{ik}$, light, ${}_{light}T_{ik}$ (positive pressure, implying a negative sign), and a further quantity having the structure of a vacuum, that is, of a cosmological term [10] ${}_{(vacuum)}T_{ik}$, displaying a negative pressure too.

(4.2)

$$ {}_{AQIs}T^i_k = \begin{bmatrix} \rho & & & \\ & \frac{\rho}{3} & & \\ & & \frac{\rho}{3} & \\ & & & \frac{\rho}{3} \end{bmatrix} = \begin{bmatrix} {}_{dust}\rho & & & \\ & 0 & & \\ & & 0 & \\ & & & 0 \end{bmatrix} + \begin{bmatrix} {}_{light}\rho & & & \\ & -\frac{{}_{light}\rho}{3} & & \\ & & -\frac{{}_{light}\rho}{3} & \\ & & & -\frac{{}_{light}\rho}{3} \end{bmatrix} + \begin{bmatrix} \Lambda_{eff} & & & \\ & \Lambda_{eff} & & \\ & & \Lambda_{eff} & \\ & & & \Lambda_{eff} \end{bmatrix} $$

According to the form of Eq. (4.2) for the tensor ${}_{AQI}T_{ik}$, there are two relations for the three parameters ${}_{dust}\rho$, ${}_{light}\rho$, and Λ_{eff}:

(4.3) $$\rho = {}_{dust}\rho + {}_{light}\rho + \Lambda_{eff}$$

(4.4) $$\rho/3 = - {}_{light}\rho/3 + \Lambda_{eff}$$

Obviously, these equations would not make sense without the Λ_{eff} term. The energy density of light would be negative, clearly an absurd consequence.

Let q denote the ratio of ${}_{dust}\rho + {}_{light}\rho$ and Λ_{eff}:

(4.5a) $$q = ({}_{dust}\rho + {}_{light}\rho)/ \Lambda_{eff}$$

(4.5b) $$\Lambda_{eff} = \rho/(q+1)$$

According to

(4.6a) $$\rho = - {}_{light}\rho + 3 \Lambda_{eff} = - {}_{light}\rho + 3 \rho/(q+1)$$

we may write ${}_{light}\rho$ as

(4.6b) $${}_{light}\rho = \rho (-1 + 3/q+1)) = \rho (3-q-1)/(q+1) = \rho (2-q)/q+1))$$

In a similar way, we obtain

(4.7) $$\rho = {}_{dust}\rho + \rho (2-q)/q+1)) + \rho/(q+1)$$

and thus

(4.8) $${}_{dust}\rho = \rho - \rho (2-q)/q+1) - \rho/(q+1) = \rho [q + 1 - 2 + q - 1] /(q+1) = 2 \rho [q - 1] /(q+1)$$

Since ρ, ${}_{light}\rho$, and ${}_{dust}\rho$ are positive, it follows from

(4.9a) $${}_{light}\rho = \rho (2-q)/(q+1)$$

(4.9b) $${}_{dust}\rho = 2 \rho (q - 1)/(q+1) ,$$

that q is bounded according to

(4.10) $$1 \leq q \leq 2$$

This means that Λ_{eff} has an energy density at least as large as that of matter and light combined, yet not exceeding twice that sum. This results in the correct order of magnitude of the cosmological term (or a time-dependent "effective cosmological constant").

4.2. The explanation of the "Dark Matter"

In a further approximation to the reality, the initial idealization of a "cosmic fluid", entailing the model of a uniformly expanding cosmos, can be refined as to assuming spatial variability. Then there may be non-uniform distributions of the three parts ${}_{dust}\rho$, ${}_{light}\rho$, and Λ_{eff} in the cosmos as fluctuations around a mean value, entailing local gravitational effects.[13] The protyposis part Λ_{eff} – as distinguished from dust (stars, planets, Black Holes) and light – does not appear as matter or light. However, as with matter and light one may assume local density variations of Λ_{eff} as well. Hence, the Λ_{eff} part of the cosmic content can be interpreted as part of the "Dark Matter" gravitation. The gravitational effect of the Black Holes, that is, a non-luminescent "dust" part of the protyposis not appearing in the form of "quantum particles" will be discussed in Sec. 6 below.

The protyposis concept dispenses with the two and a half thousand years old idea of atomism that ultimately the entire content of the cosmos should be reducible to "spatially smallest" building blocks.

5. A rationalization of Einstein's equations

The truly simplest structures consistent with physical and mathematical arguments are the protyposis AQI bits. As a necessary consequence of their existence, they cause a gravitational effect, as gravitation is the reaction of the cosmic evolution, caused by the increasing number of the AQI bits, to the content of the cosmos and the processes in it.

A cosmos with only a few AQI bits has a small radius and a large energy density (see Eq. 3.9), one with many AQI bits has a large radius and a small energy density. If one supposes that these relations, which apply to the content and the radius of the cosmos as a whole, can be transferred to the case of local fluctuations within the cosmos, then Einstein's equations follow as a classical approximation from the quantum-theoretical protyposis cosmology. [12 – 14]

The derivation along that path is permitted in so far as our argumentation up to here has been completely independent of Einstein's GTR equations.

The introduction of a cosmic time and, thus, of a distinguished coordinate or reference system seems to be at odds with what often is written about the GTR. There one may read: "There is no distinguished reference system." With regard the cosmology, however, this is incorrect [14]:

> "Already Dirac has referred to the fact that such a distinguished background metric remains unnoticeable as long as one is included into an "Einsteinian elevator" without any windows. The situation changes if a window to the cosmos is opened and the background radiation is included into the researches. [42] That all reference systems are of equal rights is therefore a local property in the description of localized systems with arbitrary accelerations, however, it is not necessary to claim this property for the whole cosmic space as well."

Since the cosmological model established here is a rigorous solution of Einstein's equations, one may invert the argumentation and, in an inductive manner, infer the structure of the equation from the solution:

The reasonable postulation that local density variations within the cosmos obey the same laws as those applying to the density and the radius of the entire cosmos justifies the general validity of the relations between the energy-momentum tensor and the Einstein tensor in their GTR form.

This explains why Einstein's equations allow for such an excellent description of local gravitational phenomena within the cosmos.

As to the linear approximations of the GTR, used for example to describe gravitational waves, their quantization is of course always possible, allowing for a quantum-theoretical treatment of gravitational waves in terms of gravitons.

6. A model of Black Holes

Already within the framework of Newtonian mechanics, the idea was conceived of an astronomical body having strength of gravity at its surface of such a magnitude that the theoretical escape velocity would surpass the velocity of light. According to the outcome of GTR calculations, there is a horizon enclosing a region in space from which nothing can escape. In its interior, as a consequence of the strict GTR description, all of the matter would vanish in a singular mathematical point.

Invoking for the first time quantum-theoretical aspects, Bekenstein [24, 25] and Hawking [26] could show that entropy has to be assigned to a black hole. The value of this entropy surpassed by many orders of magnitude anything that hitherto could be inferred from thermodynamically considerations for particles.

Considering AQI bits rather than particles in the treatment of Black Holes, already a simple toy model will make it plausible that Black Holes have entropies of the encountered order of magnitude.

Let us see what follows from the assumption that there are irreducible volumes – that is black hole equivalents – in the cosmos possessing a larger mass than Planck's mass. [10]

Elementary particles are associated with irreducible representations of the Poincaré group, and this mathematical concept implies the absence of any interior structure. Except for mass and spin, other physical quantities cannot be accommodated by those representations. Postulating the absence of accessible inner structure for an object with a mass exceeding Planck's mass, one obtains the black hole model. Here the event horizon precludes, on principle, any exterior knowledge of the conditions inside the black hole.

The challenge was to construct a simple model, which would allow one to illustrate the order of magnitude of the black-hole entropy.

Let n_{BH} denote the number of AQI bits within such an irreducible volume and N the total cosmic AQI bit number. The black-hole mass is larger than Planck's mass, so that $n_{BH} \geq \sqrt{(2N)}$.

Since each AQI bit has the energy $1/R$ the mass m_{BH} of the black hole is given by

(6.1) $$m_{BH} = n_{BH}\,(1/R)$$

The horizon property can be modeled by requiring that there must not be any overlap of the Black Holes within the cosmos, even if all cosmic AQI bits not assigned to the cosmic term Λ_{eff}, that is N/2 according to Eq. (4.10), participate in forming altogether z_{BH} Black Holes. Assuming that each of these z_{BH} Black Holes are of the same mass (and size), the product $z_{BH}\,n_{BH}$ satisfies

(6.2) $$z_{BH}\,n_{BH} = N/2 = R^2/4$$

Denoting the maximal extension (radius) of such an entity by r_{BH}, the horizon property requires that

(6.3) $$z_{BH}\,r_{BH} = R$$

This means that the Black Holes can be aligned next to each other, yet maintaining their mutual separation. If the sum of the black-hole radii is smaller than or at most equal to the radius R of the cosmos, then the z_{BH} Black Holes can always be arranged within the cosmos without necessitating any overlaps.

Combining the last two equations results in the relation

(6.4) $$n_{BH} = R\; r_{BH}\,/4$$

between n_{BH} and r_{BH}, which can be used in Eq. (6.1) to establish the linear relation

(6.5) $$m_{BH} = r_{BH}/4$$

for the mass of the black hole and its radius.

To address the black hole entropy, we will introduce two more quantities.

We consider black-hole inside-oscillations with an extension r_{BH} , of which each is formed by a subset of the n_{BH} black-hole AQI bits. Let k_{BH} denote the number of qubits in these subsets.

These oscillations, resulting from a coherent product of AQI oscillations, can be interpreted as the fundamental modes of the black hole.

According to Eq. (2.2) j qubits can form an j-dimensional irreducible representation of the $SU(2)$. In Appendix 1 it is mentioned that such an irreducible representation of the $SU(2)$ contains functions with wavelengths of order R/j. As the extension of these oscillations (its wavelengths) is given by r_{BH}, the following relation holds

(6.6) $$r_{BH} = R/k_{BH} \quad \text{or} \quad k_{BH} = R/r_{BH}$$

Now let us examine how many of those fundamental oscillation modes can arise in our constructs. Obviously, the fundamental modes within the black hole are analogous to the AQI bits in the cosmos as a whole. An AQI bit has a "wavelength" of the order of the cosmic radius R. To generate a much smaller "wavelength" of the order of the black-hole radius r_{BH}, it takes many, that is, k_{BH} coherent AQI oscillations.

Just as the total number of AQI bits correlates with the set of all possible states in the cosmos, the number of the fundamental modes of the black hole, to be denoted by s_{BH}, *corresponds to all possible inner states of the black hole, that is, information inaccessible from outside the black hole. Insofar* s_{BH} *is a measure of the entropy of the black hole.*

Obviously, s_{BH} satisfies the following relation

(6.7) $$n_{BH} = s_{BH} \times k_{BH} \quad [\text{or} \quad s_{BH} = n_{BH}/k_{BH}]$$

Using here Eqs. (6.4) and (6.6) for n_{BH} and k_{BH}, respectively, yields an explicit expression for s_{BH}:

(6.8) $$s_{BH} = r_{BH}^2/4 = 4\; m_{BH}^2$$

where the second equation follows from the mass-radius relation (6.5). This is essentially the black-hole entropy result according to Bekenstein [24, 25] and Hawking [26].

As these rather simplistic arguments show, the "wavelength" r_{BH} of the fundamental black-hole oscillations, each formed of k_{BH} AQI bits, corresponds to the extension of the event horizon. Like the cosmic AQI bits of wavelength R partitions the space into two halves, the "k_{BH}-modes" act as internal AQI-type bits for the irreducible black-hole volume. Apart from the number of these quasi-qubits, their actual states are unrecognizable from outside the black hole. At least the order of magnitude (up to a factor of π) of the black-hole entropy s_{BH} can be rationalized by these considerations. Note that a factor of 2 is missing in the present mass-radius relation (6.5), the correct result being $2m_{BH} = r_{BH.}$

The total number n_{BH} of the AQI bits forming the black hole is larger than its entropy s_{BH}. Entropy is to be understood as a measure of basically inaccessible information. In our black-hole model, s_{BH} is the share of the n_{BH} qubits for which the specific state cannot be recognized from outside. By contrast, it is possible, at least in principle, to come to know the states of the remaining AQI bits,

(6.9) $$n_{BH} - s_{BH} = n_{BH}(1-1/k_{BH})$$

The information on the position of a relatively small black hole within the vast cosmos is, among others, encoded in those qubits. Any localization requires much information, as already our everyday experience tells us. The more information is available, the easier a search will succeed. Much information or, likewise, a large number of AQI bits is required to determine the position of an object in the cosmos.

This shows that "entropy" cannot simply be equated with a respective number of AQI bits. In the case of a particle, for example, all the constituting AQI bits are accessible. Thus, the entropy concept does not apply to a particle in a known quantum state. Black Holes, by contrast, have inner states that, as a matter of principle, cannot be known. Accordingly, an amount of entropy is to be attributed to a black hole. The entropy of a black hole, being an object in the cosmos, is always smaller than the number of AQI bits forming the black hole.

7. The interior of Black Holes

Usually, Einstein's theory is also applied to describe Black Holes within their horizons. Quantum-theoretical aspects are considered only close to the Schwarzschild singularity.

> "The classical phase of the hole's internal evolution presents us with a problem which is mathematically quite definite and, in principle, straightforward. It is a hyperbolic initial-value problem of Cauchy's type. The evolution equations are the classical Einstein field equations. The initial data are set on or near the event horizon. The task is to evolve these data forward in time up to the point where a singularity is imminent. *(At this stage the classical evolution equations fail and the quantum regime takes over.)*" [27] (Highlighting by TG)

Postulating that the entire interior matter vanishes in a singular mathematical point is, in terms of physics, an utterly senseless statement, which certainly has contributed to the fact that until today there is serious resistance against the concept of Black Holes.

The non-physical hypothesis of a Schwarzschild singularity has to be dropped if quantum-theoretical arguments are taken into account. [28, 29] Here one has to abandon the idea that quantum theory applies only to "the small".

According to quantum theory, a limited volume, an impenetrable box, affects the ground state of the interior space.

Therefore – other than implied by the GTR – one cannot suppose that the vacuum inside and outside of the event horizon is the same.

The present simple black-hole protyposis model, yielding the correct order of magnitude of the entropy, was essentially based on the analogy between the inner state of a black hole and the cosmos in the protyposis description.

The energy density ρ_{BH} in a black hole is given by

(7.1) $$\rho_{BH} \approx m_{BH}/\, r_{BH}^{\,3} = 1/4\, r_{BH}^{\,2}$$

where the mass-radius relation (6.5) was used to obtain the second equation.

Analogous to Eq. (3.7) we find

(7.2) $$dm_{BH} + p_{BH}\, dV_{BH} \approx (1/4)\, dr_{BH} + p_{BH}\, 3\, r_{BH}^{\,2}\, dr_{BH} = 0$$

which results in the familiar equation of state

(7.3) $p_{BH} = -1/(4\times3)r_{BH}^2 = -\rho_{BH}/3$

for the interior of the black hole.

As has been elaborated elsewhere [28, 29], this equation of state for the interior of a black hole gives rise to the model of a Friedman-Robertson-Walker cosmos, which corresponds to the model of our cosmos. In view of this finding, one might be tempted to think of the possibility that our cosmos could be interpreted as the interior of a vast black hole.

8. The connection to the observations

8.1. The astrophysical data

At present, the generally accepted age of the cosmos is about 13.8 billion years, which, converted to Planck units, amounts to 8×10^{60} Planck times:

(8.1) $t_{cosmos} \approx 4.3588\times10^{17} s = 8\times10^{60}\, t_{Pl}$

Using that age as empirical input, we now may draw connections to the orders of other physical quantities.

According to our model, the present number of AQI bits is given by $N = R^2/2 = t_{cosmos}^2/2$, which amounts to

(8.2) $N = 3.2\times10^{121}$.

while the value of the cosmic radius becomes

(8.3) $R = 8\times10^{60}\, l_{Pl} = 1.28\times10^{28}\, cm$

where $l_{Pl} = 1.616\times10^{-33}\, cm$ was used in the conversion to cm.

In the $R=ct$ model the product of the Hubble parameter and the cosmic time t_{cosmos} is 1, which is in good agreement with the present findings. The Hubble parameter is defined by

(8.4a) $$H_0 = \frac{\dot{R}(t)}{R(t)}$$

which for $R(t) = ct$ becomes

(8.4b) $$H_0 = \frac{c}{ct} = \frac{1}{t}$$

According to WMAP5 [30] the current value is

(8.5) $H_0 = 70.5\ km\ s^{-1}\ Mpc^{-1} = 2.285\times10^{-18}\ s^{-1}$

and the multiplication with t_{cosmos} gives

(8.6a) $H_0 \times t_{cosmos} = 0.996.$

A recent paper [31] concludes a somewhat larger value, $H_0 = 73.1\ km\ s^{-1}\ Mpc^{-1}$, for which the product is

(8.6b) $H_0 \times t_{cosmos} = 1.033.$

Both results support the thesis of an expansion of the cosmic space at light velocity.

8.2. Extension of Einstein's equivalence to an equivalence of matter, energy, and quantum information

The formation of material and energetic quantum objects from the protyposis AQI bits suggests equivalence between matter, energy, and quantum information, extending Einstein's equivalence between matter and motion.

With Planck's units the gravitational constant G, the action quantum $\hbar$, and the velocity of light c have the value 1 and, thus, their actual role in the formulas is not manifest.

According to Eq. (3.4), the energy E_{AQI} of an AQI bit was seen to be equivalent to the inverse of the cosmic radius or, likewise, the inverse of the cosmic age. In a dimensionally correct form this relation reads

(8.7) $E_{AQI} = \hbar/t_{cosmos}$

With the values $t_{cosmos} = 4.358\times10^{17}\ s$ and $\hbar = 6.582\times10^{-16}\ eV\ s$, the AQI bit energy amounts to

(8.8) $E_{AQI} = 1.51\times10^{-33}\ eV$

or in cgs-units

(8.9) $$E_{AQI} = 0.242\times10^{-44}\ erg$$

A *10 eV* photon would be formed from about 6×10^{33} qubits, an electron, having a mass of *511 keV*, from 3×10^{38} qubits, and a proton with a rest mass of about *1 GeV* from 6×10^{41} qubits. The proton number is in the order of the value 10^{40} obtained by Weizsäcker already in 1974 as the result of rather demanding arguments. [41]

For the total cosmic energy $U \approx R/2$ we obtain the value

(8.10) $$U = 4.94\times10^{88}\ eV = 7.91\times10^{76}\ erg$$

With the value $c= 29\,979\,245\,800\ cm/s$ the total "cosmic mass" $M=U/c^2$ amounts to

(8.11) $$M = 5.5\times10^{56}\ g = 5.5\times10^{53}\ kg$$

For the energy density, $\rho = U/V = U/2\pi^2R^3$, we find

(8.12) $$\rho = 1.6\times10^{-8}\ erg/cm^3$$

and the corresponding matter density, $\rho_M = M/V = M/2\pi^2R^3$, is given by

(8.13) $$\rho_M = 1.34\times10^{-29}\ g/cm^3 = 1.34\times10^{-26}\ kg/m^3$$

The so-called critical density is defined by $\rho_c = 3H_0^2/(8\pi G)$, which amounts to $\rho_c = 0.934\times10^{-29}\ g/cm^3 = 0.934\times10^{-26}\ kg/m^3$, where the H_0 value according to Eq. (8.6a) and $G=6.674\ cm^3/g\ s^2$ have been assumed. (H_0 from (8.6b) implies $\rho_c = 1.004\times10^{-29}\ g/cm^3$) The corresponding energy density is $8.38\times10^{-9}\ erg/cm^3$.

The critical density corresponds to a flat universe. For larger values, such as those obtained in the present protyposis model, a closed cosmos with a finite volume results.

The value of the cosmological term Λ_{eff}, that is, the time-dependent "effective cosmological constant" is in the range

(8.14) $$\rho/2 \leq \Lambda_{eff} \leq 2\ \rho/3$$

where ρ is the total energy density of the cosmos. While there is abundant speculation on the ratio of the as yet completely unknown "Dark Energy" and "Dark Matter" ingredients of the cosmos, specific assessments (in erg/cm^3) of the actual total energy density are hard to come by. A vacuum energy of $10^{-10}\ erg/cm^3$ has been listed by A. Riess [32] in the Encyclopedia Britannica. For the matter density calculated values between [33] $4.7\times10^{-27}\ kg/m^3$ or [34] $8.47\times10^{-27}\ kg/m^3$ are given in Wikipedia. The uncertainties with regard to these quantities are apparent.

8.3. A test of consistency: The protyposis concept and the black hole entropy

As an interesting supplement, we perform a consistency check of the protyposis model with regard to cosmological and astrophysical data, referring here also to the GTR.

Since there is no principal theoretical upper bound for the mass of a black hole, one may suppose, as a thought experiment, that the total mass of the universe is contained within a single black hole.

According to the GTR the Schwarzschild radius is given by $r_S = 2GM/c^2$, or, in Planck units,

(8.15) $$r_S = 2\ M$$

This can be compared with the relation $R/2 = U$ of the protyposis model, to be written also as $R/2=M$ according to Einstein's relation between U and M with $c=1$. This shows that the cosmic radius and the Schwarzschild radius of the hypothetical black hole are of the same order of magnitude.

The value of Planck's mass is $m_{Pl} =2.176\times10^{-8}\ kg$. The proton mass is $m_p = 1.672621\times10^{-27}\ kg$ or $7.7\times10^{-20}\ m_{Pl}$. Now suppose a single proton drops into the mass M black hole (of $8\times10^{60}\ m_{Pl}$) so that the total mass becomes $M+m_p$.

The entropy of a black hole is given by Bekenstein [24, 25] and Hawking [26] as $S = k_Bc^3A/4\hbar G$ where $A=16\ \pi\ (GM/c^2)^2$ and k_B is Boltzmann's constant. (Measuring the temperature in erg, then $k_B=1$)

In Planck units this formula reduces to $S = 4\pi M^2$. The difference of the entropies before and after addition of the proton is given by

(8.16) $$\Delta S/4\pi = (M + m_p)^2 - M^2 \approx 2\ M\ m_p$$

Inserting the numerical values we find

(8.17) $$\Delta S = 8\times10^{60}\times7.7\times10^{-20}/4\pi = 9.8\times10^{40}$$

Thereby we have captured the order of magnitude of all AQI bits that become non-accessible via a single proton. Stated differently: *A proton "is" 10^{41} AQI bits.*

The comparison with the findings according to Eq. (8.9), by which 6×10^{41} *AQI* bits were attributed to a proton, shows that the data based on estimates of the orders of magnitude are rather consistent and quite realistic.

9. The protyposis concept and the foundation of interactions

Aspiring to be the basis for a physical modeling of the reality, the protyposis concept should also allow one to establish the mathematical form of interaction. This has in fact been achieved [35, 36], as will be briefly outlined in the following.

For a long time, the quest to unify the scientific description of nature followed the path towards ever-greater symmetries underlying the forces.

In the protyposis concept, the issue of a unification of the forces is shifted to the level of the simplest and thus fundamental structures of nature. This allows one to derive the basic forms of interactions and thereby establish the distinct interactions.

9.1. Interaction and gauge groups

As is well known, the interactions, seen as relevant in the atomic realm, can be understood in terms of local gauge groups. The compact groups $U(1)$, $SU(2)$, and $SU(3)$ are the gauge groups for the electromagnetic, the weak, and the strong interaction. Aiming at an eventual unification, one has searched for larger groups comprising these three gauge groups as subgroups. As possible candidates for such a larger group, the groups $SU(5)$, $SO(10)$, and $E(8)$ have been investigated. However, all these big groups have the undesirable feature of grossly inflating the manifold of the concomitant "fundamental quantum species", that is, of entities to be considered as simple and elementary. The corresponding hypothetical particles, which have not been found in the energy region known so far, must be expected to be extremely instable in view of their large mass, as the experimental experience suggests. This could already be seen in the case of the Higgs particle, which has a lifetime in the order of 10^{-22} *s*. Obviously, one here reaches the limit for the applicability of the notion of "existence".

Interaction is a concept deriving from classical physics. The tensor product in the quantum-theoretical composition generates "entireness without parts". Therefore interaction can only be a valid concept for systems that can be considered as consisting of separate parts.

Already in Newtonian mechanics an individual coordinate cosmos is assigned to each particle. The time-evolution of the particle state is monitored by the variation of the respective coordinates. In the quantum-theoretical description, change is associated with the momentum operators P_k, being essentially the derivatives with respect to the coordinates of the wave function,

$$P_k = -i\, \partial/\partial x_k \tag{9.1}$$

As is empirically well confirmed, the "switching on" of an interaction can be effected by replacing the partial derivatives with the covariant derivatives

$$\partial/\partial x_k \Rightarrow \partial/\partial x_k + g\, A_a^k\, \theta^a \tag{9.2}$$

Here Θ^a is a generator in the Lie algebra of the respective gauge group and g denotes a coupling constant.

9.2. Electromagnetic and weak interaction

The description of the electromagnetic and the weak interaction follows rather directly from the preceding considerations.

For a particle in Minkowski space, the interaction concept requires to choose the coordinate cosmos of a possible interacting partner. According to the protyposis cosmology this is a homogeneous space of the $U(1)$ and $SU(2)$ groups.

The operator for the motion in Minkowski space has to be augmented by the operators for the motion in the $U(1)\times SU(2)$ manifold, where a linear approximation is enabled in form of the Lie algebras $\boldsymbol{u(1)+su(2)}$. This means that the motions in the homogeneous spaces of the $U(1)$ and $SU(2)$

IOP Conf. Series: Journal of Physics: Conf. Series **1071** (2018) 012011 doi:10.1088/1742-6596/1071/1/012011

groups are approximated in the vicinity of the unit element of the respective group. An element γ of the *U(1)* group can be represented according to

(9.3) $$\gamma = exp\{i\, A\, \zeta\} \approx 1 + i\, A\, \zeta$$

and similarily for the *SU(2)* group,

(9.4) $$\gamma = exp\{i\Sigma\, B_a\, \sigma^a\} \approx 1 + i\, \Sigma\, B_a\, \sigma^a$$

Here, as usual, the expansions of the exponential functions have been truncated after the linear term. The generators σ^a of the Lie algebra ***su(2)*** are the three Pauli matrices.

In the electromagnetic and weak interaction the momentum operators P^k are extended to the well-known forms

(9.5) $$P^k \rightarrow -i\partial/\partial x_k + g_1 A^k \zeta + g_2 B^k{}_a \sigma^a$$

The two coupling constants g_1 and g_2 serve a measure of the respective charge and, thus, of the strengths of the interactions as well as the mass generated by the charge.

9.3. The strong interaction

In the foundation of the strong interaction it has to be taken into account that quantum theory requires the use of complex numbers. Quantum theory accounts for real effects deriving from possibilities. In classical physics, possibilities merely reflect the lack of knowledge on part of the describer and, thus, do not at all affect the behavior of the systems.

Being much more accurate, quantum theory captures the temporally non-local influence of future possibilities on the current process. This cannot be modeled using only real numbers.

Real and imaginary numbers satisfy the well-known relations

(9.6) $$real \times real = real, \qquad imaginary \times imaginary = real, \qquad real \times imaginary = imaginary$$

In the field of groups, these relations are mirrored by the so-called Cartan decomposition. [37]

Consider a group *G* with a subgroup *K,* and let *P* denote the corresponding coset. Supposing that the Lie algebra **G** can be decomposed according to

(9.7) $$\mathbf{G} = \mathbf{K} + \mathbf{P}$$

where **K** is the subalgebra associated with *K*, then this is referred to a Cartan decomposition of the Lie group **G** if the elements

(9.8) $$k_i \in \mathbf{K} \quad \text{and} \quad p_i \in \mathbf{P}$$

satisfy the relations

(9.9) $$[k_i, k_j] \in \mathbf{K} \qquad [p_i, p_j] \in \mathbf{K} \qquad [k_i, p_j] \in \mathbf{P}$$

Here $[a,b] = ab - ba$ denotes the commutator of *a* and *b*.

If the manifolds **K** and **P** are of same dimension, this algebraic structure is an analogue to the relations (9.4), underlying the transition from real to complex numbers, representing here $\boldsymbol{C}^1$ by $\boldsymbol{R}^2$.

To incorporate the quantic possibilities this suggests doubling the coordinate cosmos and establishing a structure that is analogous to that of Eqs. (9.6) and (9.9).

This postulate leads from *U(1)×SU(2)* directly to the *SU(3)* group.

Following a paper by Byrd [38], it can be shown that the *SU(3)* group, having *8* parameters to be denoted by the string *(α,β,γ,θ,a,b,c,φ)*, features together with the *U(1)* and *SU(2)* groups exactly the structure according to Eqs. (9.6-9.9).

As Byrd has shown, an arbitrary element of the *SU(3)* group can be represented as

(9.10) $$D(\alpha,\beta,\gamma,\theta,a,b,c,\varphi) = e^{(i\lambda_3\alpha)} e^{(i\lambda_2\beta)}\, e^{(i\lambda_3\gamma)} e^{(i\lambda_5\theta)} e^{(i\lambda_3 a)} e^{(i\lambda_2 b)} e^{(i\lambda_3 c)} e^{(i\lambda_8\varphi)}$$

Here λ_i denote 3×3 matrices of the set of the 8 Gell-Mann matrices (see Appendix 2). In particular, the matrices λ_1, λ_2, and λ_3 result from augmenting the three Pauli matrices with each one row and one column of zero elements.

The representation (9.10) is of the form

(9.11) $$D(\alpha,\beta,\gamma,\theta,a,b,c,\varphi) = D^{(2)}(\alpha,\beta,\gamma) e^{(i\lambda_5\theta)}\, D^{(2)}(a,b,c) e^{(i\lambda_8\varphi)}$$

Here $e^{(i\lambda_5\theta)}$ and $e^{(i\lambda_8\theta)}$ are elements of the *U(1)* group, and $D^{(2)}(\alpha,\beta,\gamma)$, $D^{(2)}(a,b,c)$ elements of *SU(2)*.

Thus, when the quantum-theoretical possibilities are taken into account, one retrieves, in addition to the electromagnetic and weak interaction, the structure of the strong interaction.

However, the corresponding quanta can only be structural quanta. Accordingly, quarks and gluons cannot appear as actual quantum particles in vacuum. This, however, does not prevent that they can act like particles within strongly interacting matter.

According to these insights, Eq. (9.5) is to be extended as

(9.12) $$P^k \rightarrow -i\partial/\partial x_k + g_1 A^k \zeta + g_2 B^k{}_a \sigma^a + g_3 C^k{}_a \lambda^a$$

The fact that the structural quanta of the strong interaction, quarks and gluons, cannot appear as free particles in vacuum is consistent with the empirical findings.

10. The general importance of the protyposis concept for a foundation of the natural sciences.

The goal of science is to explain the phenomena in the world that will allow us to derive instructions for action in an adapted and preferably rational way. To explain means to reduce complex structures to simple ones, unknown structures to ones already known. Such a reduction can be seen as successful, when something not understood becomes comprehensible.

Since for mathematical and physical reasons, there are no simpler structures than the protyposis AQI-bits, the latter are the logical basis for any scientific explanations. In the practical implementation it is of course necessary to interpret and render comprehensible the mathematical transitions from a lower structural level to a higher one.

Given the amount of knowledge available today, it has become possible to base the observation data upon sound theoretical foundations. Today, the evolution within the cosmos can be understood. The evolution has begun with very simple structures, from which very complex structures have evolved. This makes apparent that natural science must reduce. Here, a dualistic concept, supposing two basic substances such as mind and matter, must be ruled out.

In principle, the actual structures can be derived from the simplest structure, the protyposis. Here "in principle" means that the mathematical limits are understood and interpreted. Then it is not necessary or reasonable to treat the more complex structures at the next higher level without using the approximation methods suitable there and apply the procedures of the preceding level. A "Schrödinger equation for the cell" would not be a sensible approach.

In the cosmic evolution, after a very early and very hot phase, Black Holes are formed. Around them, in first galaxies, stars are formed from the then available hydrogen and helium. After the first supernova explosions had generated heavy elements, celestial bodies such as planets and comets can be formed.

If the conditions on a planet are suitable, life will develop. And if these conditions prevail sufficiently long, that is, if the star is not too big and thus explodes too soon and if the planet is not thrown off its course, then life will eventually develop forms of life capable of consciousness.

As has been known for a long time, quantum particles with a rest mass can take up and give off energy. Usually this is effected by absorption and emission of real and virtual photons. Being manifestations of the protyposis AQI bits, it is comprehensible that the material quanta and the photons can absorb or emit single AQI bits to be seen as meaningful. Mostly, these will be referred to as "properties" of the particle.

Living beings differ from stable objects in that they permanently face a plethora of instable situations at every organizational level – from the constituents of the cells, to cells and organs, up to the entire being. In instable situations already tiniest causes can effect an influence.

Therefore life can be characterized in that living beings are thermodynamically instable systems that stabilize themselves via the processing of information. ([44], [20] Chap. 12)

During the biological evolution, in rapidly movable life forms, such as animals, organs will develop that are specialized on information processing. Ultimately, in animals with a highly developed brain the formation of consciousness becomes possible (see [45] and the summary in [20] pp. 738).

The protyposis concept allows us to understand that an advanced system of information processing such as the brain is distinguished in that there is no strict separation between "hardware" and "software".

As bits of quantum information, the AQI-bits of the protyposis are to be related – if an every-day metaphor is suitable here – to our thoughts rather than to our body. With the protyposis as fundamental substance, it has become possible to explain the genesis of matter and the evolution, from the cosmos to life, up to the human psyche and consciousness, in a unifying way. [20]

With the humans, who not only can enjoy and admire the creations of nature but, moreover, are destined to comprehend all that, the arc beginning with the big bang of the $R=ct$ universe comes to a conclusion.

Acknowledgments
I am very grateful to Jochen Schirmer for various and very helpful comments and suggestions.

I thank the referee for useful comments.

Appendix 1

An element g of the group $SU(2)$ has the form

(A1-1) $$g = \begin{pmatrix} x+iy & u+iv \\ -u+iv & x-iy \end{pmatrix} \quad \text{with} \quad \det|g| = x^2+y^2+u^2+v^2 = 1$$

With the Euler angles the usual parameterization of the S^3 with radius 1 results in

(A1-2) $$g(\varphi,\psi,\theta) = \begin{pmatrix} e^{i(\varphi+\psi)}\cos\theta & ie^{i(\varphi-\psi)}\sin\theta \\ ie^{-i(\varphi-\psi)}\sin\theta & e^{-i(\varphi+\psi)}\cos\theta \end{pmatrix}$$

The Hilbert-space $L^2(S^3)$ of the square integrable functions $f(\varphi, \psi, \theta)$ over S^3 is the representation space for the regular representation of $SU(2)$.

An irreducible subrepresentation of "spin j" has a (2j+1)-dimensional representation space. In this space a basis is given by the functions $D_j^{kl}(\varphi, \psi, \theta)$. (see e.g. [43])

(A1-3) $$D^j(\varphi,\psi,\theta)_l^k = e^{-i(k\varphi+l\psi)}d^j(\theta)_l^k$$

The functions $d^j(\theta)^k{}_l$ are the Jacobian polynoms

(A1-4) $$d^j(\theta)_l^k = \sqrt{\frac{(j+k)!(j-k)!}{(j+l)!(j-l)!}}\left(\cos\frac{\theta}{2}\right)^{k+l}\left(\sin\frac{\theta}{2}\right)^{l-k} P_{j-k}^{k-l,k+l}(\cos\theta)$$

In the special cases $k=l$ Jacobian Polynoms transform into Legendre polynoms $P_j^k(\cos\theta)$

(A1-5) $$P_{j-k}^{k,k}(z) = (-2)^k\frac{j!}{(j-k)!}(1-z^2)^{-k/2}P_j^k(z)$$

In present connections the shortest wavelengths are of interest. For the trigonometric functions of φ resp. ψ the shortest wavelengths result for l resp. $k=j$, for the Legendre polynoms $P_j^k(\cos\theta)$ for $k=0$.

The wavelengths are of order $1/j$ resp. $2\pi/j$. For the case $j=30$ an illustration is given for $\cos(30x)$ und $P_{30}(\cos x)$

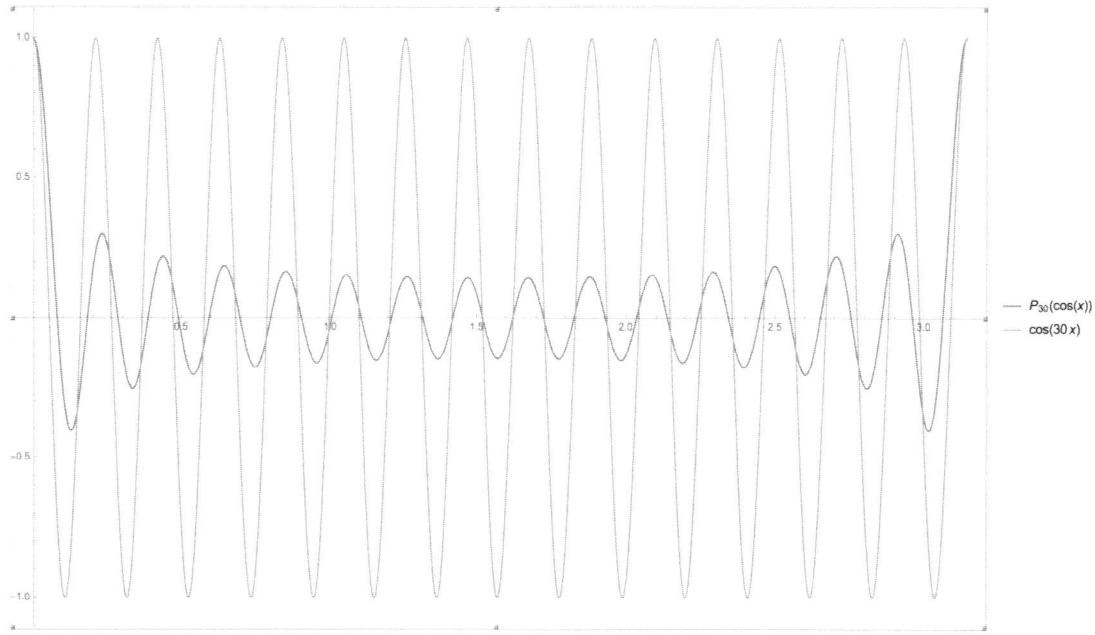

Fig. 5: The Graphs of the functions cos(30x) and P_{30}(cosx) between 0 and 2π.

Appendix 2

$$\lambda_1 = \begin{pmatrix} 0 & 1 & 0 \\ 1 & 0 & 0 \\ 0 & 0 & 0 \end{pmatrix}, \quad \lambda_2 = \begin{pmatrix} 0 & -i & 0 \\ i & 0 & 0 \\ 0 & 0 & 0 \end{pmatrix}, \quad \lambda_3 = \begin{pmatrix} 1 & 0 & 0 \\ 0 & -1 & 0 \\ 0 & 0 & 0 \end{pmatrix},$$

$$\lambda_4 = \begin{pmatrix} 0 & 0 & 1 \\ 0 & 0 & 0 \\ 1 & 0 & 0 \end{pmatrix}, \quad \lambda_5 = \begin{pmatrix} 0 & 0 & -i \\ 0 & 0 & 0 \\ i & 0 & 0 \end{pmatrix}, \quad \lambda_6 = \begin{pmatrix} 0 & 0 & 0 \\ 0 & 0 & 1 \\ 0 & 1 & 0 \end{pmatrix},$$

$$\lambda_7 = \begin{pmatrix} 0 & 0 & 0 \\ 0 & 0 & -i \\ 0 & i & 0 \end{pmatrix}, \quad \lambda_8 = \frac{1}{\sqrt{3}} \begin{pmatrix} 1 & 0 & 0 \\ 0 & 1 & 0 \\ 0 & 0 & -2 \end{pmatrix}.$$

Fig. 6: The 8 generators of the Lie algebra of the *SU(3)* group in the so-called Gell-Mann representation. λ_1, λ_2 and λ_3 are extensions of the Pauli matrices; together with λ_8 they generate the *U(2)* group.

	λ_1	λ_2	λ_3	λ_4	λ_5	λ_6	λ_7	λ_8
λ_1	0	$2i\lambda_3$	$-2i\lambda_2$	$i\lambda_7$	$-i\lambda_6$	$i\lambda_5$	$-i\lambda_4$	0
λ_2	$-2i\lambda_3$	0	$2i\lambda_1$	$i\lambda_6$	$i\lambda_7$	$-i\lambda_4$	$-i\lambda_5$	0
λ_3	$2i\lambda_2$	$-2i\lambda_1$	0	$i\lambda_5$	$-i\lambda_4$	$-i\lambda_7$	$i\lambda_6$	0
λ_4	$-i\lambda_7$	$-i\lambda_6$	$-i\lambda_5$	0	$i\lambda_3$ $+i\sqrt{3}\lambda_8$	$i\lambda_2$	$i\lambda_1$	$-i\sqrt{3}\lambda_5$
λ_5	$i\lambda_6$	$-i\lambda_7$	$i\lambda_4$	$-i\lambda_3$ $-i\sqrt{3}\lambda_8$	0	$-i\lambda_1$	$-i\lambda_2$	$i\sqrt{3}\lambda_4$
λ_6	$-i\lambda_5$	$i\lambda_4$	$i\lambda_7$	$-i\lambda_2$	$i\lambda_1$	0	$-i\lambda_3$ $+i\sqrt{3}\lambda_8$	$-i\sqrt{3}\lambda_7$
λ_7	$i\lambda_4$	$i\lambda_5$	$-i\lambda_6$	$-i\lambda_1$	$i\lambda_2$	$i\lambda_3$ $-i\sqrt{3}\lambda_8$	0	$i\sqrt{3}\lambda_6$
λ_8	0	0	0	$i\sqrt{3}\lambda_5$	$-i\sqrt{3}\lambda_4$	$i\sqrt{3}\lambda_7$	$-i\sqrt{3}\lambda_6$	0

Fig. 7: Product table for the λ_i matrices.

References

[1] van Melsen A G M 2016 Atomism, https://www.britannica.com/topic/atomism, dated 03/21/2018
[2] López-Corredoira M *et al* 2016 *Intern. J. of Modern Physics D* **25**, doi: 10.1142/S0218271816500607
[3] Melia F and López-Corredoira M 2015 arXiv:1503.05052v3 [astro-ph.CO]
[4] Wei J-J *et al* 2015 *Astron. J.* **149**, 102
[5] Melia F 2012 *Astron. J.* **144**, 110
[6] Melia F and Maier R S 2013 *Mon. Not. Roy. Astron. Soc.* **432**, 2669
[7] Nielsen J T, Guffanti A and Sarkar S 2016 *Scientific Reports* | 6:35596 | DOI: 10.1038/srep35596
[8] Melia F Jan. 2017, private Communication
[9] Görnitz T 1986 *Intern. J. Theoret. Phys.* **25**, 897
[10] Görnitz T 1988a *Intern. J. Theoret. Phys.* **27**, 527
[11] Görnitz T 1988b *Intern. J. Theoret. Phys.* **27**, 659
[12] Görnitz T 2009 https://arxiv.org/pdf/0904.1784v1.pdf
[13] Görnitz T 2010 https://arxiv.org/pdf/1008.4558v1
[14] Görnitz T 2011 *Adv. Sci. Lett.* **4** 577
[15] Zeiliger Group (2017) Experimental Quantum Foundations and Applications. http://www.iqoqi-vienna.at/research/zeilinger-group/satellite-based-quantum-communication/ Access 21.03.2018
[16] Görnitz T, Graudenz D and Weizsäcker C F v 1992 *Intern. J. Theoret. Phys.* **31**, 1929
[17] Görnitz T and Schomäcker U 2012 *Journ of Physics: Conference Series* **380** 012025
[18] Görnitz T and Görnitz B 2002, 2006, 2013 *Der kreative Kosmos* (Heidelberg, Spektrum)
[19] Görnitz T and Görnitz B 2008, 2009 *Die Evolution des Geistigen* (Göttingen, Vandenhoeck&Ruprecht)
[20] Görnitz T and Görnitz B 2016 *Von der Quantenphysik zum Bewusstsein – Kosmos, Geist und Materie* (Heidelberg, Springer)
[21] Weizsäcker, C F v (1955) *Naturwissenschaften* **42** (1955) 521, 545
ScheibeE, Suessmann G and Weizsäcker C F v 1958 *Zeitschrift für Naturforschung*, **13a**, 705
[22] Weizsäcker C F v 1985 *Aufbau der Physik* (München, Hanser)
[23] Weizsäcker C F v 2006 *The Structure of Physics,* edited, revised and enlarged by Görnitz T and Lyre H (Heidelberg, Springer)
[24] Bekenstein J D 1973 *Phys. Rev. D* **7**, 2333
[25] Bekenstein J D 1974 *Phys. Rev. D* **9**, 3292

[26] Hawking S W 1975 *Comm. in Math. Phys.* **43**, 199
[27] Israel W 1998 *Internal Stucture of Black Holes* (Black holes: theory and observation) ed. Hehl F W, Kiefer, C and Metzler R J K, (Heidelberg, Berlin, Springer) p. 365
[28] Görnitz T and Ruhnau E 1989 *Intern. J. Theoret. Phys.* **28**, 651
[29] Görnitz T 2013 *Quantum Matter*, **2**, 21
[30] Komatsu E *et al* 2009 *The Astrophysical Journal Supplement Series* **180**, 330
[31] Wong K C *et al* 2016 *MNRAS* **4**, 465 (*Preprint* arXiv:1607.01403)
[32] Riess A 2016 *Dark energy* (Encyclopedia Britannica) https://www.britannica.com/topic/dark-energy; dated 12/30/2016
[33] https://de.wikipedia.org/wiki/Universum; dated 12/30/2016 [use english pages]
[34] https://de.wikipedia.org/wiki/Lambda-CDM-Modell; dated 12/30/2016 [dto.]
[35] Görnitz T 2014 *Rev. in Theoret. Science* **2,** 289
[36] Görnitz T and Schomäcker U 2018 *Found Sci* **23**, 51
[37] Böhm M 2011 *Lie-Gruppen und Lie-Algebren in der Physik* (Heidelberg, Springer) pp. 377
[38] Byrd M 1998 The Geometry of SU(3) *Preprint* UTEXAS-HEP-97-18, arXiv:math-ph/9807032v1
[39] Woodward J *Scientific Explanation* (The Stanford Encyclopedia of Philosophy: Fall 2017 Edition) ed Zalta N: URL = <https://plato.stanford.edu/archives/fall2017/entries/scientific-explanation/>.
[40] Blair A 1990 *Journ of the History of Ideas*, **51**, 355
[41] Weizsäcker C F v 1974 *Nova acta Leopoldina*, N. F. 212, **39**, 61 (Leipzig,Barth)
[42] Dirac P A M 1980 *Why We Believe in The Einstein Theory* (Symmetries in Science) ed. Gruber B and Millman B R S (New York, London: Plenum Press) p. 1-11
[43] Tung Wu-Ki 1985 *Group Theory for Physics* (Singapore, World Scientific) pp 141
[44] Görnitz T *Found Sci* 2014 **19** 257
[45] Görnitz T *Found Sci* 2017 https://doi.org/10.1007/s10699-017-9536-9

IOP Conf. Series: Journal of Physics: Conf. Series **1071** (2018) 012012 doi:10.1088/1742-6596/1071/1/012012

Another Introduction to Geometric Algebra with some Comments on Moore-Penrose Inverses

Martin Erik Horn

HWR/BSEL – Hochschule für Wirtschaft und Recht Berlin / Berlin School of Economics and Law, Badensche Str. 52, D – 10825 Berlin, Germany

E-mail: e_hornm@doz.hwr-berlin.de mail@martinerikhorn.de

Abstract. With his theory of extensions Hermann Grassmann gave algebra a substantial different shape: He indeed had invented a generalized version of Pauli and Dirac Algebra which can be applied not only to re-model our mathematical view on space and time, but also to re-model mathematics – and especially the relation between geometry and algebra – itself. Therefore this didactical version of generalized Pauli and Dirac Algebra is called Geometric Algebra.

In the following an introduction to Geometric Algebra is given and applications of Geometric Algebra in physics and mathematics are discussed. Special emphasis is given to the mathematics of pure and mixed sandwich products and to a different view on solving systems of linear equations. At the end matrix inverses of non-square matrices (e.g. Moore-Penrose generalized inverses) are discussed.

1. Time scales of mathematical inventions: The history of solving systems of linear equations

One of the earliest cultural techniques mankind had invented has been mathematics: "as far as our evidence goes, 'mathematics' precedes writing" [1]. And some of the earliest mathematical techniques mankind had invented have been strategies to solve systems of equations: systems of linear equations and mixed systems of linear and quadratic equations as well.

The main tool of solving systems of linear equations in these early times has been the algorithm we nowadays call "Gaussian elimination" [2, 3]. This indicates that time scales of mathematical inventions can sometimes be very or even very, very long: Several thousands of years ago mathematicians may already have been able to solve simple systems of equations, but next to nothing is known about this time as writing was invented only later.

Then at about two thousand years BC mathematicians of the Old Babylonian Period fixed the solution of systems of equations at cuneiform tablets for the first time. After that we again had to wait for another two thousand years before new progress was made: At about 0 BC Diophantus and his colleagues (not even the century is known in which Diophantus lived [4]) invented symbolic algebra and the use of variables.

And the next step of mathematical progress in solving systems of linear equations came again only nearly two thousand years later. After the invention of determinants by Seki (in Edo/Tokyo), who "was not only the discoverer but … had a much broader idea than that of his great German contemporary" [36] Leibniz (in Hanover), Cramer wrote down his rule in 1750 and opened the route to Grassmann and to "another annus mirabilis, 1844, the birthyear of the Ausdehnungslehre, one of the supreme landmarks in the history of the human mind" [5].

2. The theory of extensions: What has Grassmann said?

The way Grassmann solved systems of linear equations is of enormous conceptual importance. And he surely was right, telling his readers that "es werde ... die Algebra eine wesentlich veränderte Gestalt gewinnen" [6] – Algebra will gain a substantial different shape by using outer products.

Therefore it makes sense to start this introduction to Geometric Algebra by analyzing what Grassmann had to say and how he explained his solution strategy. Read the classics! If we only look on copies, which are carefully cleaned, polished, sometimes dramatically modified or even rebuild in a totally different way, we will only get a blurred and hazy look on the supreme landmarks in the history of the human mind. To really understand these landmarks, we should directly confront ourselves with them.

So let us start with a system of three linear equations with three unknown variables x, y, z

$$\begin{aligned} a_1\, x + b_1\, y + c_1\, z &= r_1 \\ a_2\, x + b_2\, y + c_2\, z &= r_2 \\ a_3\, x + b_3\, y + c_3\, z &= r_3 \end{aligned} \qquad (1)$$

(in contrast to Grassmann who immediately discussed the general case with an arbitrary number n of variables) and let us try to follow Grassmann's explanation in German: "Hier können wir die Zahlenkoefficienten, welche verschiedenen Gleichungen angehören, sofern wir diese Verschiedenheit an ihrem Begriff noch festhalten, als verschiedenartig ansehen, und zwar alle als an sich verschiedenartig, d. h. als unabhängig in dem Sinne unserer Wissenschaft, die einer und derselben Gleichung als unter sich in derselben Beziehung gleichartig" [6].

Even as a native German speaker it is extremely difficult to grasp the meaning of this one sentence as Grassmann's train of thought not only seems to have been light-years away from standard mathematical reasoning of most contemporaries of his time, but also from standard mathematical reasoning we are used to today.

Perhaps a translation into English will help to clarify his ideas: "Here we can consider the coefficients which are part of different equations as different, if we still grasp this difference by its inner idea and indeed all of them as inherently different, this means, as independent in the sense of our science, (and the coefficients) of one and the same equation as of the same kind with respect to the same relation."

A shorter version of these lines might reveal the core of the theory of extensions: **Coefficients of the same equation (coefficients of a row) are of the same kind. And coefficients of different equations (coefficients of a column) are of different kind.** But what is the inner meaning of this statement? What has Grassmann done by implementing this concept? The answer is quite simple: Grassmann anticipated geometrization. Grassmann, as a physicist, did what later Albert Einstein and Hermann Minkowski did to describe special relativity. Grassmann pre-invented the mathematics of relativity by attaching entities of different kind, which we nowadays call directions, to different equations.

Thus a shorter version of his explanation can be in modern words: **Coefficients of a row point into identical directions, and coefficients of a column point into different directions.**

Therefore the first equation of (1) is connected with a first direction, the second equation of (1) is connected with a second direction, and the third equation of (1) is connected with a third direction. Usually we describe these directional relations today by multiplying the equations by different base vectors.

$$\begin{aligned} &(a_1\, x + b_1\, y + c_1\, z)\,(\text{base vector into x-direction}) = r_1\,(\text{base vector into x-direction}) \\ &(a_2\, x + b_2\, y + c_2\, z)\,(\text{base vector into y-direction}) = r_2\,(\text{base vector into y-direction}) \\ &(a_3\, x + b_3\, y + c_3\, z)\,(\text{base vector into z-direction}) = r_3\,(\text{base vector into z-direction}) \end{aligned} \qquad (2)$$

But what are base vectors? How can we describe them mathematically? In the course of time Grassmann developed a clear and very helpful picture of base vectors: "Ich will den Verein dreier Strecken e_1, e_2, e_3, die zu einander senkrecht sind, und deren Länge und deren äusseres Produkt $[e_1e_2e_3]$ gleich Eins sind, einen Normalverein nennen" [7]. "I would like to call the (set) of three line segments e_1, e_2, e_3 that are perpendicular to each other and whose lengths and exterior product $[e_1e_2e_3]$ equal one a normal set" [8]. These are old words, and a "Verein" is a typical German construction, which can be translated as "union" or more mathematically as "set". They form the basic building blocks of what we nowadays call a coordinate system with three axes which point into the directions of the line segments e_1, e_2, e_3. Thus eq. (2) can simply be written as:

$$\begin{aligned} (a_1\,x + b_1\,y + c_1\,z)\,e_1 &= r_1\,e_1 \\ (a_2\,x + b_2\,y + c_2\,z)\,e_2 &= r_2\,e_2 \\ (a_3\,x + b_3\,y + c_3\,z)\,e_3 &= r_3\,e_3 \end{aligned} \qquad (3)$$

From the perspective of physics, these eqs. (3) show an old-fashioned and really trumpery way of writing systems of linear equations embedded in a coordinate system which was trumped up by algebra fixed fossils ignoring or even denying geometry. According to Hestenes (and many others) geometry links the algebra to the physical world, and "the power of GA" – Geometric Algebra based on Grassmann's ideas – "derives from
- the simplicity of the grammar,
- the geometric meaning of multiplication,
- the way geometry links the algebra to the physical world" [9].

The simplicity of the grammar has already been noticed and was emphasized by Grassmann who in later papers gave his "Bedingungsgleichungen" – "conditional equations" [10] describing base vectors of three-dimensional space

$$e_1e_2 = -\,e_2e_1\,, \qquad e_2e_3 = -\,e_3e_2\,, \qquad e_3e_1 = -\,e_1e_3$$
$$e_1^2 = e_2^2 = e_3^2 = [e_r \mid e_r] = 1 \qquad (4)$$

He then proposed to create "das mittlere Product, als aus dem äusseren und inneren zusammengesetzt" [10] – "the middle product, which is composed of outer and inner product". This middle product is identical to the geometric product of Timerding [11] and Hestenes [9] who described the "canonical decomposition" of the product of two vectors $\mathbf{a} = a_1e_1 + a_2e_2 + a_3e_3$ and $\mathbf{b} = b_1e_1 + b_2e_2 + b_3e_3$ as

$$\mathbf{a\,b} = \mathbf{a} \bullet \mathbf{b} + \mathbf{a} \wedge \mathbf{b} \qquad (5)$$

The geometric meaning of multiplication is visualized in fig. 1. If two different vectors are multiplied, we will get an oriented parallelogram which consists of a symmetric scalar part or inner product

$$\mathbf{a} \bullet \mathbf{b} = \mathbf{b} \bullet \mathbf{a} = \frac{1}{2}\,(\mathbf{a\,b} + \mathbf{b\,a}) = [\mathbf{a} \mid \mathbf{b}] \quad \text{(Grassmann's style of writing on the right-hand side)}, \qquad (6)$$

and an anti-symmetric bivector part or outer product

$$\mathbf{a} \wedge \mathbf{b} = -\,\mathbf{b} \wedge \mathbf{a} = \frac{1}{2}\,(\mathbf{a\,b} - \mathbf{b\,a}) = [\mathbf{a\,b}] \quad \text{(Grassmann's style of writing on the right-hand side)}. \qquad (7)$$

Thus the product of two different base vectors results in a base bivector, which represents an oriented unit square (or an oriented unit rectangle, e.g. $\tfrac{1}{2}e_1 2e_2$) with disappearing inner product $e_1 \bullet e_2 = 0$

$$e_1e_2 = e_1 \bullet e_2 + e_1 \wedge e_2 = e_1 \wedge e_2 \qquad (8)$$

having right angles (fig. 1).

Of course the orientation then depends on the order of the multiplication. If we first make a step into the direction of the x-axis and then a step into the direction of the y-axis, we will get a positive or counter-clockwise orientation. If we first make a step into the direction of the y-axis instead and then a step into the direction of the x-axis, we will get a negative or clockwise orientation. This change of orientation is codified in the negative signs of eqs. (4).

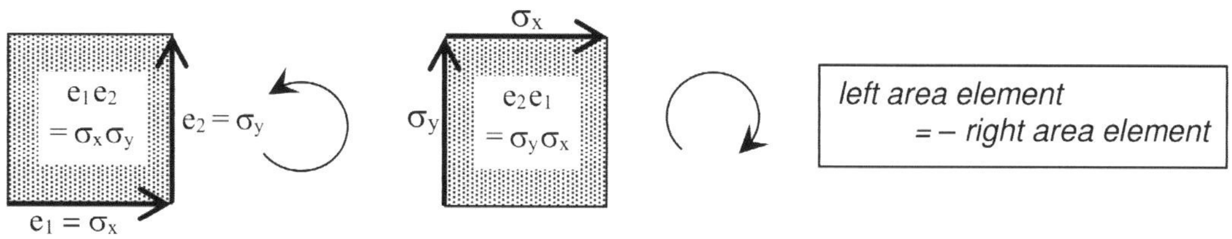

Figure 1. Visualization of a base bivector as the geometric product of two base vectors.

Again it is time for the bizarre announcement of Cambridge scientists Gull, Lasenby, and Doran: "We have now reached the point which is liable to cause the greatest intellectual shock" [12]. Grassmann's equations (4) are identical to the algebra of Pauli matrices:

$$\sigma_x \sigma_y = -\sigma_y \sigma_x\,, \qquad \sigma_y \sigma_z = -\sigma_z \sigma_y\,, \qquad \sigma_z \sigma_x = -\sigma_x \sigma_z$$

$$\sigma_x^2 = \sigma_y^2 = \sigma_z^2 = [\sigma_i \mid \sigma_i] = 1 \tag{9}$$

The historical conclusion is clear and conclusive: Grassmann had already invented Pauli Algebra (and Dirac Algebra by the way). Therefore eqs. (3) can be written in a more physics oriented, modern form as:

$$\begin{aligned}(a_1\, x + b_1\, y + c_1\, z)\, \sigma_x &= r_1\, \sigma_x \\ (a_2\, x + b_2\, y + c_2\, z)\, \sigma_y &= r_2\, \sigma_y \\ (a_3\, x + b_3\, y + c_3\, z)\, \sigma_z &= r_3\, \sigma_z\end{aligned} \tag{10}$$

Grassmann's next step to solve this system of linear equations has been, to add all equations and to rearrange the different terms.

$$(a_1\, x + b_1\, y + c_1\, z)\, \sigma_x + (a_2\, x + b_2\, y + c_2\, z)\, \sigma_y + (a_3\, x + b_3\, y + c_3\, z)\, \sigma_z = r_1\, \sigma_x + r_2\, \sigma_y + r_3\, \sigma_z$$

$$(a_1\, \sigma_x + a_2\, \sigma_y + a_3\, \sigma_z)\, x + (b_1\, \sigma_x + b_2\, \sigma_y + b_3\, \sigma_z)\, y + (c_1\, \sigma_x + c_2\, \sigma_y + c_3\, \sigma_z)\, z = r_1\, \sigma_x + r_2\, \sigma_y + r_3\, \sigma_z \tag{11}$$

Now the three coefficient vectors

$$\mathbf{a} = a_1\, \sigma_x + a_2\, \sigma_y + a_3\, \sigma_z \qquad \mathbf{b} = b_1\, \sigma_x + b_2\, \sigma_y + b_3\, \sigma_z \qquad \mathbf{c} = c_1\, \sigma_x + c_2\, \sigma_y + c_3\, \sigma_z \tag{12}$$

and the resulting vector of constant terms

$$\mathbf{r} = r_1\, \sigma_x + r_2\, \sigma_y + r_3\, \sigma_z \tag{13}$$

form the system of linear equations (11)

$$\mathbf{a}\, x + \mathbf{b}\, y + \mathbf{c}\, z = \mathbf{r} \tag{14}$$

As the outer product of two identical or two parallel vectors disappears, a simple outer multiplication by two of the three coefficient vectors will yield an equation which possesses one unknown variable only. Or in Grassmann's voluptuous German: "Aus dieser Gleichung, welche die Stelle jener n Gleichungen vertritt, lässt sich nun auf der Stelle jede der Unbekannten (…) finden, wenn wir die beiden Seiten mit dem äusseren Produkte aus den Koefficienten der übrigen Unbekannten äusserlich multipliciren (…). Da nämlich, wenn man die Glieder der linken Seite einzeln multiplicirt, nach dem Begriff des äusseren Produktes (…) alle Produkte wegfallen, welche zwei gleiche Faktoren enthalten" [6].

$$(\mathbf{a} \wedge \mathbf{b} \wedge \mathbf{c})\, x = \mathbf{r} \wedge \mathbf{b} \wedge \mathbf{c}$$
$$(\mathbf{a} \wedge \mathbf{b} \wedge \mathbf{c})\, y = \mathbf{a} \wedge \mathbf{r} \wedge \mathbf{c} \tag{15}$$
$$(\mathbf{a} \wedge \mathbf{b} \wedge \mathbf{c})\, z = \mathbf{a} \wedge \mathbf{b} \wedge \mathbf{r}$$

The solution of the system of linear equations can then be found by dividing by the oriented volume of the mathematical object already identified by Grassmann ("begrifflich, dass [**a b c**] gleich dem Parallelepipedon (Spat) ist, in welchem 3 sich aneinander schließende Kanten gleich **a**, **b** und **c** sind" [10]) as a parallelepiped

$$\mathbf{a} \wedge \mathbf{b} \wedge \mathbf{c} = \det \mathbf{A}\ \sigma_x\sigma_y\sigma_z \tag{16}$$

which some mathematicians also call determinant [13]. If the vectors **a**, **b**, **c** are linearly independent and **a**, **b**, **c**, **r** are linearly dependent, pre-division and post-division will get identical scalar results:

$$x = (\mathbf{a} \wedge \mathbf{b} \wedge \mathbf{c})^{-1}\,(\mathbf{r} \wedge \mathbf{b} \wedge \mathbf{c}) \qquad x = (\mathbf{r} \wedge \mathbf{b} \wedge \mathbf{c})\,(\mathbf{a} \wedge \mathbf{b} \wedge \mathbf{c})^{-1}$$
$$y = (\mathbf{a} \wedge \mathbf{b} \wedge \mathbf{c})^{-1}\,(\mathbf{a} \wedge \mathbf{r} \wedge \mathbf{c}) \qquad y = (\mathbf{a} \wedge \mathbf{r} \wedge \mathbf{c})\,(\mathbf{a} \wedge \mathbf{b} \wedge \mathbf{c})^{-1} \tag{17}$$
$$z = (\mathbf{a} \wedge \mathbf{b} \wedge \mathbf{c})^{-1}\,(\mathbf{a} \wedge \mathbf{b} \wedge \mathbf{r}) \qquad z = (\mathbf{a} \wedge \mathbf{b} \wedge \mathbf{r})\,(\mathbf{a} \wedge \mathbf{b} \wedge \mathbf{c})^{-1}$$

These are Grassmann's results of § 45 [6], or later reformulated on page 385 of [10]. And please kindly note that eq. (16) includes a definition for determinants of non-square matrices

$$\det \mathbf{A} = (\mathbf{a} \wedge \mathbf{b} \wedge \mathbf{c} \wedge \ldots \wedge \mathbf{n})\ \sigma_i \ldots \sigma_3\sigma_2\sigma_1 \tag{18}$$

which perfectly agrees with Arnold's emotional call to define determinants geometrically as "the oriented volume of the parallelepiped whose edges are its columns … a secret … carefully hidden in purified algebraic education" [13].

To teach outer products (or in other words: determinants) is an important part of teaching Geometric Algebra. And to teach inner products is an important part of teaching Geometric Algebra. But of similar or of even higher importance might be to teach how to avoid outer and inner products at all as the following section suggests.

3. Intermezzo: Mathematical viruses

How can we deal with mathematical viruses? David Hestenes identified several severe viruses "which can infect the mind – the mind of anyone doing mathematics, from young student to professional mathematician. (…) A mathematical virus (MV) is a preconception about the structure, function or method of mathematics which impairs one's ability to do mathematics. Just as a CV is program which impairs the operating system of a computer, an MV is an idea which impairs the conceptualization of mathematics in the mind" [14].

One of these mathematical viruses Hestenes called "Grassmann Virus (MV/G), because it is a distortion of Grassmann's own view" [14]. This virus has infected mathematicians who think that calculations using only outer products are more fundamental than calculations using the complete geometric product[1]. Or in short: "Grassmann Algebra is more fundamental than Clifford Algebra," which simply is wrong.

Thus an anti-MV/G vaccine is required. Such a vaccine might have the following didactical core: Let us replace outer products by geometric products. To prevent students from believing that calculations using only outer products are more fundamental than calculations using the complete geometric product, it is helpful to discuss the solution of systems of linear equations using complete geometric products.

---

[1] Physicists seem to be more immune to MV/G compared to mathematicians, as they surely admire Feynman who once commented: „So there is a connection, ultimately, between algebra and geometry. (…) the most remarkable formula in mathematics: $e^{i\Theta} = \cos\Theta + i \sin\Theta$. This is our jewel" [15]. This jewel can be read as a simpler expression of the geometric product $\mathbf{a}\,\mathbf{b} = \mathbf{a} \bullet \mathbf{b} + \mathbf{a} \wedge \mathbf{b}$ (5). And a jewel will not be split into pieces.

The solutions (17) of the system of linear equations (1) or (14) will then be presented as

$$\begin{aligned} x &= (\mathbf{c\,a\,b} - \mathbf{b\,a\,c})^{-1}(\mathbf{c\,r\,b} - \mathbf{b\,r\,c}) \qquad & x &= (\mathbf{c\,r\,b} - \mathbf{b\,r\,c})(\mathbf{c\,a\,b} - \mathbf{b\,a\,c})^{-1} \\ y &= (\mathbf{a\,b\,c} - \mathbf{c\,b\,a})^{-1}(\mathbf{a\,r\,c} - \mathbf{c\,r\,a}) \qquad & y &= (\mathbf{a\,r\,c} - \mathbf{c\,r\,a})(\mathbf{a\,b\,c} - \mathbf{c\,b\,a})^{-1} \\ z &= (\mathbf{b\,c\,a} - \mathbf{a\,c\,b})^{-1}(\mathbf{b\,r\,a} - \mathbf{a\,r\,b}) \qquad & z &= (\mathbf{b\,r\,a} - \mathbf{a\,r\,b})(\mathbf{b\,c\,a} - \mathbf{a\,c\,b})^{-1} \end{aligned} \tag{19}$$

Mixed sandwich products indeed play a special role in these equations. Therefore it is worth the trouble to take a good look at sandwich products.

4. Pure sandwich products – What is stronger: physics or mathematics?

If a mathematical object is pre- and post-multiplied by a second mathematical object, the result can be called a sandwich product. The first mathematical object is sandwiched by the second from left and right.

As an example we can sandwich the system of linear equations (14) by the geometric product of two coefficient vectors and its reversed geometric product to reach yet another path towards a solution

$$\begin{aligned} &\mathbf{bc\,r\,bc} - \mathbf{cb\,r\,cb} \quad \Rightarrow \quad x = (\mathbf{bcabc} - \mathbf{cbacb})^{-1}(\mathbf{bcrbc} - \mathbf{cbrcb}) \\ &\mathbf{ca\,r\,ca} - \mathbf{ac\,r\,ac} \quad \Rightarrow \quad y = (\mathbf{cabca} - \mathbf{acbac})^{-1}(\mathbf{carca} - \mathbf{acrac}) \\ &\mathbf{ab\,r\,ab} - \mathbf{ba\,r\,ba} \quad \Rightarrow \quad z = (\mathbf{abcab} - \mathbf{bacba})^{-1}(\mathbf{abrab} - \mathbf{barba}) \end{aligned} \tag{20}$$

In Geometric Algebra pure sandwich products by k-vectors (by multivectors of constant grade k) can be used to model a very basic mathematical operation: reflections! I do not know whether it is allowed to call the sandwich products of eqs. (20) reflections in a parallelogram. But it surely opens new conceptual and new didactical views on a subject, if these subjects are renamed in an unorthodox manner.

The idea that nature might follow reflection symmetry, is a basic physical concept. Therefore the mathematics of reflections is important, and I simply will repeat the Geometric Algebra equations describing reflections [16], [17] in the following.

	Reflection in a point (represented by scalar ℓ):	Reflection in a line (represented by vector $\mathbf{n}$):	Reflection in a plane (represented by bivector $\mathbf{N}$):
Scalars:	$k_{ref} = \ell\, k\, \ell^{-1}$	$k_{ref} = \mathbf{n}\, k\, \mathbf{n}^{-1}$	$k_{ref} = \mathbf{N}\, k\, \mathbf{N}^{-1}$
Vectors:	$\mathbf{r}_{ref} = -\ell\, \mathbf{r}\, \ell^{-1}$	$\mathbf{r}_{ref} = \mathbf{n\, r\, n}^{-1}$	$\mathbf{r}_{ref} = -\mathbf{N\, r\, N}^{-1}$
Bivectors:	$\mathbf{A}_{ref} = \ell\, \mathbf{A}\, \ell^{-1}$	$\mathbf{A}_{ref} = \mathbf{n\, A\, n}^{-1}$	$\mathbf{A}_{ref} = \mathbf{N\, A\, N}^{-1}$
Trivectors:	$\mathbf{V}_{ref} = -\ell\, \mathbf{V}\, \ell^{-1}$	$\mathbf{V}_{ref} = \mathbf{n\, V\, n}^{-1}$	$\mathbf{V}_{ref} = -\mathbf{N\, V\, N}^{-1}$
Quadvectors:	$\boldsymbol{Q}_{ref} = \ell\, \boldsymbol{Q}\, \ell^{-1}$	$\boldsymbol{Q}_{ref} = \mathbf{n}\, \boldsymbol{Q}\, \mathbf{n}^{-1}$	$\boldsymbol{Q}_{ref} = \mathbf{N}\, \boldsymbol{Q}\, \mathbf{N}^{-1}$
Pentavectors:	$\boldsymbol{P}_{ref} = -\ell\, \boldsymbol{P}\, \ell^{-1}$	$\boldsymbol{P}_{ref} = \mathbf{n}\, \boldsymbol{P}\, \mathbf{n}^{-1}$	$\boldsymbol{P}_{ref} = -\mathbf{N}\, \boldsymbol{P}\, \mathbf{N}^{-1}$

(21)

	Reflection in a 3d space or a reduced 3d spacetime (represented by trivector $\mathbf{T}$):	Reflection in a 4d hyper-space or spacetime (represented by quadvector Q):	Reflection in a 5d hyper-space or spacetimevelocity[2] (represented by pentavector P):
Scalars:	$k_{ref} = \mathbf{T}\, k\, \mathbf{T}^{-1}$	$k_{ref} = \mathsf{Q}\, k\, \mathsf{Q}^{-1}$	$k_{ref} = \mathsf{P}\, k\, \mathsf{P}^{-1}$
Vectors:	$\mathbf{r}_{ref} = \mathbf{T\, r\, T}^{-1}$	$\mathbf{r}_{ref} = -\mathsf{Q}\, \mathbf{r}\, \mathsf{Q}^{-1}$	$\mathbf{r}_{ref} = \mathsf{P}\, \mathbf{r}\, \mathsf{P}^{-1}$
Bivectors:	$\mathbf{A}_{ref} = \mathbf{T\, A\, T}^{-1}$	$\mathbf{A}_{ref} = \mathsf{Q}\, \mathbf{A}\, \mathsf{Q}^{-1}$	$\mathbf{A}_{ref} = \mathsf{P}\, \mathbf{A}\, \mathsf{P}^{-1}$
Trivectors:	$\mathbf{V}_{ref} = \mathbf{T\, V\, T}^{-1}$	$\mathbf{V}_{ref} = -\mathsf{Q}\, \mathbf{V}\, \mathsf{Q}^{-1}$	$\mathbf{V}_{ref} = \mathsf{P}\, \mathbf{V}\, \mathsf{P}^{-1}$
Quadvectors:	$\boldsymbol{Q}_{ref} = \mathbf{T}\, \boldsymbol{Q}\, \mathbf{T}^{-1}$	$\boldsymbol{Q}_{ref} = \mathsf{Q}\, \boldsymbol{Q}\, \mathsf{Q}^{-1}$	$\boldsymbol{Q}_{ref} = \mathsf{P}\, \boldsymbol{Q}\, \mathsf{P}^{-1}$
Pentavectors:	$\boldsymbol{P}_{ref} = \mathbf{T}\, \boldsymbol{P}\, \mathbf{T}^{-1}$	$\boldsymbol{P}_{ref} = -\mathsf{Q}\, \boldsymbol{P}\, \mathsf{Q}^{-1}$	$\boldsymbol{P}_{ref} = \mathsf{P}\, \boldsymbol{P}\, \mathsf{P}^{-1}$

(22)

Reflections in higher dimensional spaces or spacetimes (or spacetimevelocities[2]) can be written in a similar way. Thus three quarters of all reflection formulae are positive sandwich products, and a quarter of all reflection formulae are negative sandwich products.

As an amateur philosopher I ask amateur questions: What is stronger: physics or mathematics? The idea, that nature might follow reflection symmetry, is a basic physical concept. The idea, that the left-hand side of an equation will remain identical to the right-hand side of this equation, if we multiply both sides of the equation by the same factors, is a basic mathematical concept. So nature sometimes has to make a decision if naïve physicists try to do experiments with reflections. Should nature follow the idea of reflection symmetry or should nature follow the idea of mathematical consistency and equivalence, if multivectors are transformed?

The most general multivector **M** existing in three-dimensional space is the following linear combination:

$$\mathbf{M} = \mathrm{k} + \mathbf{r} + \mathbf{A} + \mathbf{V} \tag{23}$$

(with $\mathrm{k} = \mathrm{k}\cdot 1$; $\mathbf{r} = r_1\,\sigma_x + r_2\,\sigma_y + r_3\,\sigma_z$ (13); $\mathbf{A} = A_{12}\,\sigma_x\sigma_y + A_{23}\,\sigma_y\sigma_z + A_{31}\,\sigma_z\sigma_x$; $\mathbf{V} = V\,\sigma_x\sigma_y\sigma_z$)

Surely the experimenting physicist will find it possible to get a sandwich product of multivector **M** if she or he experiments with reflections in an axis as the ‘mathematical’ and the ‘physical’ outcome should be identical.

$$\mathbf{M}_{\mathrm{sand}} = \mathbf{n}\,\mathbf{M}\,\mathbf{n}^{-1} = \mathbf{n}\,\mathrm{k}\,\mathbf{n}^{-1} + \mathbf{n}\,\mathbf{r}\,\mathbf{n}^{-1} + \mathbf{n}\,\mathbf{A}\,\mathbf{n}^{-1} + \mathbf{n}\,\mathbf{V}\,\mathbf{n}^{-1} = \mathrm{k}_{\mathrm{ref}} + \mathbf{r}_{\mathrm{ref}} + \mathbf{A}_{\mathrm{ref}} + \mathbf{V}_{\mathrm{ref}} = \mathbf{M}_{\mathrm{ref}} \tag{24}$$

But what will happen, if an experimenting physicist experiments with reflections in a plane? Now nature has to make a decision as ‘mathematical’ and ‘physical’ outcome clearly are different.

$$\mathbf{M}_{\mathrm{sand}} = \mathbf{N}\,\mathbf{M}\,\mathbf{N}^{-1} = \mathbf{N}\,\mathrm{k}\,\mathbf{N}^{-1} + \mathbf{N}\,\mathbf{r}\,\mathbf{N}^{-1} + \mathbf{N}\,\mathbf{A}\,\mathbf{N}^{-1} + \mathbf{N}\,\mathbf{V}\,\mathbf{N}^{-1} = \mathrm{k}_{\mathrm{ref}} - \mathbf{r}_{\mathrm{ref}} + \mathbf{A}_{\mathrm{ref}} - \mathbf{V}_{\mathrm{ref}} \neq \mathbf{M}_{\mathrm{ref}} \tag{25}$$

Will nature decide for mathematical consistency and confuse physicists with additional minus signs by flipping over vectors and trivectors? Or will nature decide for unlimited and unrestricted perfect reflection symmetry at the experiments of physicists?

There are clear indications and an obituary postcard by Pauli (“Es ist uns eine bange Pflicht, bekannt zu geben, dass unsere langjährige, liebe Freundin PARITY am 19. Januar 1957 nach kurzem Leiden bei weiteren experimentellen Eingriffen sanft entschlafen ist.” [20]) which make it reasonable to suppose that nature has decided for mathematical consistency and against absolute reflection symmetry.

Unfortunately there is a parity confusion in the literature. Some authors explain parity as as reflection in a two-dimensional plane, e.g. Penrose: “This operation of spatial reflection (reflection in a mirror) is referred as P (which stands for *parity*)” [21]. Other authors explain parity as a reflection in a point, e.g. Kirch et al: “Die Paritätstransformation (*P*) entspricht einer Punktspiegelung am Ursprung des dreidimensionalen Raums” [22].

Of course all this is confusing. The reflection in a point is different from a reflection in an axis. And they are different from a reflection in a mirror. And all these reflections are of course different from a reflection in a three-dimensional space or an n-dimensional hyper-space. Therefore we should invent and use different names for these different reflections. If they are all called parity, then different parities of different types exist.

But good news of course is that nature also has decided to model Lorentz transformations by spacetime rotations and thus by two succeeding reflections. Therefore the minus signs will cancel in special relativistic experiments with Lorentz boosts:

$$\mathbf{M}_{\mathrm{double\text{-}sand}} = \mathbf{N}_2\,\mathbf{N}_1\,\mathbf{M}\,\mathbf{N}_1^{-1}\,\mathbf{N}_2^{-1} = \mathrm{k}_{\mathrm{rot}} + \mathbf{r}_{\mathrm{rot}} + \mathbf{A}_{\mathrm{rot}} + \mathbf{V}_{\mathrm{rot}} = \mathbf{M}_{\mathrm{rot}} \tag{26}$$

[2] Five-dimensional spacetimevelocity of Carmeli’s cosmological special relativity models our world by three spacelike dimensions x, y, z and two timelike dimensions τv, ct with Hubble time constant $\tau = 1/H_0$ [18, 19].

5. Mixed sandwich products

If a mathematical object is pre-multiplied by a second mathematical object and post-multiplied by a third mathematical object, this operation can be called a mixed sandwich product. The first mathematical object is sandwiched by two different mathematical objects from left and right.

While three quarters of all pure sandwich products can be interpreted as reflections and a quarter of all pure sandwich products as anti-reflections (a reflection followed by a change of signs), mixed sandwich products can be interpreted as sort of interchanges (German: Vertauschung) – or more precisely as exchanges (German: Austauschung) or replacements (German: Ersetzung), respectively. These are no perfect interchanges as the result depends on the order of pre- and post-multiplied factors.

As a simple example we first exchange or replace the x-axis by the y-axis by pre-multiplying a sandwiched vector **r** by σ_x and post-multiplying it by σ_y (27). And we exchange or replace the y-axis by the x-axis by pre-multiplying a sandwiched vector **r** by σ_y and post-multiplying it by σ_x (28).

$$\sigma_x \mathbf{r} \sigma_y = \sigma_x (r_1 \sigma_x + r_2 \sigma_y + r_3 \sigma_z) \sigma_y = r_2 \sigma_x + r_1 \sigma_y - r_3 \sigma_x\sigma_y\sigma_z \tag{27}$$

$$\sigma_y \mathbf{r} \sigma_x = \sigma_y (r_1 \sigma_x + r_2 \sigma_y + r_3 \sigma_z) \sigma_x = r_2 \sigma_x + r_1 \sigma_y + r_3 \sigma_x\sigma_y\sigma_z \tag{28}$$

Both transformations interchange the x- and y-axes. But these operations have different effects on axes perpendicular to σ_x and σ_y. Perpendicular vectorial axes become trivectorial axes with different orientations. The right-handed Euclidean coordinate system of the original vector **r** with base units squaring to one (9) becomes a left-handed (27) or right-handed (28) hyperbolic Pseudo-Euclidean coordinate system with two real and one imaginary coordinate axes as the base units of (27) & (28) now square differently (29).

$$\sigma_x^2 = \sigma_y^2 = 1 \qquad (\sigma_x\sigma_y\sigma_z)^2 = -1 \tag{29}$$

This change of dimensionality can now be used to solve systems of linear equations. To solve a system of three linear equations, vector **r** is now seen as a linear combination of coefficient vectors **a**, **b**, **c** (14). Subtracting mixed sandwich multiplications of **r** with two of the three coefficient vectors (e.g. **b** and **c**)

$$\mathbf{c\,r\,b} = \mathbf{c}\,(\mathbf{a}\,x + \mathbf{b}\,y + \mathbf{c}\,z)\,\mathbf{b} = \mathbf{c\,a\,b}\,x + \mathbf{c}\,b^2 y + \mathbf{b}\,c^2 z \qquad \mathbf{b\,r\,c} = \mathbf{b\,a\,c}\,x + \mathbf{c}\,b^2 y + \mathbf{b}\,c^2 z \tag{30}$$

immediately results in the solution equations already given in (19).

But a more interesting conceptual feature of mixed sandwich products is important: If mixed sandwich products change the dimensionality, we should not only look on vectors, but on complete multivectors. Exchanging vectorial axes then is only one possible feature. It might also be possible to exchange or replace two objects which have different dimensions.

A one-dimensional axis might be exchanged or replaced by a two-dimensional plane (31). Or a three-dimensional volume might be exchanged or replaced by a zero-dimensional point (32). Multivector **M** (23) will then be transformed into:

$$\mathbf{M}_{\mathbf{n}\to\mathbf{N}} = \mathbf{n\,M\,N} = \mathbf{n}\,(k + \mathbf{r} + \mathbf{A} + \mathbf{V})\,\mathbf{N} \quad \text{or} \quad \mathbf{M}_{\mathbf{N}\to\mathbf{n}} = \mathbf{N\,M\,n} = \mathbf{N}\,(k + \mathbf{r} + \mathbf{A} + \mathbf{V})\,\mathbf{n} \tag{31}$$

$$\mathbf{M}_{\mathbf{T}\to\ell} = \mathbf{T\,M}\,\ell = \mathbf{T}\,(k + \mathbf{r} + \mathbf{A} + \mathbf{V})\,\ell \quad \text{or} \quad \mathbf{M}_{\ell\to\mathbf{T}} = \ell\,\mathbf{M\,T} = \ell\,(k + \mathbf{r} + \mathbf{A} + \mathbf{V})\,\mathbf{T} \tag{32}$$

These mixed sandwich products are combined exchanges and dilations. If the magnitudes of pre- and post-multiplied factors equal one ($\ell = 1$; $|\mathbf{n}| = 1$; $|\mathbf{N}| = 1$; $|\mathbf{T}| = 1$), the mixed sandwich products describe exchanges perfectly.

What will then be the structural interpretation of reflected multivector $\mathbf{M}_{ref} = \mathbf{n\,M\,n}^{-1}$ of eq. (24)? This reflection can be split into two exchanges or replacements of unit vector $\mathbf{n} = \mathbf{n}^{-1}$ (or $|\mathbf{n}| = 1$) by unit scalar $\ell = 1$:

$$\mathbf{M}_{ref} = \mathbf{n\,M\,n} = 1\,(\mathbf{n\,M}\,1)\,\mathbf{n} = (\mathbf{M}_{\mathbf{n}\to 1})_{1\to\mathbf{n}} \tag{33}$$

Thus the reflection in an axis pointing into the direction of **n** can be modeled as an exchange or replacement of the vectorial axis pointing into the direction of **n** by the scalar axis pointing into the dimensionless direction of $\ell = 1$, followed by a second exchange or replacement of the scalar axis pointing into the dimensionless direction of $\ell = 1$ by the vectorial axis pointing into the direction of **n**.

Or in general: The reflection in an arbitrary volume (or in a hyper-volume) can be modeled as an exchange of the scalar axis with the volume of reflection (or the hyper-volume of reflection), followed by an exchange of the volume of reflection (or the hyper-volume of reflection) with the scalar axis.

6. Rotating axes in special relativity: Lorentz transformations, as usual

While Pauli matrices (9) represent base vectors of three-dimensional Euclidean space, Dirac matrices $\gamma_t, \gamma_x, \gamma_y, \gamma_z$ with

$$\gamma_x\gamma_y = -\gamma_y\gamma_x\,,\quad \gamma_y\gamma_z = -\gamma_z\gamma_y\,,\quad \gamma_z\gamma_x = -\gamma_x\gamma_z\,,\quad \gamma_x\gamma_t = -\gamma_t\gamma_x\,,\quad \gamma_y\gamma_t = -\gamma_t\gamma_y\,,\quad \gamma_z\gamma_t = -\gamma_t\gamma_z$$

$$\gamma_t^2 = 1, \qquad \gamma_x^2 = \gamma_y^2 = \gamma_z^2 = -1 \tag{34}$$

represent base vectors of four-dimensional Pseudo-Euclidean spacetime [23, 24, 25, 26, 27]. It is well known (see references [23–27]), that a spacetime rotation of a coordinate system with axes pointing into the directions of the base vectors $\gamma_t, \gamma_x, \gamma_y, \gamma_z$ into a new coordinate system with axes pointing into the directions of $\gamma_{t\text{-new}}, \gamma_{x\text{-new}}, \gamma_y, \gamma_z$ equals a Lorentz transformation. This Lorentz transformation can be modeled as a first spacetime reflection in an axis pointing into the direction of the bisector **n** of the angle between old and new time axes

$$\gamma_{t\text{-old}} = \gamma_t \qquad \gamma_{t\text{-new}} = \cosh\alpha\,\gamma_t + \sinh\alpha\,\gamma_x \qquad (\gamma_t^2 = \gamma_{t\text{-new}}^2 = \cosh^2\alpha - \sinh^2\alpha = 1) \tag{35}$$

$$\mathbf{n} = \mathbf{n}^{-1} = \frac{\gamma_t + \gamma_{t\text{-new}}}{\sqrt{(\gamma_t + \gamma_{t\text{-new}})^2}} = \frac{(1+\cosh\alpha)\,\gamma_t + \sinh\alpha\gamma_x}{\sqrt{2+2\cosh\alpha}} = \cosh\frac{\alpha}{2}\,\gamma_t + \sinh\frac{\alpha}{2}\,\gamma_x \tag{36}$$

followed by a second spacetime reflection in the new time axis

$$\mathbf{m}_1 = \mathbf{m}_1^{-1} = \gamma_{t\text{-new}} = \cosh\alpha\,\gamma_t + \sinh\alpha\,\gamma_x \tag{37}$$

Thus the four base vectors pointing into the directions of the coordinate axes are Lorentz transformed as usual [28] according to

$$\gamma_{t\text{-new}} = \mathbf{m}_1\,\mathbf{n}\,\gamma_t\,\mathbf{n}^{-1}\,\mathbf{m}_1^{-1} = \cosh\alpha\,\gamma_t + \sinh\alpha\,\gamma_x \tag{38}$$

$$\gamma_{x\text{-new}} = \mathbf{m}_1\,\mathbf{n}\,\gamma_x\,\mathbf{n}^{-1}\,\mathbf{m}_1^{-1} = \sinh\alpha\,\gamma_t + \cosh\alpha\,\gamma_x \tag{39}$$

$$\gamma_{y\text{-new}} = \mathbf{m}_1\,\mathbf{n}\,\gamma_y\,\mathbf{n}^{-1}\,\mathbf{m}_1^{-1} = \gamma_y \tag{40}$$

$$\gamma_{z\text{-new}} = \mathbf{m}_1\,\mathbf{n}\,\gamma_z\,\mathbf{n}^{-1}\,\mathbf{m}_1^{-1} = \gamma_z \tag{41}$$

Vectors transform into vectors, and this is important: In contrast to that it will be shown later, that other transformations might transform vectors into linear combinations of vectors and geometric objects of other dimensions.

To avoid confusion it is helpful **not** to speak about active transformations (… a coordinate system remains unchanged while the event at position vector **r** rotates actively…) and **not** to speak about passive transformations (… the coordinate system is rotated into a new one while the position vector **r** passively remains at the same spot…). These bracketed statements are not very helpful as always everything is transformed: An old coordinate system is transformed into a new one while the old position vector is transformed into a new one (fig. 2).

Therefore it makes more sense to speak about forward and backward transformations instead. The transformation just described transforms the original coordinate system into a new one with a negative, anti-clockwise rotation and thus will be a backward transformation.

If the first spacetime reflection in an axis pointing into the direction of the bisector **n** is now followed by a second reflection in the old time axis

$$\mathbf{m}_2 = \mathbf{m}_2^{-1} = \gamma_t \tag{42}$$

the new coordinate axes will be rotated back into the original ones, now describing the inverse of the previous backward transformations of eqs. (38) – (41), thus giving a forward transformation:

$$\gamma_t = \mathbf{m}_2\, \mathbf{n}\, \gamma_{t\text{-new}}\, \mathbf{n}^{-1}\, \mathbf{m}_2^{-1} = \cosh\alpha\, \gamma_{t\text{-new}} - \sinh\alpha\, \gamma_{x\text{-new}} \tag{43}$$

$$\gamma_x = \mathbf{m}_2\, \mathbf{n}\, \gamma_{x\text{-new}}\, \mathbf{n}^{-1}\, \mathbf{m}_2^{-1} = -\sinh\alpha\, \gamma_{t\text{-new}} + \cosh\alpha\, \gamma_{x\text{-new}} \tag{44}$$

$$\gamma_y = \mathbf{m}_2\, \mathbf{n}\, \gamma_{y\text{-new}}\, \mathbf{n}^{-1}\, \mathbf{m}_2^{-1} = \gamma_{y\text{-new}} \tag{45}$$

$$\gamma_z = \mathbf{m}_2\, \mathbf{n}\, \gamma_{z\text{-new}}\, \mathbf{n}^{-1}\, \mathbf{m}_2^{-1} = \gamma_{z\text{-new}} \tag{46}$$

Therefore the position vector **r** of an event

$$\mathbf{r} = ct\, \gamma_t + x\, \gamma_x + y\, \gamma_y + z\, \gamma_z \tag{47}$$

is either transformed backwards into

$$\begin{aligned}\mathbf{r}_{back} &= \mathbf{m}_1\, \mathbf{n}\, \mathbf{r}\, \mathbf{n}^{-1}\, \mathbf{m}_1^{-1} \\ &= (ct \cosh\alpha + x \sinh\alpha)\, \gamma_t + (x \cosh\alpha + ct \sinh\alpha)\, \gamma_x + y\, \gamma_y + z\, \gamma_z \\ &= ct_{back}\, \gamma_t + x_{back}\, \gamma_x + y\, \gamma_y + z\, \gamma_z\end{aligned} \tag{48}$$

or forwards into

$$\begin{aligned}\mathbf{r}_{for} &= \mathbf{m}_2\, \mathbf{n}\, \mathbf{r}\, \mathbf{n}^{-1}\, \mathbf{m}_2^{-1} \\ &= (ct \cosh\alpha - x \sinh\alpha)\, \gamma_t + (x \cosh\alpha - ct \sinh\alpha)\, \gamma_x + y\, \gamma_y + z\, \gamma_z \\ &= ct_{for}\, \gamma_t + x_{for}\, \gamma_x + y\, \gamma_y + z\, \gamma_z\end{aligned} \tag{49}$$

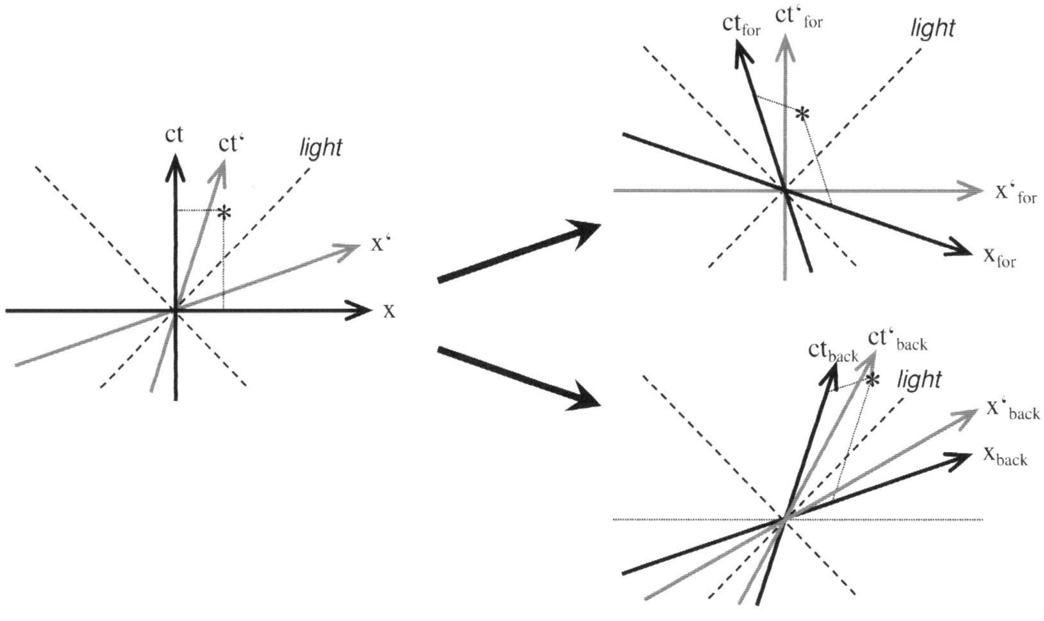

Figure 2. Sketch of forward Lorentz transformation with positive, anti-clockwise rotation of the time axis (top) and backward Lorentz transformation with negative, clockwise rotation (at the bottom).

resulting into the well-known coordinate changes

$$ct_{new} = ct_{back} = ct \cosh\alpha - x \sinh\alpha \qquad x_{new} = x_{back} = x \cosh\alpha - ct \sinh\alpha \qquad y_{new} = y \qquad z_{new} = z \tag{50}$$

As an observer of the old coordinate system will see a person, who is motionless in his own new coordinate system (x = y = z = 0), moving with constant velocity

$$\frac{v}{c} = \frac{\Delta x}{c\,\Delta t} = \frac{\sinh\alpha}{\cosh\alpha} = \tanh\alpha \tag{51}$$

the Lorentz factor will be as usual

$$\cosh\alpha = \frac{1}{\sqrt{1-\tanh^2\alpha}} = \frac{1}{\sqrt{1-\frac{v^2}{c^2}}} \qquad \sinh\alpha = \frac{\tanh\alpha}{\sqrt{1-\tanh^2\alpha}} = \frac{\frac{v}{c}}{\sqrt{1-\frac{v^2}{c^2}}} \tag{52}$$

and the transformation formulas (50) will become

$$ct_{new} = \frac{ct - \frac{v}{c}x}{\sqrt{1-\frac{v^2}{c^2}}} \qquad x_{new} = \frac{x - vt}{\sqrt{1-\frac{v^2}{c^2}}} \qquad y_{new} = y \qquad z_{new} = z \tag{53}$$

Of course these Lorentz transformations can now be interpreted as one rotation. Or they can be interpreted as two successive reflections. Or they can be seen as four exchanges of axes, e.g.

$$\begin{aligned}\mathbf{r}_{new} &= \mathbf{m\,n\,r\,n}^{-1}\,\mathbf{m}^{-1} = \mathbf{m\,n\,r\,n\,m}\\ &= \mathbf{m}\,(\mathbf{n\,r}\,1)\,\mathbf{n\,m}\\ &= 1\,(\mathbf{m}\,(1\,(\mathbf{n\,r}\,1)\,\mathbf{n}\,)\,1)\,\mathbf{m}\end{aligned} \tag{54}$$

At the beginning an axis pointing into the direction of unit vector **n** is exchanged or replaced by a scalar axis pointing into the non-existing direction of unit scalar 1. Then a scalar axis pointing into the non-existing direction of unit scalar 1 is exchanged or replaced by an axis pointing into the direction of **n**. Then an axis pointing into the direction of unit vector **m** is exchanged or replaced by a scalar axis pointing into the non-existing direction of unit scalar 1. And finally a scalar axis pointing into the non-existing direction of unit scalar 1 is exchanged or replaced by an axis pointing into the direction of **m**.

Are these intermediate steps only artificial mathematical subtleties? Are they mathematical inventions which do not represent real situations?

Or do these intermediate equations have physical significance? Do they describe situations, which might happen in nature? Are they real? Is it possible, that a second observer will look at nature from a coordinate system which is constructed by exchanging or replacing axes of the coordinate system of a first observer?

The following sections will perhaps show that the answer of the last questions might be: Yes, it is.

7. Creating something out of nothing

A coordinate system in which events can be measured at position $\mathbf{r} = ct\,\gamma_t + x\,\gamma_x + y\,\gamma_y + z\,\gamma_z$ (47) is considered as a coordinate system of empty space in special relativity, because squaring eq. (47)

$$\mathbf{r}^2 = (ct)^2 - x^2 - y^2 + z^2 \quad \text{or} \quad dr^2 = \mathbf{r}^2 = c^2\,dt^2 - dx^2 - dy^2 + dz^2 \tag{55}$$

clearly shows that non-diagonal elements of the metric tensor disappear

$$g_{\mu\nu} = 0 \text{ if } \mu \neq \nu \quad \text{and} \quad g_{00} = \gamma_t^{\,2} = c^2 \quad g_{11} = g_{22} = g_{33} = \gamma_x^{\,2} = \gamma_y^{\,2} = \gamma_z^{\,2} = -1 \tag{56}$$

and the diagonal elements show the hyperbolic structure of empty space.

In the following, axes will be exchanged or replaced and it will be seen, that this event, which took place in an empty space, will look like an event which takes place at a space which is filled with sort of a field. An empty space seen from a first observer will look like a space with a field seen from a second observer. By exchanging or replacing axes we might create (or destroy sometimes) fields.

As already said in the last section: Vectors transform into vectors, bivectors transform into bivectors, trivectors transform into trivectors, and scalars transform into scalars if spacetime rotations are modeled according to Lorentz transformations.

But now we have a different tool on hand: We are able now to exchange or replace axes. As a first try we will exchange or replace the old time axis which points into the direction of unit vector γ_t by a new time axis, which points into the direction of unit vector $\gamma_{t\text{-new}}$ (35), (38) given above. Position vector $\mathbf{r}$ will then be transformed into

$$\mathbf{r}_{ex} = \gamma_t \, \mathbf{r} \, \gamma_{t\text{-new}} \tag{57}$$

and the four base vectors pointing into the directions of the coordinate axes will be transformed into

$$\gamma_{t\text{-ex}} = \gamma_t \, \gamma_t \, \gamma_{t\text{-new}} = \cosh\alpha \, \gamma_t + \sinh\alpha \, \gamma_x \tag{58}$$

$$\gamma_{x\text{-ex}} = \gamma_t \, \gamma_x \, \gamma_{t\text{-new}} = -\cosh\alpha \, \gamma_x - \sinh\alpha \, \gamma_t \tag{59}$$

$$\gamma_{y\text{-ex}} = \gamma_t \, \gamma_y \, \gamma_{t\text{-new}} = -\cosh\alpha \, \gamma_y - \sinh\alpha \, \gamma_x\gamma_y\gamma_t \tag{60}$$

$$\gamma_{z\text{-ex}} = \gamma_t \, \gamma_z \, \gamma_{t\text{-new}} = -\cosh\alpha \, \gamma_z - \sinh\alpha \, \gamma_x\gamma_z\gamma_t \tag{61}$$

Now vectors transform into vectors (58), (59) **and into linear combinations of vectors and trivectors** (60), (61). This is an interesting feature. To compare this strange behavior with Lorentz transformations, it makes sense to transform the now left-handed coordinate system (58) – (61) into a right-handed coordinate system and to get rid of the minus signs of eqs. (59), (60), (61). Therefore the exchange or replacement (57) will be followed by a reflection at the three-dimensional new space which is represented by the oriented volume element

$$\begin{aligned}\gamma_{x\text{-new}}\gamma_{y\text{-new}}\gamma_{z\text{-new}} &= -\cosh\alpha\,(\cosh^2\alpha + 3\sinh^2\alpha)\,\gamma_x\gamma_y\gamma_z - \sinh\alpha\,(3\cosh^2\alpha + \sinh^2\alpha)\,\gamma_y\gamma_z\gamma_t \\ &= -\cosh\alpha\,(2\cosh(2\alpha) - 1)\,\gamma_x\gamma_y\gamma_z - \sinh\alpha\,(\cosh(2\alpha) + 1)\,\gamma_y\gamma_z\gamma_t\end{aligned} \tag{62}$$

which of course is a pure trivector. The now complete exchange transformation is

$$\mathbf{r}_{EX} = \gamma_{x\text{-new}}\gamma_{y\text{-new}}\gamma_{z\text{-new}} \, \mathbf{r}_{ex} \, \gamma_{x\text{-new}}\gamma_{y\text{-new}}\gamma_{z\text{-new}} \tag{63}$$

and the backward transformations of the four base vectors will result in the following base vectors of the right-handed final coordinate system:

$$\gamma_{t\text{-EX}} = \gamma_{x\text{-new}}\gamma_{y\text{-new}}\gamma_{z\text{-new}} \, \gamma_{t\text{-ex}} \, \gamma_{x\text{-new}}\gamma_{y\text{-new}}\gamma_{z\text{-new}} = \cosh\alpha \, \gamma_t + \sinh\alpha \, \gamma_x \tag{64}$$

$$\gamma_{x\text{-EX}} = \gamma_{x\text{-new}}\gamma_{y\text{-new}}\gamma_{z\text{-new}} \, \gamma_{x\text{-ex}} \, \gamma_{x\text{-new}}\gamma_{y\text{-new}}\gamma_{z\text{-new}} = \cosh\alpha \, \gamma_x + \sinh\alpha \, \gamma_t \tag{65}$$

$$\gamma_{y\text{-EX}} = \gamma_{x\text{-new}}\gamma_{y\text{-new}}\gamma_{z\text{-new}} \, \gamma_{y\text{-ex}} \, \gamma_{x\text{-new}}\gamma_{y\text{-new}}\gamma_{z\text{-new}} = \cosh\alpha \, \gamma_y + \sinh\alpha \, \gamma_x\gamma_y\gamma_t \tag{66}$$

$$\gamma_{z\text{-EX}} = \gamma_{x\text{-new}}\gamma_{y\text{-new}}\gamma_{z\text{-new}} \, \gamma_{z\text{-ex}} \, \gamma_{x\text{-new}}\gamma_{y\text{-new}}\gamma_{z\text{-new}} = \cosh\alpha \, \gamma_z + \sinh\alpha \, \gamma_x\gamma_z\gamma_t \tag{67}$$

What are coordinates? What are base vectors? Can these four entities be called base vectors? And can be coefficients of these four entities be called coordinates? The first two vectors (64) & (65) surely are base vectors as they square in time-like manner to plus one or in space-like manner to minus one.

$$\gamma_{t\text{-EX}}^2 = 1 \qquad \gamma_{x\text{-EX}}^2 = -1 \tag{68}$$

The last two entities (66) & (67) are linear combinations of vectors and trivectors, quite similar to a complex number which has a scalar component squaring to a positive value and an imaginary component squaring to a negative value. But now both components square to minus one (which is called pseudo-complex by some authors)

$$\gamma_y^2 = \gamma_z^2 = (\gamma_x\gamma_y\gamma_t)^2 = (\gamma_x\gamma_z\gamma_t)^2 = -1 \tag{69}$$

and the unit vectors itself square to a value which can be considered as pseudo-real

$$\gamma_{y\text{-EX}}^2 = -\cosh^2\alpha - \sinh^2\alpha\ \gamma_x\gamma_y\gamma_t + 2\cosh\alpha\sinh\alpha\ \gamma_x\gamma_t = -\cosh(2\alpha) + \sinh(2\alpha)\ \gamma_x\gamma_t \tag{70}$$

$$\gamma_{z\text{-EX}}^2 = -\cosh^2\alpha - \sinh^2\alpha\ \gamma_x\gamma_z\gamma_t + 2\cosh\alpha\sinh\alpha\ \gamma_x\gamma_t = -\cosh(2\alpha) + \sinh(2\alpha)\ \gamma_x\gamma_t \tag{71}$$

because the base units of all terms square to plus one:

$$1^2 = (\gamma_x\gamma_t)^2 = 1 \tag{72}$$

Are these base entities acting as base vectors? Of course they do: Their inner products disappear.

$$\gamma_{x\text{-EX}} \bullet \gamma_{t\text{-EX}} = \gamma_{y\text{-EX}} \bullet \gamma_{t\text{-EX}} = \gamma_{z\text{-EX}} \bullet \gamma_{t\text{-EX}} = \gamma_{x\text{-EX}} \bullet \gamma_{y\text{-EX}} = \gamma_{x\text{-EX}} \bullet \gamma_{z\text{-EX}} = \gamma_{y\text{-EX}} \bullet \gamma_{z\text{-EX}} = 0 \tag{73}$$

Therefore they can be considered as orthogonal to each other. All geometric products of these base vectors are pure bivectors, only the last one $\gamma_{y\text{-EX}}\ \gamma_{z\text{-EX}}$ is a linear combination of a bivector and a quadvector.

The inverse transformations from the exchanged coordinate system back to the original coordinate system then are:

$$\gamma_t = \cosh\alpha\ \gamma_{t\text{-EX}} - \sinh\alpha\ \gamma_{x\text{-EX}} \tag{74}$$

$$\gamma_x = \cosh\alpha\ \gamma_{x\text{-EX}} - \sinh\alpha\ \gamma_{t\text{-EX}} \tag{75}$$

$$\gamma_y = \cosh\alpha\ \gamma_{y\text{-EX}} - \sinh\alpha\ \gamma_{x\text{-EX}}\gamma_{y\text{-EX}}\gamma_{t\text{-EX}} \tag{76}$$

$$\gamma_z = \cosh\alpha\ \gamma_{z\text{-EX}} - \sinh\alpha\ \gamma_{x\text{-EX}}\gamma_{z\text{-EX}}\gamma_{t\text{-EX}} \tag{77}$$

Thus the complete forward transformations can be found again by a first transformation (57), followed now by a reflection at the old time axis pointing into the direction of base vector γ_t

$$\gamma_{t\text{-for}} = \gamma_t\ \gamma_{t\text{-ex}}\ \gamma_t = \cosh\alpha\ \gamma_t - \sinh\alpha\ \gamma_x \tag{78}$$

$$\gamma_{x\text{-for}} = \gamma_t\ \gamma_{x\text{-ex}}\ \gamma_t = \cosh\alpha\ \gamma_x - \sinh\alpha\ \gamma_t \tag{79}$$

$$\gamma_{y\text{-for}} = \gamma_t\ \gamma_{y\text{-ex}}\ \gamma_t = \cosh\alpha\ \gamma_y - \sinh\alpha\ \gamma_x\gamma_y\gamma_t \tag{80}$$

$$\gamma_{z\text{-for}} = \gamma_t\ \gamma_{z\text{-ex}}\ \gamma_t = \cosh\alpha\ \gamma_z - \sinh\alpha\ \gamma_x\gamma_z\gamma_t \tag{81}$$

An event, which takes place at position vector **r** will now be measured in the forwards or backwards exchanged coordinate systems at

$$\begin{aligned}\mathbf{r}_{back} = \mathbf{r}_{EX} &= \gamma_{x\text{-new}}\gamma_{y\text{-new}}\gamma_{z\text{-new}}\ \gamma_t\ \mathbf{r}\ \gamma_{t\text{-new}}\ \gamma_{x\text{-new}}\gamma_{y\text{-new}}\gamma_{z\text{-new}} \\ &= (ct\cosh\alpha + x\sinh\alpha)\ \gamma_t + (x\cosh\alpha + ct\sinh\alpha)\ \gamma_x \\ &\quad + y\cosh\alpha\ \gamma_y + y\sinh\alpha\ \gamma_x\gamma_y\gamma_t + z\cosh\alpha\ \gamma_z + z\sinh\alpha\ \gamma_x\gamma_z\gamma_t \\ &= ct_{back}\ \gamma_t + x_{back}\ \gamma_x + y_{back}\ \gamma_y + z_{back}\ \gamma_z + Y_{back}\ \gamma_x\gamma_y\gamma_t + Z_{back}\ \gamma_x\gamma_z\gamma_t\end{aligned} \tag{82}$$

or at

$$\begin{aligned}\mathbf{r}_{for} &= \gamma_t\ \gamma_t\ \mathbf{r}\ \gamma_{t\text{-new}}\ \gamma_t = \mathbf{r}\ \gamma_{t\text{-new}}\ \gamma_t \\ &= (ct\cosh\alpha - x\sinh\alpha)\ \gamma_t + (x\cosh\alpha - ct\sinh\alpha)\ \gamma_x \\ &\quad + y\cosh\alpha\ \gamma_y - y\sinh\alpha\ \gamma_x\gamma_y\gamma_t + z\cosh\alpha\ \gamma_z - z\sinh\alpha\ \gamma_x\gamma_z\gamma_t \\ &= ct_{for}\ \gamma_t + x_{for}\ \gamma_x + y_{for}\ \gamma_y + z_{for}\ \gamma_z + Y_{for}\ \gamma_x\gamma_y\gamma_t + Z_{for}\ \gamma_x\gamma_z\gamma_t\end{aligned} \tag{83}$$

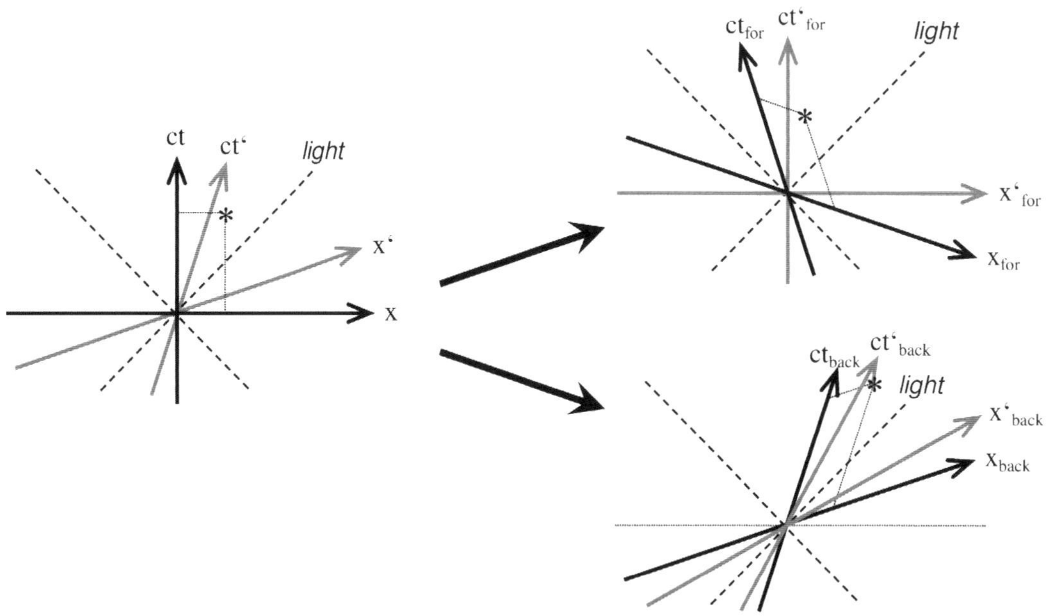

Figure 3. Sketch of forward exchange transformation with positive, anti-clockwise exchange of the time axis (top) and backward exchange transformation (at the bottom).

The components of the time directions and of the x-directions of exchange transformations of eqs. (82) & (83) are identical to the equivalent components of Lorentz transformations of eqs. (48) & (49). Therefore the sketches of the ct-x-plane of these transformations (y = z = 0), which are shown in figures 2 and 3, are completely identical.

But the components perpendicular to this plane are of course different, and a relevant question with respect to these components of the y- and z-directions should be discussed: What are coordinates? The four coordinates which are connected with the usual base vectors γ_t, γ_x, γ_y, γ_z now change according to

$$ct_{for} = ct \cosh\alpha - x \sinh\alpha \qquad x_{for} = x \cosh\alpha - ct \sinh\alpha \qquad y_{for} = y \cosh\alpha \qquad z_{for} = z \cosh\alpha \tag{84}$$

But now two additional terms pop up! There are more "coordinates":

$$Y_{for} = -\,y \sinh\alpha \qquad Z_{for} = -\,z \sinh\alpha \tag{85}$$

An observer will now measure one time-related **and** three spatial distance-related **and** two spacetime volume-related coordinates. She or he will measure two additional entities which should be considered as fields. They are part of a complete description of the event seen from the forwards transformed coordinate system and they have an effect on the physics of this event.

Again a motionless, stationary person with x = y = z = 0 will be seen by a second observer from the other coordinate system having a constant velocity. The hyperbolic transformation factors are then again identical to the Lorentz factors:

$$\tanh\alpha = \frac{v}{c} \qquad \cosh\alpha = \frac{1}{\sqrt{1-\tanh^2\alpha}} = \frac{1}{\sqrt{1-\frac{v^2}{c^2}}} \qquad \sinh\alpha = \frac{\tanh\alpha}{\sqrt{1-\tanh^2\alpha}} = \frac{\frac{v}{c}}{\sqrt{1-\frac{v^2}{c^2}}} \tag{86}$$

Therefore the transformations (84) & (85) will become

$$ct_{for} = \frac{ct - \frac{v}{c}x}{\sqrt{1-\frac{v^2}{c^2}}} \qquad x_{for} = \frac{x - vt}{\sqrt{1-\frac{v^2}{c^2}}} \qquad y_{for} = \frac{y}{\sqrt{1-\frac{v^2}{c^2}}} \qquad z_{for} = \frac{z}{\sqrt{1-\frac{v^2}{c^2}}} \tag{87}$$

$$Y_{for} = \frac{-\frac{v}{c}y}{\sqrt{1-\frac{v^2}{c^2}}} \qquad Z_{for} = \frac{-\frac{v}{c}z}{\sqrt{1-\frac{v^2}{c^2}}}$$

As a consequence of these transformations (87), time dilation, length contraction and additional fields will be observed. But this time no boost is the reason for these effects. They are caused by exchanging a small fraction of the x-direction into the time-direction.

8. Creating sort of an electric field (first try)

As discussed four years ago in Bregenz [27] 3d spatial vectors or 4d spacetime bivectors

$$\mathbf{E}_x = E_x\,\sigma_x = E_x\,\gamma_x\gamma_t \qquad \mathbf{E}_y = E_y\,\sigma_y = E_y\,\gamma_y\gamma_t \qquad \mathbf{E}_z = E_z\,\sigma_z = E_z\,\gamma_z\gamma_t \tag{88}$$

can be used to describe an electrical field and 3d spatial bivectors or 4d spacelike bivectors

$$\mathbf{B}_x = B_x\,\sigma_y\sigma_z = -\,B_x\,\gamma_y\gamma_z \qquad \mathbf{B}_y = B_y\,\sigma_z\sigma_x = -\,B_y\,\gamma_z\gamma_x \qquad \mathbf{B}_z = B_z\,\sigma_x\sigma_y = -\,B_z\,\gamma_x\gamma_y \tag{89}$$

can be used to describe a magnetic field.

To transform an empty spacetime (55), (56) into a spacetime filled with an electrical field, the old time axis represented by unit vector γ_t will be exchanged or replaced by a new axis which is a linear combination of a time-related vector and an electric bivector.

$$\gamma_{t\text{-old}} = \gamma_t \qquad \gamma_{t\text{-new}} = \cos\alpha\;\gamma_t + \sin\alpha\;\gamma_x\gamma_t \tag{90}$$

As now both base units are time-like base units squaring to one

$$\gamma_t^2 = (\gamma_x\gamma_t)^2 = 1 \qquad \Rightarrow \qquad \gamma_{t\text{-old}}^2 = \gamma_{t\text{-new}}^2 = 1 \tag{91}$$

the linear combination should be composed of sine and cosine terms to get a new unit vector again squaring to one. Position vector **r** will then be transformed analogously to eq. (57) into

$$\mathbf{r}_{el} = \gamma_t\,\mathbf{r}\,\gamma_{t\text{-new}} \tag{92}$$

and the four base vectors pointing into the directions of the coordinate axes will be transformed into

$$\gamma_{t\text{-el}} = \gamma_t\,\gamma_t\,\gamma_{t\text{-new}} = \cos\alpha\;\gamma_t + \sin\alpha\;\gamma_x\gamma_t = -\,\gamma_x\gamma_t\,\gamma_{x\text{-el}} \tag{93}$$

$$\gamma_{x\text{-el}} = \gamma_t\,\gamma_x\,\gamma_{t\text{-new}} = -\cos\alpha\;\gamma_x - \sin\alpha \tag{94}$$

$$\gamma_{y\text{-el}} = \gamma_t\,\gamma_y\,\gamma_{t\text{-new}} = -\cos\alpha\;\gamma_y - \sin\alpha\;\gamma_x\gamma_y = \gamma_x\gamma_y\,\gamma_{x\text{-el}} \tag{95}$$

$$\gamma_{z\text{-el}} = \gamma_t\,\gamma_z\,\gamma_{t\text{-new}} = -\cos\alpha\;\gamma_z - \sin\alpha\;\gamma_x\gamma_z = \gamma_x\gamma_z\,\gamma_{x\text{-el}} \tag{96}$$

Now vectors transform into linear combinations of vectors, bivectors, or scalars. But the minus signs of theses eqs. (93) – (96) show, that a left-handed coordinate system has emerged. To transform it again into a right-handed, backwards transformed coordinate system, a reflection at the three-dimensional new space which is represented by the oriented volume element

$$\gamma_{x\text{-el}}\gamma_{y\text{-el}}\gamma_{z\text{-el}} = -\cos\alpha\;\gamma_x\gamma_y\gamma_z - \sin\alpha\;\gamma_y\gamma_z = \gamma_y\gamma_z\,\gamma_{x\text{-el}} \tag{97}$$

will cancel the negative signs and will result in the following base vectors of the right-handed final coordinate system:

$$\gamma_{t\text{-}EL} = \cos\alpha\ \gamma_t + \sin\alpha\ \gamma_x\gamma_t\ = \gamma_x\gamma_t\ \gamma_{x\text{-}el} \quad (98)$$

$$\gamma_{x\text{-}EL} = \cos\alpha\ \gamma_x + \sin\alpha \quad (99)$$

$$\gamma_{y\text{-}EL} = \cos\alpha\ \gamma_y + \sin\alpha\ \gamma_x\gamma_y = \gamma_x\gamma_y\ \gamma_{x\text{-}el} \quad (100)$$

$$\gamma_{z\text{-}EL} = \cos\alpha\ \gamma_z + \sin\alpha\ \gamma_x\gamma_z = \gamma_x\gamma_z\ \gamma_{x\text{-}el} \quad (101)$$

And the inverse, forward transformations will be constructed again by reflecting eq. (92) at the old time axis into the direction of base vector γ_t.

$$\gamma_{t\text{-}for} = \gamma_t\ \gamma_{t\text{-}el}\ \gamma_t = \cos\alpha\ \gamma_t - \sin\alpha\ \gamma_x\gamma_t \quad (102)$$

$$\gamma_{x\text{-}for} = \gamma_t\ \gamma_{x\text{-}el}\ \gamma_t = \cos\alpha\ \gamma_x - \sin\alpha \quad (103)$$

$$\gamma_{y\text{-}for} = \gamma_t\ \gamma_{y\text{-}el}\ \gamma_t = \cos\alpha\ \gamma_y - \sin\alpha\ \gamma_x\gamma_y \quad (104)$$

$$\gamma_{z\text{-}for} = \gamma_t\ \gamma_{z\text{-}el}\ \gamma_t = \cos\alpha\ \gamma_z - \sin\alpha\ \gamma_x\gamma_z \quad (105)$$

An event, which takes place at position vector **r** will now be measured in both coordinate systems at

$$\begin{aligned}\mathbf{r}_{back} = \mathbf{r}_{EL} &= \gamma_{x\text{-}el}\gamma_{y\text{-}el}\gamma_{z\text{-}el}\ \gamma_t\ \mathbf{r}\ \gamma_{t\text{-}el}\ \gamma_{x\text{-}el}\gamma_{y\text{-}el}\gamma_{z\text{-}el} \\ &= \cos\alpha\ (ct\ \gamma_t + x\ \gamma_x + y\ \gamma_y + z\ \gamma_z) + \sin\alpha\ (ct\ \gamma_x\gamma_t + x + y\ \gamma_x\gamma_y + z\ \gamma_x\gamma_z) \quad (106) \\ &= ct_{back}\ \gamma_t + x_{back}\ \gamma_x + y_{back}\ \gamma_y + z_{back}\ \gamma_z + E_{x\text{-}back}\ \gamma_x\gamma_t + X_{back} + B_{z\text{-}back}\ \gamma_x\gamma_y + B_{y\text{-}back}\ \gamma_x\gamma_z\end{aligned}$$

or at

$$\begin{aligned}\mathbf{r}_{for} &= \gamma_t\ \gamma_t\ \mathbf{r}\ \gamma_{t\text{-}el}\ \gamma_t \\ &= \cos\alpha\ (ct\ \gamma_t + x\ \gamma_x + y\ \gamma_y + z\ \gamma_z) - \sin\alpha\ (ct\ \gamma_x\gamma_t + x + y\ \gamma_x\gamma_y + z\ \gamma_x\gamma_z) \quad (107) \\ &= ct_{for}\ \gamma_t + x_{for}\ \gamma_x + y_{for}\ \gamma_y + z_{for}\ \gamma_z + E_{x\text{-}for}\ \gamma_x\gamma_t + X_{for} + B_{z\text{-}for}\ \gamma_x\gamma_y + B_{y\text{-}for}\ \gamma_x\gamma_z\end{aligned}$$

The four coordinates which are connected with the usual base vectors γ_t, γ_x, γ_y, γ_z now change according to

$$ct_{new} = ct\ \cos\alpha \qquad x_{new} = x\ \cos\alpha \qquad y_{new} = y\ \cos\alpha \qquad z_{new} = z\ \cos\alpha \quad (108)$$

And again additional terms pop up. These three bivector and one scalar coordinates of eqs. (107) & (108) will be named courageously as:

$$E_x = \pm\ ct\ \sin\alpha \qquad X_{EL} = \pm\ x\ \sin\alpha \qquad B_y = \pm\ y\ \sin\alpha \qquad B_z = \pm\ z\ \sin\alpha \quad (109)$$

These additional components cannot be seen and cannot be measured by a person in the forwards transformed coordinate system

$$\mathbf{r}_{for} = ct\ \gamma_{t\text{-}for} + x\ \gamma_{x\text{-}for} + y\ \gamma_{y\text{-}for} + z\ \gamma_{z\text{-}for} + 0 \quad (110)$$

or by a person in the backwards transformed coordinate system

$$\mathbf{r}_{back} = ct\ \gamma_{t\text{-}back} + x\ \gamma_{x\text{-}back} + y\ \gamma_{y\text{-}back} + z\ \gamma_{z\text{-}back} + 0 \quad (111)$$

measuring only the usual coordinates of empty spacetime. But they can be seen and they can be measured by an observer of the original coordinate system with axes pointing into the directions of the base vectors γ_t, γ_x, γ_y, and γ_z when this observer is looking at the new coordinate systems (106 & 107).

These additional coordinates might represent an electrical field bivector, an interesting scalar part and two magnetic field bivectors. And the dynamical behavior of these additional components is surprising: The electrical field bivector is connected with the time coordinate t. An observer simply waiting at the origin at $x = y = z = 0$ will observe an electrical field which becomes stronger and stronger with increasing time t. This is indeed an unexpected feature of transformations of coordinates.

9. Creating sort of an electric field (second try)

To transform an empty spacetime (55), (56) into a spacetime filled with an electrical field in an alternative way, the old x-axis represented by unit vector γ_x will be exchanged or replaced by a new axis which is a linear combination of a space-related vector and an electric bivector.

$$\gamma_{x\text{-old}} = \gamma_x \qquad \gamma_{x\text{-new}} = \cosh\alpha\ \gamma_x + \sinh\alpha\ \gamma_x\gamma_t \tag{112}$$

As now one base unit is a space-like base unit squaring to minus one and the other base unit is a timelike base unit squaring to minus one

$$\gamma_x^2 = -1 \text{ and } (\gamma_x\gamma_t)^2 = 1 \qquad \Rightarrow \qquad \gamma_{x\text{-old}}^2 = \gamma_{x\text{-new}}^2 = -1 \tag{113}$$

the linear combination should be composed as usual of hyperbolic sine and hyperbolic cosine terms to get a new unit vector again squaring to minus one. Position vector **r** will then be transformed into

$$\mathbf{r}_{el} = \gamma_x\ \mathbf{r}\ \gamma_{x\text{-new}} \tag{114}$$

and the four base vectors pointing into the directions of the coordinate axes will be transformed into

$$\gamma_{t\text{-el}} = \gamma_x\ \gamma_t\ \gamma_{x\text{-new}} = \cosh\alpha\ \gamma_t + \sinh\alpha \tag{115}$$

$$\gamma_{x\text{-el}} = \gamma_x\ \gamma_x\ \gamma_{x\text{-new}} = -\cosh\alpha\ \gamma_x - \sinh\alpha\ \gamma_x\gamma_t = -\gamma_x\gamma_t\ \gamma_{t\text{-el}} \tag{116}$$

$$\gamma_{y\text{-el}} = \gamma_x\ \gamma_y\ \gamma_{x\text{-new}} = \cosh\alpha\ \gamma_y + \sinh\alpha\ \gamma_y\gamma_t = \gamma_y\gamma_t\ \gamma_{t\text{-el}} \tag{117}$$

$$\gamma_{z\text{-el}} = \gamma_x\ \gamma_z\ \gamma_{x\text{-new}} = \cosh\alpha\ \gamma_z + \sinh\alpha\ \gamma_z\gamma_t = \gamma_z\gamma_t\ \gamma_{t\text{-el}} \tag{118}$$

Now vectors transform again into linear combinations of vectors, bivectors, or scalars. Reflecting again at the axis pointing into the old x-direction represented by γ_x will reverse the sign first negative sign of eq. (116) giving:

$$\gamma_{t\text{-EL}} = \gamma_x\ \gamma_{t\text{-el}}\ \gamma_x = \cosh\alpha\ \gamma_t - \sinh\alpha \tag{119}$$

$$\gamma_{x\text{-EL}} = \gamma_x\ \gamma_{x\text{-el}}\ \gamma_x = \cosh\alpha\ \gamma_x - \sinh\alpha\ \gamma_x\gamma_t \tag{120}$$

$$\gamma_{y\text{-EL}} = \gamma_x\ \gamma_{y\text{-el}}\ \gamma_x = \cosh\alpha\ \gamma_y - \sinh\alpha\ \gamma_y\gamma_t \tag{121}$$

$$\gamma_{z\text{-EL}} = \gamma_x\ \gamma_{z\text{-el}}\ \gamma_x = \cosh\alpha\ \gamma_z - \sinh\alpha\ \gamma_z\gamma_t \tag{122}$$

An event, which takes place at position vector **r** will now be measured at

$$\begin{aligned} \mathbf{r}_{EL} &= \gamma_x\ \gamma_x\ \mathbf{r}\ \gamma_{x\text{-new}}\ \gamma_x = -\mathbf{r}\ \gamma_{x\text{-new}}\ \gamma_x \\ &= \cosh\alpha\ (ct\ \gamma_t + x\ \gamma_x + y\ \gamma_y + z\ \gamma_z) - \sinh\alpha\ (ct + x\ \gamma_x\gamma_t + y\ \gamma_y\gamma_t + z\ \gamma_z\gamma_t) \\ &= ct_{EL}\ \gamma_t + x_{EL}\ \gamma_x + y_{EL}\ \gamma_y + z_{EL}\ \gamma_z + cT + E_x\ \gamma_x\gamma_t + E_y\ \gamma_y\gamma_t + E_z\ \gamma_z\gamma_t \end{aligned} \tag{123}$$

The four coordinates which are connected with the usual base vectors γ_t, γ_x, γ_y, γ_z now change according to

$$ct_{EL} = ct\cosh\alpha \qquad x_{EL} = x\cosh\alpha \qquad y_{EL} = y\cosh\alpha \qquad z_{EL} = z\cosh\alpha \tag{124}$$

And again additional terms pop up. These three bivector and one scalar coordinates of eq. (123)

$$cT = -ct\sinh\alpha \qquad E_x = -x\sinh\alpha \qquad E_y = -y\sinh\alpha \qquad E_z = -z\sinh\alpha \tag{125}$$

are indeed electrical field components only and it is reasonable to conclude, that with this second try sort of an electric field has indeed been constructed now.

These additional components cannot be seen and cannot be measured by a person in a coordinate system with axes pointing into the directions of $\gamma_{t\text{-EL}}$, $\gamma_{x\text{-EL}}$, $\gamma_{y\text{-EL}}$, $\gamma_{z\text{-EL}}$. But they can be seen and they can be measured by an observer in the original coordinate system with time and space axes pointing into the directions of of γ_t, γ_x, γ_y, γ_z (123).

10. Generalized matrix inverses

As announced in the abstract, matrix inverses of non-square matrices (e.g. Moore-Penrose generalized inverses) will be discussed at the end, which now will be reached.

Let's go back to solution (17) of an exactly constrained system of three linear equations with three unknown variables (1). This solution on the left side of eq. (17) can be split into a matrix multiplication because the resulting vector **r** is a linear combination of the three base vectors (13).

$$\begin{aligned} x &= (\mathbf{a}\wedge\mathbf{b}\wedge\mathbf{c})^{-1}(\sigma_x\wedge\mathbf{b}\wedge\mathbf{c})\,r_1 + (\mathbf{a}\wedge\mathbf{b}\wedge\mathbf{c})^{-1}(\sigma_y\wedge\mathbf{b}\wedge\mathbf{c})\,r_2 + (\mathbf{a}\wedge\mathbf{b}\wedge\mathbf{c})^{-1}(\sigma_x\wedge\mathbf{b}\wedge\mathbf{c})\,r_3 \\ y &= (\mathbf{a}\wedge\mathbf{b}\wedge\mathbf{c})^{-1}(\mathbf{a}\wedge\sigma_x\wedge\mathbf{c})\,r_1 + (\mathbf{a}\wedge\mathbf{b}\wedge\mathbf{c})^{-1}(\mathbf{a}\wedge\sigma_y\wedge\mathbf{c})\,r_2 + (\mathbf{a}\wedge\mathbf{b}\wedge\mathbf{c})^{-1}(\mathbf{a}\wedge\sigma_z\wedge\mathbf{c})\,r_3 \\ z &= (\mathbf{a}\wedge\mathbf{b}\wedge\mathbf{c})^{-1}(\mathbf{a}\wedge\mathbf{b}\wedge\sigma_x)\,r_1 + (\mathbf{a}\wedge\mathbf{b}\wedge\mathbf{c})^{-1}(\mathbf{a}\wedge\mathbf{b}\wedge\sigma_y)\,r_2 + (\mathbf{a}\wedge\mathbf{b}\wedge\mathbf{c})^{-1}(\mathbf{a}\wedge\mathbf{b}\wedge\sigma_z)\,r_3 \end{aligned} \tag{126}$$

These solution equations can be expressed with the scheme of Falk (127), clearly identifying the lead matrix as the usual inverse $\mathbf{A}^{-1}$ of the coefficient matrix **A**.

$$\begin{array}{ccc|c} & & & r_1 \\ & & & r_2 \\ & & & r_3 \\ \hline (\mathbf{a}\wedge\mathbf{b}\wedge\mathbf{c})^{-1}(\sigma_x\wedge\mathbf{b}\wedge\mathbf{c}) & (\mathbf{a}\wedge\mathbf{b}\wedge\mathbf{c})^{-1}(\sigma_y\wedge\mathbf{b}\wedge\mathbf{c}) & (\mathbf{a}\wedge\mathbf{b}\wedge\mathbf{c})^{-1}(\sigma_z\wedge\mathbf{b}\wedge\mathbf{c}) & x \\ (\mathbf{a}\wedge\mathbf{b}\wedge\mathbf{c})^{-1}(\mathbf{a}\wedge\sigma_x\wedge\mathbf{c}) & (\mathbf{a}\wedge\mathbf{b}\wedge\mathbf{c})^{-1}(\mathbf{a}\wedge\sigma_y\wedge\mathbf{c}) & (\mathbf{a}\wedge\mathbf{b}\wedge\mathbf{c})^{-1}(\mathbf{a}\wedge\sigma_z\wedge\mathbf{c}) & y \\ (\mathbf{a}\wedge\mathbf{b}\wedge\mathbf{c})^{-1}(\mathbf{a}\wedge\mathbf{b}\wedge\sigma_x) & (\mathbf{a}\wedge\mathbf{b}\wedge\mathbf{c})^{-1}(\mathbf{a}\wedge\mathbf{b}\wedge\sigma_y) & (\mathbf{a}\wedge\mathbf{b}\wedge\mathbf{c})^{-1}(\mathbf{a}\wedge\mathbf{b}\wedge\sigma_z) & z \end{array} \tag{127}$$

All this happens in a three-dimensional space. What will now happen if the coefficient vectors **a**, **b**, **c** and the resulting vector **r** which possess three components pointing into the three directions x, y, z are placed into a higher dimensional space and are rotated into an oblique position, now pointing into the four directions w, x, y, z?

Of course all relations between them remain unchanged if all vectors are rotated in the same way. Especially eq. (14) $\mathbf{a}\,x + \mathbf{b}\,y + \mathbf{c}\,z = \mathbf{r}$ will not change. But the coefficient vectors and the resulting vector will now have four components, thus describing an overconstrained system of linear equations of three variables with four linear equations.

Therefore the solution equations (17) will remain unchanged, too. The only difference is now, that the decomposition of the solution into a matrix multiplication will have more components, and the leading matrix will have more columns than rows. This can be expressed again with the scheme of Falk (128).

$$\begin{array}{cccc|c} & & & & r_1 \\ & & & & r_2 \\ & & & & r_3 \\ & & & & r_4 \\ \hline (\mathbf{a}\wedge\mathbf{b}\wedge\mathbf{c})^{-1}(\sigma_w\wedge\mathbf{b}\wedge\mathbf{c}) & (\mathbf{a}\wedge\mathbf{b}\wedge\mathbf{c})^{-1}(\sigma_x\wedge\mathbf{b}\wedge\mathbf{c}) & (\mathbf{a}\wedge\mathbf{b}\wedge\mathbf{c})^{-1}(\sigma_y\wedge\mathbf{b}\wedge\mathbf{c}) & (\mathbf{a}\wedge\mathbf{b}\wedge\mathbf{c})^{-1}(\sigma_z\wedge\mathbf{b}\wedge\mathbf{c}) & x \\ (\mathbf{a}\wedge\mathbf{b}\wedge\mathbf{c})^{-1}(\mathbf{a}\wedge\sigma_w\wedge\mathbf{c}) & (\mathbf{a}\wedge\mathbf{b}\wedge\mathbf{c})^{-1}(\mathbf{a}\wedge\sigma_x\wedge\mathbf{c}) & (\mathbf{a}\wedge\mathbf{b}\wedge\mathbf{c})^{-1}(\mathbf{a}\wedge\sigma_y\wedge\mathbf{c}) & (\mathbf{a}\wedge\mathbf{b}\wedge\mathbf{c})^{-1}(\mathbf{a}\wedge\sigma_z\wedge\mathbf{c}) & y \\ (\mathbf{a}\wedge\mathbf{b}\wedge\mathbf{c})^{-1}(\mathbf{a}\wedge\mathbf{b}\wedge\sigma_w) & (\mathbf{a}\wedge\mathbf{b}\wedge\mathbf{c})^{-1}(\mathbf{a}\wedge\mathbf{b}\wedge\sigma_x) & (\mathbf{a}\wedge\mathbf{b}\wedge\mathbf{c})^{-1}(\mathbf{a}\wedge\mathbf{b}\wedge\sigma_y) & (\mathbf{a}\wedge\mathbf{b}\wedge\mathbf{c})^{-1}(\mathbf{a}\wedge\mathbf{b}\wedge\sigma_z) & z \end{array} \tag{128}$$

The lead matrix can be again clearly identified as the inverse $\mathbf{A}^{-1}$ of the now non-square coefficient matrix **A**. As this matrix is found by using Geometric Algebra or generalized Pauli algebra, it might be called the Pauli algebra generalized matrix inverse.

Summarizing and generalizing these results, it is possible to say: The matrix inverse of a square matrix with dimension n x n can be found with the Geometric Algebra equivalent of Cramer's rule by substituting coefficient vector a_j ($a_1 = \mathbf{a}$; $a_2 = \mathbf{b}$; $a_3 = \mathbf{c}$; etc.) of eq. (18) by unit vectors σ_i

$$a_1 \wedge a_2 \wedge a_3 \wedge \ldots \wedge a_{j-1} \wedge \sigma_i \wedge a_{j+1} \wedge \ldots \wedge a_n = \det \mathbf{A}_{ij}\, \sigma_1\sigma_2\sigma_3 \ldots \sigma_n \tag{129}$$

The oriented volumes of these hyper-parallelepipeds (or parallelotops) thus represent the determinants of the transposed submatrices $\mathbf{A}_{ij}$ which are already attached with the correct cofactor signs, usually called adjoint matrices. Therefore the elements of the matrix inverse can be written as quotient of eqs. (129) and (18), resulting in the following matrix inverse:

$$(\mathbf{A}^{-1})_{ij} = (a_1 \wedge a_2 \wedge a_3 \wedge \ldots \wedge a_n)^{-1}\,(a_1 \wedge a_2 \wedge a_3 \wedge \ldots \wedge a_{j-1} \wedge \sigma_i \wedge a_{j+1} \wedge \ldots \wedge a_n) \tag{130}$$

Obviously inverses of underonstrained (more variables than equations, $m < n$) systems of simultaneous linear equations do not exist, as coefficient vectors a_i then are linearly dependent and therefore the outer product (18) has to be zero.

For this reason it makes sense to only discuss matrix inverses of overconstrained and consistent systems of simultaneous linear equations ($m > n$, $\det \mathbf{A} \neq 0$).

The elements of the matrix inverse of a non-square matrix [29, 30, 31] with dimension m x n can then be found again by an equation nearly identical to eq. (130)

$$(\mathbf{A}^{-1})_{ij} = (a_1 \wedge a_2 \wedge a_3 \wedge \ldots \wedge a_n)^{-1}\,(a_1 \wedge a_2 \wedge a_3 \wedge \ldots \wedge a_{j-1} \wedge \sigma_i \wedge a_{j+1} \wedge \ldots \wedge a_n) \tag{131}$$

now resulting in a non-square matrix inverse $\mathbf{A}^{-1}$ with dimension n x m, because $i \in \{1, 2, 3, \ldots, m\}$ and $j \in \{1, 2, 3, \ldots, n\}$ have different values now.

Alternatively the right side of eq. (17) can be used to construct Pauli algebra generalized matrix inverses. A post-division by the oriented volumes of the coefficient vector parallelepipeds

$$\begin{array}{ccc|c} & & & r_1 \\ & & & r_2 \\ & & & r_3 \\ \hline (\sigma_x \wedge \mathbf{b} \wedge \mathbf{c})\,(\mathbf{a} \wedge \mathbf{b} \wedge \mathbf{c})^{-1} & (\sigma_y \wedge \mathbf{b} \wedge \mathbf{c})\,(\mathbf{a} \wedge \mathbf{b} \wedge \mathbf{c})^{-1} & (\sigma_z \wedge \mathbf{b} \wedge \mathbf{c})\,(\mathbf{a} \wedge \mathbf{b} \wedge \mathbf{c})^{-1} & x \\ (\mathbf{a} \wedge \sigma_x \wedge \mathbf{c})\,(\mathbf{a} \wedge \mathbf{b} \wedge \mathbf{c})^{-1} & (\mathbf{a} \wedge \sigma_y \wedge \mathbf{c})\,(\mathbf{a} \wedge \mathbf{b} \wedge \mathbf{c})^{-1} & (\mathbf{a} \wedge \sigma_z \wedge \mathbf{c})\,(\mathbf{a} \wedge \mathbf{b} \wedge \mathbf{c})^{-1} & y \\ (\mathbf{a} \wedge \mathbf{b} \wedge \sigma_x)\,(\mathbf{a} \wedge \mathbf{b} \wedge \mathbf{c})^{-1} & (\mathbf{a} \wedge \mathbf{b} \wedge \sigma_y)\,(\mathbf{a} \wedge \mathbf{b} \wedge \mathbf{c})^{-1} & (\mathbf{a} \wedge \mathbf{b} \wedge \sigma_z)\,(\mathbf{a} \wedge \mathbf{b} \wedge \mathbf{c})^{-1} & z \end{array} \tag{132}$$

will result again in a square matrix inverse which is identical to the lead matrix of eq. (127).

But the overconstrained, consistent system of linear equations will then have a slightly different Pauli algebra generalized matrix inverse.

$$\begin{array}{cccc|c} & & & & r_1 \\ & & & & r_2 \\ & & & & r_3 \\ & & & & r_4 \\ \hline (\sigma_w\wedge\mathbf{b}\wedge\mathbf{c})(\mathbf{a}\wedge\mathbf{b}\wedge\mathbf{c})^{-1} & (\sigma_x\wedge\mathbf{b}\wedge\mathbf{c})(\mathbf{a}\wedge\mathbf{b}\wedge\mathbf{c})^{-1} & (\sigma_y\wedge\mathbf{b}\wedge\mathbf{c})(\mathbf{a}\wedge\mathbf{b}\wedge\mathbf{c})^{-1} & (\sigma_z\wedge\mathbf{b}\wedge\mathbf{c})(\mathbf{a}\wedge\mathbf{b}\wedge\mathbf{c})^{-1} & x \\ (\mathbf{a}\wedge\sigma_w\wedge\mathbf{c})(\mathbf{a}\wedge\mathbf{b}\wedge\mathbf{c})^{-1} & (\mathbf{a}\wedge\sigma_x\wedge\mathbf{c})(\mathbf{a}\wedge\mathbf{b}\wedge\mathbf{c})^{-1} & (\mathbf{a}\wedge\sigma_y\wedge\mathbf{c})(\mathbf{a}\wedge\mathbf{b}\wedge\mathbf{c})^{-1} & (\mathbf{a}\wedge\sigma_z\wedge\mathbf{c})(\mathbf{a}\wedge\mathbf{b}\wedge\mathbf{c})^{-1} & y \\ (\mathbf{a}\wedge\mathbf{b}\wedge\sigma_w)(\mathbf{a}\wedge\mathbf{b}\wedge\mathbf{c})^{-1} & (\mathbf{a}\wedge\mathbf{b}\wedge\sigma_x)(\mathbf{a}\wedge\mathbf{b}\wedge\mathbf{c})^{-1} & (\mathbf{a}\wedge\mathbf{b}\wedge\sigma_y)(\mathbf{a}\wedge\mathbf{b}\wedge\mathbf{c})^{-1} & (\mathbf{a}\wedge\mathbf{b}\wedge\sigma_z)(\mathbf{a}\wedge\mathbf{b}\wedge\mathbf{c})^{-1} & z \end{array} \tag{133}$$

The lead matrix can be again clearly identified as an inverse $\underline{\mathbf{A}}^{-1}$ of the now non-square coefficient matrix $\mathbf{A}$. Of course the scalar parts of the pre-multiplied non-square matrix inverse $\mathbf{A}^{-1}$ and the post-multiplied non-square matrix inverse $\underline{\mathbf{A}}^{-1}$ are identical. But the bivector parts of the two non-square matrix inverses $\mathbf{A}^{-1}$ and $\underline{\mathbf{A}}^{-1}$ have reversed signs.

In a similar way the elements of post-multiplied Pauli algebra generalized matrix inverses $(\underline{\mathbf{A}}^{-1})_{ij}$ of higher-dimensional non-square coefficient matrices $\mathbf{A}$ can be constructed.

It can easily be shown that Pauli algebra generalized matrix inverses are left sided matrix inverses only:

$$\mathbf{A}^{-1}\,\mathbf{A} = \mathrm{I} \quad \text{and} \quad \underline{\mathbf{A}}^{-1}\,\mathbf{A} = \mathrm{I} \tag{134}$$

$$\mathbf{A}\,\mathbf{A}^{-1} \neq \mathrm{I} \quad \text{and} \quad \mathbf{A}\,\underline{\mathbf{A}}^{-1} \neq \mathrm{I} \tag{135}$$

A post-multiplication of Pauli algebra generalized matrix inverses from the right (135) will not result in an identity matrix I. (And by the way: the identity matrices I of eqs. (134) & (135) have different dimensions, as **A** now is a non-square coefficient matrix.)

Sandwiching or transposing eqs. (134) will then give the following three different algebraic relations:

$$\mathbf{A}\,\mathbf{A}^{-1}\,\mathbf{A} = \mathbf{A}\,\mathrm{I} = \mathbf{A} \quad \text{and} \quad \mathbf{A}\,\underline{\mathbf{A}}^{-1}\,\mathbf{A} = \mathbf{A}\,\mathrm{I} = \mathbf{A} \tag{136}$$

$$\mathbf{A}^{-1}\,\mathbf{A}\,\mathbf{A}^{-1} = \mathrm{I}\,\mathbf{A}^{-1} = \mathbf{A}^{-1} \quad \text{and} \quad \underline{\mathbf{A}}^{-1}\,\mathbf{A}\,\underline{\mathbf{A}}^{-1} = \mathrm{I}\,\underline{\mathbf{A}}^{-1} = \underline{\mathbf{A}}^{-1} \tag{137}$$

$$(\mathbf{A}^{-1}\,\mathbf{A})^{\mathrm{T}} = \mathrm{I}^{\mathrm{T}} = \mathrm{I} = \mathbf{A}^{-1}\,\mathbf{A} \quad \text{and} \quad (\underline{\mathbf{A}}^{-1}\,\mathbf{A})^{\mathrm{T}} = \mathrm{I}^{\mathrm{T}} = \mathrm{I} = \underline{\mathbf{A}}^{-1}\,\mathbf{A} \tag{138}$$

But unfortunately there is:

$$(\mathbf{A}\,\mathbf{A}^{-1})^{\mathrm{T}} \neq \mathbf{A}\,\mathbf{A}^{-1} \quad \text{and} \quad (\mathbf{A}\,\underline{\mathbf{A}}^{-1})^{\mathrm{T}} \neq \mathbf{A}\,\underline{\mathbf{A}}^{-1} \tag{139}$$

Nevertheless, the last relation (139) can be transformed into an identity if only the identical scalar parts of $\mathbf{A}^{-1}$ or $\underline{\mathbf{A}}^{-1}$ are taken into account (see section 11).

Multiplicativity does not hold either. If the product of two non-square matrices **A** and **B**

$$\mathbf{A}\,\mathbf{B} = \mathbf{D} \tag{140}$$

is given, the product of the Pauli algebra generalized matrix inverses $\mathbf{B}^{-1}$ and $\mathbf{A}^{-1}$ will not result in the Pauli algebra generalized matrix inverse $\mathbf{D}^{-1}$ of the resulting matrix **D**.

$$\mathbf{B}^{-1}\,\mathbf{A}^{-1} \neq \mathbf{D}^{-1} \tag{141}$$

A simple counter-example will be presented as attachment (see section 12).

11. Moore-Penrose generalized matrix inverses

More and more introductory textbooks of business mathematics and of mathematical economics discuss Moore-Penrose generalized matrix inverses $\mathbf{A}^{+}$ as elementary part of the foundations of matrix algebra [34, 35]. These Moore-Penrose generalized matrix inverses are defined and discussed in introductory books usually only with respect to the four Moore-Penrose conditions (142) to (145).

$$\mathbf{A}\,\mathbf{A}^{+}\,\mathbf{A} = \mathbf{A} \tag{142}$$

$$\mathbf{A}^{+}\,\mathbf{A}\,\mathbf{A}^{+} = \mathbf{A}^{+} \tag{143}$$

$$(\mathbf{A}\,\mathbf{A}^{+})^{\mathrm{T}} = \mathbf{A}\,\mathbf{A}^{+} \tag{144}$$

$$(\mathbf{A}^{+}\,\mathbf{A})^{\mathrm{T}} = \mathbf{A}^{+}\,\mathbf{A} \tag{145}$$

This starting point is didactically problematic: Students then develop a strong algebraic perspective of Moore-Penrose generalized matrix inverses. But they fail to develop a geometric perspective of generalized matrix inverses.

The geometric perspective is inherent to the Geometric Algebra interpretation of generalized Pauli and Dirac algebras. If this perspective is not mentioned and not discussed, a relevant part of mathematics education is missing. Mathematics should be considered as composed of geometry and algebra. If the geometric part of mathematics is neglected or even denied, the picture of mathematics, students will get, will be partly incomplete. This is a conceptual shortcoming and a severe didactical deficit.

Therefore another didactical path to Moore-Penrose generalized matrix inverses is proposed in the following: Moore-Penrose matrix inverses should not (or not only) be defined by the four Moore-

Penrose conditions. With respect to what Grassmann had achieved it is more helpful to first develop a complete Geometric Algebra picture and to discuss Pauli algebra generalized matrix inverses first. Then the definition of Moore-Penrose generalized matrix inverses will be straightforward and very simple: **Moore-Penrose generalized matrix inverses are the scalar part of Pauli algebra generalized matrix inverses.** The scalar parts of Pauli algebra generalized matrix inverses then automatically comply with the four Moore-Penrose conditions (142) to (145).

To get the scalar parts of $\mathbf{A}^{-1}$ or $\underline{\mathbf{A}}^{-1}$ it is possible to just throw away the bivector parts.

$$\mathbf{A}^{+} = \langle \mathbf{A}^{-1} \rangle_0 = \langle \underline{\mathbf{A}}^{-1} \rangle_0 \tag{146}$$

But it will be more elegant to define a positive and a negative superposition of pre- and post-multiplied Pauli algebra generalized matrix inverses $\mathbf{A}^{-1}$ and $\underline{\mathbf{A}}^{-1}$.

$$\mathbf{A}^{+} = \frac{1}{2}(\mathbf{A}^{-1} + \underline{\mathbf{A}}^{-1}) \tag{147}$$

$$\mathbf{A}^{-} = \frac{1}{2}(\mathbf{A}^{-1} - \underline{\mathbf{A}}^{-1}) \tag{148}$$

$\mathbf{A}^{+}$ will then be the usual Moore-Penrose generalized matrix inverse with scalar elements which solves an overconstrained, consistent system of linear equations

$$\mathbf{A}\,\mathbf{q} = \mathbf{r} \tag{149}$$

by a simple pre-multiplication from the left side

$$\mathbf{A}^{+}\,\mathbf{r} = \mathbf{A}^{-1}\,\mathbf{r} = \underline{\mathbf{A}}^{-1}\,\mathbf{r} = \mathbf{q} \tag{150}$$

while $\mathbf{A}^{-}$ will be a non-square matrix with bivector elements which is orthogonal to the resulting vector $\mathbf{r}$ of constant terms.

$$\mathbf{A}^{-}\,\mathbf{r} = 0 \tag{151}$$

Thus from a geometric viewpoint Moore-Penrose generalized matrix inverses are not complete. Geometrically complete generalized matrix inverses are composed of the scalar and bivector parts of Pauli algebra generalized matrix inverses:

$$\mathbf{A}^{-1} = \mathbf{A}^{+} + \mathbf{A}^{-} \quad \text{and} \quad \underline{\mathbf{A}}^{-1} = \mathbf{A}^{+} - \mathbf{A}^{-} \tag{152}$$

Even if these equations are not jewels (compared to the one of the first footnote), but more ordinary stones only, we should not split them. They tell us a lot about symmetry – and symmetry we will lose if we neglect the bivector (or in 3d quaternion) part $\mathbf{A}^{-}$.

Appendix: Counter-example to multiplicativity

According to the Babylonian way of teaching which Feynman[3] described [32, 33], students even today may learn by discussing and analyzing examples.

Therefore a very simple counter-example to multiplicativity will be presented in the following. It will show that the Pauli algebra generalized matrix inverse $\mathbf{D}^{-1}$ is not identical to the product of the Pauli algebra generalized matrix inverses $\mathbf{B}^{-1}$ and $\mathbf{A}^{-1}$ (141) even if $\mathbf{A}\,\mathbf{B} = \mathbf{D}$ (140).

[3] "... die alten Babylonier (besaßen) nicht die Fähigkeit, das in mathematischer Form auszudrücken. – Heute besitzen wir nicht die Fähigkeit, einem Studenten zu zeigen, wie er Physik *physikalisch* verstehen kann. (...) Mangels einer Ausdrucksmöglichkeit ist der einzige Weg, Physik physikalisch zu verstehen, auch heute noch der langweilige Babylonische Weg, viele Beispiele zu machen." [33].

The non-square matrices $\mathbf{A} = \begin{bmatrix} 1 & 0 & 0 \\ 0 & 1 & 0 \\ 0 & 0 & 1 \\ 1 & 0 & 0 \end{bmatrix}$, $\mathbf{B} = \begin{bmatrix} 1 & 0 \\ 0 & 1 \\ 1 & 0 \end{bmatrix}$, and $\mathbf{A\,B} = \mathbf{D} = \begin{bmatrix} 1 & 0 \\ 0 & 1 \\ 1 & 0 \\ 1 & 0 \end{bmatrix}$ are given.

The Geometric Algebra coefficient vectors of these matrices then are:

$$\mathbf{a}_1 = \sigma_1 + \sigma_4 \qquad \mathbf{a}_2 = \sigma_2 \qquad \mathbf{a}_3 = \sigma_3 \tag{153}$$

$$\mathbf{b}_1 = \sigma_1 + \sigma_3 \qquad \mathbf{b}_2 = \sigma_2 \tag{154}$$

$$\mathbf{d}_1 = \sigma_1 + \sigma_3 + \sigma_4 \qquad \mathbf{d}_2 = \sigma_2 \tag{155}$$

The oriented volumes and reciprocal oriented volumes are:

$$\mathbf{a}_1 \wedge \mathbf{a}_2 \wedge \mathbf{a}_3 = \sigma_1\sigma_2\sigma_3 + \sigma_2\sigma_3\sigma_4 \qquad (\mathbf{a}_1 \wedge \mathbf{a}_2 \wedge \mathbf{a}_3)^{-1} = -\frac{1}{2}(\sigma_1\sigma_2\sigma_3 + \sigma_2\sigma_3\sigma_4) \tag{156}$$

$$\mathbf{b}_1 \wedge \mathbf{b}_2 = \sigma_1\sigma_2 - \sigma_2\sigma_3 \qquad (\mathbf{b}_1 \wedge \mathbf{b}_2)^{-1} = -\frac{1}{2}(\sigma_1\sigma_2 - \sigma_2\sigma_3) \tag{157}$$

$$\mathbf{d}_1 \wedge \mathbf{d}_2 = \sigma_1\sigma_2 - \sigma_2\sigma_3 - \sigma_2\sigma_4 \qquad (\mathbf{d}_1 \wedge \mathbf{d}_2)^{-1} = -\frac{1}{3}(\sigma_1\sigma_2 - \sigma_2\sigma_3 - \sigma_2\sigma_4) \tag{158}$$

The adjoint oriented volumes are:

$$\sigma_1 \wedge \mathbf{a}_2 \wedge \mathbf{a}_3 = \sigma_1\sigma_2\sigma_3 \qquad \sigma_2 \wedge \mathbf{a}_2 \wedge \mathbf{a}_3 = 0 \qquad \sigma_3 \wedge \mathbf{a}_2 \wedge \mathbf{a}_3 = 0 \qquad \sigma_4 \wedge \mathbf{a}_2 \wedge \mathbf{a}_3 = \sigma_2\sigma_3\sigma_4 \tag{159}$$

$$\mathbf{a}_1 \wedge \sigma_1 \wedge \mathbf{a}_3 = \sigma_3\sigma_4\sigma_1 \qquad \mathbf{a}_1 \wedge \sigma_2 \wedge \mathbf{a}_3 = \sigma_1\sigma_2\sigma_3 + \sigma_2\sigma_3\sigma_4 \qquad \mathbf{a}_1 \wedge \sigma_3 \wedge \mathbf{a}_3 = 0 \qquad \mathbf{a}_1 \wedge \sigma_4 \wedge \mathbf{a}_3 = -\sigma_3\sigma_4\sigma_1 \tag{160}$$

$$\mathbf{a}_1 \wedge \mathbf{a}_2 \wedge \sigma_1 = -\sigma_4\sigma_1\sigma_2 \qquad \mathbf{a}_1 \wedge \mathbf{a}_2 \wedge \sigma_2 = 0 \qquad \mathbf{a}_1 \wedge \mathbf{a}_2 \wedge \sigma_3 = \sigma_1\sigma_2\sigma_3 + \sigma_2\sigma_3\sigma_4 \qquad \mathbf{a}_1 \wedge \mathbf{a}_2 \wedge \sigma_4 = \sigma_4\sigma_1\sigma_2 \tag{161}$$

$$\sigma_1 \wedge \mathbf{b}_2 = \sigma_1\sigma_2 \qquad \sigma_2 \wedge \mathbf{b}_2 = 0 \qquad \sigma_3 \wedge \mathbf{b}_2 = -\sigma_2\sigma_3 \tag{162}$$

$$\mathbf{b}_1 \wedge \sigma_1 = \sigma_3\sigma_1 \qquad \mathbf{b}_1 \wedge \sigma_2 = \sigma_1\sigma_2 - \sigma_2\sigma_3 \qquad \mathbf{b}_1 \wedge \sigma_3 = -\sigma_3\sigma_1 \tag{163}$$

$$\sigma_1 \wedge \mathbf{d}_2 = \sigma_1\sigma_2 \qquad \sigma_2 \wedge \mathbf{d}_2 = 0 \qquad \sigma_3 \wedge \mathbf{d}_2 = -\sigma_2\sigma_3 \qquad \sigma_4 \wedge \mathbf{d}_2 = -\sigma_2\sigma_4 \tag{164}$$

$$\mathbf{d}_1 \wedge \sigma_1 = -\sigma_1\sigma_3 - \sigma_1\sigma_4 \qquad \mathbf{d}_1 \wedge \sigma_2 = \sigma_1\sigma_2 - \sigma_2\sigma_3 - \sigma_2\sigma_4 \qquad \mathbf{d}_1 \wedge \sigma_3 = \sigma_1\sigma_3 - \sigma_3\sigma_4 \qquad \mathbf{d}_1 \wedge \sigma_4 = \sigma_1\sigma_4 + \sigma_3\sigma_4 \tag{165}$$

The Pauli algebra generalized matrix inverses of the lead matrix **A** then are:

$$\mathbf{A}^{-1} = \frac{1}{2}\begin{bmatrix} 1 - \sigma_1\sigma_4 & 0 & 0 & 1 + \sigma_1\sigma_4 \\ 0 - \sigma_1\sigma_2 - \sigma_2\sigma_4 & 2 & 0 & 0 + \sigma_1\sigma_2 + \sigma_2\sigma_4 \\ 0 - \sigma_1\sigma_3 - \sigma_3\sigma_4 & 0 & 2 & 0 + \sigma_1\sigma_3 + \sigma_3\sigma_4 \end{bmatrix} \tag{166}$$

$$\underline{\mathbf{A}}^{-1} = \frac{1}{2}\begin{bmatrix} 1 + \sigma_1\sigma_4 & 0 & 0 & 1 - \sigma_1\sigma_4 \\ 0 + \sigma_1\sigma_2 + \sigma_2\sigma_4 & 2 & 0 & 0 - \sigma_1\sigma_2 - \sigma_2\sigma_4 \\ 0 + \sigma_1\sigma_3 + \sigma_3\sigma_4 & 0 & 2 & 0 - \sigma_1\sigma_3 - \sigma_3\sigma_4 \end{bmatrix} \tag{167}$$

Check of the result:

				1	0	0
				0	1	0
				0	0	1
				1	0	0
$1-\sigma_1\sigma_4$	0	0	$1+\sigma_1\sigma_4$	2	0	0
$0-\sigma_1\sigma_2-\sigma_2\sigma_4$	2	0	$0+\sigma_1\sigma_2+\sigma_2\sigma_4$	0	2	0
$0-\sigma_1\sigma_3-\sigma_3\sigma_4$	0	2	$0+\sigma_1\sigma_3+\sigma_3\sigma_4$	0	0	2

The Moore-Penrose generalized matrix inverse then is:

$$\mathbf{A}^{+}=\frac{1}{2}\begin{bmatrix}1 & 0 & 0 & 1\\ 0 & 2 & 0 & 0\\ 0 & 0 & 2 & 0\end{bmatrix} \tag{168}$$

The orthogonal bivector matrix to the Moore-Penrose generalized matrix inverse then is:

$$\mathbf{A}^{-}=\frac{1}{2}\begin{bmatrix}-\sigma_1\sigma_4 & 0 & 0 & \sigma_1\sigma_4\\ -\sigma_1\sigma_2-\sigma_2\sigma_4 & 0 & 0 & \sigma_1\sigma_2+\sigma_2\sigma_4\\ -\sigma_1\sigma_3-\sigma_3\sigma_4 & 0 & 0 & \sigma_1\sigma_3+\sigma_3\sigma_4\end{bmatrix} \tag{169}$$

The Pauli algebra generalized matrix inverses of the lag matrix **B** then are:

$$\mathbf{B}^{-1}=\frac{1}{2}\begin{bmatrix}1-\sigma_1\sigma_3 & 0 & 1+\sigma_1\sigma_3\\ 0-\sigma_1\sigma_2-\sigma_2\sigma_3 & 2 & 0+\sigma_1\sigma_2+\sigma_2\sigma_3\end{bmatrix} \tag{170}$$

$$\underline{\mathbf{B}}^{-1}=\frac{1}{2}\begin{bmatrix}1+\sigma_1\sigma_3 & 0 & 1-\sigma_1\sigma_3\\ 0+\sigma_1\sigma_2+\sigma_2\sigma_3 & 2 & 0-\sigma_1\sigma_2-\sigma_2\sigma_3\end{bmatrix} \tag{171}$$

Check of the result:

			1	0
			0	1
			1	0
$1-\sigma_1\sigma_3$	0	$1+\sigma_1\sigma_3$	2	0
$0-\sigma_1\sigma_2-\sigma_2\sigma_3$	2	$0+\sigma_1\sigma_2+\sigma_2\sigma_3$	0	2

The Moore-Penrose generalized matrix inverse then is:

$$\mathbf{B}^{+}=\frac{1}{2}\begin{bmatrix}1 & 0 & 1\\ 0 & 2 & 0\end{bmatrix} \tag{172}$$

The orthogonal bivector matrix to the Moore-Penrose generalized matrix inverse then is:

$$\mathbf{B}^{-}=\frac{1}{2}\begin{bmatrix}-\sigma_1\sigma_3 & 0 & \sigma_1\sigma_3\\ -\sigma_1\sigma_2-\sigma_2\sigma_3 & 0 & \sigma_1\sigma_2+\sigma_2\sigma_3\end{bmatrix} \tag{173}$$

The product of both Pauli algebra generalized matrix inverses then is:

$$\mathbf{B}^{-1}\mathbf{A}^{-1}=\frac{1}{2}\begin{bmatrix}1-\sigma_1\sigma_3-\sigma_1\sigma_4-\sigma_3\sigma_4 & 0 & 1+\sigma_1\sigma_3 & \sigma_1\sigma_4+\sigma_3\sigma_4\\ 0-2\sigma_1\sigma_2-2\sigma_2\sigma_4 & 2 & 0+\sigma_1\sigma_2+\sigma_2\sigma_3 & \sigma_1\sigma_2-\sigma_2\sigma_3+2\sigma_2\sigma_4\end{bmatrix} \tag{174}$$

Check of the result:

$$\begin{array}{cccc|cc} & & & & 1 & 0\\ & & & & 0 & 1\\ & & & & 1 & 0\\ & & & & 1 & 0\\ \hline 1-\sigma_1\sigma_3-\sigma_1\sigma_4-\sigma_3\sigma_4 & 0 & 1+\sigma_1\sigma_3 & 0+\sigma_1\sigma_4+\sigma_3\sigma_4 & 2 & 0\\ 0-2\sigma_1\sigma_2-2\sigma_2\sigma_4 & 2 & 0+\sigma_1\sigma_2+\sigma_2\sigma_3 & 0+\sigma_1\sigma_2-\sigma_2\sigma_3+2\sigma_2\sigma_4 & 0 & 2\end{array}$$

The Pauli algebra generalized matrix inverses of the resulting matrix **D** then are:

$$\mathbf{D}^{-1}=\frac{1}{3}\begin{bmatrix}1-\sigma_1\sigma_3-\sigma_1\sigma_4 & 0 & 1+\sigma_1\sigma_3-\sigma_3\sigma_4 & 1+\sigma_1\sigma_4+\sigma_3\sigma_4\\ 0-2\sigma_1\sigma_2-\sigma_2\sigma_3-\sigma_2\sigma_4 & 3 & 0+\sigma_1\sigma_2+2\sigma_2\sigma_3-\sigma_2\sigma_4 & 0+\sigma_1\sigma_2-\sigma_2\sigma_3+2\sigma_2\sigma_4\end{bmatrix} \tag{175}$$

$$\underline{\mathbf{D}}^{-1}=\frac{1}{3}\begin{bmatrix}1+\sigma_1\sigma_3+\sigma_1\sigma_4 & 0 & 1-\sigma_1\sigma_3+\sigma_3\sigma_4 & 1-\sigma_1\sigma_4-\sigma_3\sigma_4\\ 0+2\sigma_1\sigma_2+\sigma_2\sigma_3+\sigma_2\sigma_4 & 3 & 0-\sigma_1\sigma_2-2\sigma_2\sigma_3+\sigma_2\sigma_4 & 0-\sigma_1\sigma_2+\sigma_2\sigma_3-2\sigma_2\sigma_4\end{bmatrix} \tag{176}$$

Check of the result:

$$\begin{array}{cccc|cc} & & & & 1 & 0\\ & & & & 0 & 1\\ & & & & 1 & 0\\ & & & & 1 & 0\\ \hline 1-\sigma_1\sigma_3-\sigma_1\sigma_4 & 0 & 1+\sigma_1\sigma_3-\sigma_3\sigma_4 & 1+\sigma_1\sigma_4+\sigma_3\sigma_4 & 3 & 0\\ 0-2\sigma_1\sigma_2-\sigma_2\sigma_3-\sigma_2\sigma_4 & 3 & 0+\sigma_1\sigma_2+2\sigma_2\sigma_3-\sigma_2\sigma_4 & 0+\sigma_1\sigma_2-\sigma_2\sigma_3+2\sigma_2\sigma_4 & 0 & 3\end{array}$$

The Moore-Penrose generalized matrix inverse then is:

$$\mathbf{D}^{+}=\frac{1}{3}\begin{bmatrix}1 & 0 & 1 & 1\\ 0 & 3 & 0 & 0\end{bmatrix} \tag{177}$$

The orthogonal bivector matrix to the Moore-Penrose generalized matrix inverse then is:

$$\mathbf{D}^{-}=\frac{1}{3}\begin{bmatrix}-\sigma_1\sigma_3-\sigma_1\sigma_4 & 0 & \sigma_1\sigma_3-\sigma_3\sigma_4 & \sigma_1\sigma_4+\sigma_3\sigma_4\\ -2\sigma_1\sigma_2-\sigma_2\sigma_3-\sigma_2\sigma_4 & 0 & \sigma_1\sigma_2+2\sigma_2\sigma_3-\sigma_2\sigma_4 & \sigma_1\sigma_2-\sigma_2\sigma_3+2\sigma_2\sigma_4\end{bmatrix} \tag{178}$$

As $\mathbf{B}^{-1}\mathbf{A}^{-1}$ of eq. (174) is not identical to $\mathbf{D}^{-1}$ of eq. (175) mulitplicativity fails and the example shows that statement (141) $\mathbf{B}^{-1}\ \mathbf{A}^{-1} \neq \mathbf{D}^{-1}$ is indeed correct.

References

[1] Hodgkin L 2005 *A History of Mathematics. From Mesopotamia to Modernity* (Oxford, New York: Oxford University Press) p 16

[2] Grcar J F 2011 Mathematicians of Gaussian Elimination *Notices of the American Mathemaitcal Society* 58 (6) pp 782–792

[3] Grcar J F 2011 How Ordinary Elimination Became Gaussian Elimination *Historia Mathematica* 38 (2) pp 163–218

[4] Derbyshire J 2006 *Unknown Quantity – A Real and Imaginary History of Algebra* (Washington, DC: Joseph Henry Press) p 34

[5] Sarton G 1944 Grassmann – 1844 *Isis* 35 (4) pp 326–330

[6] Grassmann H 1844 *Die Wissenschaft der extensiven Grösse oder die Ausdehnungslehre, eine neue mathematische Disciplin. Erster Theil, die lineale Ausdehnungslehre enthaltend* (Leipzig: Verlag von Otto Wigand) § 45 pp 70–72

[7] Grassmann H 1877 Die Mechanik nach den Principien der Ausdehnungslehre *Mathematische Annalen* 12 (2) pp 222–240 Reprint F Engel (Ed) 1902 *Hermann Grassmanns gesammelte mathematische und physikalische Werke. Zweiten Bandes zweiter Theil* (Leipzig: Verlag von B G Teubner) pp 46–72

[8] Grassmann H 1877 Mechanics, according to the principles of the theory of extensions. Translation of [7] by D H Delphenich (www.neo-classical-physics.info/theoretical-mechanics.html)

[9] Hestenes D 2003 Oersted Medal Lecture 2002: Reforming the Mathematical Language of Physics *American Journal of Physics* 71 (2) pp 104–121

[10] Grassmann H 1877 Der Ort der Hamilton'schen Quaternionen in der Ausdehnungslehre *Mathematische Annalen* 12 (3) pp 375–386 Reprint F Engel (Ed) 1904 *Hermann Grassmanns gesammelte mathematische und physikalische Werke. Zweiten Bandes erster Theil* (Leipzig: Verlag von B G Teubner) pp 268–282

[11] Schweitzer A R 1910 On the Genesis of the Middle Product in Grassmann's Extensive Algebra *Mathematische Annalen* 69 (4) pp 580–585

[12] Gull S, Lasenby A and Doran C 1993 Imaginary Numbers are not Real – The Geometric Algebra of Spacetime *Foundations of Physics* **23** (9) pp 1175–1201

[13] Arnold V I 1998 On Teaching Mathematics *Uspekhi Mat. Nauk* 53, 229-234, English translation *Russian Math. Surveys* 53, 229-236 (http://pauli.uni-muenster.de/~munsteg/arnold.htm)

[14] Hestenes D 1992 Mathematical Viruses In A Micali, R Boudet and J Helmstetter *Clifford Algebras and their Applications in Mathematical Physics* (Dordrecht: Kluwer Academic Publishers) pp 3–16

[15] Feynman R P, Leighton R B and Sands M 1977 *The Feynman Lectures on Physics, Vol I, Mainly Mechanics, Radiation, and Heat* 6th printing (Reading, Massachusetts: Addison-Wesley) p 22-10

[16] Vince J 2009 *Geometric Algebra: An Algebraic System for Computer Games and Animation* (London: Springer Verlag) p 112, eqs. 8.42 & 8.43

[17] Horn M E 2015 Sandwich Products and Reflections *PhyDid B – Didaktik der Physik* Beiträge zur DPG-Frühjahrstagung des Fachverbands Didaktik der Physik in Wuppertal 2015 (www.phydid.de/index.php/phydid-b/article/view/642) DD 17.07 eqs. {1}–{64}

[18] Carmeli M 2002 *Cosmological Special Relativity. The Large-Scale Structure of Space, Time*

and Velocity 2nd edition (Singapore: World Scientific)

[19] Horn M E 2012 Translating Cosmological Special Relativity into Geometric Algebra In S Sivasundaram (Eds) *Proceedings of the 9th International Conference on Mathematical Problems in Engineering, Aerospace and Sciences (ICNPAA) in Vienna* (Melville, New York: AIP Conference Proceedings Vol 1493) pp 492–498

[20] Pauli W 1994 The Violation of Reflection Symmetries in the Laws of Atomic Physics In C P Enz and K von Meyenn (Eds) *Wolfgang Pauli – Writings on Physics and Philosophy* (Berlin, Heidelberg: Springer-Verlag) pp 183–192 & obituary postcard at p 192

[21] Penrose R 2004 *The Road to Reality. A Complete Guide to the Laws of the Universe* (London: Jonathan Cape) p 638

[22] Kirch K, Pretz J and Wirzba A 2017 Elektrische Dipolmomente gesucht *Physik Journal* 16 (11) pp 41–46

[23] Hestenes D 2003 Spacetime Physics with Geometric Algebra *American Journal of Physics* 71 (7) pp 691–714

[24] Doran C and Lasenby A 2003 *Geometric Algebra for Physicists* (Cambridge: Cambridge University Press)

[25] Snygg J 1997 *Clifford Algebra. A Computational Tool for Physicists* (New York: Oxford University Press)

[26] Hestenes D 2002 *New Foundations for Classical Mechanics* 2nd edition (New York: Kluwer Academic Publishers) chap 9: Relativistic Mechanics

[27] Horn M E 2014 An Introduction to Geometric Algebra with some Preliminary Thoughts on the Geometric Meaning of Quantum Mechanics In D Schuch and M Ramek (Eds): Symmetries in Science XVI, Proceedings of the International Symposium in Bregenz 2013 *Journal of Physics: Conference Series* Vol 538 (2014) 012010 (Bristol: IOP Science)

[28] Landau L D and Lifschitz E M 1984 *Lehrbuch der Theoretischen Physik II, Klassische Feldtheorie* 9th edition (Berlin: Akademie-Verlag) § 4, pp 10–14

[29] Horn M E 2016 Inverse von Rechteck-Matrizen. In: *BzMU – Beiträge zum Mathematik-unterricht*. Conference Proceedings of Annual Meeting of the Society of Mathematics Education (GDM) in Heidelberg 2016, Band 1 (Münster: WTM-Verlag) pp 457-460

[30] Horn M E 2016 *More Examples of Non-Square Matrix Inverses* – Translated and Extended version of paper [29] published as attachmend of Horn M E 2016 Die Geometrische Algebra im Schnelldurchgang *PhyDid B – Didaktik der Physik* Beiträge zur DPG-Frühjahrstagung des Fachverbands Didaktik der Physik in Hannover 2016 (http://www.phydid.de/index.php/phydid-b/article/view/723/868) DD 05.19

[31] Horn M E 2016 Moderne Lineare Algebra im wirtschaftsmathematischen Kontext. In W Paravicini, J Schnieder (Eds) *Hanse-Kolloquium zur Hochschuldidaktik der Mathematik 2015*. Beiträge zum gleichnamigen Symposium an der Universität zu Lübeck (Münster: WTM-Verlag) pp 103–129

[32] Feynman R P 1967 *The Character of Physical Law* First MIT Press Paperback Edition (Cambridge, Massachusetts: MIT Press) chap 2 p 46

[33] Feynman R P 2006 *Physik »The Lost Lectures«* (Munich: Pearson Studium) pp 69-70

[34] Schmidt K and Trenkler G 2015 *Einführung in die Moderne Matrix-Algebra* Mit Anwendungen in der Statistik. 3rd edition (Berlin, Heidelberg: Springer Gabler) chap 6

[35] Dadkhah K 2011 *Foundations of Mathematical and Computational Economics*. 2nd edition (Berlin, Heidelberg: Springer-Verlag) chap 7

[36] Smith D E and Mikami Y 1914 *A History of Japanese Mathematics* (Chicago: The Open Court Publishing Company) p 125

IOP Conf. Series: Journal of Physics: Conf. Series **1071** (2018) 012013 doi:10.1088/1742-6596/1071/1/012013

Excited state quantum phase transitions and monodromy

Francesco Iachello

Center for Theoretical Physics, Sloane Laboratory, Yale University,
New Haven, Connecticut 06520-8120, USA

E-mail: francesco.iachello@yale.edu

Abstract. Quantum phase transitions (QPT) and excited state quantum phase transitions (ESQPT) in algebraic models are briefly discussed, in particular those occurring in the U(3) model of bending vibrations of molecules. Experimental evidence in the water molecule, H_2O, is presented.

1. Introduction

Quantum phase transitions (QPT) are phase transitions that occur as a function of a coupling constant, ξ, called control parameter, that appears in the quantum Hamiltonian, H, that describes the system

$$H=\varepsilon\left[(1-\xi)H_1+\xi H_2\right], \qquad 0\leq\xi\leq 1. \qquad (1)$$

Associated with QPT there are quantal order parameters, the expectation values of suitable chosen operators that describe the state of the system, $<O>$. Introduced in the 1970's [1], they have become of great importance in a variety of systems. QPT are also called ground state phase transitions and/or zero temperature phase transitions [1]. Excited state quantum phase transitions (ESQPT) are phase transitions that occur as a function of excitation energy, E_x, for fixed value of the control parameter, ξ [2]. ESQPT are intimately connected with quantum monodromy. ESQPT and quantum monodromy have been recently observed in molecular physics [3-5].

QPT and ESQPT can be conveniently studied within the framework of algebraic models [6]. For these models, one can do both the semi-classical and the quantal analysis, and thus study both semi-classical and quantum monodromy. Also, in many-body systems, finite size scaling (1/N expansion) can be easily investigated. The latter point is particularly important in applications to finite systems: nuclei, molecules, finite polymers, photonic crystals, optical lattices, etc.

In this contribution, after a brief review of algebraic models and their semi-classical limit, an experimental example of ESQPT and monodromy will be given. A full account of QPT and ESQPT, especially for applications to nuclear physics, is given in several review papers [7-10].

2. Algebraic models

2.1. *Algebraic properties*. An algebraic model is an expansion of the Hamiltonian and other operators in terms of elements, $G_{\alpha\beta}$, of an algebra (often a Lie algebra, g, or a contraction of it). The algebra g is called the spectrum generating algebra (SGA),

$$H = E_0 + \sum_{\alpha\beta} \varepsilon_{\alpha\beta} G_{\alpha\beta} + \sum_{\alpha\beta\gamma\delta} u_{\alpha\beta\gamma\delta} G_{\alpha\beta} G_{\gamma\delta} + \ldots$$
$$T = t_0 + \sum_{\alpha\beta} t_{\alpha\beta} G_{\alpha\beta} + \ldots \qquad (2)$$

An interesting situation occurs when H does not contain all elements of g, but only the invariant (Casimir) operators of a chain of algebras, $g \supset g' \supset g'' \supset \ldots$

$$H = E_0 + \alpha C(g) + \alpha' C(g') + \ldots \qquad (3)$$

called a dynamic symmetry (DS). In this case, the energy eigenvalues can be written explicitly in terms of quantum numbers labelling the representations of g

$$E = \langle H \rangle = E_0 + \alpha \langle C(g) \rangle + \alpha' \langle C(g') \rangle + \ldots \qquad (4)$$

It is convenient to write the elements $G_{\alpha\beta} \in g$ as bilinear products of creation an annihilation operators (Jordan-Schwinger realization). For bosonic systems

$$G_{\alpha\beta} = b_\alpha^\dagger b_\beta; \quad \alpha,\beta = 1,\ldots n\,. \qquad (5)$$

From

$$\left[b_\alpha, b_\beta^\dagger\right] = \delta_{\alpha\beta} \quad , \quad \left[b_\alpha, b_\beta\right] = \left[b_\alpha^\dagger, b_\beta^\dagger\right] = 0 \qquad (6)$$

One obtains the commutation relations

$$\left[G_{\alpha\beta}, G_{\gamma\delta}\right] = \delta_{\beta\gamma} G_{\alpha\delta} - \delta_{\alpha\delta} G_{\gamma\delta} \qquad (7)$$

which define the real form of $g=u(n)$. The basis upon which the elements act is the totally symmetric representation,

$$|N\rangle \equiv \square\square\ldots\square \quad , \quad |N\rangle = \frac{1}{\sqrt{N!}} \left(b_\alpha^\dagger\right)^{n_\alpha} \left(b_{\alpha'}^\dagger\right)^{n_{\alpha'}} \ldots |0\rangle \qquad (8)$$

characterized by the total number of bosons N.

Fermionic systems can also be treated algebraically in terms of bilinear products of anti-commuting operators

$$G_{ij} = a_i^\dagger a_j \ , \quad i,j = 1,\ldots,m\,, \qquad (9a)$$

$$\left\{a_i, a_j^\dagger\right\} = \delta_{ij} \quad , \quad \left\{a_i, a_j\right\} = \left\{a_i^\dagger, a_j^\dagger\right\} = \delta_{ij} \qquad (9b)$$

spanning the Lie algebra $u(m)$. Mixtures of bosonic and fermionic systems can be treated in terms of the realization

$$\begin{pmatrix} b_\alpha^\dagger b_\beta & b_\alpha^\dagger a_i \\ a_i^\dagger b_j & a_i^\dagger a_j \end{pmatrix} \qquad (10)$$

spanning the *graded* Lie algebra (also called superalgebra) $u(n/m)$. (Fermionic and mixed Bose-Fermi systems will not be discussed here).

Table 1. List of bosonic models

U(2)	Lipkin model: Schematic model of many-body systems [12] 1-dim Vibron Model (VM): Stretching vibrations of molecules [13]
U(3)	2-dim Vibron Model (VM): Bending vibrations of molecules [14]
U(4)	3-dim Vibron Model (VM): Rotations-vibrations of molecules [15]
U(6)	Interacting Boson Model (IBM): Rotations-vibrations of atomic nuclei [16]

A list [11] of algebraic models extensively investigated is given in Table 1. These provide a description of many-body systems with *f=n-1* degrees of freedom.
A convenient realization of these models is with a scalar boson, called *s* and another boson b_m with $m=2l+1$ components. The integer or half-integer number l=0, 1/2, 1, 3/2, 2, ... is called 'quasi-spin'. The use of both integer and half-integer numbers allows one to treat problems in both odd and even dimensions.

2.2. *Geometric properties*. Geometry can be associated to algebraic models with algebra *g* by constructing appropriate coset spaces, obtained by splitting *g* into

$$g = h \oplus p \tag{11}$$

where *h* is a subalgebra of *g* called stability algebra, and *p* is the remainder not closed with respect to commutation. This decomposition is also called a Cartan decomposition. For models with *u(n)* structure the appropriate coset (*g/h*) is

$$u(n)/u(n-1) \oplus u(1) \tag{12}$$

where *u(n-1)* is the maximal stability subalgebra. This space is a globally symmetric Riemannian space [6, Ch.5] with dimension *2(n-1)*. For bosonic models, the algebra *h* can be constructed by selecting one boson, b_1, and choosing

$$h \doteq b_1^\dagger b_1, b_\alpha^\dagger b_\beta \ , \quad \alpha,\beta=2,\ldots,n \tag{13a}$$

$$p \doteq b_1^\dagger b_\alpha, b_\alpha^\dagger b_1 \ , \quad \alpha=2,\ldots,n \ . \tag{13b}$$

Associated with the Cartan decomposition there are geometric variables η_i defined by

$$|\eta_i\rangle = \exp\left[\sum_i \eta_i p_i\right] |\Lambda_{ext}\rangle \ , \quad p_i \in p \ , \tag{14}$$

where Λ_{ext} is called extremal state. For bosonic systems

$$|N;\eta_\alpha\rangle = \left[\exp\left(\eta_\alpha b_\alpha^\dagger b_1 - \eta^*_\alpha b_1^\dagger b_\alpha\right)\right] \frac{1}{\sqrt{N!}} \left(b_1^\dagger\right)^N |0\rangle . \tag{15}$$

For systems with fixed number of particles, *N*, it is convenient to introduce projective coherent states in terms of projective variables, θ_α,

$$|N;\theta_\alpha\rangle = \frac{1}{\sqrt{N!}} \left[b_1^\dagger + \theta_\alpha b_\alpha^\dagger\right]^N |0\rangle \ . \tag{16}$$

For algebraic models written in terms of boson operators, s, b_m, it is convenient to rewrite the coherent state as

$$|N;\alpha\rangle = \frac{1}{\sqrt{N!}} \left(s^\dagger + \sum_m \alpha_m b_m^\dagger\right)^N |0\rangle \tag{17}$$

with normalization

$$\langle N;\alpha|N;\alpha\rangle = \left(1 + \sum_m |\alpha_m|^2\right)^N . \tag{18}$$

The semi-classical Hamiltonian associated with the quantum algebraic Hamiltonian is

$$H_{cl}(\alpha) = \frac{\langle N;\alpha|H|N;\alpha\rangle}{\langle N;\alpha|N;\alpha\rangle} \ . \tag{19}$$

This Hamiltonian depends on the complex coordinates α_m which can be split into real coordinates, q_m, and momenta, p_m.

2.3. *Quantum phase transitions (QPT and ESQPT)*. QPT and ESQPT can be studied within the framework of algebraic models both quantally and semi-classically. For QPT, in the quantum analysis, one evaluates the ground state energy, E_0 , as a function of the control parameter, ξ. At the same time, one studies the behavior of the order parameter as a function of ξ. For bosonic systems, a convenient order parameter is the expectation value of the boson number operator n_b in the ground state, normalized to N, n_b/N. The behavior of E_0 and $<n_b>$ determines whether or not there is a precursor of a phase transition. Since the calculation is done for a finite value of N, the phase transition appears as a sudden change in the quantity itself or in its slope, and not as a discontinuity. For ESQPT, in the quantum analysis, one constructs the energy spectrum as a function of the control parameter ξ, called a correlation diagram. ESQPT appear then as an increase in level density for fixed ξ as a function of excitation energy, E_x. For QPT in the semi-classical analysis, one evaluates the energy functional $H_{cl}(\alpha)=E(\alpha)$, Eq.(19). By minimizing $E(\alpha)$ with respect to the variables α one obtains an approximation to the ground-state energy which is good to order $1/N$. Denoting by α_e the equilibrium value of the variable α, one then calculates $E(\alpha_e)$ and its derivatives $\partial E(\alpha_e)/\partial\xi$, $\partial^2 E(\alpha_e)/\partial\xi^2$, … as a function of ξ. These quantities may have discontinuities. The order of the phase transition, using Ehrenfest criterion, is related to the order of the derivative that is discontinue, i.e. if $E(\alpha_e)$ is discontinue, the transition is zeroth order, if $\partial E(\alpha_e)/\partial\xi$ is discontinue, it is first order, and so on. For ESQPT, in the semi-classical analysis, one calculates the semi-classical action $S(\xi;E)$ and studies its behavior. ESQPT correspond to divergence of the level density $\rho(\xi;E)$ at some critical value $E=E_{xc}$. A similar semi-classical analysis can be also be done for the order parameter n_b/N.

3. An example of ESQPT: Bending vibrations of molecules

As an example of ESQPT, we consider here the bending vibrations of polyatomic molecules. This is a two-dimensional problem. A quantum description of this problem was given in [14]. Introducing Cartesian boson creation and annihilation operators, $\tau_x^\dagger, \tau_y^\dagger, \tau_x, \tau_y$ plus a scalar $\sigma^\dagger, \sigma$ conveniently combined into circular bosons,

$$\tau_\pm^\dagger = \mp\frac{\tau_x^\dagger \pm i\tau_y^\dagger}{\sqrt{2}} \qquad \tau_\pm = \mp\frac{\tau_x \mp \tau_y}{\sqrt{2}} , \tag{20}$$

One can write down the elements of u(3) as

$$\begin{aligned} &\hat{n} = \tau_+^\dagger\tau_+ + \tau_-^\dagger\tau_- && \hat{D}_\pm = \sqrt{2}\left(\pm\tau_\pm^\dagger\sigma \mp \sigma^\dagger\tau_\mp\right) \\ &\hat{n}_\sigma = \sigma^\dagger\sigma && \hat{Q}_\pm = \sqrt{2}\tau_\pm^\dagger\tau_\mp \\ &\hat{\ell} = \tau_+^\dagger\tau_+ - \tau_-^\dagger\tau_- && \hat{R}_\pm = \sqrt{2}\left(\tau_\pm^\dagger\sigma + \sigma^\dagger\tau_\mp\right) \end{aligned} . \tag{21}$$

The algebra of u(3) has two dynamical symmetries corresponding to the breaking into subalgebras that contain the two-dimensional rotation algebra, so(2).

$$\begin{aligned} &u(3) \supset u(2) \supset so(2) && (I) \\ &u(3) \supset so(3) \supset so(2) && (II) \end{aligned} . \tag{22}$$

For each of the two symmetries, states can be labelled by the quantum numbers denoting the irreducible representations of the appropriate group,

$$\left| \begin{matrix} u(3) & \supset & u(2) & \supset & so(2) \\ \downarrow & & \downarrow & & \downarrow \\ [N] & & n & & \ell \end{matrix} \right\rangle \tag{23}$$

With $n=N,N-1,\dots,1$ or 0 and $l=n,n-2,\dots,1$ or 0 for n=odd or even, and

$$\left| \begin{matrix} u(3) & \supset & so(3) & \supset & so(2) \\ \downarrow & & \downarrow & & \downarrow \\ [N] & & v & & \ell \end{matrix} \right\rangle \tag{24}$$

With $v=0,1,\dots,(N-1)/2$ or $N/2$ for N=odd or even, and $l=0, \pm1,\dots,\pm(N-2v)$, where N is called vibron number [15].

In order to describe QPT and ESQPT in this case, one can introduce an algebraic Hamiltonian [17]

$$\hat{H} = \varepsilon\left[(1-\xi)\hat{n} + \frac{\xi}{N}\hat{P}\right] \qquad . \qquad (25)$$
$$\hat{P} = N(N+1) - \left[\frac{1}{2}\left(\hat{D}_+\hat{D}_- + \hat{D}_-\hat{D}_+\right) + \ell^2\right]$$

This Hamiltonian admits two exactly solvable cases, called dynamical symmetries (or phases), for which the eigenvalue problem can be solved in explicit analytic form

$$E^{(I)}(n) = E_0 + \varepsilon n \qquad (26)$$
$$E^{(II)}(v) = E'_0 + 4\varepsilon\left[v\left(1+\frac{1}{2N}\right) - \frac{1}{N}v^2\right]$$

corresponding to symmetry (I) for ξ=0, and to symmetry (II) for ξ=1. Note that a single quantum number cannot be globally defined, due to the different dynamic symmetries of the two phases.

Quantal analysis. The correlation diagram obtained by numerical diagonalization of the Hamiltonian, Eq.(25), is shown in Figure 1.

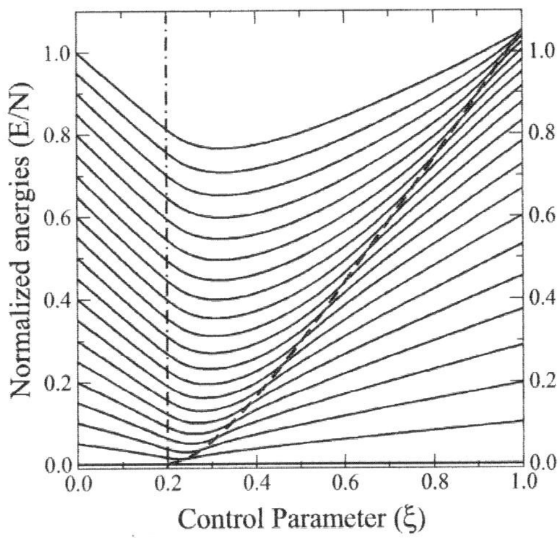

Figure 1. Correlation diagram for N=40 and vibrational angular momentum l=0. The normalized energy values E/N are plotted against ξ as full lines. The critical point ξ =0.2 is marked with a vertical dotted-dashed line. The separatrix, Eq.(27), for N=40 is shown with a dashed line.

Here the energy spectrum of l=0 states is shown as a function of the control parameter, ξ, for N=40. One can see that for ξ=0 the spectrum is that corresponding to the dynamic symmetry (I), u(2), with equidistant energy levels as in Eq. (26) (I), for ξ=1 the spectrum is that corresponding to the dynamic symmetry (II), so(3), with energies given by Eq.(26) (II). If one traverses the correlation diagram at zero energy as a function of ξ, one finds a change at a critical value $\xi=\xi_c$ from one symmetry to the other. The same change appears also as a function of energy for fixed value of ξ. The former change represents a QPT while the latter represents an ESQPT. The two symmetries (phases) are separated by a separatrix, with equation, in the limit $N \to \infty$,

$$f(\xi) = \frac{25\xi^2 - 10\xi + 1}{16\xi} \qquad . \qquad (27)$$

Also, in the limit $N \to \infty$, the value of ξ_c is ξ_c=0.2. For finite N, a modification both in Eq.(27) and in the value of ξ is needed, which amount to a multiplication by a factor (N+1)/(N-1).
In order to illustrate further the occurrence of ESQPT, it is convenient to plot the energy levels for fixed ξ as a function of l. In Figure 2, the so called quantum monodromy diagram [18] is shown. One can see that there is a change in the behaviour with l

$$
\begin{aligned}
E^{(I)}(l) &= E_0 + bl \qquad u > u_c \\
E^{(II)}(l) &= E_0 + al^2 \qquad u < u_c
\end{aligned}
\tag{28}
$$

where u numbers the states and u_c is the value of u at which the separatrix is crossed, also called monodromy region.

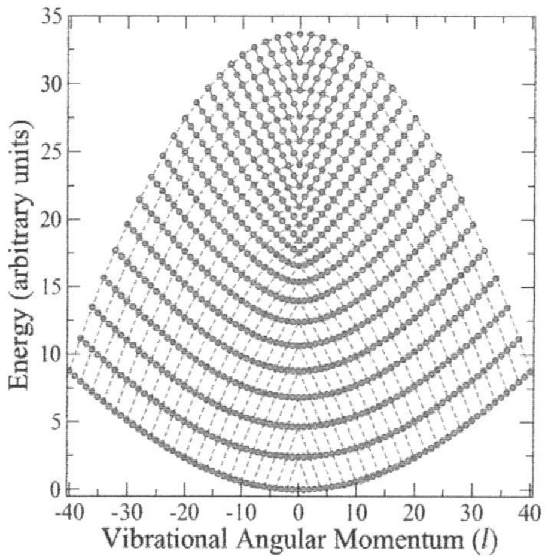

Figure 2. Energy spectrum for N=40 and ξ=0.6 as a function of the vibrational angular momentum *l*. Points linked by full lines correspond to the same value of *v*, while points joint by dashed lines correspond to the same value of *n*. In both cases one can see the transition from smooth to warped variations at *l*=0 when crossing the monodromy region.

3.1. *Semi-classical analysis*. For the semi-classical analysis, it is convenient to go from Cartesian coordinates, x,y, to polar coordinates, r, θ, and momenta p_r, p_θ. Putting all momenta equal to zero, the coherent states can be written as $|[N];r,\theta\rangle$, and the semi-classical effective potential as

$$
V(r,\theta) = \frac{\langle [N];r,\theta | H | [N];r,\theta \rangle}{\langle [N];r,\theta | [N];r,\theta \rangle} \; . \tag{29}
$$

For the Hamiltonian of Eq.(25), the effective semi-classical potential is then

$$
V(r) = \varepsilon \left[(1-\xi)\frac{r^2}{1+r^2} + \xi \left(\frac{1-r^2}{1+r^2} \right)^2 \right] . \tag{30}
$$

It is instructive to plot the potential as a function of the bending angle for various values of ξ, ξ=0 (linear), ξ=0.2 (quasi-linear), ξ=0.6 (quasi-bent) and ξ=1 (bent), Figure 3. One can see a progressive change from a potential with one minimum at φ=0 degrees (linear) with a potential with two minima at φ=±1 radiant (bent). A potential of the type quadratic-quartic, $V(r)=ar^2+br^4$ is sometimes referred to as a "sombrero potential" or a "champagne bottle potential" [18].

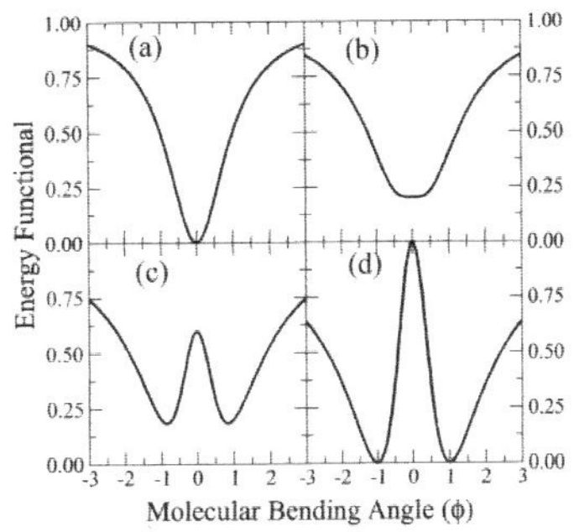

Figure 3. Energy functional (potential), for linear-to-bent transition in molecules, as a function of bending angle, φ, for four different values of ξ, (a) rigidly linear (ξ=0.0), (b) quasi-linear (ξ=0.2), (c) quasi-bent (ξ=0.6), and (d) rigidly bent (ξ=1.0).

3.2. *Experimental evidence for ESQPT and monodromy in molecules.* QPT and ESQPT have been recently observed in the bending vibrations of molecules [3-5]. The Hamiltonian used in the analysis is given in Eq.(25). There are two phases, Eq.(22), (I) linear and (II) bent. One of the best examples of ESQPT and monodromy is provided by the water molecule, H_2O, which is bent in the ground state and becomes linear at E_x~11000cm^{-1} [21]. In Figure 4, a comparison between observed and calculated energy levels is shown [4].

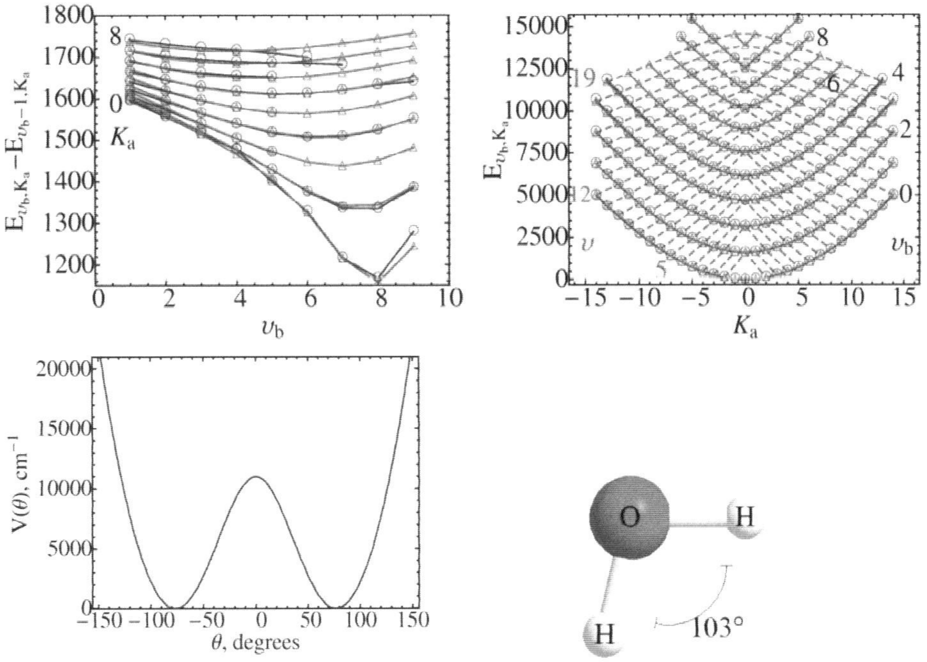

Figure 4. Birge-Spooner plot as a function of v_b (top left) and monodromy plot (top right) for H_2O. Red triangles are theory and blue circles are experiment. The bottom left panel shows the potential function and the bottom right panel shows the calculated equilibrium structure.

In the top left panel, the differences in energies $E_{vb,Ka} - E_{vb-1,Ka}$ are plotted versus the vibrational quantum number v_b for fixed K_a. At the monodromy point, this quantity should go to zero, and its inverse, related to the level density, should diverge. One observes in experiment a dip, called Dixon dip, at v_b =8. The ESQPT is smoothed out by finite N effects but it still clearly appears in the figure. In the top right panel, the energies $E_{vb,Ka}$ are plotted versus K_a=l, the angular momentum. One can again see in this plot the change from quadratic to linear behaviour at v =8. The bottom left panel shows the corresponding double well potential, and the bottom right part the equilibrium configuration of the ground state of the water molecule.

4. Summary and conclusions

QPT, ESQPT and monodromy have been studied in algebraic models of many-body systems both quantally and semi-classically. The case of u(3), Eq. (22), has been shown. Experimental evidence for ESQPT in the bending vibrations of the water molecule, H_2O, has been presented.

ESQPT and monodromy are related to properties of the potential. The analysis presented here applies equally well to all problems of the type

$$\begin{aligned} &u(n) \supset u(n-1) \\ &u(n) \supset so(n) \end{aligned} \tag{31}$$

which have a 2nd order phase transition from the symmetry *u(n-1)* to *so(n)*. These transitions have been investigated extensively and have the same universal behaviour. For algebraic models that support 1st order transitions, the structure of the spectrum and its semi-classical analysis is much more complex, since it involves several control parameters and several order parameters. The potential now depends on several coordinates. An example is the Interacting Boson Model (IBM) [16] that describes rotations and vibrations of a liquid drop with ellipsoidal shape. This model has two control parameters and three phases,

$$\begin{aligned} &u(6) \supset u(5) \\ &u(6) \supset su(3) \\ &u(6) \supset so(6) \end{aligned} \quad . \tag{32}$$

The potential depends on two intrinsic variables, β,γ. A complete analysis of classical and quantum monodromy for algebraic structures that support 1st order transitions is currently being done, but it will not be reported here.

Finally, it should be noted that the dynamics of algebraic models and its behaviour around the monodromy point has been investigated [2] for all models of the type of Eq.(31). Monodromy is also a subject of great interest in mathematics. Here I have concentrated my attention on the physical applications. For a detailed account of the mathematical aspects of monodromy, see [22], and references therein.

Acknoledgements

This work was performed in part under D.O.E. Grant No. DE-FG-02-91ER40608.

References

[1] Gilmore R 1979 *J. Math. Phys.* **20**, 891
[2] Caprio MA, Cejnar P and Iachello F 2008 *Ann. Phys. (N.Y.)* **323**, 1106
[3] Larese D and Iachello F 2011 *J. Mol. Struct.* **1006**, 611
[4] Larese D, Perez-Bernal F and Iachello F 2013 *J. Mol. Struct.* **1051**, 310
[5] Larese D, Caprio MA, Perez-Bernal F and Iachello F 2014 *J. Chem. Phys.* **140**, 014304
[6] Iachello F 2015 *Lie Algebras and Applications*, 2nd ed. (Berlin:Springer-Verlag)
[7] Iachello F and Caprio MA, in *Understanding Quantum Phase Transitions*, Carr LD ed. 2011 (Boca Raton:CRC Press)
[8] Iachello F 2011 *Rivista del Nuovo Cimento* **34**, 617
[9] Cejnar P and Jolie J 2009 *Prog. Part. Nucl. Phys.* **62**, 210
[10] Cejnar P, Jolie J, and Casten RF 2010 *Rev. Mod. Phys.* **82**, 2150

[11] Cejnar P and Iachello F 2007 *J. Phys. A: Math. Theor.* **40**, 581
[12] Lipkin HJ, Meshkov N and Glick N 1965 *Nucl. Phys. A* **62**, 188, 199, 211
[13] Van Roosmalen OS 1982 *Algebraic description of nuclear and molecular rotation-vibration spectra* (University of Groningen, The Netherlands: Ph. D. Thesis)
[14] Iachello F and Oss S 1996 *J. Chem. Phys.* **104**, 6956
[15] Iachello F and Levine RD 1995 *Algebraic Theory of Molecules* (Oxford: Oxford University Press)
[16] Iachello F and Arima A 1987 *The Interacting Boson Model* (Cambridge: Cambridge University Press)
[17] Perez-Bernal F and Iachello F 2008 *Phys. Rev. A* **77**, 032115
[18] Child MS 1998 *J. Phys. A: Math. Gen.* **31**, 657
[19] Sadowskii DA and Zhilinskii B 1999 *Phys. Lett. A* **256**, 235
[20] Cushman R and Sadovskii DA 2000 *Physica D* **142**, 166
[21] Child MS, Weston T and Tennyson J 1999 *Mol. Phys.* **96**, 371
[22] Walkens H and Dullin HR 2002 *Ann. Phys. (N.Y.)* **295**, 81

IOP Conf. Series: Journal of Physics: Conf. Series **1071** (2018) 012014 doi:10.1088/1742-6596/1071/1/012014

Coexisting partial dynamical symmetries and multiple shapes

A Leviatan and N Gavrielov

Racah Institute of Physics, The Hebrew University, Jerusalem 91904, Israel

E-mail: ami@phys.huji.ac.il, noam.gavrielov@mail.huji.ac.il

Abstract. We present an algebraic procedure for constructing Hamiltonians with several distinct partial dynamical symmetries (PDSs), of relevance to shape-coexistence phenomena. The procedure relies on a spectrum generating algebra encompassing several dynamical symmetry (DS) chains and a coherent state which assigns a particular shape to each chain. The PDS Hamiltonian maintains the DS solvability and quantum numbers in selected bands, associated with each shape, and mixes other states. The procedure is demonstrated for a variety of multiple quadrupole shapes in the framework of the interacting boson model of nuclei.

1. Introduction

During the past several decades, the concept of dynamical symmetry (DS) has become the cornerstone of algebraic modeling of dynamical systems [1–5]. Its basic paradigm is a chain of nested algebras,

$$G_{\rm dyn} \supset G_1 \supset G_2 \supset \cdots \supset G_{\rm sym} \qquad |\lambda_{\rm dyn},\, \lambda_1,\, \lambda_2,\, \ldots,\, \lambda_{\rm sym}\rangle \ , \tag{1}$$

and an Hamiltonian of the system written in terms of the Casimir operators, $\hat{C}[G]$, of the algebras in the chain

$$\hat{H}_{\rm DS} = \sum_G a_G\, \hat{C}[G] \ . \tag{2}$$

In such a case, the spectrum is completely solvable, the energies and eigenstates are labeled by quantum numbers $(\lambda_{\rm dyn},\, \lambda_1,\, \lambda_2,\, \ldots,\, \lambda_{\rm sym})$, which are the labels of irreducible representations (irreps) of the algebras in the chain. In Eq. (1), $G_{\rm dyn}$ is the dynamical (spectrum generating) algebra of the system such that operators of all physical observables can be written in terms of its generators and $G_{\rm sym}$ is the symmetry algebra. The DS spectrum exhibits a hierarchy of splitting but no mixing. A given $G_{\rm dyn}$ can encompass several DS chains, each providing characteristic analytic expressions for observables and definite selection rules.

A notable example of such algebraic setting is the interacting boson model (IBM) [2], widely used in the description of low-lying quadrupole collective states in nuclei in terms of N monopole (s) and quadrupole (d) bosons, representing valence nucleon pairs. The model is based on $G_{\rm dyn} = {\rm U}(6)$ and $G_{\rm sym} = {\rm SO}(3)$. The Hamiltonian is expanded in terms of the generators of U(6), $\{s^\dagger s,\, s^\dagger d_m,\, d^\dagger_m s,\, d^\dagger_m d_{m'}\}$, and consists of Hermitian, rotational-scalar interactions which

Table 1. Eigenvalues of the Casimir operators, $\hat{C}_k[G]$ of order $k = 1, 2$, for the leading sub-algebras (G_1) of the DS-chains (3). The equilibrium deformations $(\beta_{\rm eq}, \gamma_{\rm eq})$ define the quadrupole shape associated with each chain and determine the G_1-symmetry of $|\beta_{\rm eq}, \gamma_{\rm eq}; N\rangle$, Eq. (4). The latter is an extremal state in a particular irrep $(\lambda = \Lambda_0)$ of G_1, and serves as an intrinsic state for the respective ground-band.

Algebra G_1	Eigenvalues of $\hat{C}_k[G_1]$	Equilibrium deformations $(\beta_{\rm eq}, \gamma_{\rm eq})$	G_1-symmetry of $\|\beta_{\rm eq}, \gamma_{\rm eq}; N\rangle$ $\lambda_1 = \Lambda_0$
U(5)	n_d	$\beta_{\rm eq} = 0$	$n_d = 0$
SU(3)	$\lambda^2 + (\lambda + \mu)(\mu + 3)$	$(\beta_{\rm eq} = \sqrt{2}, \gamma_{\rm eq} = 0)$	$(\lambda, \mu) = (2N, 0)$
$\overline{\rm SU(3)}$	$\bar{\lambda}^2 + (\bar{\lambda} + \bar{\mu})(\bar{\mu} + 3)$	$(\beta_{\rm eq} = \sqrt{2}, \gamma_{\rm eq} = \pi/3)$	$(\bar{\lambda}, \bar{\mu}) = (0, 2N)$
SO(6)	$\sigma(\sigma + 4)$	$(\beta_{\rm eq} = 1, \gamma_{\rm eq}$ arbitrary$)$	$\sigma = N$

conserve the total number of s- and d- bosons, $\hat{N} = \hat{n}_s + \hat{n}_d = s^\dagger s + \sum_m d_m^\dagger d_m$. The solvable limits of the IBM correspond to the following DS chains

$$\begin{array}{llll} {\rm U(6)} \supset {\rm U(5)} \supset {\rm SO(5)} \supset {\rm SO(3)} & |N,\, n_d,\, \tau,\, n_\Delta,\, L\rangle & \text{spherical vibrator ,} & \text{(3a)} \\ {\rm U(6)} \supset {\rm SU(3)} \supset {\rm SO(3)} & |N,\, (\lambda,\mu),\, K,\, L\rangle & \text{prolate}-\text{deformed rotor ,} & \text{(3b)} \\ {\rm U(6)} \supset \overline{\rm SU(3)} \supset {\rm SO(3)} & |N,\, (\bar{\lambda},\bar{\mu}),\, \bar{K},\, L\rangle & \text{oblate}-\text{deformed rotor ,} & \text{(3c)} \\ {\rm U(6)} \supset {\rm SO(6)} \supset {\rm SO(5)} \supset {\rm SO(3)} & |N,\, \sigma,\, \tau,\, n_\Delta,\, L\rangle & \gamma-\text{unstable deformed rotor .} & \text{(3d)} \end{array}$$

Here $N, n_d, (\lambda, \mu), (\bar{\lambda}, \bar{\mu}), \sigma, \tau, L$, label the relevant irreps of U(6), U(5), SU(3), $\overline{\rm SU(3)}$, SO(6), SO(5), SO(3), respectively, and $n_\Delta, K, \bar{K}$ are multiplicity labels. The indicated basis states are eigenstates of the Casimir operators in the chain, with eigenvalues listed in Table 1 for the leading sub-algebras G_1, and eigenvalues $L(L + 1)$ $[\tau(\tau + 3)]$ for SO(3) [SO(5)]. The resulting DS spectra of the above chains resemble known paradigms of nuclear collective structure, as mentioned in Eq. (3), involving vibrations and rotations of a quadrupole shape.

Geometry is introduced in the algebraic model by means of a coset space $U(6)/U(5) \otimes U(1)$ and a 'projective' coherent state [6,7],

$$|\beta, \gamma; N\rangle = (N!)^{-1/2}(b_c^\dagger)^N\, |0\,\rangle \;, \tag{4a}$$
$$b_c^\dagger = (1 + \beta^2)^{-1/2}[\beta \cos\gamma d_0^\dagger + \beta \sin\gamma (d_2^\dagger + d_{-2}^\dagger)/\sqrt{2} + s^\dagger] \;, \tag{4b}$$

from which an energy surface is derived by the expectation value of the Hamiltonian,

$$E_N(\beta, \gamma) = \langle \beta, \gamma; N|\hat{H}|\beta, \gamma; N\rangle \;. \tag{5}$$

Here (β, γ) are quadrupole shape parameters whose values, $(\beta_{\rm eq}, \gamma_{\rm eq})$, at the global minimum of $E_N(\beta, \gamma)$ define the equilibrium shape for a given Hamiltonian. The shape can be spherical $(\beta = 0)$ or deformed $(\beta > 0)$ with $\gamma = 0$ (prolate), $\gamma = \pi/3$ (oblate), $0 < \gamma < \pi/3$ (triaxial), or γ-independent. The coherent state with the equilibrium deformations, $|\beta_{\rm eq}, \gamma_{\rm eq}; N\rangle$, serves as an intrinsic state for the ground band, whose rotational members are obtained by angular momentum projection. The equilibrium deformations associated with the DS limits, Eq. (3), are listed in Table 1 and conform with their geometric interpretation. For these values, the ground-band intrinsic state, $|\beta_{\rm eq}, \gamma_{\rm eq}; N\rangle$, becomes a lowest (or highest) weight state in a particular irrep $(\lambda_1 = \Lambda_0)$ of the leading sub-algebra G_1, as disclosed in Table 1.

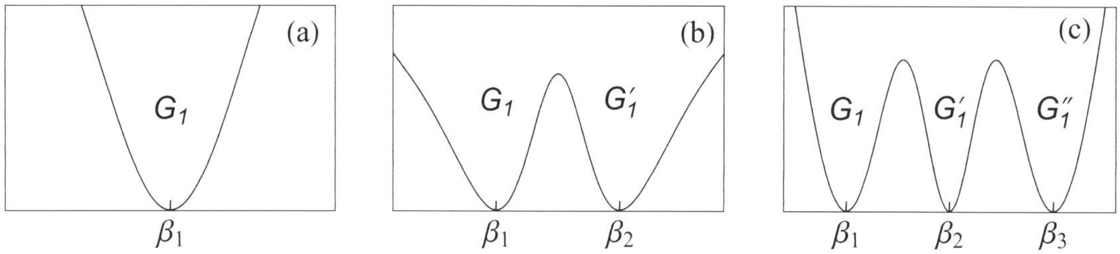

Figure 1. Energy surfaces accommodating: (a) single minimum, (b) double minima, (c) triple minima, associated with (a) G_1-DS or a single G_1-PDS, (b) coexisting G_1-PDS and G_1'-PDS, (c) coexisting G_1-PDS, G_1'-PDS and G_1''-PDS.

A dynamical symmetry corresponds to a single structural phase with a particular shape $(\beta_{\rm eq}, \gamma_{\rm eq})$. The DS Hamiltonians support a single minimum in their energy surface, hence serve as benchmarks for the dynamics of a single shape. Coexistence of different shapes in the same system is a ubiquitous phenomena in many-body systems, such as nuclei [8]. It involves the occurrence in the spectrum of several states (or bands of states) at similar energies with distinct properties, reflecting the nature of their dissimilar dynamics. The relevant Hamiltonians, by necessity, contain competing terms with incompatible (non-commuting) symmetries. The corresponding energy surface accommodates multiple minima, with different symmetry character (denoted by G_1, G_1', G_1'' in Fig. 1) of the dynamics in their vicinity. In such circumstances, exact DSs are broken, and any remaining symmetries can at most be shared by only a subset of states. To address the persisting regularities in such circumstances, amidst a complicated environment of other states, one needs to enlarge the traditional concept of exact dynamical symmetry. In the present contribution, we consider such an extended notion of symmetry, called partial dynamical symmetry (PDS) [9] and show its potential role in formulating algebraic benchmarks for the dynamics of multiple shapes.

2. Partial dynamical symmetries

A dynamical symmetry (DS) is characterized by *complete* solvability and good quantum numbers for *all* states. Partial dynamical symmetry (PDS) [9–11] is a generalization of the latter concept, and corresponds to a particular symmetry breaking for which only *some* of the states retain solvability and/or have good quantum numbers. Such generalized forms of symmetries are manifested in nuclear structure, where extensive tests provide empirical evidence for their relevance to a broad range of nuclei [9,11–28]. In addition to nuclear spectroscopy, Hamiltonians with PDS have been used in the study of quantum phase transitions [29–32] and of systems with mixed regular and chaotic dynamics [33,34]. In what follows, we present concrete algorithms for constructing Hamiltonians with single and multiple PDSs, and show explicit examples of Hamiltonians with such property, in the framework of the IBM.

The construction employs an intrinsic-collective resolution of the Hamiltonian [35–38],

$$\hat{H}' \;=\; \hat{H} + \hat{H}_c \; . \tag{6}$$

The intrinsic part $(\hat{H})$ determines the energy surface and band structure, while the collective part $(\hat{H}_c)$ is composed of kinetic rotational terms and determines the in-band structure. For a given shape, specified by the equilibrium deformations $(\beta_{\rm eq}, \gamma_{\rm eq})$, the intrinsic Hamiltonian is required to annihilate the equilibrium intrinsic state of Eq. (4),

$$\hat{H}|\beta_{\rm eq}, \gamma_{\rm eq}; N\rangle = 0 \; . \tag{7}$$

Since the Hamiltonian is rotational-invariant, this condition is equivalent to the requirement that $\hat{H}$ annihilates the states of good angular momentum L projected from $|\beta_{\rm eq},\gamma_{\rm eq};N\rangle$

$$\hat{H}|\beta_{\rm eq},\gamma_{\rm eq};N,x,L\rangle = 0 \ . \tag{8}$$

Here x denotes additional quantum numbers needed to characterize the states. Symmetry considerations enter when $(\beta_{\rm eq},\gamma_{\rm eq})$ coincide with the equilibrium deformations of the DS chains listed in Table 1. In this case, $|\beta_{\rm eq},\gamma_{\rm eq};N\rangle$ becomes an extremal state in a particular irrep of the leading sub-algebra in the DS-chain and, consequently, the states projected from it have good symmetry.

2.1. Construction of Hamiltonians with a single PDS

The starting point in constructing an Hamiltonian with a single PDS, is a dynamical symmetry chain,

$$\mathrm{U(6)} \supset \mathrm{G}_1 \supset \mathrm{G}_2 \supset \ldots \supset \mathrm{SO(3)} \qquad |N,\,\lambda_1,\,\lambda_2,\,\ldots,\,L\rangle \qquad (\beta_{\rm eq},\gamma_{\rm eq}) \ , \tag{9}$$

with leading sub-algebra G_1, related basis $|N,\,\lambda_1,\,\lambda_2,\,\ldots,\,L\rangle$ and associated shape $(\beta_{\rm eq},\gamma_{\rm eq})$ [see Eq. (3) and Table 1]. The PDS Hamiltonian of Eq. (6) is obtained in a two-step process. First, one identifies an intrinsic Hamiltonian $(\hat{H})$ with partial G_1-symmetry and then one identifies a collective Hamiltonian $(\hat{H}_c)$, which ensures that the complete Hamiltonian $(\hat{H}')$ has G_1-PDS. For that purpose, the intrinsic Hamiltonian is required to satisfy

$$\hat{H}|\beta_{\rm eq},\gamma_{\rm eq};N,\lambda_1\!=\!\Lambda_0,\lambda_2,\ldots,L\rangle = 0 \ . \tag{10}$$

The set of zero-energy eigenstates in Eq. (10) are basis states of a particular G_1-irrep, $\lambda_1=\Lambda_0$, with good G_1 symmetry, and are specified by the quantum numbers of the algebras in the chain (9). For a positive-definite $\hat{H}$, they span the ground band of the equilibrium shape and can be obtained by L-projection from the corresponding intrinsic state, $|\beta_{\rm eq},\gamma_{\rm eq};N\rangle$ of Eq. (4). $\hat{H}$ itself, however, need not be invariant under G_1 and, therefore, has partial-G_1 symmetry. According to the PDS algorithms [10, 19], the construction of number-conserving Hamiltonians obeying the condition of Eq (10), is facilitated by writing them in normal-order form,

$$\hat{H} \ = \ \sum_{\alpha,\beta} u_{\alpha\beta}\hat{T}^{\dagger}_{\alpha}\hat{T}_{\beta} \ , \tag{11}$$

in terms of n-particle creation and annihilation operators satisfying

$$\hat{T}_{\alpha}|\beta_{\rm eq},\gamma_{\rm eq};N,\lambda_1\!=\!\Lambda_0,\lambda_2,\ldots,L\rangle = 0 \ . \tag{12}$$

The collective part $(\hat{H}_c)$ is identified with the Casimir operators of the remaining sub-algebras of G_1 in the chain (9),

$$\hat{H}_c \ = \ \sum_{G_i\subset G_1} a_{G_i}\hat{C}[G_i] \ . \tag{13}$$

For this choice, the degeneracy of the above set of states is lifted, and they remain solvable eigenstates of the complete Hamiltonian $\hat{H}'$ (6). The latter, by definition, has G_1-PDS. It is interesting to note that $\hat{H}' = \hat{H} + \hat{H}_c$ has a form similar to the DS Hamiltonian, Eq. (2), with the intrinsic part $\hat{H}$ replacing the Casimir operator of the leading sub-algebra, $\hat{C}[G_1]$ and $\hat{H}_c$ is diagonal and leads to splitting but no mixing. The difference, however, is that $\hat{C}[G_1]$ has all members of the DS basis $|N,\, \lambda_1,\, \lambda_2,\, \ldots,\, L\rangle$ as eigenstates, while $\hat{H}$ has only a subset of these basis states as eigenstates.

To demonstrate the above procedure, consider the SU(3)-DS chain of Eq. (3b). The two-boson pair operators

$$\begin{aligned} P^{\dagger}_{0} &= d^{\dagger}\cdot d^{\dagger} - 2(s^{\dagger})^2 \;, &\text{(14a)}\\ P^{\dagger}_{2m} &= 2d^{\dagger}_{m}s^{\dagger} + \sqrt{7}\,(d^{\dagger}\,d^{\dagger})^{(2)}_{m} \;, &\text{(14b)}\end{aligned}$$

are $(\lambda,\mu) = (0,2)$ tensors of SU(3) and annihilate the states of the SU(3) ground band,

$$\begin{aligned} P_0\,|N,\,(\lambda,\mu)\!=\!(2N,0),\,K\!=\!0,\,L\rangle &= 0 \;, \qquad L = 0,2,4\ldots,2N\\ P_{2m}\,|N,\,(\lambda,\mu)\!=\!(2N,0),\,K\!=\!0,\,L\rangle &= 0 \;. \end{aligned} \tag{15}$$

The operators of Eq. (15) correspond to $\hat{T}_{\alpha}$ of Eq. (12). The intrinsic Hamiltonian reads

$$\hat{H} = h_0\,P^{\dagger}_{0}P_0 + h_2\,P^{\dagger}_{2}\cdot\tilde{P}_2 \;, \tag{16}$$

where $\tilde{P}_{2m} = (-)^{m}P_{2,-m}$ and the centered dot denotes a scalar product. $\hat{H}$ of Eq. (16) has partial SU(3) symmetry. For $h_2\!=\!h_0$ it is related to the quadratic Casimir operator of SU(3),

$$\hat{\theta}_2 \equiv P^{\dagger}_{0}P_0 + P^{\dagger}_{2}\cdot\tilde{P}_2 = -\hat{C}_2[\mathrm{SU(3)}] + 2\hat{N}(2\hat{N}+3) \;. \tag{17}$$

The collective Hamiltonian, $\hat{H}_c$, is composed of $\hat{C}_2[\mathrm{SO(3)}]$. The complete Hamiltonian $\hat{H}'$ (6) has SU(3)-PDS, shown to be relevant to the spectroscopy of rare earth and actinide nuclei [11–15].

A similar procedure can be adapted to the $\overline{\mathrm{SU(3)}}$-DS chain of Eq. (3c). The two-boson pair operators $P^{\dagger}_{0}$ of Eq. (14a) and

$$\bar{P}^{\dagger}_{2m} = -2d^{\dagger}_{m}s^{\dagger} + \sqrt{7}\,(d^{\dagger}\,d^{\dagger})^{(2)}_{m} \;, \tag{18}$$

are $(\bar{\lambda},\bar{\mu}) = (2,0)$ tensors of $\overline{\mathrm{SU(3)}}$ and annihilate the states of the $\overline{\mathrm{SU(3)}}$ ground band,

$$\begin{aligned} P_0\,|N,\,(\bar{\lambda},\bar{\mu})\!=\!(0,2N),\,\bar{K}\!=\!0,\,L\rangle &= 0 \;, \qquad L = 0,2,4\ldots,2N\\ \bar{P}_{2m}\,|N,\,(\bar{\lambda},\bar{\mu})\!=\!(0,2N),\,\bar{K}\!=\!0,\,L\rangle &= 0 \;. \end{aligned} \tag{19}$$

The intrinsic Hamiltonian reads

$$\hat{H} = t_0\,P^{\dagger}_{0}P_0 + t_2\,\bar{P}^{\dagger}_{2}\cdot\tilde{\bar{P}}_2 \;, \tag{20}$$

and has partial $\overline{\mathrm{SU(3)}}$ symmetry. For $t_2 = t_0$ it is related to the Casimir operator of $\overline{\mathrm{SU(3)}}$,

$$\hat{\bar{\theta}}_2 \equiv P^{\dagger}_{0}P_0 + \bar{P}^{\dagger}_{2}\cdot\tilde{\bar{P}}_2 = -\hat{C}_2[\overline{\mathrm{SU(3)}}] + 2\hat{N}(2\hat{N}+3) \;. \tag{21}$$

Occasionally, the condition of Eq. (10) necessitates higher-order terms in the Hamiltonian. This is the case for the SO(6)-DS of Eq. (3d). Here the two-boson pair operator

$$R^{\dagger}_{0} = d^{\dagger}\cdot d^{\dagger} - (s^{\dagger})^2 \;, \tag{22}$$

is a scalar ($\sigma = 0$) under SO(6) and annihilates the states in the SO(6) ground-band,

$$R_0\,|N,\sigma=N,\tau,n_{\Delta},L\rangle = 0 \;, \qquad \tau = 0,1,2,\ldots,N \tag{23}$$

However, in this case, the following two-body term

$$\hat{\theta}_0 \equiv R^{\dagger}_{0}R_0 = -\hat{C}_2[\mathrm{SO(6)}] + \hat{N}(\hat{N}+4) \;. \tag{24}$$

is related to the Casimir operator of SO(6), hence is SO(6)-invariant. A genuine SO(6)-PDS can be realized by including cubic terms in the intrinsic Hamiltonian,

$$\hat{H} = r_0\,R^{\dagger}_{0}\hat{n}_sR_0 + r_2\,R^{\dagger}_{0}\hat{n}_dR_0 \;. \tag{25}$$

The collective Hamiltonian, $\hat{H}_c$, is composed of $\hat{C}_2[\mathrm{SO(5)}]$ and $\hat{C}_2[\mathrm{SO(3)}]$. The complete Hamiltonian $\hat{H}'$ (6) has SO(6)-PDS, shown to be relevant to the spectroscopy of ^{196}Pt [19].

2.2. Construction of Hamiltonians with several PDSs

The procedure described in the previous section was oriented towards constructing Hamiltonians with a single PDS. A large number of such PDS Hamiltonians [11–21] have been constructed in this manner. They provide a valuable addition to the arsenal of DS Hamiltonians, suitable for describing systems with a single shape. To allow for a description of multiple shapes in the same system, requires an extension of the above procedure to encompass a construction of Hamiltonians with several distinct PDSs [29–32]. This is the subject matter of the present contribution. We focus on the dynamics in the vicinity of the critical point, where the corresponding multiple minima in the energy surface are near-degenerate and the structure changes most rapidly.

For that purpose, consider two different shapes specified by equilibrium deformations (β_1, γ_1) and (β_2, γ_2) whose dynamics is described, respectively, by the following DS chains

$$\mathrm{U}(6) \supset \mathrm{G}_1 \supset \mathrm{G}_2 \supset \ldots \supset \mathrm{SO}(3) \qquad |N,\, \lambda_1,\, \lambda_2,\, \ldots,\, L\rangle \qquad (\beta_1, \gamma_1)\ , \tag{26a}$$

$$\mathrm{U}(6) \supset \mathrm{G}'_1 \supset \mathrm{G}'_2 \supset \ldots \supset \mathrm{SO}(3) \qquad |N,\, \sigma_1,\, \sigma_2,\, \ldots,\, L\rangle \qquad (\beta_2, \gamma_2)\ , \tag{26b}$$

with different leading sub-algebras $(G_1 \neq G'_1)$ and associated bases. As portrayed in Fig. 1, at the critical point, the corresponding minima representing the two shapes, and respective ground bands are degenerate. Accordingly, we require the intrinsic critical-point Hamiltonian to satisfy simultaneously the following two conditions

$$\hat{H}|\beta_1, \gamma_1; N, \lambda_1 = \Lambda_0, \lambda_2, \ldots, L\rangle \;=\; 0\ , \tag{27a}$$

$$\hat{H}|\beta_2, \gamma_2; N, \sigma_1 = \Sigma_0, \sigma_2, \ldots, L\rangle \;=\; 0\ . \tag{27b}$$

The states of Eq. (27a) reside in the $\lambda_1 = \Lambda_0$ irrep of G_1, are classified according to the DS-chain (26a), hence have good G_1 symmetry. Similarly, the states of Eq. (27b) reside in the $\sigma_1 = \Sigma_0$ irrep of G'_1, are classified according to the DS-chain (26b), hence have good G'_1 symmetry. Although G_1 and G'_1 are incompatible, both sets are eigenstates of the same Hamiltonian. When the latter is positive definite, the two sets span the ground bands of the (β_1, γ_1) and (β_2, γ_2) shapes, respectively. In general, $\hat{H}$ itself is not necessarily invariant under G_1 nor under G_2 and, therefore, its other eigenstates can be mixed with respect to both G_1 and G'_1. Identifying the collective part of the Hamiltonian with the Casimir operator of SO(3) (as well as with the Casimir operators of additional algebras which are common to both chains), the two sets of states remain (non-degenerate) eigenstates of the complete Hamiltonian (6), which then has both G_1-PDS and G'_1-PDS. The case of triple (or multiple) shape coexistence, associated with three (or more) incompatible DS-chains is treated in a similar fashion. In the following sections, we apply the above procedure to a variety of coexisting shapes in the IBM framework, examine the spectral properties of the derived PDS Hamiltonians, and highlight their potential to serve as benchmarks for describing multiple shapes in nuclei.

3. Simultaneous occurrence of a spherical shape and a deformed shape

A particular type of shape coexistence present in nuclei involves spherical and quadrupole-deformed shapes, *e.g.*, in neutron-rich Sr isotopes [39,40], ^{96}Zr [41] and near ^{78}Ni [42,43]). A PDS Hamiltonian appropriate to simultaneously occurring spherical and axially-deformed prolate shapes, can be obtained from Eq. (16) by setting $h_0 = 0$. This Hamiltonian was studied in great detail in [29,30], and shown to have coexisting U(5)-PDS and SU(3)-PDS. In a similar manner, a PDS Hamiltonian with coexisting U(5)-PDS and $\overline{\mathrm{SU}(3)}$-PDS, appropriate to simultaneously occurring spherical and axially-deformed oblate shapes, can be obtained from Eq. (20) by setting $t_0 = 0$. The γ degree of freedom and triaxiality can play an important role in the occurrence of multiple shapes in nuclei [44]. In what follows, we consider a case study of PDS Hamiltonians relevant to a coexistence of spherical and non-axial γ-unstable deformed shapes.

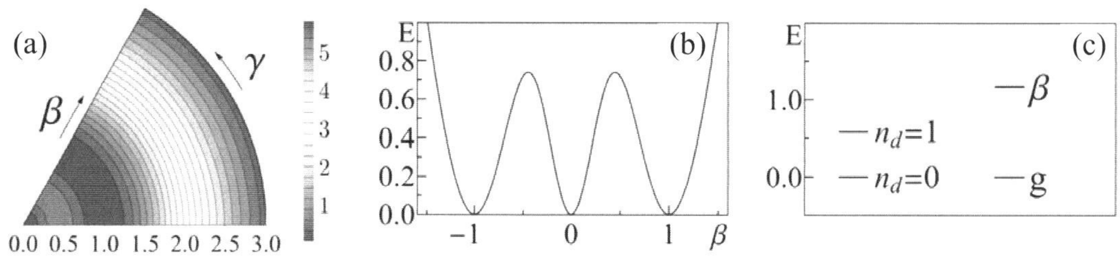

Figure 2. Spherical and γ-unstable deformed (S-G) shape coexistence. (a) Contour plots of the γ-independent energy surface (30), (b) $\gamma=0$ sections, and (c) bandhead spectrum, for the Hamiltonian $\hat{H}'$ (32) with parameters $r_2=10$, $\rho_5=\rho_3=0$ and $N=20$.

3.1. *Coexisting U(5)-PDS and SO(6)-PDS*

The U(5)-DS limit of Eq. (3a) is appropriate to the dynamics of a spherical shape. For a given N, the allowed U(5) and SO(5) irreps are $n_d=0,1,2,\ldots,N$ and $\tau=n_d,\, n_d{-}2,\ldots 0$ or 1, respectively. The values of L contained in a given τ-irrep follow the SO(5) $\supset$ SO(3) reduction rules [2]. The U(5)-DS spectrum resembles that of an anharmonic spherical vibrator, composed of U(5) n_d-multiplets whose spacing is governed by $\hat{C}_1[\mathrm{U}(5)]=\hat{n}_d$, and the splitting is generated by the SO(5) and SO(3) terms. The lowest U(5) multiplets involve the ground state with quantum numbers ($n_d=0,\,\tau=0,\,L=0$) and excited states with quantum numbers ($n_d=1,\,\tau=1,\,L=2$), ($n_d=2:\,\tau=0,\,L=0;\,\tau=2,\,L=2,4$) and ($n_d=3:\,\tau=3,\,L=6,4,3,0;\,\tau=1,\,L=2$).

The SO(6)-DS limit of Eq. (3d) is appropriate to the dynamics of a γ-unstable deformed shape. For a given N, the allowed SO(6) and SO(5) irreps are $\sigma=N,\,N-2,\ldots 0$ or 1, and $\tau=0,\,1,\,\ldots\sigma$, respectively. The SO(5) $\supset$ SO(3) reduction is the same as in the U(5)-DS chain. The SO(6)-DS spectrum resembles that of a γ-unstable deformed rotovibrator, composed of SO(6) σ-multiplets forming rotational bands, with $\tau(\tau+3)$ and $L(L+1)$ splitting generated by $\hat{C}_2[\mathrm{SO}(5)]$ and $\hat{C}_2[\mathrm{SO}(3)]$, respectively. The lowest irrep $\sigma=N$ contains the ground (g) band of a γ-unstable deformed nucleus. The first excited irrep $\sigma=N-2$ contains the β-band. The lowest members in each band have quantum numbers ($\tau=0,\,L=0$), ($\tau=1,\,L=2$), ($\tau=2,\,L=2,4$) and ($\tau=3,\,L=0,3,4,6$).

Following the procedure outlined in Eq. (27), the intrinsic part of the critical-point Hamiltonian, relevant to spherical and γ-unstable deformed (S-G) shape-coexistence, is required to satisfy

$$\hat{H}|N,\,\sigma=N,\,\tau,\,L\rangle \;=\; 0\,, \tag{28a}$$
$$\hat{H}|N,\,n_d=0,\,\tau=0,\,L=0\rangle \;=\; 0\,. \tag{28b}$$

Equivalently, $\hat{H}$ annihilates both the deformed intrinsic state of Eq. (4) with ($\beta=1,\gamma$ arbitrary), which is the lowest weight vector in the SO(6) irrep $\sigma=N$, and the spherical intrinsic state with $\beta=0$, which is the single basis state in the U(5) irrep $n_d=0$. The resulting intrinsic Hamiltonian is found to be that of Eq. (25) with $r_0=0$,

$$\hat{H} = r_2\, R_0^\dagger \hat{n}_d R_0\;, \tag{29}$$

and $R_0^\dagger$ given in Eq. (22). The energy surface, $E_N(\beta,\gamma)=N(N-1)(N-2)\tilde{E}(\beta,\gamma)$, is given by

$$\tilde{E}(\beta) = r_2\beta^2(\beta^2-1)^2(1+\beta^2)^{-3}\,. \tag{30}$$

The surface is an even sextic function of β and is independent of γ, in accord with the SO(5) symmetry of the Hamiltonian. For $r_2>0$, $\hat{H}$ is positive definite and $\tilde{E}(\beta)$ has two degenerate

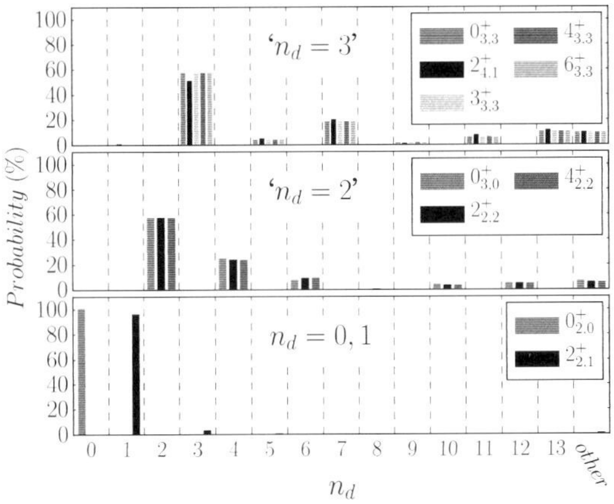

Figure 3. U(5) n_d-decomposition for spherical states, eigenstates of the Hamiltonian $\hat{H}'$ (32) with parameters as in Fig. 2, resulting in spherical and γ-unstable deformed (S-G) shape coexistence. The column 'other' depicts a sum of probabilities, each less than 5%. The spherical states are dominated by a single n_d component, in marked contrast to the deformed states exhibiting a broad n_d-distribution.

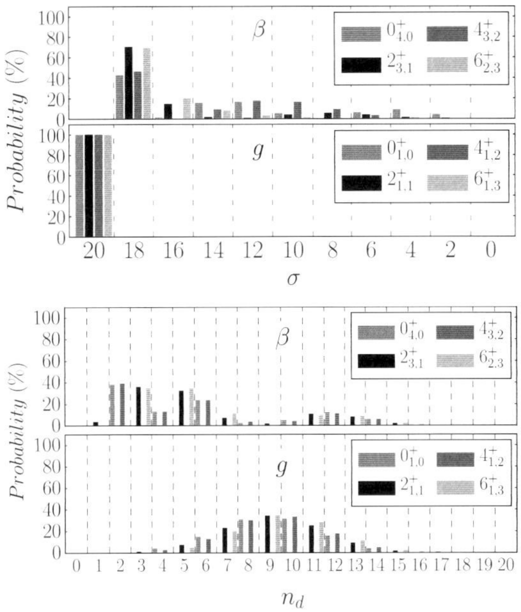

Figure 4. SO(6) σ- and U(5) n_d- decomposition for members of the deformed ground (g) and β bands, eigenstates of $\hat{H}'$ (32) as in Fig. 2.

global minima, $\beta = 0$ and $\beta^2 = 1$, at $\tilde{E} = 0$. A local maximum at $\beta_*^2 = \frac{1}{5}$ creates a barrier of height $\tilde{E} = \frac{2}{27} r_2$, separating the two minima, as seen in Fig. 2. For large N, the normal modes shown schematically in Fig. 2(c), involve β vibrations about the deformed minima, with frequency ϵ_β, and quadrupole vibrations about the spherical minimum, with frequency ϵ, respectively,

$$\epsilon_\beta = 2 r_2 N^2 \,, \tag{31a}$$
$$\epsilon = r_2 N^2 \,. \tag{31b}$$

Identifying the collective part of the Hamiltonian with the Casimir operators of the common SO(5) $\supset$ SO(3) segment of the chains (3a) and (3d), we arrive at the following complete Hamiltonian

$$\hat{H}' = r_2 R_0^\dagger \hat{n}_d R_0 + \rho_5 \hat{C}_2[\mathrm{SO}(5)] + \rho_3 \hat{C}_2[\mathrm{SO}(3)] \,. \tag{32}$$

The added rotational terms generate an exact $\rho_5 \tau(\tau+3) + \rho_3 L(L+1)$ splitting without affecting the wave functions. In particular, the solvable subset of eigenstates, Eq. (28), remain intact. Since both SO(5) and SO(3) are preserved by the Hamiltonian, its eigenstates have good (τ, L) quantum numbers and can be labeled as $L_{i,\tau}^+$, where the ordinal number i enumerates the occurrences of states with the same (τ, L), with increasing energy. The nature of the Hamiltonian eigenstates can be inferred from the probability distributions, $P_{n_d}^{(N,\tau,L)} = |C_{n_d}^{(N,\tau,L)}|^2$ and $P_\sigma^{(N,\tau,L)} = |C_\sigma^{(N,\tau,L)}|^2$, obtained from their expansion coefficients in the U(5) and SO(6) bases, Eqs (3a) and (3d). In general, the low lying spectrum of $\hat{H}'$ (32) exhibits two distinct classes of states. The first class consists of (τ, L) states arranged in n_d-multiplets of a spherical vibrator. Fig. 3 shows the U(5) n_d-decomposition of such spherical states, characterized by a narrow n_d-distribution. The lowest spherical state, $L = 0_{2,0}^+$, is the solvable U(5) state of Eq. (28b) with U(5) quantum number $n_d = 0$. The $L = 2^+_{2,1}$ state has $n_d = 1$ to a good approximation. The upper panels of Fig. 3 display the next spherical-type of multiplets ($L = 0^+_{3,0}$, $2^+_{2,2}$, $4^+_{2,2}$) and ($L = 6^+_{3,3}$, $4^+_{3,3}$, $3^+_{3,3}$, $0^+_{3,3}$, $2^+_{4,1}$), which have a somewhat less pronounced (60%) single n_d-component, with $n_d = 2$ and $n_d = 3$, respectively.

A second class consists of (τ, L) states arranged in bands of a γ-unstable deformed rotor. The SO(6) σ-decomposition of such states, in selected bands, are shown in the upper panel of Fig. 4. The ground band is seen to be pure with $\sigma = N$ SO(6) character, and coincides with the solvable band of Eq. (28a). In contrast, the non-solvable β-band (and higher β^n-bands) show considerable SO(6)-mixing. The deformed nature of these SO(5)-rotational states is manifested in their broad n_d-distribution, shown in the lower panel of Fig. 4. The above analysis demonstrates that although the critical-point Hamiltonian (32) is not invariant under U(5) nor SO(6), some of its eigenstates have good U(5) symmetry, some have good SO(6) symmetry and all other states are mixed with respect to both U(5) and SO(6). These are precisely the defining attributes of U(5)-PDS coexisting with SO(6)-PDS.

Since the wave functions for the solvable states, Eqs. (28), are known, one has at hand closed form expressions for related spectroscopic observables. Consider the $E2$ operator,

$$T(E2) = e_B\,\Pi^{(2)} = e_B\,(d^\dagger s + s^\dagger \tilde{d}) \;, \tag{33}$$

where $\Pi^{(2)}$ is a generator of SO(6) and e_B an effective charge. $T(E2)$ obeys the SO(5) selection rules $\Delta\tau = \pm 1$ and, consequently, all (τ, L) states have vanishing quadrupole moments. The $B(E2)$ values for intraband $(g \to g)$ transitions between states of the ground band, Eq. (28a), are given by the known SO(6)-DS expressions [2]. For example,

$$B(E2;\, g,\, \tau+1,\, L' = 2\tau+2 \to g,\, \tau,\, L = 2\tau) \;=\; e_B^2\,\tfrac{\tau+1}{2\tau+5}(N-\tau)(N+\tau+4)\;, \tag{34a}$$

$$B(E2;\, g,\, \tau+1,\, L' = 2\tau \to g,\, \tau,\, L = 2\tau) \;=\; e_B^2\,\tfrac{4\tau+2}{(2\tau+5)(4\tau-1)}(N-\tau)(N+\tau+4)\;. \tag{34b}$$

Similarly, the $E2$ rates for the transition connecting the pure spherical states, $(n_d = \tau = 1, L = 2)$ and $(n_d = \tau = 0, L = 0)$, satisfy the U(5)-DS expression [2]

$$B(E2; n_d = 1, L = 2 \to n_d = 0, L = 0) = e_B^2 N \;. \tag{35}$$

Member states of the deformed ground band (28a) span the entire $\sigma = N$ irrep of SO(6) and are not connected by $E2$ transitions to the spherical states since $\Pi^{(2)}$, as a generator of SO(6), cannot connect different σ-irreps of SO(6). The weak spherical $\to$ deformed $E2$ transitions persist also for a more general E2 operator obtained by adding $(d^\dagger \tilde{d})^{(2)}$ to $T(E2)$, since the latter term, as a generator of U(5), cannot connect different n_d-irreps of U(5). By similar arguments, there are no $E0$ transitions involving these spherical states, since the $E0$ operator, $T(E0) \propto \hat{n}_d$, is diagonal in n_d. These symmetry-based selection rules result in strong electromagnetic transitions between states in the same class, associated with a given shape, and weak transitions between states in different classes.

The above discussion has focused on the dynamics in the vicinity of the critical point, where the spherical and deformed configurations are degenerate. The evolution of structure away from the critical point, can be studied by adding to $\hat{H}'$ (32) the Casimir operators of U(5) and SO(6), still retaining the desired SO(5) symmetry. Adding an $\epsilon\hat{n}_d$ term, will leave the energy of the spherical $n_d = 0$ state unchanged, but will shift the deformed γ-unstable ground band to higher energy of order $\epsilon N/2$. Similarly, adding a small $\alpha\hat{\theta}_0$ term (24), will leave the solvable SO(6) $\sigma = N$ ground band unchanged, but will shift the spherical ground state $(n_d = L = 0)$ to higher energy of order αN^2. The resulting topology of the energy surfaces with such modifications are shown at the bottom row of Fig. 5. If these departures from the critical points are small, the

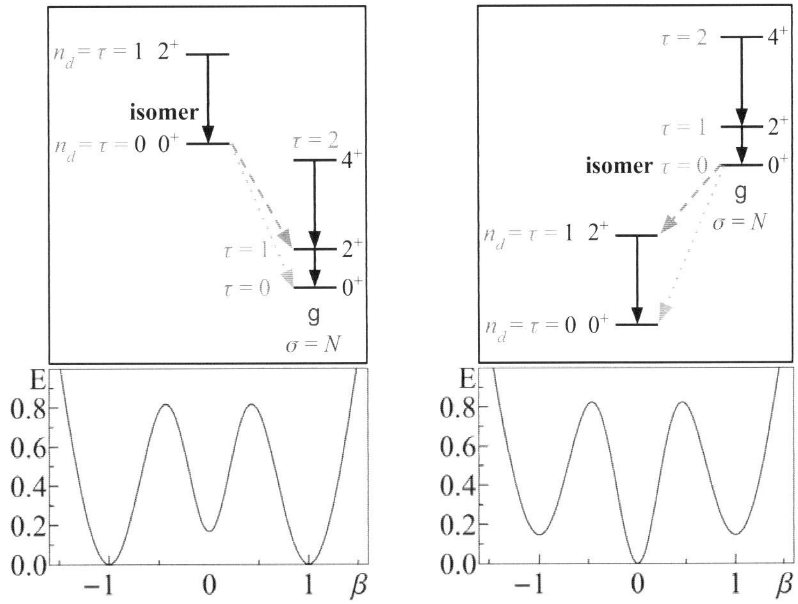

Figure 5. Energy-surface sections and level schemes corresponding to departures from the critical point of S-G shape coexistence, for $\hat{H}'$ (32) with parameters as in Fig. 2. Left panels: a spherical isomeric state ($\alpha\hat{\theta}_0$ term (24), with $\alpha=3$, added to $\hat{H}'$). Right panels: a γ-unstable deformed isomeric state ($\epsilon\hat{n}_d$ term, with $\epsilon=200$, added to $\hat{H}'$). Retarded $E2$ (dashed lines) and $E0$ (dotted lines) decays identify the isomeric states.

wave functions decomposition of Figs. 3-4 remain intact and the above analytic expressions for $E2$ observables and selection rules are still valid to a good approximation. In such scenarios, the lowest $L=0$ state of the non-yrast configuration will exhibit retarded $E2$ and $E0$ decays, hence will have the attributes of an isomer state, as depicted schematically on the top row of Fig. 5.

4. Simultaneous occurrence of two deformed shapes

Shape coexistence in nuclei can involve two deformed shapes (*e.g.*, prolate and oblate) as encountered in Kr [45], Se [46] and Hg isotopes [47]. In what follows, we study PDS Hamiltonians relevant to such axially-deformed shapes with $\gamma = 0, \pi/3$ and equal β deformation.

4.1. Coexisting SU(3)-PDS and $\overline{\text{SU(3)}}$-PDS

The DS limits appropriate to prolate and oblate shapes correspond to the chains (3b) and (3c), respectively. For a given N, the allowed SU(3) [$\overline{\text{SU(3)}}$] irreps are $(\lambda,\mu) = (2N-4k-6m, 2k)$ $[(\bar{\lambda},\bar{\mu}) = (2k, 2N-4k-6m)]$ with k, m, non-negative integers. The multiplicity label K ($\bar{K}$) corresponds geometrically to the projection of the angular momentum (L) on the symmetry axis. The basis states are eigenstates of the Casimir operator $\hat{C}_2[\text{SU(3)}]$ or $\hat{C}_2[\overline{\text{SU(3)}}]$, with eigenvalues listed in Table 1. Specifically, $\hat{C}_2[\text{SU(3)}] = 2Q^{(2)}\cdot Q^{(2)} + \frac{3}{4}L^{(1)}\cdot L^{(1)}$, $Q^{(2)} = d^\dagger s + s^\dagger\tilde{d} - \frac{1}{2}\sqrt{7}(d^\dagger\tilde{d})^{(2)}$, $L^{(1)} = \sqrt{10}(d^\dagger\tilde{d})^{(1)}$, and $\hat{C}_2[\overline{\text{SU(3)}}]$ is obtained by replacing $Q^{(2)}$ by $\bar{Q}^{(2)} = d^\dagger s + s^\dagger\tilde{d} + \frac{1}{2}\sqrt{7}(d^\dagger\tilde{d})^{(2)}$. The generators of SU(3) and $\overline{\text{SU(3)}}$, $Q^{(2)}$ and $\bar{Q}^{(2)}$, and corresponding basis states, are related by a change of phase $(s^\dagger, s) \to (-s^\dagger, -s)$, induced by the operator $\mathcal{R}_s = \exp(i\pi\hat{n}_s)$, with $\hat{n}_s = s^\dagger s$. The DS spectrum resembles that of an axially-deformed rotovibrator composed of SU(3) [or $\overline{\text{SU(3)}}$] multiplets forming rotational bands, with $L(L+1)$-splitting generated by $\hat{C}_2[\text{SO(3)}] = L^{(1)}\cdot L^{(1)}$. In the SU(3) [or $\overline{\text{SU(3)}}$] DS limit, the lowest irrep $(2N, 0)$ [or $(0, 2N)$]

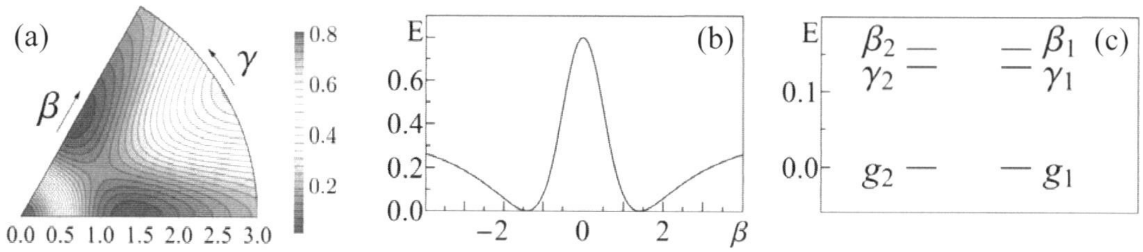

Figure 6. Prolate-oblate (P-O) shape coexistence. (a) Contour plots of the energy surface (38), (b) $\gamma=0$ sections, and (c) bandhead spectrum, for the Hamiltonian $\hat{H}'$ (40) with parameters $h_0=0.2$, $h_2=0.4$, $\eta_3=0.571$, $\alpha=0.018$, $\rho=0$ and $N=20$.

contains the ground band $g(K=0)$ [or $g(\bar{K}=0)$] of a prolate [oblate] deformed nucleus. The first excited irrep $(2N-4,2)$ [or $(2,2N-4)$] contains both the $\beta(K=0)$ and $\gamma(K=2)$ [or $\beta(\bar{K}=0)$ and $\gamma(\bar{K}=2)$] bands. Henceforth, we denote such prolate and oblate bands by (g_1,β_1,γ_1) and (g_2,β_2,γ_2), respectively. Since $\mathcal{R}_s Q^{(2)}\mathcal{R}_s^{-1}=-\bar{Q}^{(2)}$, the SU(3) and $\overline{\mathrm{SU(3)}}$ DS spectra are identical and the quadrupole moments of corresponding states differ in sign.

Following the procedure of Eq. (27), the intrinsic part of the critical-point Hamiltonian, relevant to prolate-oblate (P-O) coexistence, is required to satisfy

$$\hat{H}|N,\,(\lambda,\mu)=(2N,0),\;K=0,\;L\rangle \;=\; 0\;, \tag{36a}$$
$$\hat{H}|N,\,(\bar{\lambda},\bar{\mu})=(0,2N),\;\bar{K}=0,\;L\rangle \;=\; 0\;. \tag{36b}$$

Equivalently, $\hat{H}$ annihilates the intrinsic states of Eq. (4), with $(\beta=\sqrt{2},\gamma=0)$ and $(\beta=-\sqrt{2},\gamma=0)$, which are the lowest- and highest-weight vectors in the irreps $(2N,0)$ and $(0,2N)$ of SU(3) and $\overline{\mathrm{SU(3)}}$, respectively. The resulting intrinsic Hamiltonian is found to be [31],

$$\hat{H} = h_0\,P^{\dagger}_0\hat{n}_sP_0 + h_2\,P^{\dagger}_0\hat{n}_dP_0 + \eta_3\,G^{\dagger}_3\cdot\tilde{G}_3\;, \tag{37}$$

where $G^{\dagger}_{3m}=\sqrt{7}[(d^{\dagger}d^{\dagger})^{(2)}d^{\dagger}]^{(3)}_m=(d^{\dagger}P^{\dagger}_2)^{(3)}_m=(d^{\dagger}\bar{P}^{\dagger}_2)^{(3)}_m$ and $P^{\dagger}_0$, $P^{\dagger}_{2m}$, $\bar{P}^{\dagger}_{2m}$, are defined in Eqs. (15) and (18). The corresponding energy surface, $E_N(\beta,\gamma)=N(N-1)(N-2)\tilde{E}(\beta,\gamma)$, is given by

$$\tilde{E}(\beta,\gamma)=\left\{(\beta^2-2)^2\left[h_0+h_2\beta^2\right]+\eta_3\beta^6(1-\Gamma^2)\right\}(1+\beta^2)^{-3}\;. \tag{38}$$

The surface is an even function of β and $\Gamma=\cos 3\gamma$. For $h_0,h_2,\eta_3\geq 0$, $\hat{H}$ is positive definite and $\tilde{E}(\beta,\gamma)$ has two degenerate global minima, $(\beta=\sqrt{2},\gamma=0)$ and $(\beta=\sqrt{2},\gamma=\pi/3)$ [or equivalently $(\beta=-\sqrt{2},\gamma=0)$], at $\tilde{E}=0$. $\beta=0$ is always an extremum, which is a local minimum (maximum) for $h_2-4h_0>0$ $(h_2-4h_0<0)$, at $\tilde{E}=4h_0$. Additional extremal points include saddle points at $[\beta_1>0,\gamma=0,\pi/3]$, $[\beta_2>0,\gamma=\pi/6]$ and a local maximum at $[\beta_3>0,\gamma=\pi/6]$. The saddle points, when exist, support a barrier separating the various minima, as seen in Fig. 6. For large N, the normal modes involve β and γ vibrations about the respective deformed minima, with frequencies

$$\epsilon_{\beta 1}=\epsilon_{\beta 2}=\frac{8}{3}(h_0+2h_2)N^2\;, \tag{39a}$$
$$\epsilon_{\gamma 1}=\epsilon_{\gamma 2}=4\eta_3N^2\;. \tag{39b}$$

The members of the prolate and oblate ground-bands, Eq. (36), are zero-energy eigenstates of $\hat{H}$ (37), with good SU(3) and $\overline{\mathrm{SU(3)}}$ symmetry, respectively. The Hamiltonian is invariant

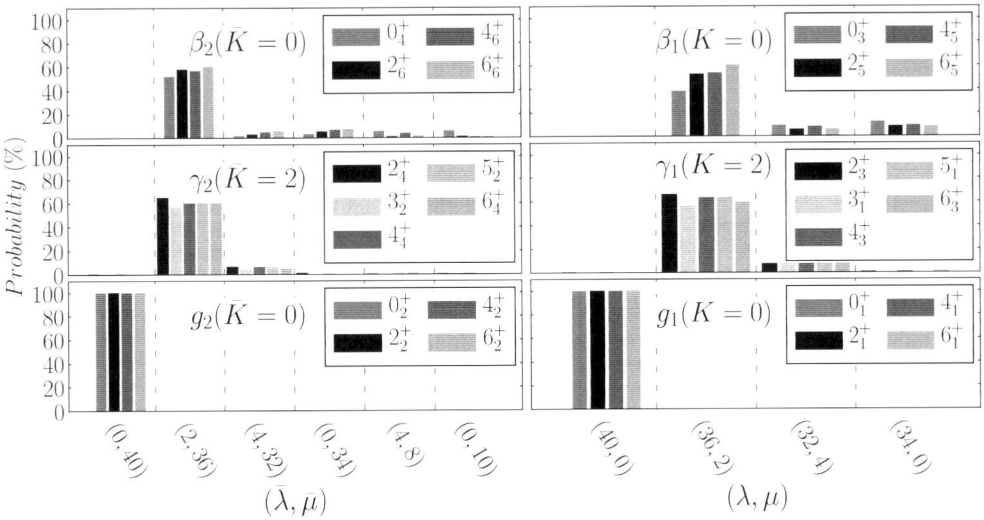

Figure 7. SU(3) (λ,μ)- and $\overline{\text{SU(3)}}$ $(\bar{\lambda},\bar{\mu})$-decompositions for members of the prolate (g_1,β_1,γ_1) and oblate (g_2,β_2,γ_2) bands, eigenstates of $\hat{H}'$ (40) with parameters as in Fig. 6, resulting in prolate-oblate (P-O) shape coexistence. Shown are probabilities larger than 5%.

under a change of sign of the s-bosons, hence commutes with the $\mathcal{R}_s$ operator mentioned above. Consequently, all non-degenerate eigenstates of $\hat{H}$ have well-defined s-parity. This implies vanishing quadrupole moments for the $E2$ operator (33), which is odd under such sign change. To overcome this difficulty, we introduce a small s-parity breaking term, $\alpha\hat{\theta}_2$ (17), which contributes to $\tilde{E}(\beta,\gamma)$ a component $\tilde{\alpha}(1+\beta^2)^{-2}[(\beta^2-2)^2+2\beta^2(2-2\sqrt{2}\beta\Gamma+\beta^2)]$, with $\tilde{\alpha}=\alpha/(N-2)$. The linear Γ-dependence distinguishes the two deformed minima and slightly lifts their degeneracy, as well as that of the normal modes (39). Identifying the collective part with $\hat{C}_2[\text{SO(3)}]$, we arrive at the following complete Hamiltonian

$$\hat{H}' \;=\; h_0\,P_0^\dagger\hat{n}_sP_0 + h_2\,P_0^\dagger\hat{n}_dP_0 + \eta_3\,G_3^\dagger\cdot\tilde{G}_3 + \alpha\,\hat{\theta}_2 + \rho\,\hat{C}_2[\text{SO(3)}]\;. \tag{40}$$

The prolate g_1-band remains solvable with energy $E_{g1}(L)=\rho L(L+1)$. The oblate g_2-band experiences a slight shift of order $\frac{32}{9}\alpha N^2$ and displays a rigid-rotor like spectrum. Replacing $\hat{\theta}_2$ by $\hat{\bar{\theta}}_2$ (21), reverses the sign of the linear Γ term in the energy surface and leads to similar effects, but interchanges the role of prolate and oblate bands. The SU(3) and $\overline{\text{SU(3)}}$ decompositions in Fig. 7 demonstrate that these bands are pure DS basis states, with $(2N,0)$ and $(0,2N)$ character, respectively, while excited β and γ bands exhibit considerable mixing. The critical-point Hamiltonian thus has a subset of states with good SU(3) symmetry, a subset of states with good $\overline{\text{SU(3)}}$ symmetry and all other states are mixed with respect to both SU(3) and $\overline{\text{SU(3)}}$. These are precisely the defining ingredients of SU(3)-PDS coexisting with $\overline{\text{SU(3)}}$-PDS.

Since the wave functions for the members of the g_1 and g_2 bands are known, one can derive analytic expressions for their quadrupole moments and $E2$ rates. For the $E2$ operator of Eq. (33), the quadrupole moments are found to have equal magnitudes and opposite signs,

$$Q_L \;=\; \mp e_B\sqrt{\tfrac{16\pi}{40}}\,\tfrac{L}{2L+3}\,\tfrac{4(2N-L)(2N+L+1)}{3(2N-1)}\;, \tag{41}$$

where the minus (plus) sign corresponds to the prolate-g_1 (oblate-g_2) band. The $B(E2)$ values

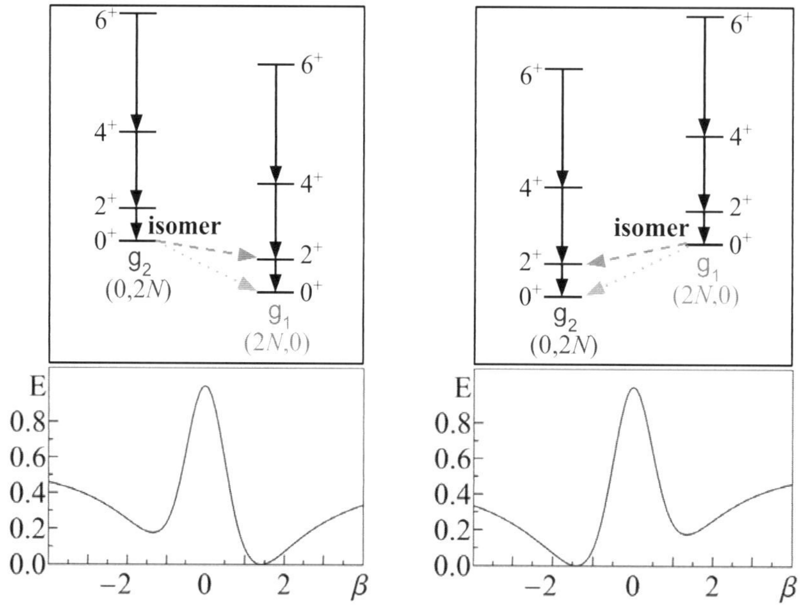

Figure 8. Energy-surface sections and level schemes corresponding to departures from the critical point of P-O shape coexistence, for $\hat{H}'$ (40) with parameters h_0, h_2, η_3 as in Fig. 6. Left panels: an oblate isomeric state ($\alpha\hat{\theta}_2$ term (17), with $\alpha = 0.9$, added to $\hat{H}'$). Right panels: a prolate isomeric state ($\bar{\alpha}\hat{\bar{\theta}}_2$ term (21), with $\bar{\alpha} = 0.9$, added to $\hat{H}'$). Retarded $E2$ (dashed lines) and $E0$ (dotted lines) decays identify the isomeric states.

for intraband ($g_1 \to g_1$, $g_2 \to g_2$) transitions,

$$B(E2; g_i,\, L+2 \to g_i,\, L) = e_B^2\, \frac{3(L+1)(L+2)}{2(2L+3)(2L+5)} \frac{(4N-1)^2(2N-L)(2N+L+3)}{18(2N-1)^2}\,, \tag{42}$$

are the same. These properties are ensured by $\mathcal{R}_s T(E2) \mathcal{R}_s^{-1} = -T(E2)$. Interband ($g_2 \leftrightarrow g_1$) $E2$ transitions, are extremely weak. This follows from the fact that the L-states of the g_1 and g_2 bands exhaust, respectively, the $(2N,0)$ and $(0,2N)$ irrep of SU(3) and $\overline{\text{SU}(3)}$. $T(E2)$ contains a $(2,2)$ tensor under both algebras, hence can connect the $(2N,0)$ irrep of g_1 only with the $(2N-4,2)$ component in g_2, which is vanishingly small. The selection rule $g_2 \not\leftrightarrow g_1$ is valid also for a more general $E2$ operator, obtained by including in it the operators $Q^{(2)}$ or $\bar{Q}^{(2)}$, since the latter, as generators, cannot mix different irreps of SU(3) or $\overline{\text{SU}(3)}$. By similar arguments, $E0$ transitions in-between the g_1 and g_2 bands are extremely weak, since the relevant operator, $T(E0) \propto \hat{n}_d$, is a combination of $(0,0)$ and $(2,2)$ tensors under both algebras. In contrast to g_1 and g_2, excited β and γ bands are mixed, hence are connected by $E2$ transitions to these ground bands.

Departures from the critical point, can be studied by varying the coupling constant of the $\alpha\hat{\theta}_2$ term (17) in $\hat{H}'$ (40). Taking larger values of α, will leave the prolate g_1-band unchanged, but will shift the oblate g_2-band to higher energy of order $16\alpha N^2/9$. Similar effects are obtained by varying the strength of the $\bar{\alpha}\hat{\bar{\theta}}_2$ term (21), but now the role of g_1 and g_2 is interchanged. The resulting topology of the energy surfaces with such modifications are shown at the bottom row of Fig. 8. If these departures from the critical points are small, the results of Fig. 7, Eqs. (41)-(42) and the selection rules remain valid to a good approximation. In such a case, the $L=0$ bandhead state of the higher g_i-band cannot decay by strong $E2$ or $E0$ transitions to the lower

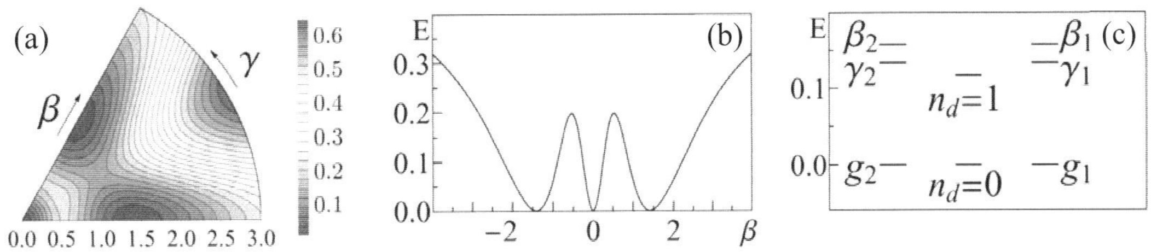

Figure 9. Spherical-prolate-oblate (S-P-O) shape coexistence. (a) Contour plots of the energy surface (45), (b) $\gamma=0$ sections, and (c) bandhead spectrum, for the Hamiltonian $\hat{H}'$ (47) with parameters $h_2=0.5$, $\eta_3=0.571$, $\alpha=0.018$, $\rho=0$ and $N=20$.

ground band, hence, as depicted schematically on the top row of Fig. 8, displays characteristic features of an isomeric state.

5. Simultaneous occurrence of spherical and two deformed shapes

Nuclei can accommodate more than two shapes simultaneously. A notable example is the observed coexistence of spherical, prolate and oblate shapes in ^{186}Pb [48]. In what follows, we consider a case study of PDS Hamiltonians relevant to such triple coexistence of a spherical shape ($\beta=0$) and two axially-deformed shapes with $\gamma=0,\pi/3$ and equal β deformations.

5.1. Coexisting U(5)-PDS, SU(3)-PDS and $\overline{\rm SU(3)}$-PDS

The DS limits relevant for spherical, prolate-deformed and oblate-deformed shapes, correspond to the chains (3a), (3b) and (3c), respectively. The intrinsic part of the critical-point Hamiltonian for triple coexistence of such shapes is now required to satisfy three conditions

$$\hat{H}|N,\, n_d=0,\, \tau=0,\, L=0\rangle \;=\; 0\;, \qquad (43a)$$
$$\hat{H}|N,\, (\lambda,\mu)=(2N,0),\, K=0,\, L\rangle \;=\; 0\;, \qquad (43b)$$
$$\hat{H}|N,\, (\bar{\lambda},\bar{\mu})=(0,2N),\, \bar{K}=0,\, L\rangle \;=\; 0\;. \qquad (43c)$$

Equivalently, $\hat{H}$ annihilates the spherical intrinsic state of Eq. (4) with $\beta=0$, which is the single basis state in the U(5) irrep $n_d=0$, and the deformed intrinsic states with $(\beta=\sqrt{2},\gamma=0)$ and $(\beta=-\sqrt{2},\gamma=0)$, which are the lowest and highest-weight vectors in the irreps $(2N,0)$ and $(0,2N)$ of SU(3) and $\overline{\rm SU(3)}$, respectively. The resulting intrinsic Hamiltonian is found to be that of Eq. (37) with $h_0=0$ [31],

$$\hat{H} = h_2\, P^{\dagger}_0 \hat{n}_d P_0 + \eta_3\, G^{\dagger}_3\cdot\tilde{G}_3\;. \qquad (44)$$

The corresponding energy surface,

$$\tilde{E}(\beta,\gamma) = \left[h_2\beta^2(\beta^2-2)^2 + \eta_3\beta^6(1-\Gamma^2)\right](1+\beta^2)^{-3}\;, \qquad (45)$$

has now three degenerate global minima: $\beta=0$, $(\beta=\sqrt{2},\gamma=0)$ and $(\beta=\sqrt{2},\gamma=\pi/3)$ [or equivalently $(\beta=-\sqrt{2},\gamma=0)$], at $\tilde{E}=0$, separated by barriers as seen in Fig. 9. In addition to the deformed β- and γ modes of Eq. (39) with $h_0=0$, there are now also spherical modes, involving quadrupole vibrations about the spherical minimum, with frequency

$$\epsilon = 4h_2N^2\;. \qquad (46)$$

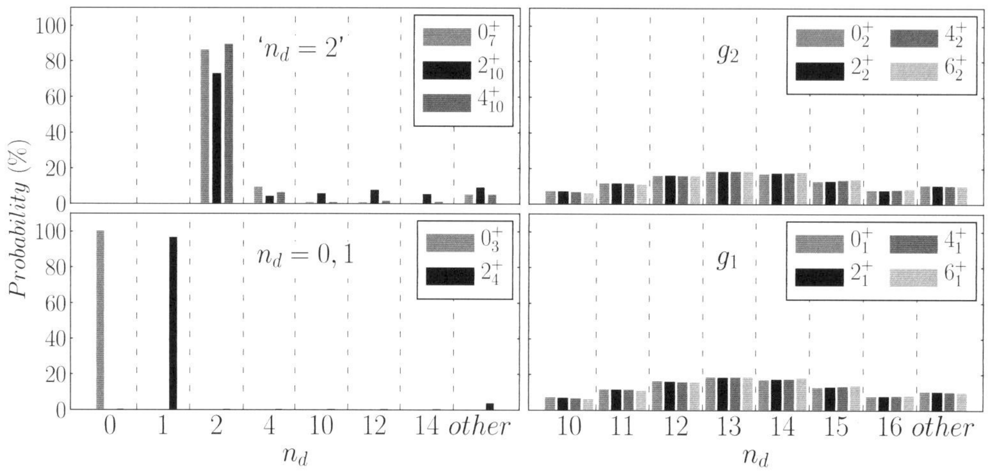

Figure 10. U(5) n_d-decomposition for spherical states (left panels) and for members of the deformed prolate (g_1) and oblate (g_2) ground bands (right panels), eigenstates of $\hat{H}'$ (47) with parameters as in Fig. 9, resulting in spherical-prolate-oblate (S-P-O) shape coexistence. The column 'other' depicts a sum of probabilities, each less than 5%.

For the same arguments as in the analysis of P-O coexistence in Section 4, the complete Hamiltonian is taken to be that of Eq. (40) with $h_0 = 0$,

$$\hat{H}' \;=\; h_2\, P_0^\dagger \hat{n}_d P_0 + \eta_3\, G_3^\dagger \cdot \tilde{G}_3 + \alpha\, \hat{\theta}_2 + \rho\, \hat{C}_2[\mathrm{SO}(3)] \; . \tag{47}$$

The deformed bands show similar rigid-rotor structure as in the P-O case. In particular, the prolate g_1-band and oblate g_2-band have good SU(3) and $\overline{\mathrm{SU(3)}}$ symmetry, respectively, while excited β and γ bands exhibit considerable mixing, with similar decompositions as in Fig. 7. A new aspect in the present S-P-O analysis, is the simultaneous occurrence in the spectrum [see Fig. 9(c)] of spherical type of states, whose wave functions are dominated by a single n_d component. As shown in Fig. 10, the lowest spherical states have quantum numbers ($n_d\!=\!L\!=\!0$) and ($n_d = 1, L = 2$), hence coincide with pure U(5) basis states, while higher spherical states have a pronounced ($\sim$70%) $n_d = 2$ component. This structure should be contrasted with the U(5) decomposition of deformed states (belonging to the g_1 and g_2 bands) which, as shown in Fig. 10, have a broad n_d-distribution. The purity of selected sets of states with respect to SU(3), $\overline{\mathrm{SU(3)}}$ and U(5), in the presence of other mixed states, are the hallmarks of coexisting partial dynamical symmetries.

For the $E2$ operator of Eq. (33), the quadrupole moments of states in the solvable g_1 and g_2 bands and intraband ($g_1 \to g_1$, $g_2 \to g_2$) $E2$ rates, obey the analytic expressions of Eqs. (41) and (42), respectively. The same selection rules depicted in Fig. 8, resulting in retarded $E2$ and $E0$ interband ($g_2 \to g_1$) decays, still hold. Furthermore, in the current S-P-O case, since $T(E2)$ obeys the selection rule $\Delta n_d\!=\!\pm 1$, the spherical states, ($n_d\!=\!L\!=\!0$) and ($n_d\!=\!1, L\!=\!2$), have no quadrupole moment and the $B(E2)$ value for their connecting transition, obeys the U(5)-DS expression of Eq. (35). These spherical states have very weak $E2$ transitions to the deformed ground bands, because they exhaust the ($n_d\!=\!0,1$) irreps of U(5), and the $n_d\!=\!2$ component in the ($L = 0,2,4$) states of the g_1 and g_2 bands is extremely small, of order $N^3 3^{-N}$. There are also no $E0$ transitions involving these spherical states, since $T(E0)$ is diagonal in n_d.

In the above analysis, the spherical and deformed minima were assumed to be degenerate. If the spherical minimum is only local, one can use the Hamiltonian of Eq. (40) with the condition

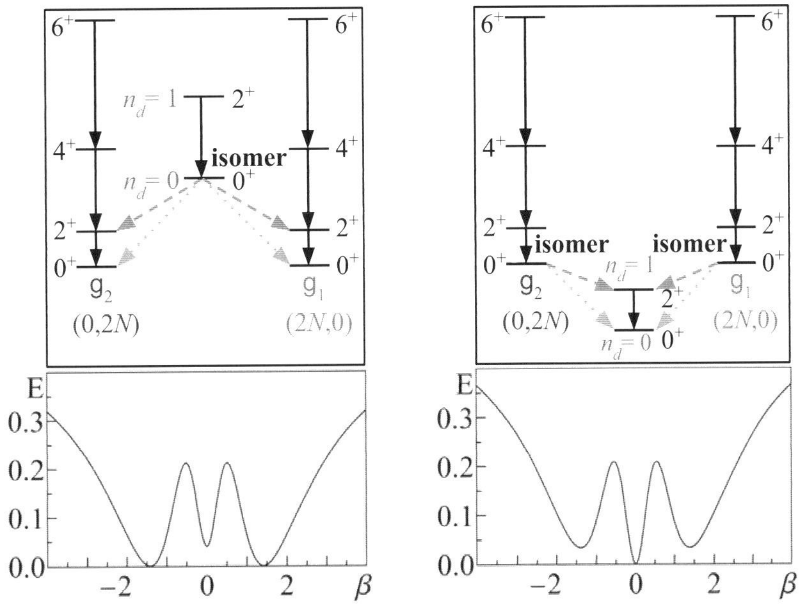

Figure 11. Energy-surface sections and level schemes corresponding to departures from the critical point of S-P-O shape coexistence, for $\hat{H}'$ (47) with $h_2=0.5$, $\eta_3=0.571$, $\alpha=0.018$ and $N=20$. Left panels: a spherical isomeric state ($h_0P^{\dagger}P_0$ term (37), with $h_0=0.01$, added to $\hat{H}'$). Right panels: P-O deformed isomeric states ($\epsilon\hat{n}_d$ term, with $\epsilon=10$, added to $\hat{H}'$). Retarded $E2$ (dashes lines) and $E0$ (dotted lines) decays identify the isomeric states.

$h_2 > 4h_0$, for which the spherical ground state ($n_d = L = 0$) experiences a shift of order $4h_0N^3$. Similarly, if the deformed minima are only local, adding an $\epsilon\hat{n}_d$ term to $\hat{H}'$ (47), will leave the $n_d = 0$ spherical ground state unchanged, but will shift the prolate and oblate bands to higher energy of order $2\epsilon N/3$. In both scenarios, the lowest $L = 0$ state of the non-yrast configuration will exhibit retarded $E2$ and $E0$ decays, hence will have the character of an isomer state, as depicted schematically in Fig. 11.

6. Concluding remarks

We have presented an algebraic symmetry-based approach for describing properties of multiple shapes in dynamical systems. The main ingredients of the approach are: (i) a spectrum generating algebra encompassing several lattices of dynamical symmetry (DS) chains. (ii) An associated geometric space, realized by means of coherent states, which assign a particular shape to a given DS chain. (iii) An intrinsic-collective resolution of the Hamiltonian. The approach involves the construction of a single number-conserving, rotational-invariant Hamiltonian which captures the essential features of the dynamics near the critical point, where two (or more) shapes coexist. The Hamiltonian conserves the dynamical symmetry (DS) for selected bands of states, associated with each shape. Since different structural phases correspond to incompatible (non-commuting) dynamical symmetries, the symmetries in question are shared by only a subset of states, and are broken in the remaining eigenstates of the Hamiltonian. The resulting structure is, therefore, that of coexisting multiple partial dynamical symmetries (PDSs).

An explicit algorithm for constructing Hamiltonians with several distinct PDS was presented and the approach was applied to a variety of coexisting quadrupole shapes, in the framework of the interacting boson model (IBM) of nuclei. The multiple PDSs and shape-coexistence scenarios considered include (i) Coexisting U(5) and SU(3) PDSs, in relation to spherical and prolate-deformed shapes. (ii) Coexisting U(5) and $\overline{\mathrm{SU(3)}}$ PDSs, in relation to spherical and oblate-deformed shapes. (iii) Coexisting U(5) and SO(6) PDSs, in relation to spherical and γ-unstable deformed shapes. (iv) Coexisting SU(3) and $\overline{\mathrm{SU(3)}}$ PDSs, in relation to prolate and oblate deformed shapes. (v) Coexisting U(5), SU(3) and $\overline{\mathrm{SU(3)}}$ PDSs, in relation to spherical, prolate-deformed and oblate-deformed shapes.

In each of the cases considered, the underlying energy surface exhibits multiple minima which are near degenerate. As shown, the constructed Hamiltonian has the capacity to have distinct families of states whose properties reflect the different nature of the coexisting shapes. Selected sets of states within each family, retain the dynamical symmetry associated with the given shape. This allows one to obtain closed expressions for quadrupole moments and transition rates, which are the observables most closely related to the nuclear shape. The resulting analytic expressions are parameter-free predictions, except for a scale, and can be used to compare with measured values of these observables and to test the underlying partial symmetries. The purity and good quantum numbers of selected states enable the derivation of symmetry-based selection rules for electromagnetic transitions (notably, for $E2$ and $E0$ decays) and the subsequent identification of isomeric states. The evolution of structure away from the critical-point can be studied by adding to the Hamiltonian the Casimir operator of a particular DS chain, which will leave unchanged the ground band of one configuration but will shift the other configuration(s) to higher energy.

A detailed microscopic interpretation of shape coexistence in nuclei is a formidable task and computational demanding [8]. The proposed algebraic approach presents a simple alternative, as a starting point to describe this phenomena, by emphasizing the role of remaining underlying symmetries, which provide physical insight and make the problem tractable. It is gratifying to note that shape-coexistence in dynamical systems, such as nuclei, constitutes a fertile ground for the development and testing of generalized notions of symmetry.

Acknowledgments

This work is supported by the Israel Science Foundation (Grant 586/16).

[1] Bohm A, Néeman Y and Barut A O eds 1988 *Dynamical Groups and Spectrum Generating Algebras* (Singapore: World Scientific)
[2] Iachello F and Arima A 1987 *The Interacting Boson Model* (Cambridge University Press, Cambridge)
[3] Iachello F and Van Isacker P 1991 *The Interacting Boson-Fermion Model* (Cambridge: Cambridge Univ. Press)
[4] Iachello F and Levine R D 1994 *Algebraic Theory of Molecules* (Oxford: Oxford Univ. Press)
[5] Iachello F 2015 *Lie Algebras and Applications* (Berlin Heidelberg: Springer-Verlag)
[6] Ginocchio J N and Kirson M W 1980 *Phys. Rev. Lett.* **44** 1744
[7] Dieperink A E L, Scholten O and Iachello F 1980 *Phys. Rev. Lett.* **44** 1747
[8] Heyde K and Wood J L (2011) *Rev. Mod. Phys.* **83** 1467
[9] Leviatan A 2011 *Prog. Part. Nucl. Phys.* **66** 93
[10] Alhassid Y and Leviatan A 1992 *J. Phys.* A **25** L1265
[11] Leviatan A 1996 *Phys. Rev. Lett.* **77** 818
[12] Leviatan A and Sinai I 1999 *Phys. Rev.* C **60** 061301(R)
[13] Casten R F, Cakirli R B, Blaum K, and Couture A 2014 *Phys. Rev. Lett.* **113** 112501
[14] Couture A, Casten R F, and Cakirli R B 2015 *Phys. Rev.* C **91** 014312
[15] Casten R F, Jolie J, Cakirli R B, and Couture A 2016 *Phys. Rev.* C **94** 061303(R)
[16] Leviatan A and Van Isacker P 2002 *Phys. Rev. Lett.* **89** 222501
[17] Kremer C *et al.* 2014 *Phys. Rev.* C **89**, 041302(R); 2015 *Phys. Rev.* C **92** 039902
[18] Leviatan A and Ginocchio J N 2000 *Phys. Rev.* C **61** 024305
[19] García-Ramos J E , Leviatan A and Van Isacker P 2009 *Phys. Rev. Lett.* **102** 112502
[20] Leviatan A, García-Ramos J E , and Van Isacker P 2013 *Phys. Rev.* C **87** 021302(R)
[21] Van Isacker P Jolie J Thomas T and Leviatan A 2015 *Phys. Rev.* C **92** 011301(R)
[22] Van Isacker P 1999 *Phys. Rev. Lett.* **83** 4269
[23] Escher J and Leviatan A 2000 *Phys. Rev. Lett.* **84** 1866
[24] Escher J and Leviatan A 2002 *Phys. Rev.* C **65** 054309
[25] Rowe D J and Rosensteel G 2001 *Phys. Rev. Lett.* **87** 172501

IOP Conf. Series: Journal of Physics: Conf. Series **1071** (2018) 012014 doi:10.1088/1742-6596/1071/1/012014

[26] Rosensteel G and Rowe D J 2003 *Phys. Rev. C* **67** 014303
[27] Van Isacker P and Heinze S 2008 *Phys. Rev. Lett.* **100** 052501
[28] Van Isacker P and Heinze S 2014 *Ann. Phys. (N.Y.)* **349** 73
[29] Leviatan A 2007 *Phys. Rev. Lett.* **98** 242502
[30] Macek M and Leviatan A 2014 *Ann. Phys. (N.Y.)* **351** 302
[31] Leviatan A and Shapira D 2016 *Phys. Rev.* C **93** 051302(R)
[32] Leviatan A and Gavrielov N 2017 *Phys. Scr.* **92** 114005
[33] Whelan N, Alhassid Y and Leviatan A 1993 *Phys. Rev. Lett.* **71** 2208
[34] Leviatan A and Whelan N D 1996 *Phys. Rev. Lett.* **77** 5202
[35] Kirson M W and Leviatan A 1985 *Phys. Rev. Lett.* **55** 2846
[36] Leviatan A 1987 *Ann. Phys. (N.Y.)* **179** 201
[37] Leviatan A and Kirson M W 1990 *Ann. Phys. (N.Y.)* **201** 13
[38] Leviatan A 2006 *Phys. Rev. C* **74** 051301(R)
[39] Clément E *et al.* 2016 *Phys. Rev. Lett.* **116** 022701
[40] Park J *et al.* 2016 *Phys. Rev.* C **93** 014315
[41] Kremer C *et al.* 2016 *Phys. Rev. Lett.* **117** 172503
[42] Gottardo A *et al.* 2016 *Phys. Rev. Lett.* **116** 182501
[43] Yang X F *et al,* 2016 *Phys. Rev. Lett.* **116** 182502
[44] A.D. Ayangeakaa *et al.* 2016 *Phys. Lett. B* **754** 254
[45] Clément E *et al.* 2007 *Phys. Rev.* C **75** 054313
[46] Ljungvall J *et al.* 2008 *Phys. Rev. Lett.* **100** 102502
[47] Bree N *et al.* 2014 *Phys. Rev. Lett.* **112** 162701
[48] Andreyev A N *et al.* 2000 *Nature* **405** 430

IOP Conf. Series: Journal of Physics: Conf. Series **1071** (2018) 012015 doi:10.1088/1742-6596/1071/1/012015

Hidden correlations in quantum optics and quantum information

Margarita A Man'ko

Lebedev Physical Institute, Leninskii Prospect 53, Moscow 119991, Russia

E-mail: mmanko@sci.lebedev.ru

Abstract. We review the notion of quantum correlations in composite systems and show that noncomposite systems have similar correlations (which we call "hidden" correlations); these quantum correlations are analogous to the ones available in composite (multipartite) systems. We demonstrate new entropic-information inequalities characterizing the hidden correlations in noncomposite systems and present the examples of some qudit states with such correlations.

Keywords: qubit, probability distributions, quantum correlations, entropy, entanglement.

1. Introduction

The aim of this work is to review recent studies of the knew notion of hidden correlations in quantum systems, which have no subsystems and present the entropic inequalities characterizing such correlations. Before discussing quantum systems, we recall a crucial role of the notion of probability in quantum mechanics and quantum optics. This notion is related to classical systems with fluctuating random variables. The probability distributions are associated with states of classical systems; they are described by nonnegative functions satisfying the normalization conditions.

All the statistical properties of classical systems like, e.g., correlations of different degrees of freedom in systems containing two (or more) subsystems are coded by the joint probability distributions. Historical aspects and foundations of the classical probability theory can be found in the book by Kolmogorov [1]. An important notion determined by the probability distribution is entropy, which gets the name of Shannon entropy due to milestone work by Shannon [2].

In quantum mechanics and quantum optics, the states are associated with the wave function (pure states) [3] and the density matrices (mixed states) [4, 5]. The notion of quantum entropy was introduced by von Neumann, and this entropy, called von Neumann entropy, is expressed in terms of the density matrix. The quantum von Neumann entropy also characterizes the correlations in the systems containing the subsystems (e.g., two spin-1/2 systems, two-level atoms, or qubits).

Recently, the new representation of quantum states was introduced [6–8]. This representation, called the probability representation of quantum states, is based on the results where the tomographic probability distributions were measured in quantum optics experiments [9] to reconstruct the Wigner function of photon states.

In the probability representation of quantum states, fair probability distributions[1] are associated with quantum-system states. Thus, the known properties of classical probability distributions can be used to study quantum systems.

The clarification of the classical–quantum entropic relations and obtaining some new relations characterizing the quantum-system properties is one of the goals of this work based on the publications [10–18]. The procedure to obtain the entropic-information relations for the density matrix $\rho_{ss'}$ of a single qudit (single spin j or N-level atom) is based on a simple tool of applying the invertible map of the matrix indices s and s' onto the pairs of indices $s \leftrightarrow (jk)$ and $s' \leftrightarrow (j'k')$ or triples of indices $s \leftrightarrow (jk\ell)$ and $s' \leftrightarrow (j'k'\ell')$, etc.

Following this way, we reconsider the density matrix $\rho_{ss'}$ of a single qudit (single spin j or N-level atom) state as the density matrix $\rho_{jk,j'k'}$ (or $\rho_{jk\ell,j'k'\ell'}$) of artificial composite bipartite (or tripartite) system state. In view of the applied invertible map of the matrix indices, the entropic relations obvious for composite systems are transformed to the relations for the density matrices of the single qudit (single spin j) state. It is just a basis of our idea to show the entropic-information relations, which are new for the noncomposite systems (single qudits).

For composite systems, the entropic-information relations correspond to quantum correlations between the subsystem degrees of freedom. These correlations are known to be a resource for quantum technologies, e.g., correlations, which are responsible for the phenomenon of entanglement [19, 20] in multipartite quantum systems. The existence of analogous entropic-information relations (inequalities and equalities) in single qudit states (single spin-j states) reflects the existence of quantum correlations (analogous to entanglement) in noncomposite systems, which do not contain any subsystems. In principle, these correlations can provide an additional resource for quantum technologies based on manipulations with single qudit states analogous to processing the multipartite system states.

For diagonal density matrices $\rho_{ss'}$, the presented map of the matrix indices provides the possibility to map any classical probability distribution $P(s) \equiv \rho_{ss}$ onto the joint probability distribution of bipartite system $p(j,k) \equiv \rho_{jk,jk}$ or tripartite systems $p(j,k,\ell) \equiv \rho_{jk\ell,jk\ell}$. This tool can be used to describe hidden correlations in classical systems with one random variable.

This paper is organized as follows.

In section 2, we study classical hidden correlations. In section 3, we consider entropic-information inequalities for the classical coin and compass. In section 4, we present quantum subadditivity and strong subadditivity conditions and discuss hidden correlations in bipartite quantum systems in section 5. In section 6, we derive new entropic inequalities for the five-level atom and study qubit–qutrit systems in section 7 and a single qudit with spin $j = 2$ in section 8. Finally, we give our conclusions and prospectives in section 9.

2. Classical hidden correlations

We formulate the idea of our approach recalling the known relations for probability distributions of bipartite and tripartite classical systems.

If for bipartite system $(1,2)$, the joint probability distribution is given by the nonnegative function $p(j,k)$, $j = 1,2,\ldots,n$ and $k = 1,2,\ldots,m$, the known entropic inequality, called the entropic subadditivity condition for Shannon entropies of system $H(1,2)$ and its two subsystems

[1] The tomographic probability distribution is a fair (actual) probability distribution; this terminology is often used in the literature as canonical one to stress that the tomographic probability distribution is, in fact, the standard notion for probability distributions used in classical probability theory. In the quantum case, it is fair (actual) one in contrast to quasidistributions like the Wigner function, the Husimi–Kano function, and the Glauber–Sudarshan P-distribution.

$H(1)$ and $H(2)$, holds

$$-\sum_{j=1}^{n}\sum_{k=1}^{m} p(j,k)\ln p(j,k) \leq -\sum_{k=1}^{m}\left[\left(\sum_{j=1}^{n} p(j,k)\right)\ln\left(\sum_{j=1}^{n} p(j,k)\right)\right] -\sum_{j=1}^{n}\left[\left(\sum_{k=1}^{m} p(j,k)\right)\ln\left(\sum_{k=1}^{m} p(j,k)\right)\right]. \quad (1)$$

Here, the left-hand side is equal to Shannon entropy $H(1,2)$, and the right-hand side is equal to the sum of Shannon entropies $H(1)$ and $H(2)$, respectively.

For tripartite system $(1,2,3)$ described by the joint probability distribution $p(j,k,\ell)$, $j=1,2,\ldots,n$, $k=1,2,\ldots,m$, and $\ell=1,2,\ldots,s$, the entropic inequality holds

$$-\sum_{j=1}^{n}\sum_{k=1}^{m}\sum_{\ell=1}^{s} p(j,k,\ell)\ln p(j,k,\ell) - \sum_{k=1}^{m}\left(\sum_{j=1}^{n}\sum_{\ell=1}^{s} p(j,k,\ell)\right)\ln\left(\sum_{j=1}^{n}\sum_{\ell=1}^{s} p(j,k,\ell)\right)$$
$$\leq -\sum_{j=1}^{n}\sum_{k=1}^{m}\left(\sum_{\ell=1}^{s} p(j,k,\ell)\right)\ln\left(\sum_{\ell=1}^{s} p(j,k,\ell)\right) - \sum_{k=1}^{m}\sum_{\ell=1}^{s}\left(\sum_{j=1}^{n} p(j,k,\ell)\right)\ln\left(\sum_{j=1}^{n} p(j,k,\ell)\right). \quad (2)$$

Here, the left-hand side is equal to the sum of Shannon entropy $H(1,2,3)$ of tripartite system $(1,2,3)$ and Shannon entropy $H(2)$ of its second subsystem, and the right-hand side is equal to the sum of Shannon entropies $H(1,2)$ and $H(2,3)$ of subsystems $(1,2)$ and $(2,3)$, respectively. Inequality (2) is called the strong subadditivity condition.

Below we show that such inequalities exist also for systems with one random variable. For this aim, we use the map of indices described above. To introduce the approach, we start with a simplest example of the probability distribution of one classical coin, then consider two coins.

3. Entropy of a classical coin and entropic inequalities for two coins and compass

We consider an example of the classical bit, taking into account that it has a quantum counterpart like the two-level atom and one qubit (or spin $j=1/2$ system). By nonnegative number $p_1 \leq 1$, we describe the probability for the coin to be in state "UP," and by nonnegative number $p_2 \leq 1$ we describe the probability for the coin to be in state "DOWN" (Fig. 1). The two numbers satisfy the normalization condition $p_1 + p_2 = 1 \leftrightarrow \sum_{n=1}^{2} p_n = 1$.

Shannon entropy [2] $H = -p_1 \ln p_1 - p_2 \ln p_2 = -\sum_{n=1}^{2} p_n \ln p_n$ satisfies the inequality $\ln 2 \geq H \geq 0$. Shannon entropy $H=0$ for "complete order," i.e., either $p_1 = 1$ and $p_2 = 0$ or $p_1 = 0$ and $p_2 = 1$, and $H = \ln 2$ for $p_1 = p_2 = 1/2$.

There exists another entropy called Tsallis entropy [21] or q-entropy

$$T(q) = \frac{p_1^q + p_2^q - 1}{1-q} = -\sum_{n=1}^{2} p_n^q \,\frac{p_n^{1-q} - 1}{1-q}. \quad (3)$$

In the limit $q \to 1$, $T(1) = H$, Tsallis entropy converts to Shannon entropy. Both entropies are nonnegative. For Tsallis entropy, also there exist inequalities which can be used within the framework of our approach.

For two coins, there are four possibilities: "UP" "UP" , "UP" "DOWN" , "DOWN" "UP" , "DOWN" "DOWN" .

Figure 1. Coin.

Let the probabilities for the two coins to have these positions to be described by nonnegative numbers $1 \geq p_1,\ p_2,\ p_3,\ p_4 \geq 0$, such that $p_1 + p_2 + p_3 + p_4 = 1$ or $\sum_{n=1}^{4} p_n = 1$. On the other hand, the physical meaning of these four numbers is reflected by the notation $p(jk)$, such that one has

$$p(11) = p_1, \quad p(12) = p_2, \quad p(21) = p_3, \quad p(22) = p_4,$$

where $p(jk)$ is the joint probability distribution that the first coin is in the jth position ("UP" or "DOWN") and the second coin, in the kth position ("UP" or "DOWN"). This means that the position "UP" "UP" for two coins is described by the probability $p(11) = p_1$, and analogously for the other three positions of these coins. We see that, in fact, we use the map of the integers $1 \leftrightarrow 1\,1,\ 2 \leftrightarrow 1\,2,\ 3 \leftrightarrow 2\,1,\ 4 \leftrightarrow 2\,2$ to label the four probabilities. We can use either four numbers $p_1,\ p_2,\ p_3,\ p_4$ or the same four numbers labeled as $p(11),\ p(12),\ p(21),\ p(22)$.

We can formulate this observation as follows.

Given four nonnegative numbers $p_1,\ p_2,\ p_3,\ p_4$, which can be considered as components of the probability vector $\vec{p} = (p_1, p_2, p_3, p_4)$. These four numbers can be identified either with a joint probability distribution described by the table $p(jk)$ or as a probability distribution $p(s)$, $s = 1, 2, 3, 4$ for four events, which have four different possibilities, e.g., four different orientations of compass "North" , "South" , "East" , "West" (Fig. 2).

Figure 2. Compass.

From the viewpoint of mathematical relations between these four numbers, the relations like equalities or inequalities do not depend on the interpretation of these numbers connected either with the positions of two coins or with the compass orientation. But if we identify these four numbers with the joint probability distribution $p(jk)$, i.e., $1 \geq p(jk) \geq 0$, $\sum_{j=1}^{2}\sum_{k=1}^{2} p(jk) = 1$, we elaborate the approach to study entropies associated with the probabilities $p(jk)$.

For example, the probability distribution describing the positions "UP" or "DOWN" only for the first coin reads

$$p(1)_j = \sum_{k=1}^{2} p(jk) = \begin{cases} p(11) + p(12) \equiv p_1 + p_2, \\ p(21) + p(22) \equiv p_3 + p_4, \end{cases} \tag{4}$$

and analogously the probability distribution describing the positions "UP" or "DOWN" only for the second coin is

$$p(2)_k = \sum_{j=1}^{2} p(jk) = \begin{cases} p(11) + p(21) \equiv p_1 + p_3, \\ p(12) + p(22) \equiv p_2 + p_4. \end{cases} \tag{5}$$

Shannon entropies associated with two probability distributions $p(1)_j$, $p(2)_k$ and the entropy associated with joint probability distribution $p(jk)$ are known to satisfy the inequality (subadditivity condition) $H(1) + H(2) \geq H(12)$, where

$$\begin{aligned} H(1) &= -(p(1))_1 \ln (p(1))_1 - (p(1))_2 \ln (p(1))_2, \\ H(2) &= -(p(2))_1 \ln (p(2))_1 - (p(2))_2 \ln (p(2))_2, \\ H(1,2) &= -p(11) \ln p(11) - p(12) \ln p(12) - p(21) \ln p(21) - p(22) \ln p(22). \end{aligned} \tag{6}$$

The physical meaning of this inequality consists in the fact that the degree of "disorder" in composite system (two coins) is smaller than the sum of degrees of "disorder" in the behavior of two subsystems (first coin and second coin).

The difference $I = H(1) + H(2) - H(12) \geq 0$ is called the mutual Shannon information. If the behaviors of both coins are completely independent (there is no correlations between the subsystems – coins), information $I = 0$. If there is a strong correlation in the behaviors of two coins (i.e., the position of the first coin determines the position of the second coin), information is maximum, i.e., $I = \ln 2$.

The absence of correlations between two coins means that the probability vector $\vec{p} = (p_1, p_2, p_3, p_4)$ has the product form $\vec{p} = (xy,\, x(1-y),\, (1-x)y,\, (1-x)(1-y))$, where the numbers x and y are the probabilities to have the position of the first coin "UP" and of the second coin also "UP." In this case, the system entropy is equal the sum of entropies of subsystems, $H(1,2) = H(1) + H(2) \to I = 0$. A complete correlation means that the position of the first coin determines the position of the second coin.

If the probability distributions are such that the coin positions are completely correlated, e.g., the probability vector $\vec{p} = (p_1, p_2, p_3, p_4) = (x, y, y, x)$ has the form $(x,\, 1/2 - x,\, 1/2 - x,\, x)$, and it just describes this situation. In fact, $(p(1))_1 = 1/2$, $(p(1))_2 = 1/2$, $(p(2))_1 = 1/2$, $(p(2))_2 = 1/2$; this means that the Shannon information $I = 2\ln 2 + 2x\ln x + 2(1/2 - x)\ln(1/2 - x)$. For $x = 1/2$ (or for $x = 0$), the Shannon information is maximum, $I = \ln 2$.

Thus, one can see that for an arbitrary probability vector $\vec{p} = (p_1, p_2, p_3, p_4)$, the vector of the form $(x,\, 1/2 - x,\, 1/2 - x,\, x)$ at $x = 0$ and $x = 1/2$ provides a maximum of the function

$$\begin{aligned} I &= (p_1 \ln p_1 + p_2 \ln p_2 + p_3 \ln p_3 + p_4 \ln p_4) \\ &\quad -(p_1 + p_2)\ln(p_1 + p_2) - (p_3 + p_4)\ln(p_3 + p_4) \\ &\quad -(p_1 + p_3)\ln(p_1 + p_3) - (p_2 + p_4)\ln(p_2 + p_4). \end{aligned} \tag{7}$$

For two coins, complete correlations of the coins described by the probability vector $(x,\, 1/2 - x,\, 1/2 - x,\, x)$ means that, if the first coin is in the position "UP" the second coin is also definitely in the position "UP" (for $x = 1/2$) or, if the first coin is in the position "UP" the second coin is definitely in the position "DOWN" (for $x = 0$). This means that such distribution corresponds to the "entanglement" of classical coins.

The function I can be interpreted as Shannon information for two "artificial coins" associated with a compass. The correlations in the compass orientations analogous to the correlations in positions of two coins provides the possibility to extend the notion of mutual information known for bipartite system to the case of a single (noncomposite) system.

The systems under consideration – "two coins" and "compass" are quite different systems. The former one is a composite system containing two subsystems. The latter one is noncomposite system which does not contain subsystems. But from the viewpoint of mathematical properties, the fair probability distributions describing the results of measuring the states of these systems are identical. This means that all notions based on the properties like entropies, information, equalities and inequalities for entropies and information can be introduced for both composite ("two coins") systems and noncomposite ("compass") system. The difference in the introduced notions is in the interpretation of entropic inequalities, since the same numerical inequalities reflect the presence or absence of correlations in the systems, but the physical nature of the correlations is different.

In composite ("two coins") systems, correlations providing the nonnegativity of the mutual information are correlations between the states of the subsystems (two different coins). In noncomposite ("compass") system, "mutual information" introduced for this system describes the correlations between the states of the same system, i.e., between the "compass" orientations. Specific properties of the correlations are related to the events which correlate. In the case of compass, a pair of compass orientations is one event but, the pairs, in principle, are not independent. Formally, namely, the dependence of pairs of the compass-orientation events is mimicking correlations between the events in experiments with two coins.

Figure 3. Casino roulette.

We make another comment. Different noncomposite systems can exist; for example, in addition to the compass already considered, there exists "casino roulette" with four outputs (Fig. 4). It is different system but the probability description of the casino roulette is identical to the statistics of the compass behavior. The correlations in the casino roulette processing are described by the same entropic-information formula existing for the compass correlations and two-coin correlations.

From this discussion, one can draw the conclusion:

Entropic relations obvious for composite ("two coins") system are found for noncomposite (nondivisible) system, which has no subsystems. For example, if there is no correlations in the behavior of two coins, the entropy of this composite system is the sum of entropies of both subsystems. It is worth noting that completely analogous properties exist for noncomposite system like casino roulette.

In the case of maximum correlations in the behavior of two coins, the position of the first coin determines the position of the second coin, and the mutual information takes a maximum possible value. An analogous property for compass or casino roulette is not obvious but it also exists.

Applying the map between the system states, we can get and clarify the meaning of the inequalities, which are "new" entropic inequalities for noncomposite systems. This idea is the general idea for the approach we are presenting in this work and not only for classical systems but quantum (qudit) systems as well.

4. Quantum subadditivity and strong subadditivity conditions

In previous sections, we discussed classical probability distributions of composite and noncomposite systems and known inequalities. Now we present the relations known for quantum systems and called the subadditivity and strong subadditivity conditions. A basic concept for quantum states, e.g., the qudit state is the density matrix ρ with matrix elements $\rho_{ss'}$, where $s, s' = 1, 2, \ldots, N$. The von Neumann entropy of the state is $S = -\mathrm{Tr}\,(\rho \ln \rho)$. If the system is composite with two subsystems 1 and 2, for example, two qudits, the density matrix $\rho(1,2)$ has matrix elements $\rho_{jk,j'k'}(1,2)$, with $j = 1, 2, \ldots, m$ and $k = 1, 2, \ldots, n$.

The density matrix $\rho(1)$ of the first subsystem is given by the so-called partial tracing tool that provides the matrix elements of this matrix

$$\rho_{jj'}(1) = \sum_{k=1}^{n} \rho_{jk,j'k}(1,2), \qquad j, j' = 1, 2, \ldots, m. \tag{8}$$

For the density matrix $\rho(2)$ of the second subsystem, we have the matrix elements, in view of the partial tracing with respect to indices connected with the first subsystem, i.e.,

$$\rho_{kk'}(2) = \sum_{j=1}^{m} \rho_{jk,jk'}(1,2). \tag{9}$$

The known subadditivity condition for three entropies of composite system and its two subsystems

$$S(1) = -\mathrm{Tr}\,(\rho(1) \ln \rho(1)), \quad S(2) = -\mathrm{Tr}\,(\rho(2) \ln \rho(2)) \quad S(1,2) = -\mathrm{Tr}\,(\rho(1,2) \ln \rho(1,2)) \tag{10}$$

reads

$$S(1) + S(2) \geq S(1, 2), \tag{11}$$

which means nonnegativity of the mutual quantum information

$$I_q = S(1) + S(2) - S(1, 2) \geq 0.$$

For tripartite system with three subsystems 1,2,3, for example, three qudits, the density matrix $\rho(1, 2, 3)$ has matrix elements $\rho_{jk\ell,j'k'\ell'}(1, 2, 3)$, with $j, j' = 1, 2, \ldots, m$, $k, k' = 1, 2, \ldots, n$, and $l, l' = 1, 2, \ldots, s$. The density matrices of the subsystems have matrix elements obtained from these ones by the partial tracing tool.

The density matrix $\rho(1, 2)$ of subsystems 1 and 2 reads $\rho_{jk,j'k'}(1, 2) = \sum_{l=1}^{s} \rho_{jkl,j'k'l}(1, 2, 3)$, and $\rho(2, 3)$ of subsystems 2 and 3 is $\rho_{kl,k'l'}(2, 3) = \sum_{j=1}^{m} \rho_{jkl,jk'l'}(1, 2, 3)$; for the density matrix $\rho(2)$ of the second subsystem, we have the matrix elements $\rho_{kk'}(2) = \sum_{l=1}^{s} \rho_{kl,k'l}(2, 3)$.

The quantum strong subadditivity condition is known as the entropic inequality

$$S(1, 2) + S(2, 3) \geq S(1, 2, 3) + S(2),$$

where the von Neumann entropies are

$$\begin{aligned} S(1, 2) &= -\mathrm{Tr}\,(\rho(1, 2) \ln \rho(1, 2)), \qquad S(2, 3) = -\mathrm{Tr}\,(\rho(2, 3) \ln \rho(2, 3)), \\ S(2) &= -\mathrm{Tr}\,(\rho(2) \ln \rho(2)), \qquad S(1, 2, 3) = -\mathrm{Tr}\,(\rho(1, 2, 3) \ln \rho(1, 2, 3)). \end{aligned} \tag{12}$$

The nonnegative number $I_{\rm con} = S(1, 2) + S(2, 3) - S(1, 2, 3) - S(2)$ is called the conditional information.

Our approach is to employ the same map of indices, which we used for classical probability distributions in previous sections, but now we apply the map to the density matrix $\rho_{ss'}$ of a single qudit (noncomposite system) and obtain the corresponding entropic inequalities. This means that we interpret the matrix $\rho_{ss'}$ either as the density matrix of bipartite system, $\rho_{ss'} \leftrightarrow \rho_{jk,j'k'}$, or as the matrix of tripartite system, $\rho_{ss'} \leftrightarrow \rho_{jk\ell,j'k'\ell'}$.

In the following sections, we give particular examples of the approach.

5. Hidden correlations in bipartite systems

In previous sections, we discussed hidden correlations in systems with one random variable. We can associate the tool of partitions, which provides the possibility to map the set of indices (integers) $s = 1, 2, \ldots N$, $N = n_1 n_2$ onto pairs of indices (integers) (j, k), $j = 1, 2, \ldots, n_1$, $k = 1, 2, \ldots, n_2$, and the tool can be used to associate a bipartite system, say, of two qudits with a tripartite system (with three artificial subsystems). Then we introduce the notion of hidden correlations not for a single qudit but for a system of two qudits. The correlations existing in these bipartite systems are characterized by strong subadditivity conditions known for composite tripartite systems.

Such approach can be developed if the integer number N can be presented in two forms — one form is the mentioned product form $N = n_1 n_2$, and the other form is $N = m_1 m_2 m_3 = n_1 n_2$, where m_1, m_2, and m_3 are integers. This equality permits to consider the density matrix $\rho_{jk,j'k'}$ of two-qudit state as the density matrix $\rho_{mn\ell,m'n'\ell'}$ of three-qudit state, where three integer indices $mn\ell, m'n'\ell'$ are taken from domains $1 \leq m, m' \leq m_1$, $1 \leq n, n' \leq m_2$, and $1 \leq \ell, \ell' \leq m_3$. We can construct this map since there exists the bijective correspondence of the matrix elements $\rho_{ss'} \Leftrightarrow \rho_{jk,j'k'}$ and $\rho_{ss'} \Leftrightarrow \rho_{mn\ell,m'n'\ell'}$, where $1 \leq s, s' \leq N$, $1 \leq j, j' \leq n_1$, and $1 \leq k, k' \leq n_2$.

In view of this fact, if one has the density matrix of bipartite system $\rho_{jk,j'k'}$, this density matrix can be interpreted first as the density matrix of noncomposite system $\rho_{ss'}$, which is the same numerical $N\times N$-matrix. Then, this density matrix (again the same numerical matrix) can be interpreted as the density matrix of tripartite quantum state. For tripartite quantum states, we have the strong subadditivity condition for von Neumann entropies of the system and its three subsystems. The entropic inequality for this composite system with three artificial qudits (the initial density matrix of bipartite system is the matrix corresponding to two qudits) provides the relation (inequality) for the matrix elements of the density matrix of this two-qudit state. This inequality characterizes hidden three-partite-like correlations in bipartite quantum systems.

We give the example of such inequality for the system containing the spin-1/2 subsystem and the spin-3/2 subsystem; the density 8×8-matrix of this system has matrix elements $\rho_{m_1m_2,m_1'm_2'}$, where $m_1, m_1' = -1/2, +1/2$ and $m_2, m_2' = -3/2, -1/2, +1/2, +3/2$. We employ the map of indices $ss' \Leftrightarrow mm'$ of the form

$$\begin{array}{llll} -1/2\;\; -3/2 \leftrightarrow 1, & -1/2\;\; -1/2 \leftrightarrow 2, & -1/2\;\; +1/2 \leftrightarrow 3, & -1/2\;\; +3/2 \leftrightarrow 4, \\ +1/2\;\; -3/2 \leftrightarrow 5, & +1/2\;\; -1/2 \leftrightarrow 6, & +1/2\;\; +1/2 \leftrightarrow 7, & +1/2\;\; +3/2 \leftrightarrow 8, \end{array}$$

and interpret the matrix $\rho_{m_1m_2,m_1'm_2'}$ as the density matrix of noncomposite system $\rho_{ss'}$ corresponding to the qudit system with $j = 7/2$. Then we apply another map of indices, namely, $\rho_{ss'} \Leftrightarrow \rho_{M_1M_2M_3,M_1'M_2'M_3'}$, where $M_1, M_2, M_3, M_1', M_2'$, and M_3' take values $\pm 1/2$, i.e.,

$$\begin{array}{lll} 1 \leftrightarrow +1/2\;\; +1/2\;\; +1/2, & 2 \leftrightarrow +1/2\;\; +1/2\;\; -1/2, & 3 \leftrightarrow +1/2\;\; -1/2\;\; +1/2, \\ 4 \leftrightarrow +1/2\;\; -1/2\;\; -1/2, & 5 \leftrightarrow -1/2\;\; +1/2\;\; +1/2, & 6 \leftrightarrow -1/2\;\; +1/2\;\; -1/2, \\ & 7 \leftrightarrow -1/2\;\; -1/2\;\; +1/2, & 8 \leftrightarrow -1/2\;\; -1/2\;\; -1/2. \end{array}$$

Producing the partial trace procedure with the density matrix of three artificial qubits, we arrive at the density matrices $\rho_{M_1M_2,M_1'M_2'}(1,2)$, $\rho_{M_2M_3,M_2'M_3'}(2,3)$, and $\rho_{M_2M_2'}(2)$ of the qubit subsystems; in terms of matrix elements $\rho_{ss'}$, the matrices read

$$\begin{aligned} \rho(1,2) &= \begin{pmatrix} \rho_{11}+\rho_{22} & \rho_{13}+\rho_{24} & \rho_{15}+\rho_{26} & \rho_{17}+\rho_{28} \\ \rho_{31}+\rho_{42} & \rho_{33}+\rho_{44} & \rho_{25}+\rho_{46} & \rho_{27}+\rho_{48} \\ \rho_{51}+\rho_{62} & \rho_{52}+\rho_{64} & \rho_{55}+\rho_{66} & \rho_{57}+\rho_{68} \\ \rho_{31}+\rho_{82} & \rho_{72}+\rho_{84} & \rho_{75}+\rho_{86} & \rho_{77}+\rho_{88} \end{pmatrix}, \\ \rho(2,3) &= \begin{pmatrix} \rho_{11}+\rho_{55} & \rho_{12}+\rho_{56} & \rho_{13}+\rho_{57} & \rho_{14}+\rho_{58} \\ \rho_{21}+\rho_{65} & \rho_{22}+\rho_{66} & \rho_{23}+\rho_{67} & \rho_{24}+\rho_{68} \\ \rho_{31}+\rho_{75} & \rho_{51}+\rho_{76} & \rho_{33}+\rho_{77} & \rho_{34}+\rho_{78} \\ \rho_{41}+\rho_{85} & \rho_{42}+\rho_{86} & \rho_{43}+\rho_{87} & \rho_{44}+\rho_{88} \end{pmatrix}, \\ \rho(2) &= \begin{pmatrix} \rho_{11}+\rho_{22}+\rho_{55}+\rho_{66} & \rho_{13}+\rho_{24}+\rho_{57}+\rho_{68} \\ \rho_{11}+\rho_{42}+\rho_{75}+\rho_{86} & \rho_{33}+\rho_{44}+\rho_{77}+\rho_{98} \end{pmatrix}. \end{aligned} \tag{13}$$

The hidden correlations in the bipartite system with spin $-1/2$ and spin $-3/2$ subsystems satisfy the strong subadditivity condition

$$-\mathrm{Tr}\,\rho\ln\rho - \mathrm{Tr}\,\rho(2)\ln\rho(2) \leq -\mathrm{Tr}\,\rho(1,2)\ln\rho(1,2) - \mathrm{Tr}\,\rho(2,3)\ln\rho(2,3); \tag{14}$$

thus, we obtain the new entropic inequality for the bipartite system containing the two-level atom and the four-level atom as subsystems. The hidden correlations in this bipartite system correspond to correlations of artificial qubits. If the considered density matrix ρ describes completely chaotic states, the off-diagonal elements in matrices $\rho(1,2)$, $\rho(2,3)$, and $\rho(2)$ are equal to zero, and the inequality obtained corresponds to the inequality for classical probability distribution $P(s) = \rho_{ss}$ given explicitly as follows:

$$
\begin{aligned}
&-\sum_{s=1}^{8} \rho_{ss} \ln \rho_{ss} - (P(1)+P(2)+P(5)+P(6)) \ln (P(1)+P(2)+P(5)+P(6)) \\
&-(P(3)+P(4)+P(7)+P(8)) \ln (P(3)+P(4)+P(7)+P(8)) \\
&\leq -(P(1)+P(2)) \ln (P(1)+P(2)) - (P(3)+P(4)) \ln (P(3)+P(4)) \\
&-(P(5)+P(6)) \ln (P(5)+P(6)) - (P(7)+P(8)) \ln (P(7)+P(8)) \\
&-(P(1)+P(5)) \ln (P(1)+P(5)) - (P(2)+P(6)) \ln (P(2)+P(6)) \\
&-(P(3)+P(7)) \ln (P(3)+P(7)) - (P(4)+P(8)) \ln (P(4)+P(8)). \qquad (15)
\end{aligned}
$$

All inequalities obtained from the above one by arbitrary permutations of the integers $1, 2, 3, 4, 5, 6, 7, 8$ are also valid.

Thus, we derived some entropic inequalities valid for an arbitrary Hermitian 8×8-matrix $\rho = \rho^{\dagger}$, $\mathrm{Tr}\,\rho = 1$, and $\rho \geq 0$, but the physical meaning of the inequalities depends on the interpretation of this matrix. For the case of three-qubit states, the entropic inequality is the standard strong subadditivity condition characterizing the degree of quantum correlations of the three qubits. In the classical case, say, if one has the system – a coin plus the compass, the obtained inequality characterizes the degree of correlations of three artificial classical coins for which their joint probability distribution of three random variables is mapped onto the joint probability distribution of two random variables (one coin and one compass). The entropic inequalities obtained can be checked in experiments with superconducting circuits discussed in [22–32].

6. New entropic inequality for the five-level atom (spin j = 2)

First, we formulate, without derivation, new entropic-information inequality for qudit (spin $j = 2$) or the five-level atom. Then we show how to obtain these inequalities.

Qudit with $j = 2$ has five states. These states can be modeled by states of a five-level atom. The atomic states can be realized by superconducting circuit states based on the Josephson junction. On the other hand, these states can be considered as states of qubit–qutrit having six states but one of these six states is not occupied.

In Fig. 4, the atomic levels are labeled either by numbers $1, 2, 3, 4, 5$ or by pairs of numbers $(2, 2)$, $(2, 1)$, $(2, 0)$, $(2, -1)$, $(2, -2)$ corresponding to labeling the states of spin $j = 2$ with spin projections $m = 2, 1, 0, -1, -2$.

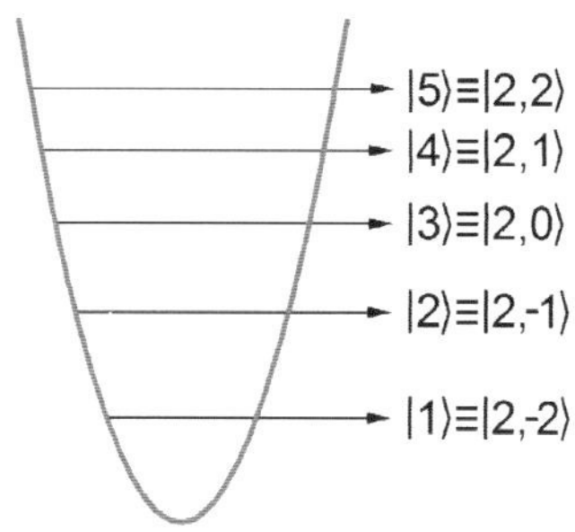

Figure 4. Qubit–qutrit model of the five-level atom $\mid n\rangle \equiv \mid j, m\rangle$, $j = 2$.

Thus, we use the invertible map of the labels

$$1 \leftrightarrow (2,2), \quad 2 \leftrightarrow (2,1), \quad 3 \leftrightarrow (2,0), \quad 4 \leftrightarrow (2,-1), \quad 5 \leftrightarrow (2,-2).$$

For the density matrix of the five-level atom, we can use both possibilities to label the matrix elements.

The qudit (spin $j = 2$) density matrix $R_{mm'}$ $(m, m' = -2, -1, 0, 1, 2)$ realized as states of the five-level atom reads

$$
R_{mm'} = \begin{pmatrix}
R_{22} & R_{21} & R_{20} & R_{2\,-1} & R_{2\,-2} \\
R_{12} & R_{11} & R_{10} & R_{1\,-1} & R_{1\,-2} \\
R_{02} & R_{01} & R_{00} & R_{0\,-1} & R_{0\,-2} \\
R_{-12} & R_{-11} & R_{-10} & R_{-1\,-1} & R_{-1\,-2} \\
R_{-22} & R_{-21} & R_{-20} & R_{-2\,-1} & R_{-2\,-2}
\end{pmatrix}
\equiv
\begin{pmatrix}
\widetilde{R}_{11} & \widetilde{R}_{12} & \widetilde{R}_{13} & \widetilde{R}_{14} & \widetilde{R}_{15} \\
\widetilde{R}_{21} & \widetilde{R}_{22} & \widetilde{R}_{23} & \widetilde{R}_{24} & \widetilde{R}_{25} \\
\widetilde{R}_{31} & \widetilde{R}_{32} & \widetilde{R}_{33} & \widetilde{R}_{34} & \widetilde{R}_{35} \\
\widetilde{R}_{41} & \widetilde{R}_{42} & \widetilde{R}_{43} & \widetilde{R}_{44} & \widetilde{R}_{45} \\
\widetilde{R}_{51} & \widetilde{R}_{52} & \widetilde{R}_{53} & \widetilde{R}_{54} & \widetilde{R}_{55}
\end{pmatrix}.
$$

Using the other notation for the matrix R and its indices, i.e., $R_{mm'} \leftrightarrow \widetilde{R}_{jk}$, the matrix $\widetilde{R}$ can be interpreted as the density matrix of the five-level atom.

One can construct two density matrices of artificial "atoms" (qubit and qutrit). In terms of matrix elements $R_{mm'}$, the density matrix of the qubit (two-level "atom") state is

$$r_{MM'} = \begin{pmatrix} R_{22} + R_{00} + R_{-2\,-2} & R_{21} + R_{1\,-1} \\ R_{12} + R_{-11} & R_{11} + R_{-1\,-1} \end{pmatrix}, \qquad M, M' = 1/2, -1/2.$$

The density matrix of the qutrit ("three-level" atom) state reads

$$\rho_{\mu\mu'} = \begin{pmatrix} R_{22} + R_{11} & R_{20} + R_{1\,-1} & R_{2\,-2} \\ R_{02} + R_{-11} & R_{00} + R_{-1\,-1} & R_{0\,-2} \\ R_{-22} & R_{-20} & R_{-2\,-2} \end{pmatrix}, \qquad \mu, \mu' = 1, 0, -1.$$

Now we are in the position to write some entropic inequalities for these three density matrices, namely, $R_{mm'}$; $m, m' = -2, -1, 0, 1, 2$ of qudit (spin $j = 2$) state realized by the state of superconducting circuit based on Josephson junction, $r_{MM'}$; $M, M' = 1/2, -1/21$ of the qubit (two-level "atom") state, and $\rho_{\mu\mu'}$; $\mu, \mu' = 1, 0, -1$ of the qutrit (three-level "atom") state.

The subadditivity condition for von Neumann entropies determined by the three density matrices reads

$$-\mathrm{Tr}\,(R \ln R) \leq -\mathrm{Tr}\,(r \ln r) - \mathrm{Tr}\,(\rho \ln \rho)\,.$$

An analog of the Araki–Lieb inequality known for bipartite quantum systems is

$$-\mathrm{Tr}\,(R \ln R) \geq \left|-\mathrm{Tr}\,(r \ln r) + \mathrm{Tr}\,(\rho \ln \rho)\right|.$$

For the mutual information of the five-level atomic system, we obtain

$$I = \mathrm{Tr}\,(R \ln R) - \mathrm{Tr}\,(r \ln r) - \mathrm{Tr}\,(\rho \ln \rho) \geq 0.$$

The inequalities can be checked for arbitrary density 5×5-matrices $\widetilde{R}_{jk}$; $j, k = 1, 2, 3, 4, 5$, for example, if it is obtained in experiments with superconducting qudits. The approach with such ideas was employed in [30–32].

Now we present another inequality, which is an analog of the conditional von Neumann information nonnegativity known for tripartite quantum systems; we write it for entropies of the following density matrices:

The density 4×4-matrix $\widetilde{R}_{\mu\mu'}$; $(\mu, \mu' = 3/2, 1/2, -1/2, -3, 2)$ of an artificial qudit (spin $j = 3/2$) expressed in terms of matrix elements $R_{mm'}$ reads

$$\widetilde{R}_{\mu\mu'} = \begin{pmatrix} R_{22} + R_{-2\,-2} & R_{21} & R_{20} & R_{2\,-1} \\ R_{12} & R_{11} & R_{10} & R_{1\,-1} \\ R_{02} & R_{01} & R_{00} & R_{0\,-1} \\ R_{-12} & R_{-11} & R_{-10} & R_{-1\,-1} \end{pmatrix}.$$

The density matrix $\widetilde{\rho}_{mm'}$; $m, m' = 1, 0, -1$ of an artificial qutrit (spin $j = 1$) expressed in terms of matrix elements $R_{mm'}$ is

$$\widetilde{r}_{mm'} = \begin{pmatrix} R_{22} + R_{11} & R_{20} + R_{1\,-1} & R_{2\,-2} \\ R_{02} + R_{-11} & R_{00} + R_{-1\,-1} & R_{0\,-2} \\ R_{-22} & R_{-20} & R_{-2\,-2} \end{pmatrix}.$$

The density matrix $\widetilde{\rho}_{mm'}$; $(m, m' = \pm 1/2)$ of an artificial qubit is

$$\widetilde{\rho}_{mm'} = \begin{pmatrix} R_{11} + R_{22} + R_{-2\,-2} & R_{20} + R_{1\,-1} \\ R_{02} + R_{-11} & R_{00} + R_{-1\,-1} \end{pmatrix}.$$

IOP Conf. Series: Journal of Physics: Conf. Series **1071** (2018) 012015 doi:10.1088/1742-6596/1071/1/012015

The new conditional information nonnegativity (an analog of the strong subadditivity condition) for the five-level atomic state reads

$$I_{\rm con} = -\mathrm{Tr}\,\widetilde{\rho}\ln\widetilde{\rho} - \mathrm{Tr}\,\widetilde{R}\ln\widetilde{R} + \mathrm{Tr}\,R\ln R + \mathrm{Tr}\,\widetilde{r}\ln\widetilde{r} \geq 0; \qquad (16)$$

it can be checked experimentally, if one measures the density matrix of five-level "atom" or qudit with $j=2$ in experiments with superconducting circuits.

For completely decoherent state with diagonal matrix R, relation (16) yields the inequality for the population of the five-level atom,

$$\begin{aligned} I_{\rm con} &= -(R_{22}+R_{11})\ln(R_{22}+R_{11}) - (R_{22}+R_{-2\,-2})\ln(R_{22}+R_{-2\,-2}) \\ &\quad +R_{22}\ln R_{22} + (R_{22}+R_{11}+R_{-2\,-2})\ln(R_{22}+R_{11}+R_{-2\,-2}) \geq 0. \end{aligned} \qquad (17)$$

One can see that, in this case, the conditional information depends only on the three parameters R_{11}, R_{22}, and $R_{-2\,-2}$. It is a new matrix inequality.

The inequalities obtained for the density matrix of the five-level atom can be formulated as mathematical inequalities for an arbitrary 5×5-matrix ρ such that $\rho^\dagger = \rho$, $\mathrm{Tr}\,\rho = 1$, and $\rho \geq 0$, i.e., eigenvalues of the matrix are nonnegative.

For example, for such 5×5-matrix ρ, the mathematical subadditivity condition reads

$$\begin{aligned} &-\mathrm{Tr}\begin{pmatrix} \rho_{11} & \rho_{12} & \rho_{13} & \rho_{14} & \rho_{15} \\ \rho_{21} & \rho_{22} & \rho_{23} & \rho_{24} & \rho_{25} \\ \rho_{31} & \rho_{32} & \rho_{33} & \rho_{34} & \rho_{35} \\ \rho_{41} & \rho_{42} & \rho_{43} & \rho_{44} & \rho_{45} \\ \rho_{51} & \rho_{52} & \rho_{53} & \rho_{54} & \rho_{55} \end{pmatrix} \ln \begin{pmatrix} \rho_{11} & \rho_{12} & \rho_{13} & \rho_{14} & \rho_{15} \\ \rho_{21} & \rho_{22} & \rho_{23} & \rho_{24} & \rho_{25} \\ \rho_{31} & \rho_{32} & \rho_{33} & \rho_{34} & \rho_{35} \\ \rho_{41} & \rho_{42} & \rho_{43} & \rho_{44} & \rho_{45} \\ \rho_{51} & \rho_{52} & \rho_{53} & \rho_{54} & \rho_{55} \end{pmatrix} \\ &\leq -\mathrm{Tr}\begin{pmatrix} \rho_{11}+\rho_{33}+\rho_{55} & \rho_{12}+\rho_{34} \\ \rho_{21}+\rho_{43} & \rho_{22}+\rho_{44} \end{pmatrix} \ln \begin{pmatrix} \rho_{11}+\rho_{33}+\rho_{55} & \rho_{12}+\rho_{34} \\ \rho_{21}+\rho_{43} & \rho_{22}+\rho_{44} \end{pmatrix} \\ &\quad -\mathrm{Tr}\begin{pmatrix} \rho_{11}+\rho_{22} & \rho_{13}+\rho_{24} & \rho_{15} \\ \rho_{31}+\rho_{24} & \rho_{33}+\rho_{44} & \rho_{35} \\ \rho_{51} & \rho_{53} & \rho_{55} \end{pmatrix} \ln \begin{pmatrix} \rho_{11}+\rho_{22} & \rho_{13}+\rho_{24} & \rho_{15} \\ \rho_{31}+\rho_{24} & \rho_{33}+\rho_{44} & \rho_{35} \\ \rho_{51} & \rho_{53} & \rho_{55} \end{pmatrix}. \end{aligned} \qquad (18)$$

The mathematical strong subadditivity condition for such 5×5-matrix ρ is

$$\begin{aligned} &-\mathrm{Tr}\begin{pmatrix} \rho_{11} & \rho_{12} & \rho_{13} & \rho_{14} & \rho_{15} \\ \rho_{21} & \rho_{22} & \rho_{23} & \rho_{24} & \rho_{25} \\ \rho_{31} & \rho_{32} & \rho_{33} & \rho_{34} & \rho_{35} \\ \rho_{41} & \rho_{42} & \rho_{43} & \rho_{44} & \rho_{45} \\ \rho_{51} & \rho_{52} & \rho_{53} & \rho_{54} & \rho_{55} \end{pmatrix} \ln \begin{pmatrix} \rho_{11} & \rho_{12} & \rho_{13} & \rho_{14} & \rho_{15} \\ \rho_{21} & \rho_{22} & \rho_{23} & \rho_{24} & \rho_{25} \\ \rho_{31} & \rho_{32} & \rho_{33} & \rho_{34} & \rho_{35} \\ \rho_{41} & \rho_{42} & \rho_{43} & \rho_{44} & \rho_{45} \\ \rho_{51} & \rho_{52} & \rho_{53} & \rho_{54} & \rho_{55} \end{pmatrix} \\ &-\mathrm{Tr}\begin{pmatrix} \rho_{11}+\rho_{22}+\rho_{55} & \rho_{13}+\rho_{24} \\ \rho_{31}+\rho_{42} & \rho_{33}+\rho_{44} \end{pmatrix} \ln \begin{pmatrix} \rho_{11}+\rho_{22}+\rho_{55} & \rho_{13}+\rho_{24} \\ \rho_{31}+\rho_{42} & \rho_{33}+\rho_{44} \end{pmatrix} \\ &\leq -\mathrm{Tr}\begin{pmatrix} \rho_{11}+\rho_{55} & \rho_{12}+\rho_{13} & \rho_{14} & \\ \rho_{21} & \rho_{22} & \rho_{23} & \rho_{24} \\ \rho_{31} & \rho_{32} & \rho_{33} & \rho_{34} \\ \rho_{41}+\rho_{42} & \rho_{43}+\rho_{44} & & \end{pmatrix} \ln \begin{pmatrix} \rho_{11}+\rho_{55} & \rho_{12}+\rho_{13} & \rho_{14} & \\ \rho_{21} & \rho_{22} & \rho_{23} & \rho_{24} \\ \rho_{31} & \rho_{32} & \rho_{33} & \rho_{34} \\ \rho_{41}+\rho_{42} & \rho_{43}+\rho_{44} & & \end{pmatrix} \\ &\quad -\mathrm{Tr}\begin{pmatrix} \rho_{11}+\rho_{22} & \rho_{13}+\rho_{24} & \rho_{15} \\ \rho_{31}+\rho_{42} & \rho_{33}+\rho_{44} & \rho_{35} \\ \rho_{51} & \rho_{53} & \rho_{55} \end{pmatrix} \ln \begin{pmatrix} \rho_{11}+\rho_{22} & \rho_{13}+\rho_{24} & \rho_{15} \\ \rho_{31}+\rho_{42} & \rho_{33}+\rho_{44} & \rho_{35} \\ \rho_{51} & \rho_{53} & \rho_{55} \end{pmatrix}. \end{aligned} \qquad (19)$$

Inequalities (18) and (19) are new matrix inequalities.

7. Qubit–qutrit state

Now we consider the qubit–qutrit system and explain how the results for qudit with $j = 2$ are obtained. The pure states $\mid m\,M\rangle$ of the system consisted of qubit (spin $s_1 = 1/2$) and qutrit (spin $s_2 = 1$), where spin projections are $m = \pm 1/2$ and $M = 1, 0, -1$, can be described also by the density 6×6-matrix $\rho(1,2)$ with matrix elements $\rho_{mM,m'M'}(1,2)$ of the form

$$\rho_{mM,m'M'}(1,2) =$$
$$\begin{pmatrix}
\rho_{\frac{1}{2}1,\frac{1}{2}1} & \rho_{\frac{1}{2}1,\frac{1}{2}0} & \rho_{\frac{1}{2}1,\frac{1}{2}-1} & \rho_{\frac{1}{2}1,-\frac{1}{2}1} & \rho_{\frac{1}{2}1,-\frac{1}{2}0} & \rho_{\frac{1}{2}1,-\frac{1}{2}-1} \\
\rho_{\frac{1}{2}0,\frac{1}{2}1} & \rho_{\frac{1}{2}0,\frac{1}{2}0} & \rho_{\frac{1}{2}0,\frac{1}{2}-1} & \rho_{\frac{1}{2}0,-\frac{1}{2}1} & \rho_{\frac{1}{2}0,-\frac{1}{2}0} & \rho_{\frac{1}{2}0,-\frac{1}{2}-1} \\
\rho_{\frac{1}{2}-1,\frac{1}{2}1} & \rho_{\frac{1}{2}-1,\frac{1}{2}0} & \rho_{\frac{1}{2}-1,\frac{1}{2}-1} & \rho_{\frac{1}{2}-1,-\frac{1}{2}1} & \rho_{\frac{1}{2}-1,-\frac{1}{2}0} & \rho_{\frac{1}{2}-1,-\frac{1}{2}-1} \\
\rho_{-\frac{1}{2}1,\frac{1}{2}1} & \rho_{-\frac{1}{2}1,\frac{1}{2}0} & \rho_{-\frac{1}{2}1,\frac{1}{2}-1} & \rho_{-\frac{1}{2}1,-\frac{1}{2}1} & \rho_{-\frac{1}{2}1,-\frac{1}{2}0} & \rho_{-\frac{1}{2}1,-\frac{1}{2}-1} \\
\rho_{-\frac{1}{2}0,\frac{1}{2}1} & \rho_{-\frac{1}{2}0,\frac{1}{2}0} & \rho_{-\frac{1}{2}0,\frac{1}{2}-1} & \rho_{-\frac{1}{2}0,-\frac{1}{2}1} & \rho_{-\frac{1}{2}0,-\frac{1}{2}0} & \rho_{-\frac{1}{2}0,-\frac{1}{2}-1} \\
\rho_{-\frac{1}{2}-1,\frac{1}{2}1} & \rho_{-\frac{1}{2}-1,\frac{1}{2}0} & \rho_{-\frac{1}{2}-1,\frac{1}{2}-1} & \rho_{-\frac{1}{2}-1,-\frac{1}{2}1} & \rho_{-\frac{1}{2}-1,-\frac{1}{2}0} & \rho_{-\frac{1}{2}-1,-\frac{1}{2}-1}
\end{pmatrix}.$$

The density 2×2-matrix of qubit $\rho_{mm'}(1)$ is obtained by the partial trace of the matrix $\rho_{mM,m'M'}(1,2)$, i.e., $\rho_{mm'}(1) = \sum_{M=-1}^{1} \rho_{mM,m'M}(1,2)$. The density 3×3-matrix of qutrit $\rho_{MM'}(2)$ is obtained by the partial trace of the matrix $\rho_{mM,m'M'}(1,2)$, i.e., $\rho_{MM'}(2) = \sum_{m=-1/2}^{1/2} \rho_{mM,mM'}(1,2)$. These matrices read

$$\rho_{mm'}(1) = \begin{pmatrix}
\rho_{\frac{1}{2}1,\frac{1}{2}1} + \rho_{\frac{1}{2}0,\frac{1}{2}0} + \rho_{\frac{1}{2}-1,\frac{1}{2}-1} & \rho_{\frac{1}{2}1,-\frac{1}{2}1} + \rho_{\frac{1}{2}0,-\frac{1}{2}0} + \rho_{\frac{1}{2}-1,-\frac{1}{2}-1} \\
\rho_{-\frac{1}{2}1,\frac{1}{2}1} + \rho_{-\frac{1}{2}0,\frac{1}{2}0} + \rho_{-\frac{1}{2}1,\frac{1}{2}-1} & \rho_{-\frac{1}{2}1,-\frac{1}{2}1} + \rho_{-\frac{1}{2}0,-\frac{1}{2}0} + \rho_{-\frac{1}{2}-1,-\frac{1}{2}-1}
\end{pmatrix},$$

$$\rho_{MM'}(2) = \begin{pmatrix}
\rho_{\frac{1}{2}1,\frac{1}{2}1} + \rho_{-\frac{1}{2}1,-\frac{1}{2}1} & \rho_{\frac{1}{2}1,\frac{1}{2}0} + \rho_{-\frac{1}{2}1,-\frac{1}{2})} & \rho_{\frac{1}{2}1,\frac{1}{2}-1} + \rho_{-\frac{1}{2}1,-\frac{1}{2}-1} \\
\rho_{\frac{1}{2}0,\frac{1}{2}1} + \rho_{-\frac{1}{2}0,-\frac{1}{2}1} & \rho_{\frac{1}{2}0,\frac{1}{2}0} + \rho_{-\frac{1}{2}0,-\frac{1}{2}0} & \rho_{\frac{1}{2}0,\frac{1}{2}-1} + \rho_{-\frac{1}{2}0,-\frac{1}{2}-1} \\
\rho_{\frac{1}{2}-1,\frac{1}{2}1} + \rho_{-\frac{1}{2}-1,-\frac{1}{2}1} & \rho_{\frac{1}{2}-1,\frac{1}{2}0} + \rho_{-\frac{1}{2}-1,-\frac{1}{2}0} & \rho_{\frac{1}{2}-1,\frac{1}{2}-1} + \rho_{-\frac{1}{2}-1,-\frac{1}{2}-1}
\end{pmatrix}.$$

The matrices $\rho(1,2)$, $\rho(1)$, and $\rho(2)$ are known to satisfy the subadditivity condition, which means the nonnegativity of the von Neumann mutual information,

$$I_q = \mathrm{Tr}\left(\rho(1,2)\ln\rho(1,2)\right) - \mathrm{Tr}\left(\rho(1)\ln\rho(1)\right) - \mathrm{Tr}\left(\rho(2)\ln\rho(2)\right) \geq 0.$$

If there is no correlations, i.e., $\rho(1,2) = \rho(1) \otimes \rho(2)$, the mutual information is equal to zero.

8. Single qudit with j = 2

Now we consider the density 5×5-matrix $\rho_{\mu\mu'} = \begin{pmatrix}
\rho_{22} & \rho_{21} & \rho_{20} & \rho_{2-1} & \rho_{2-2} \\
\rho_{12} & \rho_{11} & \rho_{10} & \rho_{1-1} & \rho_{1-2} \\
\rho_{02} & \rho_{01} & \rho_{00} & \rho_{0-1} & \rho_{0-2} \\
\rho_{-12} & \rho_{-11} & \rho_{-10} & \rho_{-1-1} & \rho_{-1-2} \\
\rho_{-22} & \rho_{-21} & \rho_{-20} & \rho_{-2-1} & \rho_{-2-2}
\end{pmatrix}$ of the single-qudit state with $j = 2$, where $\mu, \mu' = 2, 1, 0, -1, -2$, and explain how we obtained the relations of section 6. To reduce the problem to the consideration of the 6×6-matrix, we construct the 6×6 matrix $\rho_{mM,m'M'}(1,2)$ by adding a zero column and a zero row, i.e.,

$$\rho_{mM,m'M'}(1,2) = \begin{pmatrix}
\rho_{22} & \rho_{21} & \rho_{20} & \rho_{2-1} & \rho_{2-2} & 0 \\
\rho_{12} & \rho_{11} & \rho_{10} & \rho_{1-1} & \rho_{1-2} & 0 \\
\rho_{02} & \rho_{01} & \rho_{00} & \rho_{0-1} & \rho_{0-2} & 0 \\
\rho_{-12} & \rho_{-11} & \rho_{-10} & \rho_{-1-1} & \rho_{-1-2} & 0 \\
\rho_{-22} & \rho_{-21} & \rho_{-20} & \rho_{-2-1} & \rho_{-2-2} & 0 \\
0 & 0 & 0 & 0 & 0 & 0
\end{pmatrix} \equiv \begin{pmatrix} R & 0 \\ 0 & 0 \end{pmatrix}.$$

IOP Conf. Series: Journal of Physics: Conf. Series **1071** (2018) 012015 doi:10.1088/1742-6596/1071/1/012015

After this, we construct matrices $\rho_{mm'}(1)$ and $\rho_{MM'}(2)$ in the same way, as we did for composite system, and obtain the matrices which numerically are the same as we already considered. This means that matrices $\rho(1,2) \equiv R$, $\rho(1) \equiv r$, and $\rho(2) \equiv \rho$ satisfy the entropic inequality or the subadditivity condition we found.

Now we are in the position to derive another inequality by constructing 8×8-matrix, which we interpret as a specific density matrix of three artificial "qubits" by adding 3 zero columns and 3 zero rows, i.e.,

$$\rho_{m_1m_2m_3,m_1'm_2'm_3'}(1,2,3) = \begin{pmatrix} \rho_{22} & \rho_{21} & \rho_{20} & \rho_{2\,-1} & \rho_{2\,-2} & 0 & 0 & 0 \\ \rho_{12} & \rho_{11} & \rho_{10} & \rho_{1\,-1} & \rho_{1\,-2} & 0 & 0 & 0 \\ \rho_{02} & \rho_{01} & \rho_{00} & \rho_{0\,-1} & \rho_{0\,-2} & 0 & 0 & 0 \\ \rho_{-12} & \rho_{-11} & \rho_{-10} & \rho_{-1\,-1} & \rho_{-1\,-2} & 0 & 0 & 0 \\ \rho_{-22} & \rho_{-21} & \rho_{-20} & \rho_{-2\,-1} & \rho_{-2\,-2} & 0 & 0 & 0 \\ 0 & 0 & 0 & 0 & 0 & 0 & 0 & 0 \\ 0 & 0 & 0 & 0 & 0 & 0 & 0 & 0 \\ 0 & 0 & 0 & 0 & 0 & 0 & 0 & 0 \end{pmatrix},$$

where $(m_j, m_{j'} = \pm 1/2)$.

After this, we construct matrices $\rho(1,2) \equiv \widetilde{R}$, $\rho(2,3) \equiv \widetilde{r}$, and $\rho(2) \equiv \widetilde{\rho}$ of "subsystems" by partial tracing. Numerically they are the same matrices as were obtained for $j = 2$ qudit; this means that they satisfy the strong subadditivity condition, which we have already described and presented in the previous section as the nonnegativity of conditional mutual information $I_{\rm con}$ for the five-level atom.

As a particular result, the strong subadditivity condition applied to diagonal density matrix of the five-level atom yields the inequality for any three-level populations $\mathcal{P}_1$, $\mathcal{P}_2$, and $\mathcal{P}_3$ of the atom of the form

$$\begin{aligned} I_{\rm con} &= (\mathcal{P}_1 + \mathcal{P}_2)\ln(\mathcal{P}_1 + \mathcal{P}_2) + (\mathcal{P}_1 + \mathcal{P}_3)\ln(\mathcal{P}_1 + \mathcal{P}_3) \\ &\quad -\mathcal{P}_1 \ln \mathcal{P}_1 - (\mathcal{P}_1 + \mathcal{P}_2 + \mathcal{P}_3)\ln(\mathcal{P}_1 + \mathcal{P}_2 + \mathcal{P}_3) \geq 0. \end{aligned}$$

Formally this inequality means the nonnegativity of the conditional mutual information calculated for the three qubit state with a specific diagonal density 8×8-matrix, and it can be checked experimentally.

The inequality is valid for any three nonnegative numbers $\mathcal{P}_1$, $\mathcal{P}_2$, and $\mathcal{P}_3$, satisfying the inequality $\mathcal{P}_1 + \mathcal{P}_2 + \mathcal{P}_3 \leq 1$ independently of the interpretation of these numbers.

9. Conclusions

To conclude, we list the main results of our work.

First of all, it is worth pointing out that in this work we did not derive new inequalities but did use the well-known for bipartite and multipartite systems entropic-information relations to attract attention to the fact that, by employing the tool of a map of indices, these inequalities can also be applied to the system which does not contain subsystems, e.g., for a single qudit. For noncomposite systems, the relations discussed here, like the subadditivity condition, were not discussed in the literature.

The motivation of writing such entropic relations in explicit matrix forms is connected with recent discussions of experimental studies of qudit density matrices, where the entropic-information matrix formulas can be checked using the data characterizing the states of superconducting quantum circuits realized by Josephson junctions [22–32].

We reviewed the approach to the set of nonnegative numbers and Hermitian nonnegative matrices with unit trace, in view of the interpretation of the numbers and the matrices as the probability distributions and the density matrices, respectively.

We showed that the known entropic inequalities, which are applied to composite systems, both classical and quantum, can be also applied to the systems without subsystems. We obtained new entropic inequalities, in view of the interpretation of the density matrix of the system state as the density matrix of artificial bipartite or tripartite system.

In fact, the approach presented provides the possibility to extend all known entropic-information relations for classical and quantum composite systems to the case of systems without subsystems, and the relations reflect the presence of correlations either classical or quantum ones of the system degrees of freedom. The quantum correlations of the single qudit states, which we call hidden correlations [33, 34] can be used in quantum technologies, analogously to the employment of entanglement as a quantum resource.

To clarify the physical meaning of hidden correlations in noncomposite (nondivisible) systems, we define the "events" of these systems as "pairs" or "triples" of the "initial events," statistics of which we study employing either the probability distribution $p(s)$ in classical domain or the density matrix $\rho_{ss'}$ in quantum domain.

Acknowledgments
The author is grateful to the Organizers of the 2017 Symposium "Symmetries in Science" (Bregenz, Austria, July 30 – August 4, 2017) and especially to Prof. Dr. Dieter Schuch for invitation and kind hospitality.

References

[1] Kolmogorov A N 1956 *Foundations of the Theory of Probability*, 2nd ed., Chelsea Publishing Company, New York, USA
[2] Shannon C E 1948 *Bell. Syst. Tech. J.* **27** 379; 623
[3] Schrödinger E 1926 *Ann. Phys.* **79**, 361; **81**, 109
[4] Landau L D 1927 *Z. Phys.* **45** 430
[5] von Neumann J 1927 *Göttingenische Nachrichten* **11** 245
[6] Mancini S, Man'ko V I and Tombesi P 1996 *Phys. Lett. A* **213** 1
[7] Dodonov V V and Man'ko V I 1997 *Phys. Lett. A* **229** 335
[8] Man'ko V I and Man'ko O V 1997 *J. Exp. Theor. Phys.* **85** 430
[9] Smithey D T, Beck M, Raymer M G and Faridani A 1993 *Phys. Rev. Lett.* **70** 1244
[10] Manko M A and Manko V I 2011 *Found. Phys.* **41** 330
[11] Manko M A and Manko V I 2011 "Dynamic symmetries and entropic inequalities in the probability representation of quantum mechanics," *Latin-American School of Physics XI ELAF: Symmetries in Physics*, AIP Conf. Ser., Vol. 1334, pp. 217–248
[12] Manko M A and Manko V I 2012 "Tomographic entropic inequalities in the probability representation of quantum mechanics," in: R. Bijker, (Ed.), *Beauty in Physics: Theory and Experiment: in Honor of Francesco Iachello on the Occasion of His 70th Birthday, Hacienda Cocoyoc, Mexico, 14-18 May 2012*, AIP Conference Proceedings, Vol. 1488, pp. 110–121
[13] Man'ko M A 2013 *Phys. Scr.* **87** 038113
[14] Manko M A and Manko V I 2014 *Phys. Scr.* **T160**, 014030
[15] Man'ko M A, Man'ko V I, Marmo G, Simoni A and Ventriglia F 2013 *Nuovo Cim. C* **36** 163
[16] Manko M A and Manko V I 2014 *Int. J. Quantum Inf.* **12** 156006.
[17] Manko M A and Manko V I 2016 *J. Phys.: Conf. Ser.* **698** 012004
[18] Man'ko M A and Man'ko V I 2015 *Entropy* **17** 2876
[19] Shrödinger E 1935 *Naturwissenschaften* **23** 807
[20] Nielsen M A and Chuang I L 2000 *Quantum Computation and Quantum Information*, Cambridge University Press, Cambridge, UK
[21] Tsallis C 2001 "Nonextensive statistical mechanics and thermodynamics: historical background and present status," in: S. Abe and Y. Okamoto (Eds.), *Nonextensive Statistical Mechanics and Its Applications*, *Lecture Notes in Physics*, Springer, Berlin, Vol. 560, pp. 3–98
[22] Dodonov V V, Man'ko V I and Man'ko O V 1989 *J. Sov. Laser Res.* **10** 413
[23] Man'ko V I 1991 *J. Sov. Laser Res.* **12** 383
[24] Man'ko O V 1994 *J. Korean Phys. Soc.* **27** 1
[25] Pashkin Yu A, Yamamoto T, Astafiev O, Nakamura Y, Averin D V and Tsai J S 2003 *Nature* **421** 823

IOP Conf. Series: Journal of Physics: Conf. Series **1071** (2018) 012015 doi:10.1088/1742-6596/1071/1/012015

[26] Devoret M H, Wallraff A and Martinis J M 2004 "Superconducting qubits: A short review" arXiv:cond-mat/0411174v1
[27] Devoret M H and Schoelkopf R J 2013 *Science* **339** 1169
[28] Shalibo Y, Resh R, Fogel O, Shwa D, Bialczak R, Martinis J M and Katz N 2013 *Phys. Rev. Lett.* **110** 100404
[29] Braumüller J, Cramer J, Schlör S, Rotzinger H, Radtke L, Lukashenko A, Yang P, Marthaler M, Guo L, Ustinov A V and Weides M 2015 *Phys. Rev. B* **91** 054523
[30] Fedorov A K, Kiktenko E O, Man'ko O V and Man'ko V I 2015 *Phys. Rev. A* **91** 042312
[31] Kiktenko E O, Fedorov A K, Strakhov A A and Man'ko V I 2015 *Phys. Lett. A* **379** 1409
[32] Glushkov E, Glushkova A and Man'ko V I 2015 *J. Russ. Laser Res.* **36** 448
[33] Man'ko M A and Man'ko V I 2015 *J. Russ. Laser Res.* **36** 301
[34] Man'ko M A 2017 *J. Russ. Laser Res.* **38** 211

IOP Conf. Series: Journal of Physics: Conf. Series **1071** (2018) 012016 doi:10.1088/1742-6596/1071/1/012016

Quasi-exactly solvable polynomial extensions of the quantum harmonic oscillator

Christiane Quesne

Physique Nucléaire Théorique et Physique Mathématique, Université Libre de Bruxelles, Campus de la Plaine CP229, Boulevard du Triomphe, B-1050 Brussels, Belgium

E-mail: cquesne@ulb.ac.be

Abstract. The (analytic) sextic oscillator is often considered as the prototype of quasi-exactly solvable (QES) Schrödinger equations, i.e., those Schrödinger equations for which at some ad hoc couplings a finite number of eigenstates can be found explicitly by algebraic means, while the remaining ones remain unknown. Recently, a (non-analytic) QES symmetrized quartic oscillator was introduced and shown to complete the list of (analytic) QES anharmonic oscillators, which does not contain any quartic one. Here we prove that such a quartic oscillator is amenable to an sl(2, $\mathbb{R}$) description. Furthermore, we generalize it to a symmetrized sextic oscillator. The latter is obtained by parting the real line into two subintervals $\mathbb{R}^-$ and $\mathbb{R}^+$ on which the corresponding Schrödinger equations are solved by using the functional Bethe ansatz method, and the resulting wavefunctions and their first derivatives are matched at $x = 0$. Two categories of QES potentials are obtained: the first one containing the well-known analytic sextic potentials as a subset, and the second one of novel potentials with no counterpart in such a class.

1. Introduction

Exact solutions of the Schrödinger equation are known to be very useful for approximating solutions of more realistic equations appearing in practical problems, because they may give a hint for suggesting a good starting point in perturbation theory or variational calculus. Such exact solutions essentially belong to one of two categories.

The first category consists of solutions of the so-called exactly solvable (ES) Schrödinger equations, for which all the eigenstates can be algebraically determined. Among them, one finds equations containing piece-wise constant point interactions, such as the rectangular potential hole [1], for which the solutions are obtained by matching the wavefunctions and their first derivatives at the non-analyticity points of the potential. Other ES Schrödinger equations are those containing an analytic potential that can be transformed into a second-order differential equation of hypergeometric type [2]. The eigenfunctions of such equations can be constructed in closed form by using the theory of special functions [3]. Still others can be obtained from the previous ones by using higher-order supersymmetric quantum mechanics or Darboux transformations and the corresponding potentials are rational extensions of the previous potentials [4, 5, 6, 7], while their wavefunctions can be expressed in terms of exceptional orthogonal polynomials [8, 9].

The second category of exact solutions comprises solutions of the so-called quasi-exactly solvable (QES) Schrödinger equations, for which a finite number of eigenstates can be algebraically determined at some ad hoc couplings, while the remaining ones remain unknown.

IOP Conf. Series: Journal of Physics: Conf. Series **1071** (2018) 012016 doi:10.1088/1742-6596/1071/1/012016

The simplest ones, which were discovered in the 1980s, are characterized by a hidden sl(2, $\mathbb{R}$) algebraic structure [10, 11, 12, 13, 14] and are connected with polynomial solutions of the Heun equation [15]. More complicated ones, related to polynomial solutions of generalized Heun equations, can be solved by the recursion relation method [16] or by the functional Bethe ansatz method [17, 18]. In such a context, the latter has proven very effective [19, 20, 21, 22].

It was recently suggested [23, 24, 25, 26, 27] that the ES status should also be attributed to less common Schrödinger equations for which the real line of coordinates is splitted into subintervals wherein the potential admits different definitions, while being continuous on the whole line, and the wavefunctions remain piece-wise proportional to special functions, while being matched, as well as their first derivatives, at the subinterval limit points. For instance, the Schrödinger equation with the non-analytic potential $V(x) = -g^2 \exp(-|x|)$ was solved by Sasaki and Znojil in terms of Bessel functions [24].

A similar extension to the QES case was also proposed by Znojil [28], who considered a symmetrized quartic oscillator and showed by using the recursion relation method that such an oscillator completes the list of (analytic) QES anharmonic oscillators, which contains sextic polynomial oscillators [10], but no quartic one.

In a recent work [29], we proved that Znojil's symmetrized quartic oscillator is amenable to an sl(2, $\mathbb{R}$) description and we introduced a new QES symmetrized sextic oscillator, which we solved by the functional Bethe ansatz method.

It is the purpose of the present communication to review such a work.

2. Symmetrized quartic polynomial oscillator

The symmetrized quartic polynomial oscillator [28, 29] is described by the Schrödinger equation

$$\left(-\frac{d^2}{dx^2} + V(x)\right)\psi(x) = E\psi(x), \tag{1}$$

where the non-analytic potential

$$V(x) = x^4 - s|x|^3 + rx^2 - q|x| = \begin{cases} x^4 + sx^3 + rx^2 + qx & \text{if } x < 0, \\ x^4 - sx^3 + rx^2 - qx & \text{if } x > 0, \end{cases} \tag{2}$$

is defined separately on the two semi-infinite intervals $x < 0$ and $x > 0$ with $q, r, s \in \mathbb{R}$.

Physically acceptable wavefunctions should tend to zero for $x \to \pm\infty$ and be continuous everywhere, as well as their first derivatives. Furthermore, as the potential is symmetric under parity, they must also be either symmetric or antisymmetric.

We may therefore solve the Schrödinger equation in one of the intervals, for instance $x < 0$, then prescrive that in the other interval $x > 0$, $\psi(x) = \epsilon\psi(-x)$ with $\epsilon = +1$ or $\epsilon = -1$ according to whether ψ is even or odd. On the resulting wavefunctions $\psi^{(\epsilon)}(x)$ and their derivatives it then remains to impose continuity at $x = 0$, which amounts to setting either $\psi^{(+)\prime}(0) = 0$ or $\psi^{(-)}(0) = 0$.

To get QES solutions of (1), it is appropriate to assume that

$$\psi_n^{(\epsilon)}(x) = e^{W(x)}\phi_n^{(\epsilon)}(x), \qquad W(x) = -\tfrac{1}{3}|x|^3 + ax^2 - b|x|, \tag{3}$$

and

$$\phi_n^{(\epsilon)} = \begin{cases} \sum_{k=0}^n v_k x^k & \text{if } x < 0, \\ \epsilon \sum_{k=0}^n (-1)^k v_k x^k & \text{if } x > 0, \end{cases} \tag{4}$$

where a, b are two real constants and v_k $(k = 0, 1, \ldots, n)$ are some expansion coefficients to be determined. Such functions tend to zero for $x \to \pm\infty$, as it should be, and they have the right behaviour at $x = 0$ provided the additional condition

$$\begin{aligned} &v_1 + bv_0 = 0 \qquad \text{if } \epsilon = +1, \\ &v_0 = 0 \qquad \text{if } \epsilon = -1, \end{aligned} \tag{5}$$

is imposed.

On inserting potential (2) and wavefunctions (3) in (1), we obtain on the interval $x < 0$ the differential equation

$$\left(-\frac{d^2}{dx^2} - 2(x^2 + 2ax + b)\frac{d}{dx} + (q - 4ab - 2)x - b^2 - 2a \right) \phi_n(x) = E_n\phi_n(x), \tag{6}$$

where higher-order terms have been eliminated by setting

$$s = 4a \qquad \text{and} \qquad r = 4a^2 + 2b. \tag{7}$$

This is a (triconfluent) Heun equation [15]. Hence, it can be solved by an sl(2, $\mathbb{R}$) algebraic approach [14].

The first-order differential operators

$$J_n^+ = x^2\frac{d}{dx} - nx, \qquad J_n^0 = x\frac{d}{dx} - \frac{n}{2}, \qquad J_n^- = \frac{d}{dx}, \tag{8}$$

satisfying the commutation relations

$$\left[J_n^0, J_n^\pm\right] = \pm J_n^\pm, \qquad \left[J_n^+, J_n^-\right] = -2J_n^0, \tag{9}$$

provide a realization of sl(2, $\mathbb{R}$) in the $(n+1)$-dimensional space of polynomials of order not higher than n, $\mathcal{P}_{n+1} = \langle 1, x, x^2, \ldots, x^n \rangle$. Their action in such a space is given by

$$J_n^+ x^k = (k - n)x^{k+1}, \qquad J_n^0 x^k = \left(k - \frac{n}{2}\right) x^k, \qquad J_n^- x^k = kx^{k-1}. \tag{10}$$

Provided we set

$$q = 4ab + 2n + 2, \tag{11}$$

the second-order differential operator h on the left-hand side of (6) can be written as a second-degree polynomial in the sl(2, $\mathbb{R}$) generators (8),

$$h = -J_n^- J_n^- - 2J_n^+ - 4aJ_n^0 - 2bJ_n^- - 2(n+1)a - b^2. \tag{12}$$

On using relations (10), it is then straightforward to write the matrix of h in the space $\mathcal{P}_{n+1}$. The admissible energies $E_n^{(\epsilon)}$ and wavefunctions $\psi_n^{(\epsilon)}(x)$ can then be obtained by diagonalizing such a matrix and imposing constraint (5).

The simplest examples are provided by the $n = 0$ and $n = 1$ cases.

For $n = 0$, we get a 1×1 matrix for h, so that we directly obtain $E_0 = -2a - b^2$ and $\phi_0(x) = 1$ on $x < 0$. Since $v_0 = 1$ and $v_1 = 0$, it results from constraint (5) that the wavefunction on the whole real line cannot be odd, but it can be even provided the condition $b = 0$ is satisfied. Hence the potential

$$V(x) = x^4 - 4a|x|^3 + 4a^2x^2 - 2|x| \tag{13}$$

has an energy eigenvalue $E_0^{(+)}$ and associated wavefunction $\psi_0^{(+)}(x)$, given by

$$E_0^{(+)} = -2a, \qquad \psi_0^{(+)}(x) = e^{-\frac{1}{3}|x|^3+ax^2}, \tag{14}$$

corresponding to a ground state.

For $n = 1$, the matrix for h is 2×2, so that its diagonalization leads to a second-degree equation for E_1. The solutions of the latter, $E_{1\pm} = -4a - b^2 \pm 2\sqrt{a^2-b}$, are real provided $a^2 \geq b$, which can be satisfied if either $b \leq 0$ or $b > 0$ and simultaneously $a \geq \sqrt{b}$ or $a \leq -\sqrt{b}$. The corresponding eigenfunctions on $x < 0$ are $\phi_{1\pm}(x) = x + a \pm \sqrt{a^2-b}$, showing that $v_0 = a \pm \sqrt{a^2-b}$ and $v_1 = 1$. On the whole line, we can now get an odd or even wavefunction according to the parameter choice.

The wavefunction is odd if $a = \mp\sqrt{a^2-b}$, which imposes $b = 0$, $a < 0$ for ϕ_{1+}, and $b = 0$, $a > 0$ for ϕ_{1-}. We therefore obtain that the potential

$$V(x) = x^4 - 4a|x|^3 + 4a^2x^2 - 4|x| \tag{15}$$

has an energy eigenvalue $E_1^{(-)}$ and associated wavefunction $\psi_1^{(-)}(x)$, given by

$$E_1^{(-)} = -6a, \qquad \psi_1^{(-)}(x) = e^{-\frac{1}{3}|x|^3+ax^2}x, \tag{16}$$

corresponding to a first-excited state.

From (5), the wavefunction is even provided $b \neq 0$ and $a \pm \sqrt{a^2-b} = -1/b$ for $\phi_{1,\pm}$. Examination of these conditions [29] leads to the conclusion that the potential

$$V(x) = x^4 + \frac{2(b^3+1)}{b}|x|^3 + \frac{b^6+4b^3+1}{b^2}x^2 + 2(b^3-1)|x|, \qquad b \neq 0, \tag{17}$$

has an energy eigenvalue $E_1^{(+)}$ and associated wavefunction $\psi_1^{(+)}(x)$, given by

$$E_1^{(+)} = \frac{2b^3+1}{b}, \qquad \psi_1^{(+)}(x) = e^{-\frac{1}{3}|x|^3-\frac{b^3+1}{2b}x^2-b|x|}\left(|x|+\frac{1}{b}\right), \tag{18}$$

corresponding to a ground state for $b > 0$ and to a second-excited state for $b < 0$.

A similar discussion can be carried out for higher n values and, in particular, for the $n = 2$, $\epsilon = +1$ case solved by Znojil [28] using the recursion relation method.

3. Symmetrized sextic polynomial oscillator

Instead of potential (2), let us consider now the symmetrized sextic polynomial oscillator potential

$$V(x) = x^6 - u|x|^5 + tx^4 - s|x|^3 + rx^2 - q|x| \tag{19}$$

in equation (1). Such a potential depends on five real constants q, r, s, t, u instead of three.

To get QES solutions of (1), let us assume that

$$\psi_n^{(\epsilon)}(x) = e^{W(x)}\phi_n^{(\epsilon)}(x), \qquad W(x) = -\frac{1}{4}x^4 - a|x|^3 + bx^2 - c|x|, \tag{20}$$

with $a, b, c \in \mathbb{R}$, and $\phi_n^{(\epsilon)}(x)$ given by (4), with the new constraint

$$\begin{aligned} &v_1 + cv_0 = 0 \qquad \text{if } \epsilon = +1, \\ &v_0 = 0 \qquad \text{if } \epsilon = -1. \end{aligned} \tag{21}$$

By setting

$$u = -6a, \qquad t = 9a^2 - 4b, \qquad s = 12ab - 2c, \tag{22}$$

to eliminate some higher-order terms, the Schrödinger equation (1) leads to the differential equation

$$\begin{aligned}\Bigg(&-\frac{d^2}{dx^2} + 2(x^3 - 3ax^2 - 2bx - c)\frac{d}{dx} + (r - 4b^2 - 6ac + 3)x^2 + (q - 4bc - 6a)x \\ &- c^2 - 2b\Bigg)\phi_n(x) = E_n\phi_n(x),\end{aligned} \tag{23}$$

on $x < 0$. This is a special case of a generalized Heun equation, whose polynomial solutions can be found by the functional Bethe ansatz method, as previously shown by Zhang [18]. Before reviewing this method, it is worth observing that a special case of (19) and (20), corresponding to $u = s = q = a = c = 0$, gives back the analytic sextic potential considered by Turbiner and Ushveridze [10]. In such a case, equation (23) can be transformed into a Heun-type equation by making the change of variable $\xi = x^2$ and, consequently, can be dealt with by an sl(2, $\mathbb{R}$) approach.

In the functional Bethe ansatz method, one considers a second-order differential equation

$$X(z)y''(z) + Y(z)y'(z) + Z(z)y(z) = 0 \tag{24}$$

with polynomial coefficients

$$X(z) = \sum_{l=0}^{k} a_l z^l, \qquad Y(z) = \sum_{l=0}^{k-1} b_l z^l, \qquad Z(z) = \sum_{l=0}^{k-2} c_l z^l, \tag{25}$$

such that $a_l, b_l, c_l \in \mathbb{R}$, and one looks for polynomial solutions of the type

$$y(z) = \prod_{i=1}^{n}(z - z_i) \tag{26}$$

with real and distinct roots z_i, $i = 1, 2, \ldots, n$.

On inserting (25) in (24), the latter is transformed into

$$-c_0 = \left(\sum_{l=0}^{k} a_l z^l\right)\sum_{i=1}^{n}\frac{1}{z - z_i}\sum_{\substack{j=1\\ j\neq i}}^{n}\frac{2}{z_i - z_j} + \left(\sum_{l=0}^{k-1} b_l z^l\right)\sum_{i=1}^{n}\frac{1}{z - z_i} + \sum_{l=1}^{k-2} c_l z^l, \tag{27}$$

where the left-hand side is a constant, while the right-hand one is a meromorphic function with simple poles at $z = z_i$ and a singularity at $z = \infty$. Equation (27) will be satisfied provided the residues at the simple poles vanish and the coefficients of $z, z^2, \ldots, z^{k-2}$ vanish too. The first condition leads to the Bethe ansatz equations determining the roots,

$$\sum_{\substack{j=1\\ j\neq i}}^{n}\frac{2}{z_i - z_j} + \frac{\sum_{l=0}^{k-1} b_l z_i^l}{\sum_{l=0}^{k} a_l z_i^l} = 0, \qquad i = 1, 2, \ldots, n, \tag{28}$$

while the second one provides some relations between the coefficients a_l, b_l, c_l and the roots.

In the case of equation (23), we obtain the Bethe ansatz equations

$$\sum_{\substack{j=1\\ j\neq i}}^{n} \frac{1}{x_i - x_j} - x_i^3 + 3ax_i^2 + 2bx_i + c = 0, \qquad i = 1, 2, \ldots, n, \tag{29}$$

and the relations

$$\begin{aligned} r &= 4b^2 + 6ac - 2n - 3, \qquad q = -2\sum_{i=1}^{n} x_i + 6a(n+1) + 4bc, \\ E &= 2\sum_{i=1}^{n} x_i^2 - 6a\sum_{i=1}^{n} x_i - 2b(2n+1) - c^2. \end{aligned} \tag{30}$$

On the resulting $\phi_n(x)$, it then remains to impose constraint (21).

The $n = 0$ case is easy to solve because then $\phi_0(x) = 1$ on $x < 0$ and there is no root. Relations (30) then become

$$r = 4b^2 + 6ac - 3, \qquad q = 6a + 4bc, \qquad E_0 = -2b - c^2. \tag{31}$$

Since $v_0 = 1$, $v_1 = 0$, there is no odd wavefunction, but there is an even one provided $c = 0$. We then get the potential

$$V(x) = x^6 + 6a|x|^5 + (9a^2 - 4b)x^4 - 12ab|x|^3 + (4b^2 - 3)x^2 - 6a|x| \tag{32}$$

with an energy eigenvalue $E_0^{(+)}$ and associated wavefunction $\psi_0^{(+)}(x)$, given by

$$E_0^{(+)} = -2b, \qquad \psi_0^{(+)}(x) = e^{-\frac{1}{4}x^4 - a|x|^3 + bx^2}, \tag{33}$$

and corresponding to a ground state. For $a = 0$, we obtain, as a special case, the analytic sextic potential

$$V(x) = x^6 - 4bx^4 + (4b^2 - 3)x^2 \tag{34}$$

with

$$E_0^{(+)} = -2b, \qquad \psi_0^{(+)}(x) = e^{-\frac{1}{4}x^4 + bx^2}, \tag{35}$$

as found by Turbiner and Ushveridze [10].

For $n = 1$, there is a single Bethe ansatz equation

$$x_1^3 - 3ax_1^2 - 2bx_1 - c = 0 \tag{36}$$

and relations (30) become

$$r = 4b^2 + 6ac - 5, \qquad q = -2x_1 + 12a + 4bc, \qquad E = 2x_1^2 - 6ax_1 - 6b - c^2. \tag{37}$$

Since $v_0 = -x_1$ and $v_1 = 1$, we can get both an even and an odd wavefunction according to which parameters are chosen.

The wavefunction is odd if $x_1 = 0$, which imposes $c = 0$. The potential

$$V(x) = x^6 + 6a|x|^5 + (9a^2 - 4b)x^4 - 12ab|x|^3 + (4b^2 - 5)x^2 - 12a|x| \tag{38}$$

has therefore an energy eigenvalue $E_1^{(-)}$ and associated wavefunction $\psi_i^{(-)}(x)$, given by

$$E_1^{(-)} = -6b, \qquad \psi_1^{(-)}(x) = e^{-\frac{1}{4}x^4 - a|x|^3 + bx^2}x, \tag{39}$$

and corresponding to a first-excited state. The special case resulting from $a = 0$ is now the analytic sextic potential

$$V(x) = x^6 - 4bx^4 + (4b^2 - 5)x^2 \tag{40}$$

with

$$E_1^{(-)} = -6b, \qquad \psi_1^{(-)}(x) = e^{-\frac{1}{4}x^4 + bx^2} x, \tag{41}$$

in accordance with Turbiner and Ushveridze results [10].

The even wavefunction is obtained for $x_1 = 1/c$, where c is a nonvanishing solution of the quartic equation $c^4 + 2bc^2 + 3ac - 1 = 0$. For such a c value, the potential

$$\begin{aligned} V(x) &= x^6 + 6a|x|^5 + (9a^2 - 4b)x^4 - (12ab - 2c)|x|^3 + (4b^2 + 6ac - 5)x^2 \\ &\quad - \left(12a + 4bc - \frac{2}{c}\right)|x| \end{aligned} \tag{42}$$

has an eigenvalue $E_1^{(+)}$ and an eigenfunction $\psi_1^{(+)}(x)$, given by

$$E_1^{(+)} = -6b - c^2 - \frac{6a}{c} + \frac{2}{c^2}, \qquad \psi_1^{(+)}(x) = e^{-\frac{1}{4}x^4 - a|x|^3 + bx^2 - c|x|} \left(|x| + \frac{1}{c}\right), \tag{43}$$

describing a ground state for $c > 0$ and a second-excited state for $c < 0$. Such a potential has no counterpart among the analytic sextic potentials of Ref. [10].

For $n = 2$, as detailed in Ref. [29], it is also possible to find both odd and even wavefunctions. An odd wavefunction $\psi_2^{(-)}(x) \propto x\left(|x| + \frac{1}{c}\right)$ is obtained if $x_1 = 0$, $x_2 = 1/c$, and $c \neq 0$ is a solution of a quartic equation $2c^4 + 2bc^2 + 3ac - 1 = 0$. The resulting potential has no counterpart in the set of analytic sextic potentials of Ref. [10]. On the other hand, an even wavefunction $\psi_2^{(+)}(x) \propto (|x| + x_1)(|x| + x_2)$ corresponds to roots $x_1, x_2 \neq 0, 1/c$ that are solutions of

$$x_1^3 - 3ax_1^2 - 2bx_1 - c = -(x_2^3 - 3ax_2^2 - 2bx_2 - c) = \frac{1}{x_1 - x_2}, \qquad x_1 + x_2 = cx_1x_2. \tag{44}$$

This time, we get a counterpart in the set of analytic sextic potentials for $a = c = x_1 + x_2 = 0$. In this case, there indeed remains a single equation $2x_1^4 - 4bx_1^2 - 1 = 0$ with the two solutions $(x_1^2)_\pm = \frac{1}{2}(2b \pm \sqrt{4b^2 + 2})$, giving rise to two eigenvalues and eigenfunctions

$$E_{2\pm}^{(+)} = -6b \pm 2\sqrt{4b^2 + 2}, \qquad \psi_{2\pm}^{(+)}(x) = \exp\left(-\frac{1}{4}x^4 + bx^2\right)\left(x^2 - b \mp \frac{1}{2}\sqrt{4b^2 + 2}\right). \tag{45}$$

In Ref. [29], some examples of QES symmetrized sextic oscillators and associated wavefunctions have been displayed and compared with results for the analytic case whenever the latter exist.

4. Results for QES symmetrized sextic polynomial oscillators

We have shown that the set of new QES symmetrized sextic polynomial oscillators can be separated into two categories.

The first one contains those symmetrized sextic oscillators whose known wavefunction is of 'natural' parity, i.e., for which $\epsilon = (-1)^n$. These include those associated with the cases $(n, \epsilon) = (0, +1)$, $(1, -1)$, and $(2, +1)$, which we have explicitly considered. They comprise as special cases the QES analytic sextic potentials dealt with by Turbiner and Ushveridze [10], which constitute a subset.

The second category consists of those symmetrized sextic oscillators whose known wavefunction is of 'unnatural' parity, i.e., for which $\epsilon = (-1)^{n+1}$. They include, in particular, those associated with the cases $(n, \epsilon) = (0, +1)$ and $(1, -1)$, which have been explicitly studied. They do not comprise any QES analytic sextic potential, as found by Turbiner and Ushveridze [10].

In conclusion, considering symmetrized sextic oscillators gives rise to a novel type of QES potentials with no counterpart in the set of analytic sextic potentials.

References

[1] Flügge S 1971 *Practical Quantum Mechanics I* (Berlin: Springer-Verlag)
[2] Cooper F, Khare A and Sukhatme U 1995 *Phys. Rep.* **251** 267
[3] Szegö G 1939 *Orthogonal Polynomials* (New York: American Mathematical Society)
[4] Quesne C 2008 *J. Phys. A: Math. Theor.* **41** 392001
[5] Quesne C 2009 *SIGMA* **5** 084
[6] Odake S and Sasaki R 2009 *Phys. Lett.* B **679** 414
[7] Odake S and Sasaki R 2011 *Phys. Lett.* B **702** 164
[8] Gómez-Ullate D, Kamran N and Milson R 2009 *J. Math. Anal. Appl.* **359** 352
[9] Gómez-Ullate D, Kamran N and Milson R 2010 *J. Approx. Theory* **162** 987
[10] Turbiner A V and Ushveridze A G 1987 *Phys. Lett.* A **126** 181
[11] Turbiner A V 1988 *Commun. Math. Phys.* **118** 467
[12] Ushveridze A G 1994 *Quasi-Exactly Solvable Models in Quantum Mechanics* (Bristol: Institute of Physics Publishing)
[13] González-López A, Kamran N and Olver P J 1993 *Commun. Math. Phys.* **153** 117
[14] Turbiner A V 2016 *Phys. Rep.* **642** 1
[15] Ronveaux A 1995 *Heun Differential Equations* (Oxford: Oxford University Press)
[16] Ciftci H, Hall R I, Saad N and Dogu E 2010 *J. Phys. A: Math. Theor.* **43** 415206
[17] Ho C-L 2008 *Ann. Phys. (NY)* **323** 2241
[18] Zhang Y-Z 2012 *J. Phys. A: Math. Theor.* **45** 065206
[19] Agboola D and Zhang Y-Z 2012 *Mod. Phys. Lett.* A **27** 1250112
[20] Agboola D and Zhang Y-Z 2013 *Ann. Phys. (NY)* **330** 246
[21] Agboola D, Links J, Marquette I and Zhang Y-Z 2014 *J. Phys. A: Math. Theor.* **47** 395305
[22] Quesne C 2017 *J. Math. Phys.* **58** 052104
[23] Znojil M 2016 *Mod. Phys. Lett.* A **31** 1650088
[24] Sasaki R and Znojil M 2016 *J. Phys. A: Math. Theor.* **49** 445303
[25] Znojil M 2016 *Mod. Phys. Lett.* A **31** 1650195
[26] Sasaki R 2016 Confining non-analytic exponential potential $V(x) = g^2 \exp(2|x|)$ and its exact Bessel-function solvability *Preprint* arXiv:1611.02467
[27] Sasaki R 2016 Symmetric Morse potential is exactly solvable *Preprint* arXiv:1611.05952
[28] Znojil M 2016 *Phys. Lett.* A **380** 1414
[29] Quesne C 2017 *Eur. Phys. J. Plus* **132** 450

IOP Conf. Series: Journal of Physics: Conf. Series **1071** (2018) 012017 doi:10.1088/1742-6596/1071/1/012017

On local representations of rotations on discrete configuration spaces

M Reginatto[1] **and M J W Hall**[2]

[1]Physikalisch-Technische Bundesanstalt, Bundesallee 100, 38116 Braunschweig, Germany
[2]Centre for Quantum Computation and Communication Technology (Australian Research Council), Centre for Quantum Dynamics, Griffith University, Brisbane, QLD 4111, Australia

E-mail: marcel.reginatto@ptb.de

Abstract.
A quantum mechanical spin-half system may be characterized abstractly as having a set of two-valued observables which generate infinitesimal rotations in three dimensions. We consider a concrete realization of such a two-level system within the formalism of ensembles on configuration space, an approach which is not only capable of describing quantum mechanical systems but allows also for theories that are generalizations of quantum theory. Such a spin-half system may be called a rotational bit or robit, to distinguish it from the standard quantum qubit. After reviewing ensembles on configuration space and examining the example of constructing representations of the Galilean Lie algebra for the free particle, we construct probabilistic models for ensembles that consist of one and two spin-half systems. In the case of a single spin-half system, there are two main requirements: the configuration space must be a discrete set, labeling the outcomes of two-valued spin observables, and these observables must provide an algebraic representation of $so(3)$. The case of a pair of spin-half systems is more complicated, in that additional physical requirements concerning locality and subsystem independence must also be taken into account and now the observables must provide an algebraic representation of $so(3) \oplus so(3)$. We compare the resulting theories to the corresponding quantum mechanical systems.

1. Introduction

Physical systems that are composed of spin-half particles, while simple enough to be tractable, display enough complexity to bring out many of the most puzzling features of quantum mechanics. Already the relatively simple case of two spin-half particles leads to remarkable results. This became apparent after Bohm [1] provided a reformulation of the EPR paradox [2] in terms of a pair of spin-half particles. This same model would later be used by Bell [3] to formulate his famous theorem. The literature on such quantum systems is enormous and their study still remains a very active area of research to this date, particularly in quantum information theory.

A quantum mechanical spin-half system may be characterized abstractly as having a set of two-valued observables which generate infinitesimal rotations in three dimensions, as discussed in more detail in section 5. Using this abstract formulation, we consider a concrete realization of physical systems of this type within the formalism of ensembles on configuration space [4], an approach which is capable of describing quantum mechanical systems and, in addition, allows also for theories which are generalizations of quantum theory. The aim is to construct probabilistic models for ensembles that consist of one and two spin-half systems and to see whether a set of basic physical requirements still allows for generalizations of the familiar quantum systems. Thus, our starting point is not the usual formulation of spin-half systems using operators and vectors in Hilbert space [5] but the more general formulation of ensembles on configuration space [4]. We then compare the resulting theories to the corresponding quantum mechanical systems.

The essential tool that we use are representations of appropriate Lie algebras on the configuration space of the system. Thus, in the case of a single spin-half system, we consider representations of the Lie algebra $so(3)$ for a configuration space that is two-dimensional. In the case of a pair of spin-half systems, we need to deal with representations of the Lie algebra $so(3) \oplus so(3)$ for a configuration space that is four-dimensional. This case however becomes more complicated than the case of a single spin-half particle because we must take into account additional physical requirements concerning locality and subsystem independence.

In the next section, we introduce the formulation of ensembles on configuration space. Before considering the case of discrete configuration spaces, which is the one that is needed for the description of spin-half particles, we discuss the more intuitive case of ensembles on configuration space for the continuous configuration space of a single particle. In the section that follows, we consider a particular example in which we look for representations of the Galilean Lie algebra and show that non-equivalent representations can arise, leading to inequivalent theories of the free particle. The following two sections are devoted to the formulation of ensembles on configuration space on discrete configuration spaces and the description of spin-half particles. We end with some concluding remarks.

2. Ensembles on configuration space

Consider the case of a non-relativistic particle with a continuous configuration space with coordinates x. The description of physical systems in terms of ensembles on configuration space requires three basic ingredients,

- a probability density $P(x)$ on configuration space, with $P(x) \geq 0$ and $\int dx P(x) = 1$,
- a canonically conjugate quantity $S(x)$,
- an ensemble Hamiltonian $H[P, S]$,

where $H[P, S]$ is a functional of P and S that must satisfy certain requirements (these conditions, which are common to all observables, are discussed below).

The *state* of the system is fully described by the functions $P(x)$ and $S(x)$. The time evolution of the state follows Hamiltonian equations of motion determined by H which, when expressed in terms of Poisson brackets, take the form

$$\frac{\partial P}{\partial t} = \{P, H\} = \frac{\delta H}{\delta S}, \qquad \frac{\partial S}{\partial t} = \{S, H\} = -\frac{\delta H}{\delta P}. \tag{1}$$

To understand why we introduce $S(x)$ as a canonically conjugate variable to $P(x)$, it is convenient to think of the formalism as resulting from a two-step process. In the first step, we introduce a configuration space with coordinates x and probabilities $P(x)$. In the second step, we introduce dynamics. We assume that the equations of motion for $P(x)$ follow from an action principle. Then, if we do this via a Hamiltonian formalism, we are forced to introduce an auxiliary field canonically conjugate to $P(x)$. This is the field $S(x)$.

Notice that this is a very general formalism that essentially describes dynamics of probabilities. A physical interpretation is only possible if an appropriate ensemble Hamiltonian is introduced together with other appropriate observables.

We consider a couple of physically relevant examples of ensemble Hamiltonians. The ensemble Hamiltonian that describes a *classical particle* subject to a force derived from a potential term $V(x)$ is given by

$$H_C[P,S] = \int dx\, P\left[\frac{|\nabla S|^2}{2m} + V\right]. \tag{2}$$

To see this, evaluate the equations of motion,

$$\frac{\partial S}{\partial t} + \frac{|\nabla S|^2}{2m} + V = 0, \qquad \frac{\partial P}{\partial t} + \nabla \cdot \left(P\frac{\nabla S}{m}\right) = 0. \tag{3}$$

The first equation is the Hamilton-Jacobi equation, the second equation is the continuity equation which ensures that probability is preserved. In the case of a particle, the probability $P(x,t)$ is interpreted as the probability that the particle is at a given location in configuration space at a given time.

The ensemble Hamiltonian that describes a *quantum particle* subject to a force derived from a potential $V(x)$ is given by

$$H_Q[P,S] = H_C[P,S] + \frac{\hbar}{4}\int dx\, \frac{1}{P}\left(\frac{|\nabla P|^2}{2m}\right). \tag{4}$$

The corresponding equations of motion are

$$\frac{\partial S}{\partial t} + \frac{|\nabla S|^2}{2m} + V + \frac{\hbar}{2}\left(\frac{\nabla^2\sqrt{P}}{m\sqrt{P}}\right) = 0, \qquad \frac{\partial P}{\partial t} + \nabla \cdot \left(P\frac{\nabla S}{m}\right) = 0. \tag{5}$$

These equations are the Madelung equations [9] which describe a quantum particle using Madelung hydrodynamical variables P and S. To see the connection to the Schrödinger equation, introduce the complex *canonical transformation* given by $\psi = \sqrt{P}\, e^{iS/\hbar}$, $\bar{\psi} = \sqrt{P}\, e^{-iS/\hbar}$ which leads to

$$i\hbar\frac{\partial \psi}{\partial t} = \frac{-\hbar^2}{2m}\nabla^2\psi + V\psi, \tag{6}$$

the Schrödinger equation. Thus, in the formulation in terms of ensembles on configurations space, the wavefunction representation given in terms of ψ, $\bar{\psi}$ is related to the representation in terms of hydrodynamical variables P, S via a simple canonical transformation.

Observables are functionals $A[P,S]$ of P and S which satisfy the following basic requirements: an infinitesimal canonical transformation generated by A,

$$P \to P + \epsilon\frac{\delta A}{\delta S}, \qquad S \to S - \epsilon\frac{\delta A}{\delta P}, \tag{7}$$

must preserve the normalization and the positivity of the probability. This leads to [4]

$$A[P, S+c] = A[P,S], \qquad \frac{\delta A}{\delta S} = 0 \text{ if } P(x) = 0. \tag{8}$$

The first of these conditions implies that only relative values of S have physical significance. In addition, we will require that observables be functionals which are homogeneous of degree one in P, i.e., $A[\lambda P, S] = \lambda A[P,S]$, where λ is an arbitrary positive constant [4]. No further conditions are necessary. We emphasize that *the set of observables forms a closed Lie algebra under the Poisson bracket.*

Examples of particle observables include:

- the ensemble Hamiltonians introduced previously; i.e. H_C and H_Q,
- the position observable $X[P,S] := \int dx\, P\, x$,
- the momentum observable $\pi[P,S] := \int dx\, P\, \nabla S$.

Classical and quantum observables may be defined as averages. To define a classical observable, consider a function $f(x,p)$ defined on the standard phase space of classical mechanics and define the corresponding *classical observable* by:

$$C_f \;:=\; \int dx\, P\, f(x, \nabla S). \tag{9}$$

Then, one can show that the Poisson bracket for *classical observables* is isomorphic to the *phase space* Poisson bracket. To define a quantum observable, consider a Hermitian operator $\hat{M}$ defined on Hilbert space. Define the corresponding *quantum observable* by:

$$Q_{\hat{M}} := \langle\psi|\hat{M}|\psi\rangle = \int dq\, dq'\, (PP')^{1/2} e^{i(S-S')/\hbar} \langle q'|\hat{M}|q\rangle. \tag{10}$$

Then, one can show that the Poisson bracket for *quantum observables* is isomorphic to the *commutator* on Hilbert space.

Thus, we have the non-trivial result that the algebras of each of these two classes of observables are isomorphic to the algebras that arise naturally in the phase space and Hilbert space representations of classical and quantum mechanics [4].

3. Non-relativistic free particle via a representation of the Galilean group

We now look at an example of representing the action of Lie algebras within this formalism. We consider the case of a non-relativistic free particle, which may be defined in terms of the representation of the Galilean Lie algebra on a continuous configuration space of three dimensions. We will show how non-equivalent representations of this algebra can arise, leading to inequivalent theories of the free particle [4, 6]. Among them, we will recover the classical and quantum theories of the free particle.

The Galilean Lie algebra has 10 generators,

- A_i : space displacements,
- H : time displacement,
- L_i : space rotations,
- G_i : Galilean transformations ("boosts"),

which satisfy the following Lie algebra,

$$\begin{array}{rclll} \{H, A_i\} &=& 0, & \{H, L_i\} = 0, & \{H, G_i\} = -A_i, \\ \{L_i, A_j\} &=& \epsilon_{ijk} A_k, & \{L_i, L_j\} = \epsilon_{ijk} L_k, & \{L_i, G_j\} = \epsilon_{ijk} G_k, \\ \{A_i, A_j\} &=& 0, & \{A_i, G_j\} = -m\delta_{ij}, & \{G_i, G_j\} = 0, \end{array} \tag{11}$$

where m is the mass of the particle.

The configuration space of a non-relativistic particle is given by Euclidean space R^3. We will look for a representation of the generators in terms of observables. For space displacements and rotations, we set

$$A_i = \int d^3x\, P\, (\partial_i S), \qquad L_i = \int d^3x\, P\, (\epsilon_{ijk}\, x_j\, \partial_k S), \tag{12}$$

which correspond to the momentum and angular momentum observables. A natural choice for the Galilean boost transformations is given by

$$G_i = \int d^3x \; P \; (mx_i - t\partial_i S). \tag{13}$$

This follows from the definition $G_i = (mX_i - tA_i)$ where $X_i = \int d^3x \, P \, x_i$ is the position observable. Finally, the remaining generator H must be a Galilean scalar that satisfies

$$\{H, G_i\} = -m \int d^3x \frac{\delta H}{\delta S} x_i = -A_i = -\int d^3x \; P \; \partial_i S. \tag{14}$$

The condition given by Eq. (14) is not a very stringent one and it leads to a family of solutions $H^{(K)}$ labeled by a functional K,

$$H^{(K)}[P,S] = \int dx \, P \, \frac{|\nabla S|^2}{2m} + K[P,S], \tag{15}$$

where K is any observable that is invariant under translations, rotations and boosts, i.e., K is a Galilean scalar. Different choices of K lead to *different* theories of the free particle. In particular, the family of solutions includes both classical and quantum systems,

$$H_C[P,S] = \int dx \, P \frac{|\nabla S|^2}{2m}, \tag{16}$$

$$H_Q[P,S] = H_C[P,S] + \left[\frac{\hbar^2}{4} \int dx \; P \frac{|\nabla \log P|^2}{2m}\right]. \tag{17}$$

Thus, we have shown that one may represent the action of the Galilean Lie algebra in terms of an algebra of observables and that there are physically inequivalent representations of this algebra leading to distinct physical theories of the free particle.

4. Ensembles on configuration space for discrete systems

Consider now a system with a discrete configuration space. In this case, the formalism of ensembles on configuration space that results is analogous to the one that we already described for the case of continuous configuration spaces, except that we replace continuous variables by discrete variables and functionals by functions [4, 7, 8]. Thus, we need

- a probability $P = (P_1, ..., P_n)$ where n is the number of states, with $P_i \geq 0$ and $\sum_{i=1}^{n} P_i = 1$,
- canonically conjugate quantities $S_1, ..., S_n$,
- an ensemble Hamiltonian $H(P_i, S_j)$,

where $H(P_i, S_j)$ is a function of the P_i and S_j that must satisfy all the conditions that are required of observables. These are that the infinitesimal canonical transformation generated by an observable $A(P_i, S_j)$ must preserve both the normalization and the positivity of the probability and that it must satisfy the homogeneity condition $A[\lambda P_i, S_j] = \lambda A[P_i, S_j]$. The first two conditions imply

$$A(P_i, S_j + c) = A(P_i, S_j), \qquad \partial A/\partial S_i = 0 \text{ if } P_i = 0, \tag{18}$$

where c is a constant [4]. The Poisson bracket for two observables $A(P_i, S_j)$ and $B(P_i, S_j)$ is defined by

$$\{A, B\} = \sum_i \left(\frac{\partial A}{\partial P_i} \frac{\partial B}{\partial S_i} - \frac{\partial A}{\partial S_i} \frac{\partial B}{\partial P_i} \right). \tag{19}$$

IOP Conf. Series: Journal of Physics: Conf. Series **1071** (2018) 012017 doi:10.1088/1742-6596/1071/1/012017

5. Rotational bits

We now introduce the concept of a rotational bit, or *robit*. Our starting point is the observation that a spin-half system may be characterized abstractly as having a set of two-valued observables which generate infinitesimal rotations in three dimensions. Therefore, to describe a robit, we must introduce a two-dimensional configuration space and observables L_k, with $k = 1, 2, 3$, which satisfy the $so(3)$ Lie algebra,

$$\{L_j, L_k\} = \epsilon_{jkl} L_l. \tag{20}$$

The generator of rotations about a given direction will be associated with measurements of the spin in that direction.

We will first consider the case of a single robit and then the case of a physical system that consists of two robits.

5.1. One robit

In the case of a single robit, the configuration space is two-dimensional and thus the associated phase space is four-dimensional. We introduce coordinates (P_+, P_-, S_+, S_-) on this phase space.

It is convenient to define the reduced phase space with coordinates (q_1, p_1) via the canonical transformation

$$q_0 = (P_+ + P_-)/2, \qquad q_1 = (P_+ - P_-)/2, \qquad p_0 = S_+ + S_-, \qquad p_1 = S_+ - S_-. \tag{21}$$

Then, the coordinate q_0 is a constant of the motion, as the probability must be normalized, $(P_+ + P_-) = 2q_0 = 1$. Furthermore, observables can only depend on the difference $(S_+ - S_-)$, thus they cannot be functions of p_0.

We now look for a representation of the Lie algebra of $so(3)$ in this reduced phase space. The identification of generators with average values immediately fixes L_3,

$$L_3 = \tfrac{1}{2}(P_+ - P_-) = q_1. \tag{22}$$

Using Eq. (20), the general solutions for L_1, L_2, up to canonical transformations, are given by

$$L_1 = \sqrt{(1/2)^2 - q_1^2}\,\cos(p_1 + b(q_1)), \qquad L_2 = -\sqrt{(1/2)^2 - q_1^2}\,\sin(p_1 + b(q_1)), \tag{23}$$

where b is an arbitrary function. However, we can always set $b(q_1) = 0$ via the simple canonical transformation

$$q_1 \to q_1, \qquad p_1 \to p_1 - b(q_1). \tag{24}$$

Consider then the case $b(q_1) = 0$. We can now show that single robits are strictly equivalent to quantum mechanical qubits. To see this, we consider the operators $\hat{L}_k$ for spin-half, which satisfy $[\hat{L}_j, \hat{L}_k] = i\epsilon_{jkl}\hat{L}_l$ and may be represented in terms of Pauli matrices,

$$\hat{L}_1 = \frac{1}{2}\begin{pmatrix} 0 & 1 \\ 1 & 0 \end{pmatrix}, \qquad \hat{L}_2 = \frac{1}{2}\begin{pmatrix} 0 & -i \\ i & 0 \end{pmatrix}, \qquad \hat{L}_3 = \frac{1}{2}\begin{pmatrix} 1 & 0 \\ 0 & -1 \end{pmatrix}. \tag{25}$$

If we now introduce the two-component wave function

$$|\psi> = \begin{pmatrix} \psi_+ \\ \psi_- \end{pmatrix} = \begin{pmatrix} \sqrt{P_+}\, e^{iS_+} \\ \sqrt{P_-}\, e^{iS_-} \end{pmatrix}, \tag{26}$$

we find that

$$< \psi|\hat{L}_k|\psi> = L_k. \tag{27}$$

This shows that the most general theory of single robits describes quantum mechanical qubits, leading to no predictions which differ from quantum theory.

5.2. Two robits

In the case of two robits, we need to introduce the configuration space of a pair of spin-half systems, which is four-dimensional, and spin observables that provide a representation of the Lie algebra $so(3) \oplus so(3)$. What makes this case more interesting and far more complicated is that we now have to consider additional physical requirements concerning locality and subsystem independence. In particular, to get a viable physical theory, we must require that

- no signalling is possible via a rotation of either system (evolution locality),
- no signalling is possible via a measurement of either system (update locality), and
- initially independent systems remain independent under local rotations.

We introduce coordinates $(P_{++}, P_{+-}, P_{-+}, P_{--}, S_{++}, S_{+-}, S_{-+}, S_{--})$ on this phase space. Similar to the case of the single robit, we define a six-dimensional reduced phase space with coordinates $(q_1, q_2, q_3, p_1, p_2, p_3)$ via the canonical transformation

$$\begin{aligned} q_0 &= (P_{++} + P_{+-} + P_{-+} + P_{--})/2, \\ q_1 &= (P_{++} + P_{+-} - P_{-+} - P_{--})/2, \\ q_2 &= (P_{++} - P_{+-} + P_{-+} - P_{--})/2, \\ q_3 &= (P_{++} - P_{+-} - P_{-+} + P_{--})/2, \end{aligned} \tag{28}$$

and

$$\begin{aligned} p_0 &= (S_{++} + S_{+-} + S_{-+} + S_{--})/2, \\ p_1 &= (S_{++} + S_{+-} - S_{-+} - S_{--})/2, \\ p_2 &= (S_{++} - S_{+-} + S_{-+} - S_{--})/2, \\ p_3 &= (S_{++} - S_{+-} - S_{-+} + S_{--})/2. \end{aligned} \tag{29}$$

The same as in the previous case, q_0 is a constant of the motion, as the probability must be normalized. Furthermore, observables can only depend on the differences of the conjugate variables $S_{\alpha\beta}$ with $\alpha, \beta = +, -$, thus they cannot be functions of p_0.

We now look for representations of $so(3) \oplus so(3)$ in this reduced phase space. They are given in terms of six generators M_j, N_k that satisfy

$$\{M_j, M_k\} = \epsilon_{jkl} M_l, \qquad \{N_j, N_k\} = \epsilon_{jkl} N_l, \qquad \{M_j, N_k\} = 0, \tag{30}$$

for $j, k = 1, 2, 3$.

Before giving explicit representations of the generators, we introduce definitions of marginal and conditional probabilities. The marginal probabilities are given by

$$\begin{aligned} P_{\pm}^{(1)} &:= P_{\pm+} + P_{\pm-} = (q_0 \pm q_1) = (1/2 \pm q_1), \\ P_{\pm}^{(2)} &:= P_{+\pm} + P_{-\pm} = (q_0 \pm q_2) = (1/2 \pm q_2), \end{aligned} \tag{31}$$

and the conditional probabilities by

$$P_{\alpha|\beta}^{(1)} = \frac{P_{\alpha\beta}}{P_{\beta}^{(2)}}, \qquad P_{\alpha|\beta}^{(2)} = \frac{P_{\alpha\beta}}{P_{\beta}^{(1)}}. \tag{32}$$

These will be needed in the expressions that we give below.

The identification of generators with average values immediately fixes the forms of M_3 and N_3. The average value of spin measurements in the z-direction is given by

$$M_3 = \tfrac{1}{2}(P_{+}^{(1)} - P_{-}^{(1)}) = q_1, \qquad N_3 = \tfrac{1}{2}(P_{+}^{(2)} - P_{-}^{(2)}) = q_2. \tag{33}$$

To fix the functional form of the other generators, we need to consider the physical requirements concerning locality and subsystem independence. We first consider evolution and update locality requirements, i.e., that no signalling is possible via a rotation of either system (evolution locality) and no signalling is possible via a measurement of either system (update locality).

Evolution locality is guaranteed by $\{M_j, N_k\} = 0$, thus it is already satisfied if, as assumed, the generators satisfy Eq. (30).

Enforcing update locality requires substantially more effort. One can show that it requires that M_j and N_k are of the form [4]

$$M_j = P_+^{(2)} L_j(q_+^{(1)}, p_+^{(1)}) + P_-^{(2)} L_j(q_-^{(1)}, p_-^{(1)}) := M_j^+ + M_j^-, \tag{34}$$

$$N_k = P_+^{(1)} L_k(q_+^{(2)}, p_+^{(2)}) + P_-^{(1)} L_k(q_-^{(2)}, p_-^{(2)}) := N_k^+ + N_k^-, \tag{35}$$

where the L_k are *single-robit generators* and the $q_\pm^{(1)}$, $q_\pm^{(2)}$, $p_\pm^{(1)}$, $p_\pm^{(2)}$ are some appropriate functions of the reduced phase space coordinates which will have to be determined. To put M_3 and N_3 in the form of Eqs (34-35), first write them in the form

$$M_3 = \tfrac{1}{2}(P_+^{(1)} - P_-^{(1)}) = P_+^{(2)} \frac{P_{+|+} - P_{-|+}}{2} + P_-^{(2)} \frac{P_{+|-} - P_{-|-}}{2}, \tag{36}$$

$$N_3 = \tfrac{1}{2}(P_+^{(2)} - P_-^{(2)}) = P_+^{(1)} \frac{P_{+|+} - P_{+|-}}{2} + P_-^{(1)} \frac{P_{-|+} - P_{-|-}}{2}. \tag{37}$$

This leads immediately to expressions for $M_3^\pm$ and $N_3^\pm$,

$$M_3^\pm = \frac{q_1 \pm q_3}{2}, \qquad N_3^\pm = \frac{q_2 \pm q_3}{2}. \tag{38}$$

The generators $M_j^\pm$ and $N_k^\pm$ for $j, k = 1, 2$ are derived using $M_3^\pm$ and $N_3^\pm$ with the same type of computations that were carried out for the case of a single robit.

The most general solution is then of the form [4]

$$M_j = M_j^+ + M_j^-, \qquad N_k = N_k^+ + N_k^-, \tag{39}$$

where

$$\begin{aligned} M_1^\pm &= \sqrt{\left(\frac{1/2 \pm q_2}{2}\right)^2 - \left(\frac{q_1 \pm q_3}{2}\right)^2} \cos((p_1 \pm p_3) + b_M^\pm) \\ M_2^\pm &= -\sqrt{\left(\frac{1/2 \pm q_2}{2}\right)^2 - \left(\frac{q_1 \pm q_3}{2}\right)^2} \sin((p_1 \pm p_3) + b_M^\pm) \\ M_3^\pm &= \frac{q_1 \pm q_3}{2}, \end{aligned} \tag{40}$$

with similar expressions for the $N_k^\pm$ which follow from the $M_k^\pm$ via $(q_1, p_1) \to (q_2, p_2)$.

It turns out that the functional forms of $b_M^\pm$ and $b_N^\pm$ can be further specified using various symmetries which come from requiring invariance under particular types of relabeling and subsystem independence [4], leading to

$$b_M^\pm = b\left(\frac{q_1 \pm q_3}{1/2 \pm q_2}\right), \qquad b_N^\pm = b\left(\frac{q_2 \pm q_3}{1/2 \pm q_1}\right), \tag{41}$$

where b is an arbitrary function. However, all of the requirements that we have considered so far do not allow us to fix the form of b *uniquely*.

5.3. Two robits vs. two qubits: the spin correlation matrix

We now examine how a pair of robits is related to a pair of quantum mechanical qubits. In quantum mechanics, the generators of $so(3) \oplus so(3)$ are represented by 4×4 matrix operators in the tensor product space,

$$\hat{M}_1 = \frac{1}{2}\begin{pmatrix} 0 & \mathbf{1} \\ \mathbf{1} & 0 \end{pmatrix}, \quad \hat{M}_2 = \frac{1}{2}\begin{pmatrix} 0 & -i\mathbf{1} \\ i\mathbf{1} & 0 \end{pmatrix}, \quad \hat{M}_3 = \frac{1}{2}\begin{pmatrix} \mathbf{1} & 0 \\ 0 & -\mathbf{1} \end{pmatrix},$$

and

$$\hat{N}_1 = \frac{1}{2}\begin{pmatrix} \sigma_1 & 0 \\ 0 & \sigma_1 \end{pmatrix}, \quad \hat{N}_2 = \frac{1}{2}\begin{pmatrix} \sigma_2 & 0 \\ 0 & \sigma_2 \end{pmatrix}, \quad \hat{N}_3 = \frac{1}{2}\begin{pmatrix} \sigma_3 & 0 \\ 0 & \sigma_3 \end{pmatrix}.$$

When $b(q) = 0$, we have

$$< \psi|\hat{M}_j|\psi >= M_j, \qquad < \psi|\hat{N}_k|\psi >= N_k. \tag{42}$$

Thus *equivalence to a pair of quantum mechanical qubits is achieved for the choice* $b = 0$.

It is natural then to turn to the question of whether there are any further conditions that may force $b = 0$, in which case the theory of a pair of robits would be physically equivalent to the theory of a pair of quantum mechanical qubits, or whether there are choices of $b \neq 0$ which lead to viable theories that are physically distinct from quantum mechanics.

One possibility is to enlarge the algebra of observable beyond the generators M_j and N_k by adding other physically relevant observables, to see if the algebra of this larger set of observables brings any additional conditions that may lead to $b = 0$.

This can be done by considering a more general situation than the one that we have considered so far. Up to now, we have derived generators for the robits using a fixed coordinate system common to both of them, one in which the Cartesian coordinates of the first robit are aligned with the Cartesian coordinates of the second robit, thus leading to a situation where both robits lie essentially on the same axis. We now consider a more general physical situation, one in which

- the z axis for the first system is aligned with the unit vector $\vec{m}$,
- the z axis for the second system is aligned with the unit vector $\vec{n}$.

Then, one can show (see Appendix A) that

$$P_{jk}(\vec{m},\vec{n}) = \frac{1}{4}\left[1 + j\,M_3(\vec{m}) + k\,N_3(\vec{n}) + jk\,C(\vec{m},\vec{n})\right]. \tag{43}$$

where $j, k = \pm 1$ and C is the correlation matrix. As the $P_{jk}(\vec{m},\vec{n})$ are in principle measurable, and furthermore exhaust all measurable quantities, we need to consider the consequences that this has for the algebra of observables.

As $P_{jk}(\vec{m},\vec{n})$ depends on $C(\vec{m},\vec{n})$, it will be sufficient to focus on the correlation matrix. The crucial observation is that the correlation matrix C can be expressed solely in terms of *Poisson brackets* which depend on the generators M_j, N_k and q_3. If we write it down explicitly, we find

$$C = \begin{pmatrix} \{\{q_3, M_2\}, N_2\} & -\{\{q_3, M_2\}, N_1\} & -\{q_3, M_2\} \\ -\{\{q_3, M_1\}, N_2\} & \{\{q_3, M_1\}, N_1\} & \{q_3, M_1\} \\ -\{q_3, N_2\} & \{q_3, N_1\} & q_3 \end{pmatrix}$$

This suggests that we *enlarge* the algebra, which until now was assumed to be generated by the closed set of observables (M_j, N_k), to include at a minimum the set of observables (M_j, N_k, q_3) plus the other generators that correspond to the elements of the correlation matrix.

We must however require that this enlarged algebra also be closed, otherwise it will generate an ever larger set of observables which will include, in addition to the $P_{jk}(\vec{m},\vec{n})$ that we can measure, other quantities that are *not* measurable. A straightforward but tedious calculation shows that the more general algebra generated by the M_j, N_k and the elements of C only closes if $b = 0$.

Thus the more general situation that we have just considered leads to the condition $b = 0$ and thus to the equivalence of the theory of a pair of robits to that of a pair of quantum mechanical qubits. One can show that the generators of this larger algebra turn out to be linear combinations of $SU(4)$ generators, as one would expect by its equivalence to the quantum theory.

6. Concluding remarks

We have carried out a derivation of probabilistic models of ensembles consisting of one and two spin-half particles, known as robits to distinguish them from the standard quantum mechanical qubits. Our starting point was not the usual formulation of spin-half systems using operators and vectors in Hilbert space but the more general formulation of ensembles on configuration space. Thus none of the conditions used in the derivation of the models were formulated using the standard building blocks of quantum mechanics; e.g., operators acting on states in a complex Hilbert space, unitary transformations, etc. The models are based on a minimal set of assumptions:

- the configuration space is a discrete set, labelling the outcomes of two-valued spin observables,
- the spin observables provide a representation of $so(3)$ for the case of a single robit or a representation of $so(3) \oplus so(3)$ for the case of two robits,
- for the case of a pair of robits, additional physical requirements concerning locality and subsystem independence must be taken into account,
- for the case of a pair of robits, the algebra of generators must include all observables and no additional generators that are not measurable.

We have shown that this set of assumptions leads to theories which are equivalent to the quantum theory of one and two qubits. Since it is not obvious that this result is a general one that will hold when such an analysis is applied to more complicated systems, it is of of interest to extend the approach to composite systems that have more particles and/or particles of higher spin.

7. Acknowledgments

We wish to thank Gerald A. Goldin for comments and suggestions regarding this work.

Appendix A. Derivation of $P_{jk}(\vec{m},\vec{n})$

To prove Eq. (43) notice that for arbitrary directions $\vec{m}$ and $\vec{n}$ we have

$$\begin{aligned} M_3(\vec{m}) &:= P_+^{(1)}(\vec{m}) - P_-^{(1)}(\vec{m}) \\ &= P_{++}(\vec{m},\vec{n}) + P_{+-}(\vec{m},\vec{n}) - P_{-+}(\vec{m},\vec{n}) - P_{--}(\vec{m},\vec{n}), \end{aligned} \tag{A.1}$$

$$\begin{aligned} N_3(\vec{n}) &:= P_+^{(2)}(\vec{n}) - P_-^{(2)}(\vec{n}) \\ &= P_{++}(\vec{m},\vec{n}) - P_{+-}(\vec{m},\vec{n}) + P_{-+}(\vec{m},\vec{n}) - P_{--}(\vec{m},\vec{n}). \end{aligned} \tag{A.2}$$

Furthermore, the correlation is defined by

$$C(\vec{m},\vec{n}) := P_{++}(\vec{m},\vec{n}) - P_{+-}(\vec{m},\vec{n}) - P_{-+}(\vec{m},\vec{n}) + P_{--}(\vec{m},\vec{n}). \tag{A.3}$$

IOP Conf. Series: Journal of Physics: Conf. Series **1071** (2018) 012017 doi:10.1088/1742-6596/1071/1/012017

Using the additional relation

$$1 = P_{++}(\vec{m},\vec{n}) + P_{+-}(\vec{m},\vec{n}) + P_{-+}(\vec{m},\vec{n}) + P_{--}(\vec{m},\vec{n}), \tag{A.4}$$

leads to

$$P_{jk}(\vec{m},\vec{n}) = \frac{1}{4}\left[1 + j\,M_3(\vec{m}) + k\,N_3(\vec{n}) + jk\,C(\vec{m},\vec{n})\right], \tag{A.5}$$

as desired.

References

[1] Bohm D 1951 *Quantum Theory* (New York: Prentice Hall)
[2] Einstein E, Podolsky B and Rosen N 1935 *Phys. Rev.* **47**, 777
[3] Bell J S 1964 *Physics* **1** 195
[4] Hall M J W and Reginatto M 2016 *Ensembles on Configuration Space: Classical, Quantum, and Beyond* (Switzerland: Springer International Publishing)
[5] Cohen-Tannoudji C, Diu B and Laloe F 1977 *Quantum Mechanics* (New York: Wiley)
[6] Reginatto M and Hall M J W 2012 *AIP Conf. Proc.* **1443** 96
[7] Hall M J W 2004 *J. Phys. A* **27** 7799
[8] Reginatto M and Hall M J W 2013 *AIP Conf. Proc.* **1553** 246
[9] Madelung E 1926 *Z. Physik* **40** 322

IOP Conf. Series: Journal of Physics: Conf. Series **1071** (2018) 012018 doi:10.1088/1742-6596/1071/1/012018

Quantum Interference, Hidden Symmetries: Theory and Experimental Facts

Ángel S. Sanz

Department of Optics, Faculty of Physical Sciences, Universidad Complutense de Madrid, Pza. Ciencias 1, Ciudad Universitaria E-28040 Madrid, Spain

E-mail: a.s.sanz@fis.ucm.es

Abstract. The concept of quantum superposition is reconsidered and discussed from the viewpoint of Bohmian mechanics, the hydrodynamic formulation of quantum mechanics, in order to elucidate some physical consequences that go beyond the simple mathematical idea of linearly combining vectors in a Hilbert space. Specifically, the discussion turns around the connection between symmetries characterizing the wave function and the behavior in time displayed by the quantum flux when the latter is analyzed in terms of streamlines (Bohmian trajectories). This is illustrated with a series of analytical results and numerical simulations, which include Young's two-slit experiment, counter-propagating wave packets, grating diffraction and quantum carpets (e.g., Talbot carpets), and diffraction under confinement conditions. From the analysis presented it follows that quantum paradoxes appear whenever symmetries related to interference are neglected in the interpretation and understanding of the corresponding phenomena.

1. Introduction

> In making some experiments on the fringes of colours accompanying shadows, I have found so simple and so demonstrative a proof of the general law of the interference of two portions of light [...]. The proposition on which I mean to insist at present, is simply this, that fringes of colours are produced by the interference of two portions of light; [...] the assertion is proved by the experiments I am about to relate [...].

With these remarkable words, Thomas Young started *Experiments and calculations relative to physical optics* [1], published by the end of 1804, although presented to the Royal Society a year before, the 24th of November of 1803. In this work, already considered a classic of the physics literature, Young provides the first empirical evidence of the wave nature of light through one of its most recognizable traits, namely the interference phenomenon. This experiment overthrew 100 years of Newton's corpuscular conception of light[1] and strongly supported Huygens' wave view. The latter would become the dominant trend until 1905. Ironically, just a century after Young's work, Einstein went back to Newton's model, introducing the concept of quantum of light or radiation (*Lichtquant*) to explain the photoelectric effect [2].

Essentially, what Young observed is the following. Consider light coming from a single source (e.g., sunlight), passing through a hole. If an obstacle with two nearly similar apertures is positioned in the way of the light, it splits up. Light then continues traveling until it reaches

[1] Newton published his *Opticks* precisely in 1704, about a century earlier than Young's experimental finding.

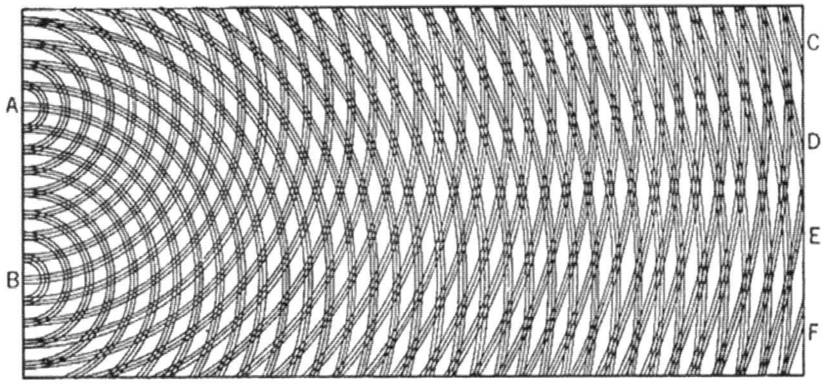

Figure 1. Young's two-slit experiment [1]. Explanation using the Huygens construction with semicircular wavefronts generated at the source slits A and B. In the outgoing or diffracted waves, maxima are denoted with a higher accumulation of semicircles (three closely packed semicircles) and minima with absence of them; the alternation of maxima and minima gives an idea of the concept of wavelength. Far from the slits, on the opposite side of the figure, the wavefronts may coincide (overlap), giving rise to interference maxima; on the contrary, they may also arrive with half a wavelength of delay one another, generating minima (at the locations indicated with the letters C, D, E and F).

another distant screen (see Fig. 1). At this screen, instead of observing essentially two spots of light, there is a redistribution of the light in the form of an alternating pattern of dark and bright spots or fringes. This pattern cannot be in any way accounted for by means of the notion of light ray from geometric optics, as happens with reflection, refraction or even mirage phenomena. To explain the observed phenomenon, Young had to reject Newton's approach and consider Huygens' his wavefront-based construction, as seen in Fig. 1. Thus, after passing through each opening, A or B, the emerging wavefronts increase their radii until they start overlapping. In some regions, the maxima of the wavefronts coincide and we observe a reinforcement of the light intensity (*constructive interference*). In other regions, maxima coincide with minima, thus leading to a cancelation of light (*destructive interference*), which is precisely what we have at the spots C, D, E and F in Fig. 1. In other words, light plus light may equal shadow.

Up to date, Young's experiment constitutes a powerful proof of the wave nature of light, which has reinforced the concept of superposition in physics, with notorious applications in optics and quantum mechanics. In the later case, it has a special practical relevance. Consider a set of solutions $|\Psi_i\rangle$, with $i = 1, 2, 3, \ldots$, all satisfying the time-dependent Schrödinger equation,

$$i\hbar \frac{\partial |\Psi_i\rangle}{\partial t} = \hat{H}|\Psi_i\rangle. \tag{1}$$

Additional solutions can be readily constructed by picking up different vectors from this set and linearly combining them in various proportions, i.e.,

$$|\Psi\rangle = c_1|\Psi_1\rangle + c_2|\Psi_2\rangle + c_3|\Psi_3\rangle + \cdots \tag{2}$$

In this linear combination, the c_i contain information about both the contribution of each $|\Psi_i\rangle$ to the superposition (through the weight $|c_i|^2$) and the mutual phase difference between every two of these vectors [through the phase $(\tan)^{-1}(\mathrm{Im}\{c_i c_j^*\}/\mathrm{Re}\{c_i c_j^*\})$, with $i \neq j$]. The sum (2) is a generalization of Young's empirical result very convenient to solve complex quantum problems by means of known (eigen)solutions (the so-called spectral methods [3], for instance, are based on this idea), and also to investigate the behavior exhibited by quantum systems acted by slits, gratings or lattices (e.g., electron, neutron or molecular diffraction, regardless of the mass and/or complexity of the probing particle [4, 5]).

Apart from its practical orientation, the above idea also introduces a rather natural and deep question: has physical superposition the same meaning as mathematical superposition? In a real experiment one always obtains results that, after analysis, can be recast or modeled in terms of sums like (2). However, the experiment itself is a whole that cannot be decomposed in the same way. A simple illustration of this fact is the experiment performed in 1989 by Akira Tonomura and his group at the Hytachi Central Research Laboratory,[2] consisting in reproducing Young's experiment with the electrons expelled by an electron microscope [11]. In this renowned experiment, electrons are detected individually after having passed through a magnetic equivalent to Young's two slits. Initially, electrons reach the detector following a sort of random distribution. However, as time proceeds and the number of electrons detected becomes larger, an interference pattern starts emerging, where alternating dark and bright fringes can be readily identified. Bright fringes denote regions of higher number of arrivals (i.e., high probability), while dark regions indicate locations that electrons tend to avoid. Accordingly, interference is only going to be observable if the sum (2), for $i = 2$, is taken as a whole, even if mathematically it does not matter to proceed with separate waves [12].

What is then special about having two slits open at the same time beyond the superposition principle? To provide an answer to this question, we are going to appeal here to the notion of symmetry, in particular to a kind of symmetry associated with the full wave function whenever it is split up, as in the case of Young's experiment, that has nothing to do with the usual ones related to conserved quantities. For instance, when matter waves are diffracted by (transmission or reflection) periodic gratings, as mentioned above, Talbot-like patterns arise in the near field (Fresnel regime) [13]. On the other hand, when dealing with scattering off periodic structures, it is known that diffraction features in the far field (Fraunhofer regime) can be directly characterized from the periodic properties of such structures [14]. This is what happens, for example, in solid state physics, where diffraction features and energy band structures are essentially obtained from symmetry properties associated with a single unit cell (Bravais lattice), that is, the unit from which the whole of an ideal crystal solid is formed by repetition (translational invariance). This constitutes the physical basis for the periodic Born-von Karman boundary conditions and Bloch's theorem [15]. Thus, these symmetries have to do precisely with the physical manifestation of the superposition (2), which avoids dealing with the system dynamics in terms of separated waves, but their combination has to be considered as a whole, as mentioned above.

The discussion here thus turns around the physical implications of the symmetries displayed or associated with coherent superpositions, i.e., wave functions that can be recast in the form of the sum (2). To this end, instead of using the typical language in terms of fields or operators, of common use in quantum mechanics, the issue will be tackled with the aid of Bohmian mechanics [16–18], i.e., its hydrodynamic formulation [19]. In it, quantum systems are described as if they were diffusing fluids in a configuration space that can be tracked by means of ensembles of streamlines or trajectories. These ensembles of trajectories are characterized by two properties relevant to our discussion: (1) their gathering and evolution in time is in compliance with the natural evolution of standard wave functions (i.e., they do not arise from any sort of

[2] It is worth noting that the tradition in event-by-event experiments (i.e., particle-by-particle in the present context), produced with dim light, precedes quantum mechanics itself, dating back to the experiments of Geoffrey Ingram Taylor in the first decade of the XXth century [6]. On the other hand, it is also to be highlighted the first realization of Young's two-slit experiment with matter waves, specifically with electrons, by Claus Jönsson [7, 8], as well as the works in the same line of Giulio Pozzi and coworkers a few years later [9], who also produced the first experimental evidence of electron-by-electron interference patterns [10].

IOP Conf. Series: Journal of Physics: Conf. Series **1071** (2018) 012018 doi:10.1088/1742-6596/1071/1/012018

approximation), and (2) this evolution enhances the role of the quantum phase as the main agent ruling quantum dynamics. Accordingly, the work has been organized as follows. The main elements of the Bohmian approach are introduced and discussed in Section 2. In Section 3 some general aspects on quantum superpositions and their analysis with Bohmian mechanics are discussed before starting a discussion on numerical results obtained for different paradigmatic examples, including Young's two-slit experiment, counter-propagating wave packets, grating diffraction and quantum carpets (e.g., Talbot carpets), and diffraction under confinement conditions. The work concludes with a series of remarks summarized in Section 4.

2. About Bohm's quantum picture: A convenient working tool

As in classical mechanics, there are different formulations equally legitimate in quantum mechanics (e.g., Schrödiner's wave mechanics, Heisenberg's matrix formulation, Feynman's path integral approach, etc.), each exploring or based on a different aspect of the quantum phenomenon or process to be analyzed. Using one or another will depend essentially on the simplicity with which we can treat the corresponding problem [19]. One of such formulations is the quantum hydrodynamic approach, also known as Bohmian mechanics. In this formulation, apart from the wave function (or wave field), there is a set of associated auxiliary trajectories that help us to understand how the system spreads throughout configuration space. Initially, such trajectories were conceived as hidden variables [16], because they allowed to monitor the evolution of the quantum system without perturbing its wave function. However, here we are going to leave aside this ontological view (interesting, although out of the scope of the current work) and focus on a more pragmatic level, regarding the trajectories just as streamlines that help us to visualize the time-evolution of the matter wave in space, in analogy to what happens in classical fluid mechanics [20]. This view arises from a joint combination of three experimental facts that the Bohmian description nicely gathers:

- Quantum phenomena occur in real time. Hence the approach to be used in their description should also include this feature, rendering a view of the phenomenon analyzed as evolving in time.
- In an experiment, quantum observables are just statistical outcomes (something particularly evident in low-intensity experiments [4, 11]), so the approach should also allow us to produce the experimental outcomes by means of a specific, non-ambiguous statistical procedure.
- Typically, in such experimental statistics, there is no time-correlation among different detections (detected events) [21], which means that the evolution of one event (a diffracted particle, for example) is unaware of the rest of events (the other diffracted particles, past and future, that contribute to the final diffraction pattern).

Simulators based on these three facts seem to be ideal candidates to explore the physics of quantum systems beyond the bare density approach typically enabled by other quantum pictures.

Now, what are the essential ingredients of Bohmian mechanics? In the case of Bohmian mechanics, the starting point, as formerly considered by Madelung [22] or Bohm [16, 17], is Schrödinger's picture, where the wave function accounting for the state of the quantum system in the position representation is recast in the form of two real-valued fields instead of a complex-valued one by virtue of the nonlinear polar transformation

$$\Psi(\mathbf{r}, t) = A(\mathbf{r}, t) e^{iS(\mathbf{r}, t)/\hbar}. \tag{3}$$

From this relation, we find that the amplitude A is directly connected to the probability density,

$$\rho = \Psi^* \Psi = A^2, \tag{4}$$

while the phase field, defined as

$$S = \frac{\hbar}{2i} \ln\left(\frac{\Psi}{\Psi^*}\right), \tag{5}$$

is going to be related to the quantum current density or quantum flux, $\mathbf{J}$, as seen below. After substitution of the ansatz (3) into the time-dependent Schrödinger equation,

$$i\hbar \frac{\partial \Psi}{\partial t} = -\frac{\hbar^2}{2m} \nabla^2 \Psi + V\Psi, \tag{6}$$

we obtain two coupled real-valued partial differential equations of motion ruling the evolution in space and time of the fields (4) and (5):

$$\frac{\partial \rho}{\partial t} + \nabla \cdot \left(\rho \frac{\nabla S}{m}\right) = 0, \tag{7}$$

$$\frac{\partial S}{\partial t} + \frac{(\nabla S)^2}{2m} + V + Q = 0. \tag{8}$$

Equation (7) is the usual continuity equation, which accounts for the conservation of the probability, while Eq. (8) was regarded by Bohm as a quantum Hamilton-Jacobi equation due to its formal similarity with its classical homolog, although the presence of the Q term, discussed below, sets a remarkable difference. In Eq. (8), the phase field S thus becomes a generalization of the classical mechanical action (or the eikonal, in the context of optics), which was used by Bohm to postulate an associated momentum according to an also generalized Jacobi's law,

$$\mathbf{p}_B = \nabla S. \tag{9}$$

In classical mechanics, this law relates the generalized momentum in phase space with the surfaces of constant action, while here it allows to define a momentum in the position representation (not to be confused with the variable defining the momentum representation), which is perpendicular to surfaces of constant phase. The quantum phase does play an important role in the dynamical behavior displayed by quantum systems, which is typically disregarded because it is not an observable in strict sense (although its effects cannot be neglected).

From Eq. (9) an equation of motion for trajectories or streamlines can be thus straightforwardly obtained:

$$\dot{\mathbf{r}} = \frac{\mathbf{p}_B}{m} = \frac{\nabla S}{m} = \frac{1}{m} \frac{\text{Re}\{\Psi^* \hat{\mathbf{p}} \Psi\}}{|\Psi|^2}, \tag{10}$$

which is what Bohm did back in 1952, showing that the concept of trajectory, obtained after integration of this equation, is not in contradiction at all with quantum mechanics. Notice in the term after the third equality that $\hat{\mathbf{p}} = -i\hbar\nabla$ is the usual momentum operator in the position representation, which means that the equation of motion can be understood as a sort of expectation value for the momentum operator for the quantum state Ψ (or, dividing by the mass m, for an associated velocity operator). Nevertheless, it is convenient to bear in mind that, although Bohm introduced the trajectories after postulating the quantum Jacobi law, this is not a necessary condition, since it can be directly generated from usual quantum mechanics through the standard definition of the quantum flux [23] by noting that

$$\mathbf{J} = \frac{1}{m} \text{Re}\{\Psi^* \hat{\mathbf{p}} \Psi\} = \frac{\nabla S}{m} \rho. \tag{11}$$

This relation establishes a bridge between standard quantum tools and Bohmian mechanics. If, according to the continuity equation, we assume the probability density is "transported" through configuration space by means of a certain velocity field,[3] thus generating a current density, i.e., $\mathbf{J} = \mathbf{v}\rho$, then we readily find that such a velocity field is proportional to Bohm's momentum, $\mathbf{v} = \mathbf{p}_B/m$, as can be seen from (11). The possibility to specify the velocity field $\mathbf{v}$ legitimizes the monitoring of the flow by means of a series of associated streamlines, just as a direct application of the theory of characteristics [27], although there is no way to connect these paths with the actual paths (if any) quantum particles would follow. At this level (and to date, with the available experimental data we have), such paths can only be taken as indicators of how the particles flow on average through configuration space.

Regarding the Q contribution to Eq. (8), it is typically called Bohm's quantum potential and has the functional form of a measure of the curvature of the wave function amplitude,

$$Q = -\frac{\hbar^2}{2m}\frac{\nabla^2 A}{A} = -\frac{\hbar^2}{4m}\left[\frac{\nabla^2\rho}{\rho} - \frac{1}{2}\left(\frac{\nabla\rho}{\rho}\right)^2\right]. \qquad (12)$$

Its identification with a potential is somehow misleading, since it does not come at all from the potential energy term in Schrödinger's equation, but from the kinetic one. Notice that, after substituting the ansatz (3) into the kinetic part of Eq. (6) and then dividing the resulting expression by Ψ, we obtain

$$-\frac{\hbar^2}{2m}\frac{\nabla^2\Psi}{\Psi} = \frac{(\nabla S)^2}{2m} + Q + \frac{\hbar}{2i}\frac{\nabla\cdot\mathbf{J}}{\rho}. \qquad (13)$$

The first two terms provide us with an idea on the system velocity (the first term) and also with how fast it spreads out (the second term), that is, they constitute some sort of kinetic and internal energies [12], respectively, which rule the dynamical behavior of the system state. The last term (the imaginary one), on the other hand, will contribute to the continuity equation, since it reflects the conservation of the probability through the divergence of the quantum flux.

3. Quantum superpositions

3.1. General aspects

Discussion around Bohmian mechanics can be put at two levels. One concerns the predictions or calculation of quantum observables. In this regard, the outcomes rendered by Bohmian mechanics coincide with those provided by any other standard quantum picture, hence they are also compatible with the experiment. So far, no difference. Differences arise in the way how observables are computed. In standard quantum approaches, the key element is the wave function or, in more general frameworks, the density operator, which provide us with a global description of the phenomenon or process described. That is, we get all accessible information (energy spectra, diffraction patterns, etc.) in one go. The Bohmian procedure, on the contrary, is closer in nature to classical-like statistical approaches (e.g., molecular dynamics or Monte Carlo samplings), where the observable is reconstructed after collecting a number of data and performing a statistically treatment over them. In other words, the observable is obtained event by event. Since experiments are typically performed in high-intensity regimes, there is no difference between the standard analysis based on assuming a continuous probability distribution and the Bohmian one, relying on a granular-like analysis in the large-number domain — actually,

[3] Transport phenomena are described by the diffusion equation. In this regard, it is interesting to note the formal analogy between Schrödinger's equation and the diffusion equation on the basis that they both are parabolic partial differential equations (see, for instance, Ch. 12 in [24]). This was an aspect already noticed by Fürth [25] or, later on, also by Comisar [26], identifying the quantity $i\hbar/2m$ with an imaginary diffusion constant. However, the analogy breaks when we noticed that the diffusion equation accounts for a conservation law, while the same does not hold for Schrödinger's equation, since the conserved quantity is the probability density and not the probability amplitude or wave function.

standard treatments are usually more efficient in this case. In the low-intensity regime, though, the Bohmian treatment may be quite convenient, because we can reproduce the experiment on the same event-by-event basis — even if each click or spot observed in the experiment cannot be uniquely associated with a single Bohmian trajectory.

The other level is the dynamical one, i.e., the way how Bohmian mechanics contributes to understand the evolution or diffusion of quantum probability in the configuration space. This is done by stressing the role of the quantum flux and, by means of it, the presence of a dynamical velocity field associated with the quantum phase, which is not the case in other more standard quantum approaches. Once we have seen the rudiments of Bohmian mechanics we shall focus on the equation of motion (10), which is going to be the key element from now on in this work concerning the relationship between quantum symmetry and quantum superposition. In this regard, let us start with the simplest case. It can easily be seen from the Schrödinger equation (6) that any of its solutions is well determined except for a constant phase factor. This means that, if Ψ is a solution, then

$$\Psi_\alpha = e^{i\alpha}\Psi \tag{14}$$

is also a valid solution for any constant α. Taking into account the definitions (4) and (5), we have

$$\rho_\alpha = \rho, \tag{15}$$
$$S_\alpha = S + \alpha. \tag{16}$$

Substituting now the relation (16) into the equation of motion (10) leads us to the invariant form

$$\dot{\mathbf{r}}_\alpha = \dot{\mathbf{r}}. \tag{17}$$

Accordingly, global constant phase factors are irrelevant regarding the dynamics displayed by the system; the trajectories associated with Ψ and with Ψ' will be indistinguishable.

Sometimes the wave function acquires a global time-dependent phase factor, which is necessary in order to keep the consistence of the solutions to Eq. (6) along time. This is the case, for example, of the time-dependent phase factor developed by Gaussian wave packets over time (the direct quantum analog of the Gouy phase in Gaussian beam optics [28]), or the geometric or Pancharatnam-Berry phase developed by energy eigenstates over the course of a cycle (in this case, ϕ would correspond to such energy eigenstates, thus being a time-independent solution to Schrödinger's equation). In general, these wave functions can be recast as

$$\Psi(\mathbf{r},t) = e^{i\varphi(t)}\phi(\mathbf{r},t). \tag{18}$$

Unlike the relation (14), here the full term on the right-hand side is a solution (ϕ is the part of the wave function globally affected by the preceding phase factor, but is not a solution itself, in general). As before, when (18) is substituted into the definitions (4) and (5), we obtain

$$\rho_\Psi = \rho_\phi, \tag{19}$$
$$S_\Psi = S_\phi + \varphi(t)\hbar, \tag{20}$$

where subscripts here denote whether the field is associated with Ψ or with ϕ. However, from the relation (20), again we obtain that the velocity field is invariant in those cases where a time-dependent phase factor is added to the phase field, since

$$\dot{\mathbf{r}}_\Psi = \dot{\mathbf{r}}_\phi. \tag{21}$$

An interesting scenario in this regard is when the wave function displays nodes, around which the phase undergoes a 2π jump once a full loop has been closed. In such cases, although the

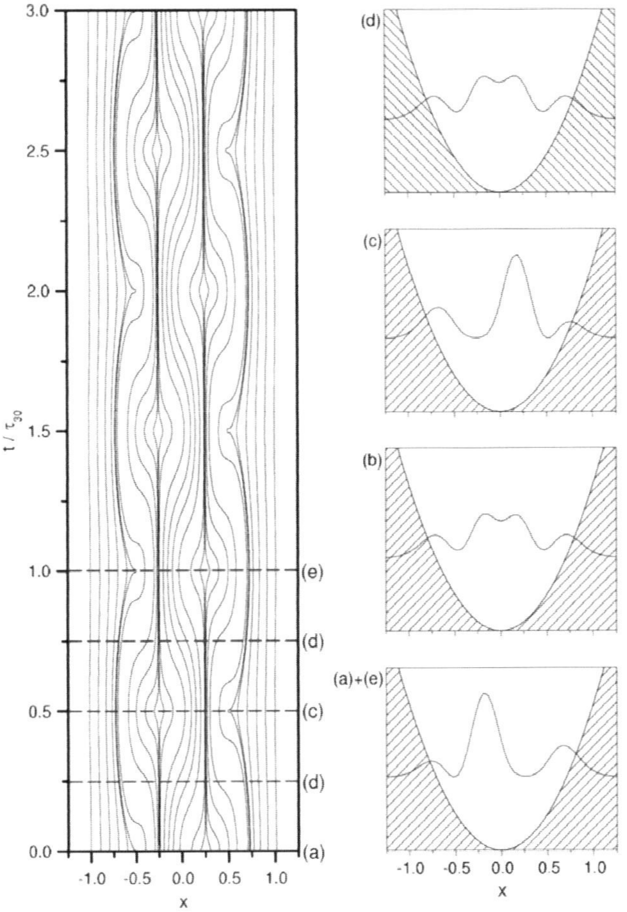

Figure 2. Time-evolution of a two-level state consisting of the superposition of the ground state of the harmonic oscillator and its third excited state. Sets of Bohmian trajectories (left) are useful to determine the periodic evolution of the flow inside the potential, while the usual probability density plots (right) render more general statistical information. Comparing both pictures, it can be noticed that, within a cycle [transition from (a) to (e)], swarms of (probability) trajectories move towards certain regions while avoiding others, this explaining the appearance of maxima in the probability density. This reasoning, close to classical statistical ones (in the end, quantum mechanics is a statistical theory), is different from the usual one based on the idea of constructive and destructive interference discussed above. [For visual purposes, the initial conditions of the trajectories are uniformly distributed along the region covered by the initial probability density, $\rho(x, t = 0)$, instead of being weighted with this density.])

flux is quantized, the trajectories moving around the nodes are not affected by the sudden jump in the phase [29, 30]. Another example is that of a coherent superposition formed by two energy eigenstates of a bound potential. Consider, for an instance, the superposition of the ground state and the third excited state of the harmonic potential, which can be recast as

$$\Psi(x,t) = c_0\phi_0(x)e^{-iE_0t/\hbar} + c_3\phi_3(x)e^{-iE_3t/\hbar} = e^{-iE_0t/\hbar}\left[c_0\phi_0(x) + c_3\phi_3(x)e^{-i\omega_{30}t}\right]. \quad (22)$$

As can be seen in Fig. 2, where a simulation in terms of trajectories (left) and probability densities (right), only the relative phase factor depending on the frequency $\omega_{30} \equiv (E_3 - E_1)/\hbar$ is dynamically relevant; the time-dependent prefactor can be neglected, although it constitutes part of the solution to the Schrödinger equation.

Bearing in mind the behavior of the trajectories displayed in Fig. 2 (left panel), before concluding this section let us consider a general coherent superposition of two states,

$$\Psi(\mathbf{r},t) = \Psi_1(\mathbf{r},t) + \Psi_2(\mathbf{r},t). \quad (23)$$

Each one of the contributing states represents a different possibility for the system, e.g., passing through one slit or another in a Young-type two-slit experiment, or choosing one path or another in Wheeler's delayed choice experiment, or just dealing with a two-level system, as in the case described above. Consider that each state is recast in polar form, as it was done with the ansatz (3). The probability density and the quantum flux can then be recast as

$$\rho = \rho_1 + \rho_2 + 2\sqrt{\rho_1\rho_2}\cos\mathcal{S}, \tag{24}$$

$$\mathbf{J} = \mathbf{J}_1 + \mathbf{J}_2 + 2\sqrt{\rho_1\rho_2}\left(\frac{\nabla\bar{S}}{m} - \frac{\mathcal{Q}}{\hbar}\tan\mathcal{S}\right)\cos\mathcal{S}, \tag{25}$$

where the definitions (4) and (11) have been taken into account, respectively, as well as the new quantities

$$\bar{S} = \frac{S_1 + S_2}{2}, \tag{26}$$

$$\mathcal{S} = \frac{S_1 - S_2}{\hbar}, \tag{27}$$

$$\mathcal{Q} = \frac{Q_1 - Q_2}{2}. \tag{28}$$

From these expressions, the associated velocity field reads as

$$\dot{\mathbf{r}} = \mathbf{v} = \frac{\rho_1}{\rho}\mathbf{v_1} + \frac{\rho_2}{\rho}\mathbf{v_2} + 2\sqrt{\frac{\rho_1\rho_2}{\rho^2}}\left(\frac{\nabla\bar{S}}{m} - \frac{\mathcal{Q}}{\hbar}\tan\mathcal{S}\right)\cos\mathcal{S}. \tag{29}$$

According to this expression, the superposition generates dynamics where each partial wave contributes with a given weight, ρ_i/ρ, with $i = 1, 2$. There is also an extra term containing the averaged phase field and the difference between quantum potentials that makes the dynamics highly nonlinear. It is by virtue of this nonlinearity that, for instance, in Young's experiment we can eventually observe interference fringes, which would never arise if only the former weighted sum would contribute [31–33]. Notice that although Eq. (29) has been derived or taken under a Bohmian view, it is just standard and simple quantum mechanics, without the artificial prejudice that trajectories are incompatible with quantum mechanics or, if trajectories are allowed, that they are associated with (the also artificial) idea of particle-wave duality.

Next we are going to discuss the physics involved in some particular illustrative examples where the above superposition holds. Further details related either to the associated models or to computational aspects can be found in the references provided in there).

3.2. Young's two-slit experiment

Apparently, we all may agree that, at least on a formal level, Young's two-slit experiment is well understood, even if the experiment itself is still considered the mystery having "in it the heart of quantum mechanics", quoting Feynman [34]. The situation changes, though, if it is reexamined from a Bohmian perspective, that is, if we switch from one quantum picture to another. Although this may sound a bit odd, the hydrodynamic language of Bohmian mechanics is able to highlight a kind of symmetry that is not apparent or easy to see with the other pictures (although, as mentioned above, it could be inferred). This symmetry is associated with the well-known *non-crossing rule* typically associated with Bohmian mechanics [12, 18].

Consider the Young two-slit experiment illustrated in Fig. 1 in terms of the Huygens' construction. In brief, in the experimental setup there are two holes or slits on a screen, such that when the matter wave reaches this screen, it undergoes diffraction at both openings. The diffracted waves (behind the screen) form a coherent superposition that can be described by the ansatz (23). If the openings are identical, and so the diffracted waves, Eq. (29) forbids trajectories started in the regions covered by each wave packet from crossing the symmetry line dividing the system into two mirror images [12, 19], as can be seen in Fig. 3. The behavior displayed by

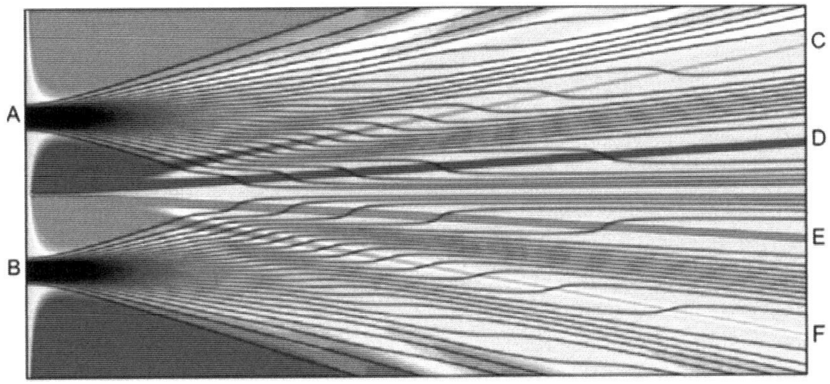

Figure 3. Bohmian explanation of the Young two-slit diagram displayed in Fig. 1. The contour-plot represents the velocity field, with the gradual transition from red to blue denoting the passage from lowest values of this field to larger ones. As initial condition a sum of two Gaussian wave packets has been considered. The trajectories shown (solid lines) illustrate how the quantum probability flows independently emanates from each slit and then, after some time, it evolves into a series of separate swarms flowing along the interference maxima directions, while avoiding others corresponding to the interference minima (regions C, D, E and F). For visual purposes, the initial conditions of the trajectories are uniformly distributed along the regions covered by the initial probability density, $\rho(x, t = 0)$, instead of being weighted with this density. (Further analytical and numerical details can be found in [19].)

the trajectories is a manifestation of the above non-crossing rule, which avoids trajectories from crossing the same point of the position space at the same time. In this particular simulation the diffracted waves have been described in terms of a coherent superposition of two identical Gaussian wave packets accounting for the transverse degree of freedom, with each wave packet having the functional form

$$\Psi(x,t) = \left(\frac{1}{2\pi\tilde{\sigma}_t^2}\right)^{1/4} e^{-(x-x_t)^2/4\sigma_0\tilde{\sigma}_t + ip(x-x_t)/\hbar + iEt/\hbar}, \tag{30}$$

where $x_t = x_0 + vt$, $p = mv$, $E = p^2/2m$, and

$$\tilde{\sigma}_t = \sigma_0 \left(1 + \frac{i\hbar t}{2m\sigma_0^2}\right). \tag{31}$$

In the case dealt with here, $v = 0$ and the initial position for each wave packet is $x_0 = x_\pm = \pm d/2$, where d the distance between their centers and $\pm$ makes reference to the relative position of each wave packet.

The symmetry mentioned above becomes apparent already with tools that we have at hand in quantum mechanics, as it is the case of the probability density [19]. A close inspection to the evolution of this quantity provide us with information on how the two diffracted beams increase their size according to the ratio

$$\frac{\sigma_t}{\sigma_0} = \sqrt{1 + \left(\frac{\hbar t}{2m\sigma_0^2}\right)^2}, \tag{32}$$

with $\sigma_t = |\tilde{\sigma}_t|$. At some time, when their width σ_t is large enough, they start overlapping and, some time later, their full overlap gives rise to the appearance of the well-know maxima and minima of interference (modulated by the corresponding single-slit diffraction pattern [35]). Because of the ideal conditions chosen on purpose in this example, the pattern in space and time displayed by the probability density is going to be symmetric with respect to the center between the two slits ($x = 0$). Now, the question that we pose here is whether this has any deeper physical implication, and the answer is yes. It is clear that the probability density by itself is not enough, because it only tells us that there are some directions along which it is more likely to find or detect the system than others, but still we can appeal to traditional arguments to explain why, as it was done in Section 1. It is precisely a need to go beyond this simple approach why we are going to consider the velocity field defined by Eq. (10) for this particular problem and the trajectories rendered after integration.

In Fig. 3, the contour-plot represents the evolution of the velocity field along time, with the transition from red to blue colors denoting the increase from negative to positive velocity values (green denotes nearly zero velocities). At $t = 0$ the Gaussian wave packet (30) is real-valued, so a superposition of two of such wave packets is also real-valued, which implies an initial zero velocity field. As time proceeds, diffraction causes that both wave packets start developing a phase, and therefore a relative phase between them starts appearing. Because both evolve exactly in the same way (notice how to the left in both cases velocities become negative and to the right, positive), the matching at the center between them becomes a sort of conflict line. As a consequence, the (quantum) flux associated with one of the wave packets (slits) cannot penetrate the region dominated by the other, an vice versa, which can be better appreciated when the picture is recast in terms of a set of streamlines or trajectories. These trajectories evolve towards the different interference maxima, as expected, but more importantly those associated with the wave packet labeled with A never cross to the domain of wave packet B, and vice versa. There is a perfectly symmetric picture for the flux, which has nothing to do with the quantum potential itself (since it is nearly zero in the far field, as also happens with the velocity field).

Bohmian trajectories themselves are not experimentally accessible, because, as mentioned above, they are just streamlines along which probability flows rather than corresponding to actual paths followed by real particles. If we are talking about particles diffracted by two slits, as in the case illustrated in Fig. 3, this means that, in principle, we do not know whether such particles travel along those trajectories, because there is no empirical evidence on that. However, we know that, on average, a number proportional to $\rho(\mathbf{r}, t)$ has passed through a region $d\mathbf{r}$ around the point $\mathbf{r}$ in the position space at a time t. Nonetheless, we *do* know that the transverse momentum in that direction is given by $\nabla S(\mathbf{r}, t)$, because this is a property linked to the statistical ensemble dynamics, which is what Schrödinger's equation eventually describes. Inspired by the Bohmian ideas, an experiment performed at the University of Toronto in 2011 by Steinberg and coworkers [36] showed that, by using *weak measurements* (a slight perturbation on the quantum system before performing the true measurement, namely a von Neumann measurement, which helps to obtain complementary information according to Vaidman *et al.* [37]), the transverse momentum could be measured on average. Although the experiment was performed with photons, i.e., zero-mass particles, and Bohmian mechanics, as it was formerly devised, was employed to describe massive ones (or, in general, any quantum system characterized by a nonzero mass), the experimental outcomes were in compliance with the expected transverse momentum, although this momentum is associated with Poynting's vector (as formerly suggested by Prosser [38] back in 1976) rather than with Bohm's momentum, and from them an ensemble of paths could be reconstructed, in good agreement with the corresponding trajectories [39–42].

3.3. General counter-propagating wave packets

The system analyzed in the previous section is a singular case of a more general scenario characterized by superpositions like (23). In particular, for the purpose here, let us consider an initial superposition of two Gaussian wave packets like (30),

$$\Psi_0(x) = c_1 e^{-(x+d/2)^2/4\sigma_{0,2}^2 + iv(x+d/2)/\hbar} + c_2 e^{-(x-d/2)^2/4\sigma_{0,1}^2 - iv(x-d/2)/\hbar}, \quad (33)$$

which represents two counter-propagating wave packets initially moving towards each other. In the case of Young's two-slit experiment, we set $v = 0$ for simplicity. Thus, only the time-evolution (spreading) of the wave packets was responsible for the generation of phase factors [see the complex spreading rate (31)] that eventually led to interference traits. In the case described by the superposition (33) there are going to be two competing phase factors [12], one coming from the spreading, and another one from the propagation speeds. Actually, the latter implies the presence of a (space-dependent) relative phase even at $t = 0$, unlike the two-slit scenario. Taking this into account, the speeds of the wave packets, v, can be selected in such a way that they cross each other and reach positions opposite to their initial ones with an almost negligible spreading. For instance, if $x_{f,1} = x_{0,2} = d/2$, the time lasted in going from $x_{0,1}$ to $x_{f,1}$ is $t = d/v$. Thus, using the expression (32), if we wish the Gaussian wave packets only increase their widths in about 10% when moving from the initial position to the opposite one (i.e., $\sigma_t = 1.1\sigma_0$), the speed has to be

$$v \approx 2.2 \times \left(\frac{\hbar d}{2m\sigma_0^2} \right), \quad (34)$$

or, equivalently,

$$\frac{v}{v_s} \approx 2.2 \times \left(\frac{d}{\sigma_0} \right), \quad (35)$$

where $v_s = \hbar/2m\sigma_0$ is the speed or rate at which the wave packet spreads out [12]. Putting some more numbers, if $\sigma_0 \approx d/20$, in order to ensure a negligible initial overlapping, then $v/v_s \approx 44$, which gives an idea of the slow wave-packet spreading along the full evolution of the system. When this conditions are fulfilled, interference is a temporary feature localized around the region where the wave packets cross each other (no traces of interference remain later on). Now, is this picture of crossing wave packets correct, even if the description of the phenomenon requires it?

Such has always been the usual approach to superpositions: the superimposed waves are distinguishable, because each contributing state is itself a separated solution of Schrödinger's equation,[4] as in Eq. (2), with a well-known time-evolution. However, according to what we have seen in the previous section, that idea may become misleading. Of course, in the case of Young's experiment it is difficult to determine what is going on after the two wave packets have reached a rather large spreading and fully overlapped. In the case dealt with here, though, interference is localized in both space and time, as said above; before and afterwards we can perfectly distinguish the propagation of both wave packets. If we appeal to the same picture, as before, we observe again the same kind of symmetry. Accordingly, the quantum flux associated with one wave packet cannot cross to the region or domain of the other, and vice versa. From this, an interesting alternative picture arises: the outgoing wave packets after interference are not the crossed wave packets, but the initial ones after having bounced backwards (at the region where interference took place). And this is so regardless of the shape of the wave packets, their propagation speeds, or their relative weights in the superposition. Nevertheless, notice that this is just a picture about the quantum flux, which, in the end of the day, is not an observable quantity [43] In other words, although the flux may be a convenient tool to ascertain certain particular properties of the theory, any further physical interpretation must be taken with care, since only a precise experiment can validate it. For example, in the experiment described at the end of Sec. 3.2, the trajectories cannot be associated with actual photon paths, because the only

[4] This notion of distinguishability (or indistinguishability) should not be confused with the one applied to quantum particles (states), defined in terms of complete sets of communing observables.

observable or measured quantities are densities, from which the transverse flow is (indirectly) obtained. Anyway, even if it is only at a formal level, if we consider the initial superposition (33), the following cases of interest can be specified:

- If the two counter-propagating wave packets have the same initial spreading ($\sigma_{0,1} = \sigma_{0,2}$) and weight in the superposition ($|c_1|^2 = |c_2|^2 = 1/2$), their speeds are exchanged after interference, in analogy to the transfer of momentum between two classical particles of mass m in a context of elastic scattering.
- Analogously, if the wave packets have different initial spreadings ($\sigma_{0,1} \neq \sigma_{0,2}$), but still the same weight in the superposition, they exchange their corresponding spreadings, giving the impression that they have crossed each other, although this is only in appearance. What actually happens is that the corresponding fluxes undergo a rearrangement (the trajectories associated with the wave packet with smaller spreading have dispersed more, while the trajectories associated with the wider wave packet have get closer). Again, this a case analogous to momentum transfer in a classical elastic scattering process between two particles with different masses.
- If the wave packets are not properly normalized or their weights are different (notice that both conditions are equivalent), part of the trajectories associated with one of the wave packets (the more prominent one in the superposition) pass, after interference, to form part of the ensemble of trajectories associated with the other wave packet. In this case, there is a gain of trajectories by one of the wave packets and a loss by the other, in analogy to a classical case of inelastic scattering.

Such behaviors are expected to survive even in those conditions where the action of the quantum potential is somehow canceled out by removing its contribution within Schrödinger's equation. This procedure leads to a nonlinear Schrödinger equation proposed by Nathan Rosen back in the mid-1960s [44–46],

$$i\hbar\frac{\partial\Psi}{\partial t} = -\frac{\hbar^2}{2m}\nabla^2\Psi + (V - Q)\,\Psi, \tag{36}$$

which gives rise to a Hamilton-Jacobi equation without the contribution of the quantum potential. This equation has recently been explored in the context of wave-packet superposition [47, 48], showing that even in the limit $\hbar \to 0$ classical mechanics is not fully recovered, but the above symmetries persist. This behavior, already discussed by Rosen in the 1980s [49], has also been commented by others, like Berry [50], in the context of the standard formulation of quantum mechanics. To understand this behavior in simple terms, notice that the evolution of wave functions take place in spaces of reduced dimensionality, that is, either described by positions or described by momenta. When the above equation is recast in the form of a continuity equation and an apparently classical Hamilton-Jacobi equation, the same property is maintained. That is, in the transformation from Ψ to (ρ, S) there is never a condition that implies unfolding positions and momenta, and therefore breaking the non-crossing rule. In classical mechanics, on the contrary, positions and momenta are independent variables, which specify an expanded space, namely the phase space.

Taking into account the above facts, one may wonder whether the analogy with a classical scattering process can be further exploited. And the answer is positive. In the classical scattering process one may simplify the description by appealing to the physics of an effective mass (the full-system reduced mass) acted by an also effective force (e.g., a central force when dealing with masses or charges). In the problem of interference (the fully symmetric case, to simplify) an analogous approach can be considered if we notice that the collision of a single wave packet with a hard (infinite) wall renders, at the time of maximal proximity to the potential, an interference pattern analogous to that displayed by the two counter-propagating wave packets at the time of

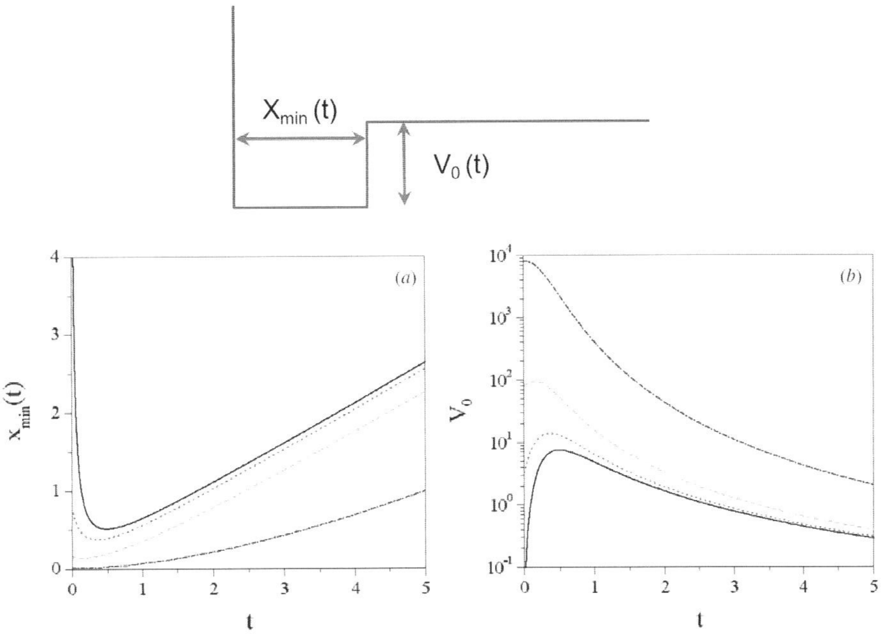

Figure 4. Simple toy-model that allows an effective substitution of a coherent superposition of two wave packets by the scattering of a single wave packet with a hard wall followed by a dynamical attractive well with time-dependent width and depth (both as being functions of the parameters defining the wave packets) [12]. Such time-dependence is explicitly shown in the two panels below for four different cases with increasing value of the propagation speed of the wave packets. The black solid like denotes a scenario recreating Young's two-slit experiment, while the blue dashed-dotted line indicates just the opposite regime, namely a scenario where the wave-packet spreading is negligible compared to the propagation speed. (Further analytical and numerical details can be found in [12].)

maximal interference. The only difference between both cases is that in the single wave-packet problem all maxima have the same width, while in the two wave-packet problem, if we divide it by a half (along the symmetry line), one of the maxima has have width. To circumvent this problem, we may assume that the collision is not just against a hard wall, but that there is a well close to the wall, such that part of the wave packet may get attached to it after the collision, with the corresponding resonance have the appropriate width, i.e., half the width of the interference maxima.

A simple toy model that gathers such features can be constructed [12], where the interference between two different wave packets is *effectively* described by the (self)interaction of a single wave packet with its own reflection. This model, sketched in Fig. 4, consists of a hard repulsive wall followed by square-well attractive region, with both the width and depth of the well being time-dependent:

$$V(t) = \begin{cases} 0 & x < x_{\rm min}(t) \\ -V_0[x_{\rm min}(t)] & x_{\rm min}(t) \leq x \leq 0 \\ \infty & 0 < x \end{cases}, \tag{37}$$

where

$$V_0[x_{\rm min}(t)] = \frac{2\hbar^2}{m} \frac{1}{x_{\rm min}^2(t)}, \tag{38}$$

$$x_{\rm min}(t) = \frac{\pi}{\dfrac{2p}{\hbar} + \dfrac{\hbar t}{2m\sigma_0^2}\dfrac{x_t}{\sigma_t^2}} = \frac{\pi\sigma_t^2}{\dfrac{2p\sigma_0^2}{\hbar} + \dfrac{\hbar t}{2m\sigma_0^2}x_0}, \tag{39}$$

with $p = mv$. The time-dependence in the width and the depth of the well is precisely related to the fact that the resonant state has to asymptotically increase its width and, at the same time, the shallowness will decrease. These two quantities are represented in the two lower panels of Fig. 4 for different values of the propagation speed of the wave packets, where the black solid line denotes a typical Young-like regime and the blue dashed-dotted line a case with negligible dispersion. As can be noticed, in the Young-like regime, the minimum width of the well is rather large, which means that the wave packet is fully inside the attractive well, while in the case with negligible dispersion the wave packet starts outside it. This already gives a clue on why in one case interference will be a persisting trait, while in the other it is only temporary (and spatially localized). Nevertheless, also notice that for the Young-type scenario the potential is rather shallow, while in the other scenario the depth is remarkable. This sort of play between width and depth can then be rather misleading: apparently meaningless potential wells may lead to important quantum effects (persistent interference), while the opposite is not necessarily true (remember that the scenario with negligible dispersion can be well approximated by a typical ray-based description, as it is commonly done when explaining atomic Mach-Zehnder interferometers [51, 52] or Wheeler's delayed choice experiment [53]).

Although such phenomena have not been experimentally confirmed yet, as it is the case of the fluxes in Young's experiment, it is worth noting that they constitute a key step in matter-wave interferometry, where an incident matter wave is split up and then, the two partial waves, are made to overlap again on some spot, as it happens in atomic Mach-Zehnder interferometers [13] or in the renowned Wheeler delayed-choice experiment [54] when both setups are left in open configuration. Although we are not going to enter into details (the interested reader is kindly addressed to the corresponding references), it is worth noting that in either case the spreading of the two split wavefronts is negligible with their passage through the interferometer and that the performance is due to a broken or restored phase symmetry at the end of the interferometer, by disregarding or inserting an element (a grating in the first case and and a beam splitter in the second case) that is able to undo the phase splitting at the entrance of the interferometer by precisely an analogous element.

3.4. Grating diffraction and quantum carpets

So far we have discussed the behavior of quantum systems described by two-state superpositions in the position representation (remember the two-state superposition with energy eigenvalues of the harmonic oscillator from Section 3.1), with Young's two-slit experiment being a paradigm of such systems and counter-propagating wave packets a generalization of such a case. What happens if we increase the number of slits or, equivalently, the number of contributing waves to the coherent superposition, trying a general superposition like (2)? This is precisely the case shown in Fig. 5, where we go from a single wave packet to a large number of them in order to illustrate a transition from single-slit diffraction to diffraction by a N-slit grating [55] (as before, we only focus on the transverse direction). As can be noticed, with increasing number of wave packets the emergence of a highly ordered pattern or structure becomes more and more apparent in the near field or Fresnel regime, i.e., the space region behind an aperture or obstacle where the matter wave still strongly feels the action of such a diffracting element, and that is characterized by a rearrangement or quantization of the momenta that later on, far away from the diffracting element, will give rise to the Fraunhofer features in the position representation [35, 56]. Such an emergent structure is called a *Talbot carpet* in analogy to the so-called Talbot phenomenon in optics [57], and that has direct atomic analogs [58].

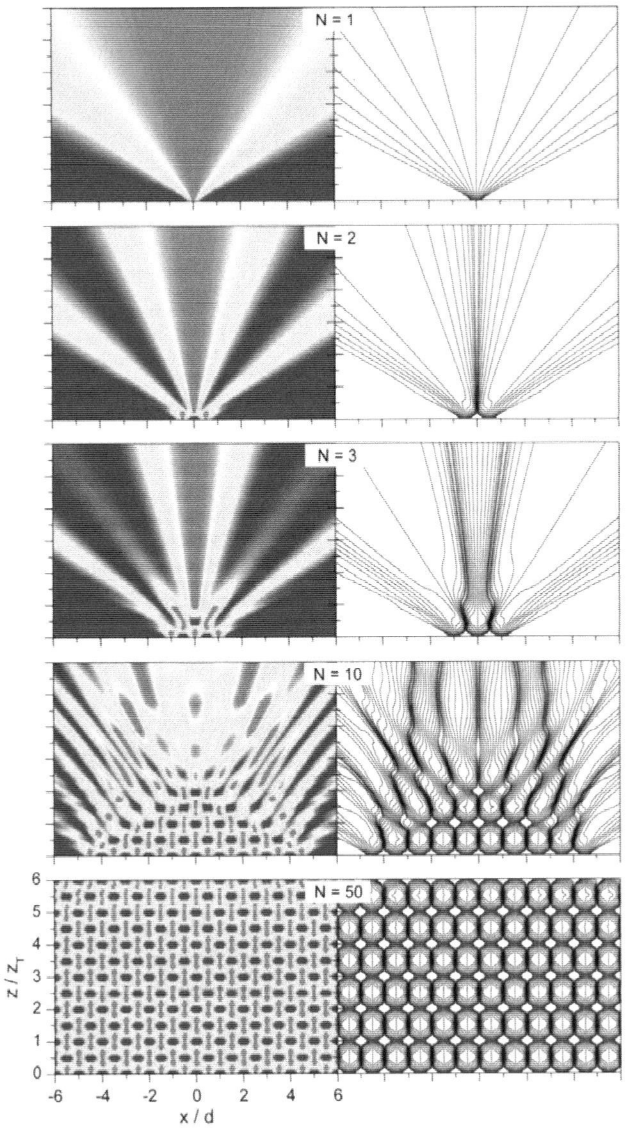

Figure 5. Emergence of a Talbot carpet as the number of slits (diffracted beams) increases. The probability density along the transverse direction (x) as a function of the longitudinal coordinate [z, normalized to twice the Talbot distance, Eq. (eq39)] is displayed in the form of contour-plots on the left column. The gradual transition from blue to red denotes increasing values of the density; for a better visualization, at any z value the maximum of the probability density has been normalized to unity. The Bohmian trajectory picture for each case is displayed on the corresponding right-column panels. Also for visual purposes, the initial conditions of the trajectories are uniformly distributed along the regions covered by the initial probability density, $\rho(x, z = 0)$, instead of being weighted with this density. (Further analytical and numerical details can be found in [55].)

In order to analyze the Talbot structure with matter waves, here we are also going to proceed as in previous sections and consider only the transverse coordinate in the description for practical purposes. So, if the distance between centers of slits (which are assumed to be characterized by a Gaussian transmission function [13]) is d, the grating is perfectly periodic (infinite extension along x), and the (de Broglie) wavelength of the incident matter wave is λ, an exact replica of the initial diffracted wave can be found at integer multiples of the distance

$$2z_T = \frac{2d^2}{\lambda}, \tag{40}$$

where z_T is called the *Talbot distance*. This distance is, actually, the one at which we can observe a full replica of the initial wave, although it is half a period of the grating (i.e., $d/2$) shifted with respect to the latter, which means that there are three space and time translational symmetries:

$$\Psi(x + \ell d, t) = \Psi(x, t), \tag{41}$$

$$\Psi(x, t + \tau_T) = \Psi(x, t), \tag{42}$$
$$\Psi(x + d/2, t + \tau_T/2) = \Psi(x, t), \tag{43}$$

with $\ell = \pm 1, \pm 2, \ldots$, and where the recurrence time $\tau_T = md^2/\pi\hbar$ is the *Talbot time*. Apart from this evident symmetry, by increasing the period of the grating it would also possible to observe other fractional recurrences [59, 60], although we are not going to enter into details in this regard, since they will be discussed later on when dealing with the problem of bound diffraction in next section.

As in previous sections, we are going to switch to the trajectory picture to elucidate whether the symmetries displayed by the Talbot carpet have further physical implications (although at this point it is already clear they will have). To this end, it is useful to define the domain of a slit in analogy to the concept of unit cell in solid state physics or condensed matter [15], which in the present context is understood as the region spanning a distance d around the center of a given slit [if the center of the slit is x_0, the slit unit cell spans the region $x \in [x_0 - d/2, x_0 + d/2)$]. Proceeding this way, we readily notice that the translational symmetry of the carpet gives rise to a kind of unbound channeling structure: although there are no physical boundaries at all (coming from a potential function or even the quantum potential, the trajectories started in the domain of a given unit cell keep evolving in a way analogous to being confined within an infinite box, without crossing to adjacent unit cells. In other words, the whole pattern can be reconstructed by repeating exactly the same structure, which provides a physical picture to a well-known result in solid state theory, namely Bloch's theorem, as well as a physical interpretation in the position representation for the Born-von Karman boundary conditions [15], thus setting a bridge between two different disciplines, as they are the solid state physics (typically understood in the reciprocal momentum and energy representations) and the matter-wave interferometry or matter-wave optics (typically explained within the position representation). As can be noticed, in the problem of Talbot carpets the wave function behind the grating can be described as a result of the linear combination of a series of diffracted waves. However, the fact that there is a periodicity implies that, if such diffracted waves are decomposed in terms of plane waves (with well-defined energies and momenta), most components must be neglected, only remaining a countably infinite set of allowed plane waves. In other words, periodicity in this context is equivalent to quantize the momentum (and energy), as also happens in solid state physics with Bloch's waves and the appearance of energy bands after introducing periodic boundary conditions. This can be better appreciated by means of a simple example. Consider that the state describing the diffraction through the slits of the grating is given by Gaussian wave functions like (30). Applying the translational invariance condition (41), such wave function becomes [55]

$$\Psi(x, t) = \sqrt{\frac{1}{d}} \left(\frac{8\pi\sigma_0^2}{d^2} \right)^{1/4} \sum_{|n|=0}^{\infty} e^{-\sigma^2 p_n^2/\hbar^2 + ip_n x/\hbar - i\omega_n t}, \tag{44}$$

where $p_n = 2\pi n\hbar/d$ and $\omega_n = 2\pi^2 n^2 \hbar^2/md^2$ are quantized quantities. Notice that, although the analysis has been restricted to what happens within a single period of the grating, the fact that the wave function (44) is given in terms of planes waves serves to replicate the behavior within a single unit cell all the way through the whole grating. If the number of slits does not go to infinity, though, then the pattern is restricted to a certain space area, showing an approximately triangular shape, as can be seen in Fig. 5 for low N (for large N we need to go beyond the frames shown in the figure); beyond this area the typical Fraunhofer grating diffraction traits start appearing, which is precisely what we observe in the far field in atomic Mach-Zehnder interferometers [13], but also in diffraction by reflection gratings, as it is the case of atom scattering off periodic surfaces [61, 62]. In such cases, Eq. (44) cannot be used and the exact form for the full superposition has to be considered.

Now, apart from such important physical consequences concerning the application of these ideas to the description and understanding of matter-wave diffraction processes and the analysis of different types of matter-wave interferometry techniques and their outcomes [13], the approach can be smartly used to devise and implement efficient numerical tools on an event-by-event basis. Observe that the calculation of the trajectories does not require either considering the full expression for the superposition with all Gaussians, but just (44), since the periodicity along the full transverse direction is already contained within this relation in a compact form (all the dynamical information encoded in a single Gaussian). This information is automatically transferred to the Bohmian trajectories. Thus, substitution of the ansatz (44) into the equation of motion (10) renders

$$\dot{x} = \frac{\sum_{i,j} p_i e^{-\sigma_0^2(p_i^2+p_j^2)/\hbar^2} \cos[(p_i - p_j)x/\hbar - (\omega_i - \omega_j)t]}{\sum_{i,j} e^{-\sigma_0^2(p_i^2+p_j^2)/\hbar^2} \cos[(p_i - p_j)x/\hbar - (\omega_i - \omega_j)t]} \tag{45}$$

with $i, j = \pm 1, \pm 2, \ldots$ This equation admits a similar analysis to Eq. (29), from which one gets an interesting answer: the periodic boundary conditions translate, in the trajectory picture, in a sort of impenetrable walls that constrain the evolution of the trajectories to a series of channels. Only if the number of openings is finite, such channels will get diluted far from the grating and will eventually disappear, letting the trajectories to start gathering towards the different Fraunhofer interference features. It can be shown [13, 55] that in the far-field limit, for a relatively large number of openings, the flux is quantized and acquires a ladder shape, with each step being associated with a diffraction order, as seen in Fig. 6 — if the number of openings considered is rather small (see upper panels of Fig. 6), then the structure of the transverse quantum flux is closer to what Steinberg and coworkers observed in the case of photons [33, 36].

3.5. Bound diffraction

In Section 3.5 we have said (and illustrated with simulations) that the channel structure found when considering perfectly periodic gratings is analogous to what we find when a true channel is analyzed, i.e., the confinement of a matter wave inside an infinite square potential. This situation essentially corresponds to diffraction under confinement, where such an analogy can be easily shown [55] if instead of considering the symmetry imposed by the boundary condition (41), we impose on the Gaussian wave packet the condition that the constituting plane waves have to be eigenstates of the infinite potential (i.e., solutions that vanish, as well as their first derivatives, at the boundaries of the potential). In this case, and assuming that the Gaussian is nearly zero close to the borders, we obtain

$$\Psi(x,t) = \sqrt{\frac{8}{d}} \left(\frac{2\pi\sigma_0^2}{d^2}\right)^{1/4} \sum_{n=0}^{\infty} e^{-\sigma_0^2 p_n^2/\hbar^2 - iE_n t} \cos(p_n x/\hbar), \tag{46}$$

where $p_n = (2n+1)\pi\hbar/d$ and $E_n = p_n^2/2m$. These solutions, analogous to the example displayed in Fig. 2 (although in that case it was for a harmonic oscillator), have full recurrences whenever the time is an integer multiple of the period

$$\tau_r = \frac{2\pi}{\omega_{1,0}} = \frac{2\pi\hbar}{E_1 - E_0} = \frac{md^2}{2\pi\hbar}, \tag{47}$$

which coincides with the time at which we observe the first Talbot recurrence in the grating (the half-a-period shifted), at $\tau_T/2$. It is worth noting here that the symmetry displayed by the wave function in this case is of the kind described by the relations (19) and (20), since

$$\Psi(x, t + \tau_r) = e^{i\varphi(t)} \Psi(x,t). \tag{48}$$

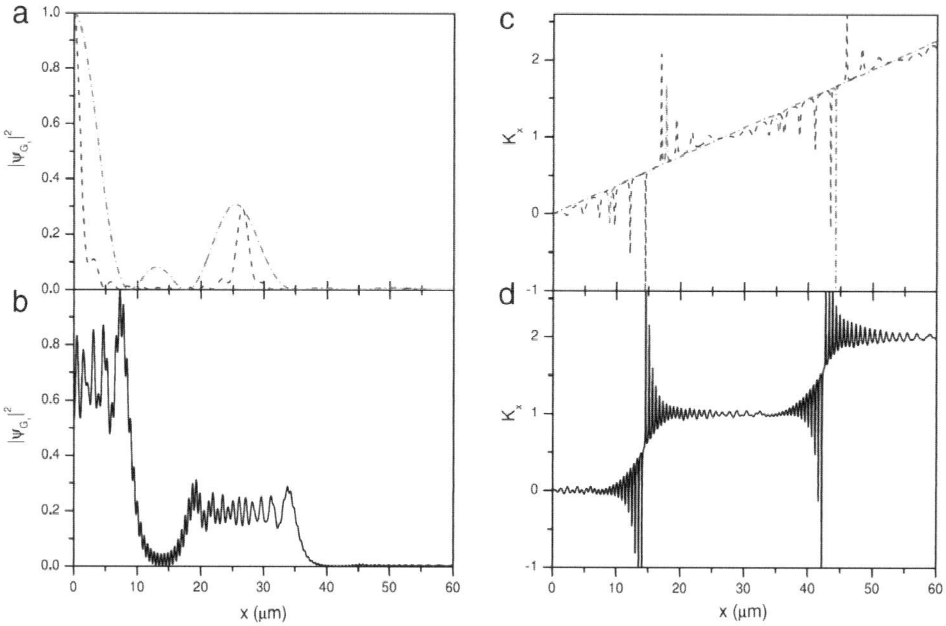

Figure 6. As the number of slits considered to numerically recreate the diffracted wave in an atomic Mach-Zehnder interferometer, just before getting diffracted again by a second grating, it observed how a few, distinctive principal interference maxima in the the probability density along the transverse direction become something closer to square waves, each progressing along a particular diffraction direction (in scattering off surfaces or lattices these directions correspond to the Bragg angles [61]). In panel (a), the blued dashed line and the red dashed-dotted one represent the probability density for 3 and 11 slits, respectively, good examples of a few-slit regime. On the contrary, the many-slit regime is represented in panel (b) with 51 slits. In the right-column panels, the corresponding transverse (Bohmian) momentum has been represented for exactly the same cases. As it can be noticed, in the few-slit regime a series of spikes appear coinciding with the minima of the density, analogous to what can be observed in Young's experiment [33, 36]. This structure evolves into an approximate step-ladder one as the number of slits becomes larger and larger, as seen in panel (d), with each step indicating a quantized value of the transverse momentum (a Bragg direction). For a better understanding of this phenomenon, the transverse momentum is normalized to $2\pi/d$. (Further analytical and numerical details can be found in [13].)

The type of symmetries or quantum carpets observed in confined systems has a very interesting limit. If the shape of the initial diffracted wave is a square instead of a Gaussian, as shown by Michael Berry [64], at some times the quantum state displays fractal features even if these carpets have the same global properties associated with diffracted states inside confined potentials [65]. These fractal states arise whenever the diffracted wave has sharp edges, i.e., it is continues everywhere, but not differenciable at some points, as it happens with a square function of width w inside a well of length $L > w$. As in Section 3.1, the wave function can be defined in terms of a linear combination of energy eigenstates,

$$\Psi(x, t; N) = \sum_{n=1}^{N} c_n \phi_n(x) e^{-iE_n t/\hbar}, \tag{49}$$

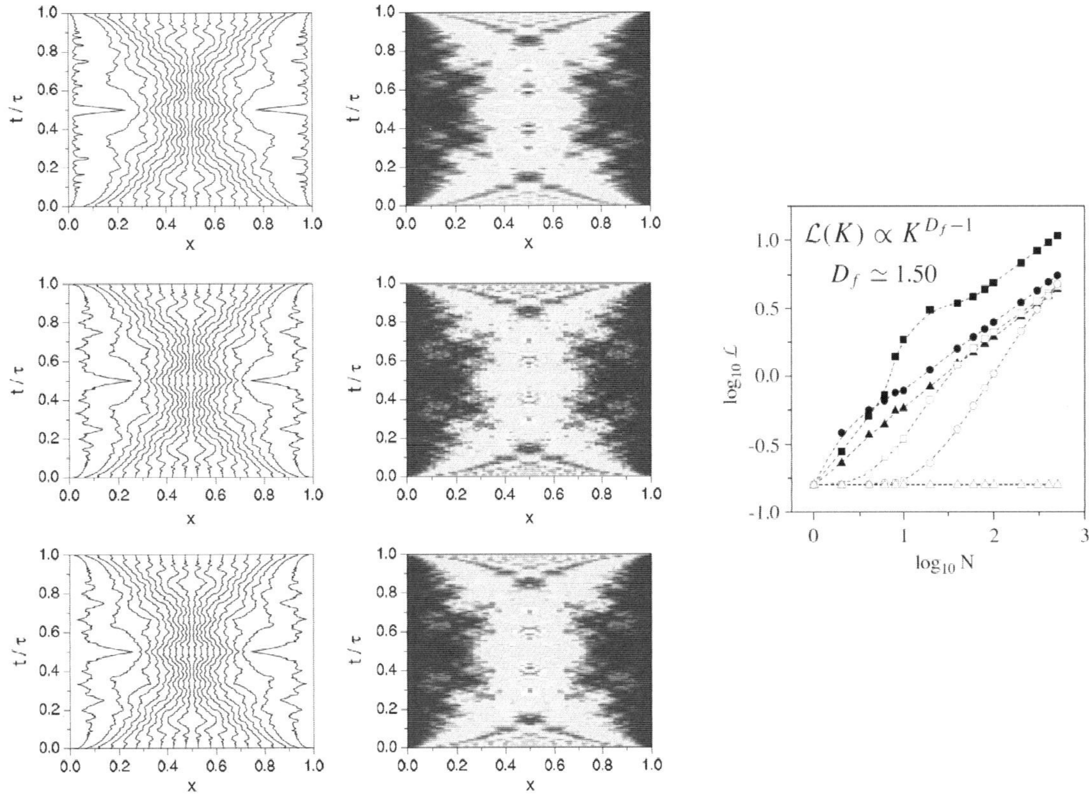

Figure 7. Time-evolution of a square wave function of width w inside an infinite well of length $L > w$. The panels on the left and central columns show the evolution of the trajectories and probability density as a function of the number of eigenstates considered, increasing from top to bottom. As it is apparent, such an increase implies a higher degree of fractality in both the trajectories and the shape displayed by the probability density. For visual purposes, the initial conditions of the trajectories are uniformly distributed along the regions covered by the initial probability density, $\rho(x, t = 0)$, instead of being weighted with this density. On the right, the panel shows the measure of the length of some selected trajectories as a function of the number of eigenstates involved in the initial superposition (both in log scale). As the initial conditions of the trajectories become closer to the edge of the initial state, the fractal dimension gets closer to that of the probability density — actually, notice that in the limit of large N, all trajectories approach the same fractal dimension. (Further analytical and numerical details on the model can be found in [63].)

although here the expectation value of the energy is going to monotonically increase with the number of eigenfunctions considered in the superposition. In this expression, $\phi_n(x)$ denote the corresponding eigenfunctions and $|c_n|^2$ the relative weights of each component in the superposition (or, equivalently, the population of each energy level). It can be shown [63] that the fractal dimension of the quantum state can be obtained from the measure

$$\mathcal{L}(\mathcal{K}) \propto \mathcal{K}^{D_f-1}, \tag{50}$$

where $\mathcal{L}$ denotes the the Euclidean length of, say, the probability density, and $\mathcal{K}$ the number of eigenfunctions considered in the superposition. As $\mathcal{K}$ increases, the length $\mathcal{L}$ also increases, and from their relation, in the limit of large $\mathcal{K}$, we obtain the fractal dimension of the state, D_f, which in the particular case displayed in Fig. 7 is approximately 1.50. The Bohmian trajectories associated with the state (49) can be obtained by means of the expression

$$\dot{x}_N = \frac{\hbar}{m} \mathrm{Im} \left\{ \frac{1}{\Psi(x,t;N)} \frac{\partial \Psi(x,t;N)}{\partial x} \right\} \tag{51}$$

with the limiting case

$$x(t) = \lim_{N \to \infty} x_N(t) \tag{52}$$

being, in general, a fractal trajectory (if the initial condition does not correspond to the central line of the state) with the same dimension of the quantum state [in this particular example, $D_f \approx 1.50$, when measured according to (50)].

4. Final remarks

The starting point of this work was a very specific question: has physical superposition the same meaning as mathematical superposition? This question was then reoriented to the presence of symmetries displayed by matter waves under different circumstances in diffraction-like experiments. To tackle the issue, apart from using standard quantum techniques and others coming from Fourier optics (spectral-like decompositions of the wave function), the Bohmian picture of quantum mechanics has also been taken as a very convenient tool to explore the physics of such phenomena beyond ordinary, well-known quantum results. In this picture, the emphasis is put on the quantum flux and the trajectories or streamlines "flying" with it, without no detriment to other more usual probability density based considerations. Notice, though, that the use of such trajectories has been taken as a probing tool to understand the "hidden" implications of interference-mediated symmetries, and not with an ontological purpose (associating them with the actual motion of real quantum particles, i.e., as "hidden variables").

Such an analysis renders a series of interesting conclusions, which can be summarized as follows:

- The symmetries displayed by the probability density, apart from being mathematically describable in terms of superpositions, are associated with distinctive non-superimposing flux dynamics that split up the configuration (position) space into separate dynamical domains.
- Those domains can be analyzed by means of effective potential models that constrain the evolution of the quantum system to them, neglecting what happens everywhere else. From these models it is then possible to get a precise and intuitive idea of what spatial coherence means, setting the grounds for future experiments to explore the "weirdness" of quantum systems and their capability to display interference.
- From the exchange symmetries displayed by coherent superpositions consisting of two counter-propagating wave packets, we have seen that they behave analogously to two classical colliding particles, identifying regimes similar to classical elastic and inelastic scattering events.
- Superpositions of periodically repeated wave functions generate densities with translational invariance, namely Talbot carpets, which are analogous to the structures we find in solid state physics and condensed matter physics due to the momentum (and energy) quantization induced on the initial wave function. Furthermore, such structures, when analyzed in terms of trajectories, seem to give rise or to be originated by repetitions of the dynamical behavior displayed by subsets of trajectories contained within individual domains or unit cells.

- The confinement or quantization observed in periodic grating diffraction processes is isomorphic with the behavior of matter waves inside bound potentials, including similar recurrence periods and trajectory behaviors.

These results, rather than constituting a conclusion from a series of works on matter-wave diffraction, they open new ways to rethink these processes and design new experiment to test the peculiarities of quantum mechanics. The experiment carried out by Steinberg and coworkers opened a new path to understand Young's two-slit experiment on a different level, but also to keep digging down in our understanding of quantum phenomena beyond elementary (and unrealistic) path-interfering arguments, as commonly done in matter-wave interferometry (Mach-Zehnder or Talbot-Lau) or to introduce "paradoxes", such as Wheeler's delayed-choice experiment. The quantum reality is far more complex (and richer) than such reasonings, which have to be overcome once for all, particularly taking into account the technological advances we have currently at hand to investigate it in the laboratory. Hence it is a matter of going beyond old-fashioned prejudices and using different alternative tools and routes to explore them. The Bohmian picture is probably a suitable tool to that end, but not necessarily the only one or the best one. At least, here it has served us to reach a deeper understanding of interference-mediated symmetries in matter-wave diffraction phenomena.

Acknowledgments
The author thanks the organizers of the International Symposium *Symmetries in Science XVII* (Bregenz, Austria, 2017) for their kind invitation to participate in the event. On a more personal side, he would also like to express his gratitude to Dieter and Ivette for their exquisite hospitality and attention during this event. The author also acknowledges financial support from the Spanish MINECO (grant reference number FIS2016-76110-P).

References
[1] Young T 1804 *Phil. Trans. R. Soc. Lond.* **94** 1–16
[2] Einstein A 1905 *Ann. Phys.* **17** 132–148
[3] Shizgal B 2015 *Spectral Methods in Chemistry and Physics: Applications to Kinetic Theory and Quantum Mechanics* Scientific Computation (Heidelberg: Springer)
[4] Juffmann T, Milic A, Müllneritsch M, Asenbaum P, Tsukernik A, Tüxen J, Mayor M, Cheshnovsky O and Arndt M 2012 *Nat. Nanotech.* **7** 297–300
[5] Arndt M and Hornberger K 2014 *Nature Phys.* **10** 271–277
[6] Taylor G I 1909 *Proc. Camb. Philos. Soc.* **15** 114–115
[7] Jönsson C 1961 *Z. Phys.* **161** 454–474
[8] Jönsson C, Brandt D and Hirschi S 1974 *Am. J. Phys.* **42** 4–11
[9] Donati O, Missiroli G F and Pozzi G 1973 *Am. J. Phys.* **41** 639–644
[10] Merli P G, Missiroli G F and Pozzi G 1976 *Am. J. Phys.* **44** 306–307
[11] Tonomura A, Endo J, Matsuda T, Kawasaki T and Ezawa H 1989 *Am. J. Phys.* **57** 117–120
[12] Sanz A S and Miret-Artés S 2008 *J. Phys. A: Math. Theor.* **41** 435303(1–23)
[13] Sanz A S, Davidović M and Božić M 2015 *Ann. Phys.* **353** 205–221
[14] Guantes R, Sanz A S, Margalef-Roig J and Miret-Artés S 2004 *Surf. Sci. Rep.* **53** 199–330
[15] Ashcroft N W and Mermin N D 1976 *Solid State Physics* (Fort Worth, USA: Harcourt College Publishers)
[16] Bohm D 1952 *Phys. Rev.* **85** 166–179
[17] Bohm D 1952 *Phys. Rev.* **85** 180–193
[18] Holland P R 1993 *The Quantum Theory of Motion* (Cambridge: Cambridge University Press)
[19] Sanz A S 2017 *arXiv:1707.00609v1* **[quant-ph]** 1–13
[20] Sanz A S 2014 *J. Phys.: Conf. Ser.* **504** 012028(1–14)
[21] Matteucci G, Pezzi M, Pozzi G, Alberghi G L, Giorgi F, Gabrielli A, Cesari N S, Villa M, Zoccoli A, Frabboni S and Gazzadi G C 2013 *Eur. J. Phys.* **34** 511–517
[22] Madelung E 1926 *Z. Phys.* **40** 322–326
[23] Schiff L I 1968 *Quantum Mechanics* 3rd ed (Singapore: McGraw-Hill)
[24] Morse P M and Feshbach H 1953 *Methods of Theoretical Physics* vol 2 (New York: McGraw-Hill)
[25] Fürth R 1933 *Z. Phys.* **81** 143–162

[26] Comisar G G 1965 *Phys. Rev.* **138** B1332–B1337
[27] Courant R and Hilbert D 1966 *Methods of Mathematical Physics* vol 2 (New York: John Wiley & Sons)
[28] Sanz A S, Davidović M and Božić M submitted, 2017 *Phys. Rev. A* 1–15
[29] Sanz A S, Borondo F and Miret-Artés S 2004 *J. Chem. Phys.* **120** 8794–8806
[30] Sanz A S, Borondo F and Miret-Artés S 2004 *Phys. Rev. B* **69** 115413(1–5)
[31] Sanz A S and Borondo F 2007 *Eur. Phys. J. D* **44** 319–326
[32] Sanz A S and Borondo F 2009 *Chem. Phys. Lett.* **478** 301–306
[33] Luis A and Sanz A S 2015 *Ann. Phys.* **357** 95–107
[34] Feynman R P, Leighton R B and Sands M 1965 *The Feynman Lectures on Physics* vol 3 (Reading, MA: Addison-Wesley)
[35] Sanz A S, Borondo F and Miret-Artés S 2002 *J. Phys.: Condens. Matter* **14** 6109–6145
[36] Kocsis S, Braverman B, Ravets S, Stevens M J, Mirin R P, Shalm L K and Steinberg A M 2011 *Science* **332** 1170–1173
[37] Aharonov Y, Albert D Z and Vaidman L 1988 *Phys. Rev. Lett.* **60** 1351–1354
[38] Prosser R D 1976 *Int. J. Theor. Phys.* **15** 169–180
[39] Davidović M, Sanz A S, Arsenović D, Božić M and Miret-Artés S 2009 *Phys. Scr.* **T135** 014009(1–5)
[40] Sanz A S, Davidović M, Božić M and Miret-Artés S 2010 *Ann. Phys.* **325** 763–784
[41] Božić M, Davidović M, Dimitrova T L, Miret-Artés S, Sanz A S and Weis A 2010 *J. Russ. Laser Res.* **31** 117–128
[42] Davidović M and Sanz A S 2013 *Europhys. News* **44** 33–36
[43] Heisenberg W 1925 *Z. Phys.* **33** 879–880
[44] Rosen N 1964 *Am. J. Phys.* **32** 377–379
[45] Rosen N 1964 *Am. J. Phys.* **32** 597–600
[46] Rosen N 1965 *Am. J. Phys.* **33** 146–150
[47] Richardson C D, Schlagheck P, Martin J, Vandewalle N and Bastin T 2014 *Phys. Rev. A* **89** 032118
[48] Chou C C 2016 *Ann. Phys.* **371** 437–459
[49] Rosen N 1986 *Found. Phys.* **16** 687–700
[50] Berry M V 1996 *Chaos and Quantum Physics* Les Houches Lecture Series LII (1989) ed Giannoni M J, Voros A and Zinn-Justin J (Amsterdam: North-Holland) pp 251–304
[51] Keith D W, Ekstrom C R, Turchette Q A and Pritchard D E 1991 *Phys. Rev. Lett.* **66** 2693–2696
[52] Cronin A D, Schmiedmayer J and Pritchard D E 2009 *Rev. Mod. Phys.* **81** 1051–1129
[53] Wheeler J A 1978 *The past and the delayed-choice double-slit experiment* Mathematical Foundations of Quantum Theory (New York: Academic Press)
[54] Sanz A S 2015 *Found. Phys.* **45** 1153–1165
[55] Sanz A S and Miret-Artés S 2007 *J. Chem. Phys.* **126** 234106(1–11)
[56] Sanz A S and Miret-Artés S 2012 *Am. J. Phys.* **80** 525–533 (*Preprint* 1104.1296)
[57] Talbot H F 1836 *Philos. Mag.* **9** 401–407
[58] Chapman M S, Ekstrom C R, Hammond T D, Schmiedmayer J, Tannian B R, Wehinger S and Pritchard D E 1995 *Phys. Rev. A* **51** R14–R17
[59] Case W B, Tomandl M, Deachapunya S and Arndt M 2009 *Opt. Express* **17** 20966–20974
[60] McMorran B J and Cronin A D 2009 *New J. Phys.* **11** 033021(1–7)
[61] Sanz A S, Borondo F and Miret-Artés S 2000 *Phys. Rev. B* **61** 7743–7751
[62] Sanz A S, Borondo F and Miret-Artés S 2001 *Europhys. Lett.* **55** 303–309
[63] Sanz A S 2005 *J. Phys. A: Math. Theor.* **38** 6037–6049
[64] Berry M V 1996 *J. Phys. A: Math. Theor.* **29** 6617–6629
[65] Tounli J, Alvarado A and Sanz A S (submitted, 2018)

IOP Conf. Series: Journal of Physics: Conf. Series **1071** (2018) 012019 doi:10.1088/1742-6596/1071/1/012019

GUP parameter and Lorentz violating terms

Fabio Scardigli[a b], **Gaetano Lambiase** [c d]

[a] Dipartimento di Matematica, Politecnico di Milano, Piazza Leonardo da Vinci 32, 20133 Milano, Italy

[b] Institute-Lorentz for Theoretical Physics, Leiden University, P.O. Box 9506, Leiden, The Netherlands

[c] Dipartimento di Fisica "E.R. Caianiello", Universita' di Salerno, 84084 Fisciano (Salerno), Italy

[d] INFN Gruppo Collegato di Salerno, Italy

E-mail: `fabio@phys.ntu.edu.tw`, `lambiase@sa.infn.it`

Abstract. Investigations on possible violation of Lorentz invariance have been widely pursued in the last decades, both from theoretical and experimental sides. A comprehensive framework to formulate the problem is the standard model extension (SME) proposed by A.Kostelecky, where violation of Lorentz invariance is encoded into specific coefficients. Here we present a procedure to link the deformation parameter β of the generalized uncertainty principle (GUP) to the SME coefficients of the gravity sector. The idea is to compute the Hawking temperature of a black hole in two different ways. The first way involves the deformation parameter β, and therefore we get a deformed Hawking temperature containing the parameter β. The second way involves a deformed Schwarzschild metric containing the Lorentz violating terms $\bar{s}^{\mu\nu}$ of the gravity sector of the SME. The comparison between the two different techniques yields a relation between β and $\bar{s}^{\mu\nu}$. In this way bounds on β transferred from $\bar{s}^{\mu\nu}$ are improved by many orders of magnitude when compared with those derived in other gravitational frameworks. Also the opposite possibility of bounds transferred from β to $\bar{s}^{\mu\nu}$ is briefly discussed.

1. Introduction

Possible breakdowns of the fundamental symmetries in Physics have received a more and more growing interest and have been studied in different areas (see for example [1, 2, 3, 4, 5, 6]). The most general setting in which they have been investigated is the Standard Model Extension (SME) [1]. The violation of the fundamental symmetries, i.e. Lorentz's and CPT symmetries, follows from the observation that the vacuum solution of the theory could spontaneously violate them, even though they are preserved by the underlying theory. Modern tests for Lorentz and CPT invariance breakdown have been discussed in [7]. More recently, the SME has been extended to incorporate the gravitational interaction [1, 9, 8]. The latter results foresee that the effective action is given by

$$S = S_{HE} + S_m + S_{LV} ,$$

where $S_{HE} = (16\pi G)^{-1} \int d^4x \sqrt{-g}(R - 2\Lambda)$ is the standard Hilbert-Einstein action of General Relativity (Λ is the cosmological constant), S_m the general matter action (which also includes Lorentz violating matter gravity coupling), while Lorentz violating gravitational couplings are included in S_{LV} [10]

$$S_{LV} = \frac{1}{16\pi G} \int d^4 e \left(-uR + s^{\mu\nu} R^T_{\mu\nu} + t^{\kappa\lambda\mu\nu} C_{\kappa\lambda\mu\nu} \right) , \tag{1}$$

where R^T is the trace-free Ricci tensor and $C_{\kappa\lambda\mu\nu}$ is the Weyl conformal tensor. The coefficients u, $s^{\mu\nu}$ and $t^{\kappa\lambda\mu\nu}$ are real and dimensionless. The coefficients $s^{\mu\nu}$ and $t^{\kappa\lambda\mu\nu}$ fulfill the Ricci and Riemann properties, respectively, and are traceless:

$$s^{\mu}{}_{\mu} = 0 , \quad t^{\kappa\lambda}{}_{\kappa\lambda} = 0 , \quad t^{\kappa}{}_{\mu\kappa\lambda} = 0 .$$

We restrict to the case $u = 0$ and $t^{\kappa\lambda\mu\nu} = 0$, therefore only the coefficients $s^{\mu\nu}$ control the Lorentz violation degrees of freedom. By varying the action S with respect to the background metric, and considering the limiting procedures detailed in [10], we obtain a two point-particles Lagrangian, which gives the effective equations of motion for the two bodies system, in the coordinate acceleration. In the hypothesis $M \gg m$ and considering the heaviest body M at rest (with respect to the test particle with mass m), this effective two bodies Lagrangian reads

$$L = \frac{1}{2} m v^2 + \frac{GmM}{r} \left(1 + \frac{3}{2} \bar{s}^{00} + \frac{1}{2} \bar{s}^{jk} \frac{x_j x_k}{r^2} \right) - \frac{GmM}{r} \left(3 \bar{s}^{0j} v_j + \bar{s}^{0j} \frac{x_j}{r} v_k \frac{x_k}{r} \right) \tag{2}$$

where we understood summation over indexes j, k, and where $v^2 = v_1^2 + v_2^2 + v_3^2$, $r^2 = x_1^2 + x_2^2 + x_3^2$, $v_k = \dot{x_k}$, and the derivative is taken in respect to the coordinate time. The $\bar{s}^{\mu\nu}$ are the Lorentz violation coefficients. We work in units with $c = 1$, and we consider particles moving slowly in respect to the speed of light, in a stationary and weak gravitational field. Therefore $v \ll 1$, so at the level of approximation we need, we can neglect the terms depending on the velocity v. The effective potential therefore reads

$$V(r) = \frac{U(r)}{m} = -\frac{GM}{r} \left(1 + \frac{3}{2} \bar{s}^{00} + \frac{1}{2} \bar{s}^{jk} \frac{x_j x_k}{r^2} \right) \tag{3}$$

As indicated in several points of Ref. [10] (see Eqs. 62, 100, 126, 137, 175) the scalar factor $(1 + 3\bar{s}^{00}/2)$ merely acts as a rescaling of the gravitational constant, and hence is unobservable in the present context. We can in fact rewrite $V(r)$ as

$$V(r) = -\frac{G_{\text{eff}} M}{r} \left(1 + \bar{s}^{jk}_{\text{eff}} \frac{x_j x_k}{r^2} \right) \tag{4}$$

where

$$G_{\text{eff}} = G \left(1 + \frac{3}{2} \bar{s}^{00} \right) \; ; \quad \bar{s}^{jk}_{\text{eff}} = \frac{\bar{s}^{jk}}{2 + 3\bar{s}^{00}} \tag{5}$$

Notice that the term $\bar{s}^{jk} x_j x_k / r^2$ cannot be reabsorbed into G_{eff} since it is an anisoptropic term, which depends upon the directions $x_j x_k / r^2$. Given the unobservability of the factors containing $\bar{s}^{00}$, in the following we shall simply rename $G_{\text{eff}} \equiv G$ and $\bar{s}^{jk}_{\text{eff}} \equiv \bar{s}^{jk}$. The parameters $s^{\mu\nu}$ have been constrained in different frameworks, see Table 1 [11].

The aim of this paper is to relate the coefficients $\bar{s}^{\mu\nu}$ to the the deforming parameter β of the generalized uncertainty principle (GUP) [16]. In fact, one of the most studied forms of the deformation of Heisenberg uncertainty principle (HUP), usually called GUP, is

$$\Delta x \, \Delta p \geq \frac{\hbar}{2} \left(1 + \beta \, \frac{4 \, \ell_{\text{p}}^2}{\hbar^2} \, \Delta p^2 \right) = \frac{\hbar}{2} \left[1 + \beta \left(\frac{\Delta p}{m_{\text{p}}} \right)^2 \right] \tag{6}$$

Table 1. Upper bounds on the $\bar{s}^{\mu\nu}$ derived in different physical frameworks (see Ref. [11] for a complete list).

$\bar{s}^{\mu\nu} <$	Physical framework	Refs.
10^{-15}	Torsion pendulum	[10]
10^{-14}	Cosmic rays	[12]
10^{-12}	Lunar laser ranging	[13]
10^{-11}	Binary pulsars	[14]
10^{-9}	Atom interferometry Perihelion precession (Solar system data)	[15] [10]

Here $\ell_{\rm p}$ is the Planck length, $m_{\rm p}$ the Planck mass, and we work in units where $2Gm_{\rm p} = \ell_{\rm p}$, $\hbar = 2m_{\rm p}\ell_{\rm p}$, $c = k_B = 1$. Typically, investigations mainly focus on understanding how gravity may affect the formulation of Heisenberg Uncertainty Principle (HUP). Given the pivotal role of gravitation in these arguments, it is not surprising that the most relevant modifications to the HUP have been proposed in string theory, loop quantum gravity, deformed special relativity, and studies of black hole physics [17, 18, 19, 20, 21, 22, 23]. In principle, the dimensionless parameter β is not fixed by the theory, although it is generally assumed to be of order one (this happens, in particular, in some models of string theory, see Ref. [17]).

In our approach we use the GUP to compute the Hawking temperature of a given black hole, which however can be also computed as well via the effective potential $V(r)$, and hence related to $\bar{s}^{\mu\nu}$. We shall find $\beta \simeq (M/m_p)^2|\bar{s}^{jk}|$, where M is the mass of the gravitational source. By making use of the most stringent bound on the parameter $\bar{s}^{jk}$ (see Table 1), we will derive an upper bound on the deformation parameter β.

2. Metric mimicking a potential corrected with Lorentz violating terms

Now we consider the effective potential produced by a metric of the very general class

$$ds^2 = F(r)dt^2 - g_{ik}(x_1, x_2, x_3)dx^i dx^k \tag{7}$$

where $r = |\mathbf{x}| = (x_1^2 + x_2^2 + x_3^2)^{1/2}$, and x_1, x_2, x_3 are the standard Cartesian coordinates. Particular cases of the metric (7) is the Schwarzschild metric, in the standard form as well as in harmonic coordinates. Once the metric is in the form (7), in Cartesian coordinates, then, with well known procedures [35], it is easy to show that the effective Newtonian potential [1] is of the form

$$V(r) \simeq \frac{1}{2}\left(F(r) - 1\right) \tag{8}$$

or, equivalently,

$$F(r) \simeq 1 + 2\,V(r)\,. \tag{9}$$

Therefore, the metric able to mimic the corrected Newtonian potential (4), containing Lorentz violating terms, will be

$$F(r) = 1 - \frac{2GM}{r}\left(1 + \bar{s}^{jk}f_{jk}(\theta,\phi)\right). \tag{10}$$

[1] The effective Newtonian potential is produced by the metric given in (7) for a point particle which moves slowly, in a stationary and weak gravitational field, i.e., quasi-Minkowskian far from the source, $r \to \infty$.

IOP Conf. Series: Journal of Physics: Conf. Series **1071** (2018) 012019 doi:10.1088/1742-6596/1071/1/012019

where we have introduced standard spherical coordinates $\mathbf{x} = r(\sin\theta\cos\phi, \sin\theta\sin\phi, \cos\theta)$ and $x_j x_k/r^2 = f_{jk}(\theta,\phi)$. Notice that the formal angular dependence on (θ,ϕ) displayed by (10), in our context does not really matter. In fact we know that $|\bar{s}^{\mu\nu}| \lesssim 10^{-10}$, $|f_{jk}(\theta,\phi)| \leq 1$ for any (θ,ϕ), and we are just interested in transferring bounds from $\bar{s}^{jk}$ to β. Hence, angular dependence does not influence our considerations.

3. Temperature from a deformed Schwarzschild metric

Let us define, for fixed θ and ϕ,

$$\epsilon(r) = -\frac{2GM}{r}\,\bar{s}^{jk} f_{jk}(\theta,\phi). \tag{11}$$

Therefore, $F(r)$ will now be of the form

$$F(r) = 1 - \frac{2GM}{r} + \epsilon(r). \tag{12}$$

Since $|f_{jk}(\theta,\phi)| \leq 1$, and we are supposing $|\bar{s}^{jk}| \ll 1$, then it is clear that $|\epsilon(r)| \ll 2GM/r$ for any $r \geq 2GM$.

We can legitimately wonder what kind of deformed Hawking temperature can be inferred from a deformed Schwarzschild metric as in (12). The deformation (12) makes sense when $|\epsilon(r)| \ll GM/r$. For computational reasons, we can introduce a regulatory small parameter ε so that we can write $\epsilon(r) \equiv \varepsilon\phi(r)$. At the end of the calculation, ε can can be sent to unity. Of course, we will look for the lowest order correction in the dimensionless parameter ε. The horizon's equation, i.e., $F(r) = 0$, now can be solved in a first approximation in ε. The solution reads

$$r_H \;=\; a - \frac{\varepsilon\, a\,\phi(a)}{1 \,+\, \varepsilon\,[\phi(a) \,+\, a\,\phi'(a)]} \tag{13}$$

where $a = 2GM$. Using Eqs.(11) and (12) for $F(r)$, and Eq.(13) for r_H, and expanding in ε, one arrives to the deformed Hawking temperature as (for details, see [25])

$$T = \hbar\frac{F'(r_H)}{4\pi} = \frac{\hbar}{4\pi a}\left\{1 + \varepsilon\left[2\phi(a) + a\phi'(a)\right] + \varepsilon^2\phi(a)\left[\phi(a) - 2a\phi'(a) - a^2\phi''(a)\right] + \dots\right\}. \tag{14}$$

It is noteworthy that the only function $\phi(r)$ that annihilates the first-order in ε temperature correction term is the solution of the differential equation $2\phi(r) + r\phi'(r) = 0$, namely $\phi(r) = A/r^2$, where A is an arbitrary constant. In particular, for the function $\phi(r) = G^2M^2/r^2$, the coefficient of ε in (14) is zero, and the coefficient of ε^2 is $-1/16$.

4. Relation between β and $\bar{s}^{jk}$

We are now in the position to compute the temperature generated by the metric (12), by simply employing (14). Therefore, the metric-deformed Hawking temperature is of the form

$$T = \frac{\hbar}{4\pi a}\left\{1 + \left[2\epsilon(a) + a\epsilon'(a)\right] + \dots\right\} \tag{15}$$

while the GUP-deformed Hawking temperature reads [2]

$$T = \frac{\hbar}{8\pi GM}\left(1 + \frac{\beta\, m_{\rm p}^2}{4\pi^2\, M^2} + \dots\right)\,, \tag{16}$$

[2] To compute the thermal GUP corrections to the Hawking spectrum, we follow the arguments of Refs. [26, 27, 28, 29, 30, 31, 32, 33].

where it is evident that to zero order in β, we recover the usual well known Hawking formula. We stress that we are assuming that the correction induced by the GUP has a thermal character, and, therefore, it can be cast in the form of a shift of the Hawking temperature. Of course, there are also different approaches, where the corrections do not respect the exact thermality of the spectrum, and thus need not be reducible to a simple shift of the temperature (an example is the corpuscular model of a black hole of Ref. [34]).

By comparing the two respective first-order correction terms in the two previous expansions, we obtain

$$\beta = \frac{4\pi^2 M^2}{m_{\rm p}^2}\left[2\epsilon(a) + a\epsilon'(a)\right] . \tag{17}$$

Using now the expression (11) for $\epsilon(r)$, we get

$$\beta = -\frac{4\pi^2 M^2}{m_{\rm p}^2}\,\bar{s}^{jk} f_{jk}(\theta,\phi) \tag{18}$$

for fixed[3] θ and ϕ.

Again, we can comment that angular dependence shown by Eq.(18) is not particularly puzzling in this context, since we are here interested in linking the magnitudes of $\bar{s}^{jk}$ and β, therefore it suffices to notice that $|f_{jk}(\theta,\phi)| \leq 1$ for any (θ,ϕ). Furthermore, the fact that β in some cases could result *negative* for positive values of $\bar{s}^{jk}$ (in general, the quantities $\bar{s}^{jk}$ can be positive or negative), shouldn't actually be a worry, since negative β can be interpreted as a signal of a lattice structure of the space-time at the Planck scale (see Refs. [36, 37]). We can get rid of angular dependence by averaging over (θ,ϕ), in fact $\langle x^j x^k\rangle = r^2\delta^{jk}/3$, and therefore setting $f_{jk} = \frac{1}{3}\delta_{jk}$, one gets (see (18))

$$\beta = \frac{4\pi^2}{3}\left(\frac{M}{m_{\rm p}}\right)^2 \delta\,, \quad \delta \equiv |\bar{s}^{11} + \bar{s}^{22} + \bar{s}^{33}| = |\bar{s}^{ii}|\,. \tag{19}$$

5. Comparison with experimental data.

5.1. Bounds from $\bar{s}^{jk}$ to β

Before proceeding further, we have to clarify what kind of mass M we have to insert in formula (19) in order to compute the relevant bounds on β. Clearly, M is the same mass which appears in the deformed metric (10). Therefore M will be chosen according to the specific gravitational experiment used to produce specific bounds (of gravitational origin) on $\bar{s}^{jk}$. Examining the experimental situations described in Refs. [10, 13, 14] (see there in particular Table VI of [10]), and the relevant gravitational bounds on $\bar{s}^{jk}$, we can list the following cases.

Lunar ranging. It is well known that lunar laser ranging is among the most sensitive tests of gravitational physics within the solar system. In this situation M is the mass of the Earth, and the distance of the Moon is probed with lasers with a precision at the centimeter level. According to Ref. [10] and its Table VI, the best attainable sensitivity is of order $|\bar{s}^{jk}| \lesssim 10^{-11}$. According to Ref.[13], we have $|\bar{s}^{jk}| \lesssim 10^{-12}$. In any case here we have $M = M_{Earth} \simeq 2.74 \times 10^{32} m_{\rm p}$.

Binary pulsars. In the case of binary pulsars we have bounds on the quantities $\bar{s}_e$, $\bar{s}_\omega$. In particular $\bar{s}_\omega \leq 10^{-11}$. According to Ref. [10], both $\bar{s}_e$, $\bar{s}_\omega$ are linear combinations of the

[3] We note that the relation (18) can be derived also in a different way. The metric (10) can be written as

$$F(r) = 1 - \frac{2GM}{r}\cdot A$$

where $A = 1+\xi$, with $\xi = \bar{s}^{jk} f_{jk}(\theta,\phi)$. The horizon $F(r) = 0$ is now $r_H = 2GMA$, and therefore the temperature is $T = \hbar\frac{F'(r_H)}{4\pi} \simeq \frac{\hbar}{8\pi GM}(1-\xi)$. Finally, on comparing with (16) we get $\beta = -\frac{4\pi^2 M^2}{m_{\rm p}^2}\,\xi$, which coincides with (18).

quantities $\bar{s}_{PP}$, $\bar{s}_{kP}$, $\bar{s}_{QQ}$, etc. (see Eq.184 of [10]), which in turns are linear combinations of the quantities $\bar{s}^{jk}$ (see Eq.185 of [10]). The coefficients entering these linear combinations are all of order 1, therefore the bounds stated for $\bar{s}_{\omega}$ can be safely transferred to $\bar{s}^{jk}$. So finally we can affirm that from binary pulsar data we get the bounds $|\bar{s}^{jk}| \leq 10^{-11}$, confirmed by Ref.[14]. Here the relevant mass is the total mass of the binary pulsar system (on this, see also Refs. [37, 38, 39]), namely $M = m_1 + m_2$. Considering the very well known system PRS B 1913+16, we have $M = m_1 + m_2 = 2.828 \times M_{\odot} = 2.55 \times 10^{38} m_{\rm p}$.

Perihelion precession. For the perihelion precession in the solar system we consider of course in particular the data from Mercury. Again we see from Ref. [10], Eq.191, that we have a bound $\bar{s}_{Mer} \leq 10^{-9}$, and $\bar{s}_{Mer}$ turns out to be a linear combination of $\bar{s}^{jk}$ with coefficients $O(1)$. So we can use that bound also for $\bar{s}^{jk}$. Obviously, the mass here involved is the mass of the Sun $M = M_{\odot} = 0.9 \times 10^{38} m_{\rm p}$.

Torsion pendulum. We conclude this inspection with the most stringent case, as for the bounds on $\bar{s}^{jk}$. In a laboratory experiment, on the Earth surface, a torsion pendulum has been considered in Ref. [10]. The bounds in principle attainable with this device on the coefficients $\bar{s}^{JK}$ are of order 10^{-15} (see again table VI of Ref. [10]). The quantities $\bar{s}^{JK}$ are here again just linear combinations of the quantities $\bar{s}^{jk}$, with coefficients of $O(1)$. So we can infer the bounds $|\bar{s}^{jk}| \lesssim 10^{-15}$. Obviously the mass to be considered here is again the Earth mass (see Eqs.119, 121, 122 of Ref. [10]), namely $M = M_{Earth} \simeq 2.74 \times 10^{32} m_{\rm p}$.

The above analysis suggests that the relevant masses to be inserted in relation (19), in order to get bounds on β, are essentially the Earth mass, $M_{Earth} \simeq 2.74 \times 10^{32} m_{\rm p}$, or the Solar mass, $M_{\odot} = 0.9 \times 10^{38} m_{\rm p}$.

As for the constraints on the parameter β, in recent years there has been a wide and lively research on this topic (see e.g. Ref.[42, 40, 41]), summarized in Tables 2 and 3. To be consistent with the logic of the argument presented here, we focus only on the bounds of β of gravitational origin, reported in Table 2. If we use now, in relation (19), the best bound on $\bar{s}^{jk}$, namely the one from Torsion Pendulum experiments, $|\bar{s}^{jk}| < 10^{-15}$, and consequently the Earth mass for $M = M_{Earth}$, then we get for the deformation parameter of GUP

$$\beta < 10^{51} \,. \tag{20}$$

As it clear from Table 2, this bound improves by many orders of magnitude almost all the bounds on β inferred in different gravitational experiments. In particular, the procedure above outlined does not involve a violation of the equivalence principle. Finally, the bound (20) is quite close to the one derived from Landau levels measurements, i.e. a non-gravitational bound.

5.2. Bounds from β to $\bar{s}^{jk}$

It is however clear that relation (19) can be formally inverted, resulting in $\delta \simeq 3(m_{\rm p}/2\pi M)^2\beta$, and suggesting therefore also the opposite path, namely the possibility to transfer bounds from β to $\bar{s}^{jk}$. Obviously, in this context we keep considering only the bounds on β displayed in Table 2, i.e. those of gravitational origin.

- If we consider the bound on β coming from the perihelion precession in the solar system, $\beta < 10^{69}$, then we use $M = M_{\odot} \simeq 10^{38} m_{\rm p}$, and from the above relation we get $|\bar{s}^{jk}| < 10^{-9}$, perfectly in line with the correspondent value displayed in Table 1.
- If we consider the bound on β coming from the gravitational wave event GW150914, $\beta < 10^{60}$, then we should use for M the total mass of the (supposed) two-black holes system, roughly $M = m_1 + m_2 \simeq 50\, M_{\odot} \simeq 5 \times 10^{39} m_{\rm p}$. From the above relation we the obtain

Table 2. Upper bounds on the GUP parameter β inferred in gravitational measurements/experiments.

$\beta <$	Physical framework	Refs.
10^{21}	Violation of equivalence principle (on Earth) Law of reciprocal action	[43] (2014)
10^{60}	GW 150914	[42]
10^{69}	Perihelion precession (Solar system data)	[37] (2015)
10^{71}	Perihelion precession (Pulsar PRS B 1913+16 data)	[37] (2015)
10^{78}	Modified mass-temperature relation Light deflection	[37] (2015)

Table 3. Upper bounds on the GUP parameter β inferred in different non-gravitational measurements/experiments.

$\beta <$	Physical framework	Refs.
10^{18}	Evolution of micro and nano mechanical oscillators (masses $\sim m_{\rm p}$)	[44] (2015)
10^{20}	Lamb shift	[45] (2011)
10^{21}	Scanning tunneling microscope	[45] (2008)
10^{33}	Gravitational bar detectors[4]	[41] (2013)
10^{34}	Electroweak measurement	[45] (2011)
10^{34}	Charmonium levels Energy difference in Hydrogen levels $1S-2S$	[45] (2011) [46] (2010)
10^{39}	^{87}Rb cold-atom-recoil experiment	[47] (2016)
10^{46}	Landau levels	[45] (2011)

$|\bar{s}^{jk}| \lesssim 10^{-21}$. Obviously, this evaluation cannot be retained as reliable as the previous one, given for example the still large uncertainties affecting the event GW150914.

• The constraint $\beta < 10^{21}$ coming from universality of free fall, or from the law of reciprocal action, is somehow problematic in this context. Such constraint is derived in Ref.[43] by postulating a deformation of the classical (covariant) Poisson brackets which resembles the deformed quantum commutator (6). As it is explicitly shown in Ref.[37], a deformation of Poisson brackets implies immediately a deformation of the equation of motion (i.e. of the geodesic equation, in the relativistic context), in such a way that the motion of a test particle depends on the mass of the particle itself. That is, a violation of the equivalence principle. This immediately reflects on the modified relations for free fall, or for reciprocal actions, from which in Ref.[43] the bound $\beta < 10^{21}$ is obtained. On the contrary, the other bounds reported in Table 2 are derived by deforming the dispersion relation (Ref.[42]), or by deforming the metric (Ref.[37]), in order to mimic the GUP-deformed Hawking temperature, but this is done always under the strict assumption of validity of equivalence principle, i.e. without deforming the geodesic equation.

Because of its origin from violation of the equivalence principle, the bound $\beta < 10^{21}$ should be therefore considered not homogeneous with the others reported in Table 2, which on the contrary respect the equivalence principle. Moreover, the bound $\beta < 10^{21}$, with the use of $M = M_{Earth}$, would imply $|\bar{s}^{jk}| < 1.01 \times 10^{-45}$. This suppresses by many orders of magnitude the current bounds reported in Table 1, and appears quite unexpected with respect to the corresponding bounds of SME coefficients in the matter sector, where the present sensitivity of experiments has not provided the evidence of such a suppression (in the SME, in fact, coefficients can be transferred from the gravitational sector to matter sector and viceversa with an appropriate change of coordinates). [5]

In this context, it is however instructive to derive an upper bound on $\bar{s}^{jk}$ by using directly the present bounds on the violation of Law of Reciprocal Action obtained from data of Lunar Laser Ranging. According to [48, 49] the active mass m_a is the source of the gravitational field ($\nabla^2 V(\mathbf{x}) = -4\pi m_a \delta\mathbf{x}$), while the passive mass is related to the response of a mass to a gravitational field, and appears in the equation of motion $m_i\ddot{\mathbf{x}} = m_p \nabla V(\mathbf{x})$, where m_i is the inertial mass. Following [43, 49], the equations of motion for a gravitationally bound system of particles, and of its center of mass $\mathbf{X}$ are

$$m_{1i}\ddot{\mathbf{x}}_1 = G m_{1p} m_{2a} \frac{\mathbf{x}_2 - \mathbf{x}_1}{|\mathbf{x}_2 - \mathbf{x}_1|^3},$$

$$m_{2i}\ddot{\mathbf{x}}_2 = G m_{2p} m_{1a} \frac{\mathbf{x}_1 - \mathbf{x}_2}{|\mathbf{x}_1 - \mathbf{x}_2|^3},$$

$$\ddot{\mathbf{X}} = G \frac{m_{1p} m_{2p}}{m_{1i} + m_{2i}} C_{21} \frac{\mathbf{x}}{|\mathbf{x}|^3}, \quad C_{21} = \frac{m_{2a}}{m_{2p}} - \frac{m_{1a}}{m_{1p}},$$

where $\mathbf{x}$ is the relative coordinate. If $C_{21} \neq 0$, then the center of mass possesses a self-acceleration. In SME model this occurs through the coefficients $\bar{s}_n^{jk}$ by assuming that they are particle depending. Here, therefore, m_{ni}, ($n = 1, 2$) is the inertial mass of particles, and $m_{na} = (1 + \bar{s}_n^{jk} \hat{\mathbf{x}}^{\mathbf{j}} \hat{\mathbf{x}}^{\mathbf{k}}) m_{ni}$ is the active mass of particles (see Eq.4). Besides, m_{np} is identified with m_{ni}. The absence of self interaction of the Moon (one considers the distribution of the main constituent of Moon, Al and Fe) yields the bound $|C_{Al-Fe}| < 7 \times 10^{-13}$, which implies $|\delta_{Al-Fe}| < 7 \times 10^{-13}$, which is in clear agreement with the expected bounds for $\bar{s}^{jk}$ of Table 1.

• In Ref. [33] the deformation parameter β has been computed by making use of the leading quantum corrections to the Newtonian potential [50, 51, 52]. The corrections to the Newtonian potential imply naturally a quantum correction to the Schwarzschild metric, and this leads to a precise numerical value for β, namely $\beta = 82\pi/5$, which is of the same order of magnitude expected from string theory. If we use this value for β, together with the Earth mass, $M \simeq 2.74 \times 10^{32} m_{\rm p}$, we get $|\bar{s}^{jk}| \simeq 5.21 \times 10^{-65}$, a value which is hugely beyond the tested experimental bounds of $\bar{s}^{jk}$. Although the actual physical relevance of this bound remains questionable, such a minute value is somehow expected, since with the above procedure we checked, for the first time, the Lorentz violating SME coefficients against a quantity, β, typically linked with Planck scale phenomena. The previous gravitational bounds on $\bar{s}^{jk}$ are all obtained in the contest of classical gravity, although post Newtonian. Here instead we face a "quantum" gravity scenario, or to be more precise, a semiclassical gravity scenario, represented by the Hawking effect. To this circumstance can be presumably traced back such a huge refinement of the value of the $\bar{s}^{jk}$ coefficients.

[5] The bound $\beta < 10^{21}$ can perhaps be made more reasonable if we consider that, according to [46], where composite systems have been investigated, to get the deformation parameter referred to particle physics one should consider an effective deformation parameter that must be multiplied with the square of the nucleon number N_{nuc} making up the gravitational source. Considering the Moon with $M_{Moon} \simeq 7.3 \times 10^{22}$Kg and the nucleon mass $m_{nuc} \simeq 1.67 \times 10^{-27}$Kg, one gets $N_{nuc} = \frac{M_{Moon}}{m_{nuc}} \simeq 4 \times 10^{49}$. The deformation parameter turns out to be $\beta = N_{nuc}^2 10^{21}$, and therefore this procedure increases the deformation parameter by many orders of magnitude, relaxing in such a way the strong suppression of the SME parameter s^{jk}. It should also be noticed, however, that the considerations proposed in Ref.[46] are themselves based on deformed Poisson brackets, namely on violation of equivalence principle.

6. Conclusions

In this paper we have derived an upper bound on the deformation parameter β of the generalized uncertainty principle (6), by relating β to the coefficients $\bar{s}^{jk}$ defined in the gravitational sector of SME. The main point of the derivation relies on the fact that we directly compute a quantum mechanical effect, the Hawking temperature, for which the GUP is necessarily relevant, without postulating a specific representation of canonical commutators. We then compute the same temperature using a deformed Schwarzschild-like metric, thereby linking together the deformed uncertainty relation, and the deformed metric. It is noteworthy that in our formalism General Relativity and standard Quantum Mechanics are recovered in the limits $s^{jk} \to 0$ and $\beta \to 0$, respectively. Our main results can be summarized in two distinct cases:

- By considering the experimental upper bounds on the parameter $|\bar{s}^{jk}| < 10^{-15}$, we infer a bound on the GUP deformation parameter $\beta < 10^{51}$, which improves by many order of magnitude the bounds obtained in gravitational frameworks compatible with the equivalence principle, and it lies quite close to the Landau level measurements, obtained with non-gravitational measurements/experiments.
- If we adopt for the parameter β the value $82\pi/5$, derived in the framework of QFT and GR, then the coefficients $\bar{s}^{jk}$ turns out to be bounded by $\bar{s}^{jk} \simeq 10^{-65}$. Such a minute value, although extremely tight because derived within a semiclassical gravity approach (Hawking effect), seems however to demand for further investigations, both on the experimental as well on the theoretical side.

There is, nowadays, a lively debate on the measurable features implied by various kinds of GUPs, and many efforts are devoted to the predictions about the size of these modifications. In this respect, several experiments have been also proposed to test GUPs in the laboratory. As shown in this paper, GUP measurements could have an interplay with the violation of the fundamental symmetries in physics, such as CPT and Lorentz invariance, through the SME. Here we focus on SME for the gravitational sector, but to understand whenever the other coefficients of the model may affect GUP, or specific representations of canonical operators, might be of great interest, especially in perspective of possible links with quantum gravity. These aspects, in turn, are particularly appealing in view of the possibility to create, in the next future, a laboratory-scale imitation of a black hole horizon, emitting analogue Hawking radiation [53].

References

[1] D. Colladay, V.A. Kostelecký, Phys. Rev. D **58**, 116002 (1998); Phys. Rev. D **55**, 6760 (1997).
[2] J. Ellis, K. Farakos, N.E. Mavromatos, V.A. Mitsou, D.V. Nanopoulos, Astrophys. J. **535**, 139 (2000). J. Ellis, N.E. Mavromatos, D.V. Nanopoulos, Phys. Rev. D **61**, 027503 (2000).
[3] J. Alfaro, H.A. Morales-Téctol, L.F. Urrutia, Phys. Rev. Lett. **84**, 2318 (2000); Phys. Rev. D **65**, 103509 (2002); Phys. Rev. D **66**, 124006 (2002). R. Gambini, J. Pullin, Phys. Rev. D **59**, 124021 (1999). A. Ashtekar, *Lectures on New Perturbative Canonical Gravity*, World Scientific, 1991. C. Rovelli, S. Speziale, Phys. Rev. D **67**, 064019 (2003).
[4] G. Amelino-Camelia, Int. J. Mod. Phys. D **11**, 35 (2002). S.M. Carroll, J. Harvey, V.A. Kostelecký, C.D. Lane, T. Okamoto, Phys. Rev. Lett. **87**, 141601 (2001).
[5] V.A. Kostelecký, R. Lehnert, M.J. Perry, Phys. Rev. D **68**, 123511 (2003). O. Bertolami, R. Lehnert, R. Potting, A. Ribeiro, Phys. Rev. D **69**, 083513 (2004).
[6] T. Jacobson, D. Mattingly, Phys. Rev. D **64**, 024028 (2001); Phys. Rev. D **70**, 024003 (2004).
[7] D. Mattingly, Living Rev. Rel. **8**, 5 (2005). H. Vucetich, gr-qc/0502093. R. Bluhm, hep-ph/0506054
[8] V.A. Kostelecký, Phys. Rev. D **69**, 105009 (2004).
[9] R. Bluhm, V.A. Kostelecký, Phys. Rev. D **71**, 065008 (2005).
[10] Q.G. Bailey, V.A. Kostelecký, Phys. Rev. D **74**, 045001 (2006).
[11] V.A. Kostelecky and N. Russell, Rev. Mod. Phys. **83**, 11 (2011).
[12] V.A. and J.D. Tasson, Phys. Lett. B **749**, 551 (2015).
[13] A. Bourgoin et al., Phys. Rev. Lett. **117**, 241301 (2016).
[14] L. Shao, Phys. Rev. Lett. **112**, 111103 (2014).

[15] J.B.R. Battat, J.F. Chandler, and C.W. Stubbs, Phys. Rev. Lett. **99**, 241103 (2007).
[16] H.S. Snyder, Phys.Rev. **71**, 38 (1947); C.N. Yang, Phys. Rev. **72**, 874 (1947); C.A. Mead, Phys. Rev. **135**, B 849 (1964); F. Karolyhazy, Nuovo Cim. A **42**, 390 (1966).
[17] D. Amati, M. Ciafaloni, G. Veneziano, Phys. Lett. B **197**, 81 (1987); D.J. Gross, P.F. Mende, Phys. Lett. B **197**, 129 (1987); D. Amati, M. Ciafaloni, G. Veneziano, Phys. Lett. B **216**, 41 (1989); K .Konishi, G. Paffuti, P. Provero, Phys. Lett. B **234**, 276 (1990); S. Capozziello, G. Lambiase, G. Scarpetta, Int. J. Theor. Phys. **39**, 15 (2000).
[18] M. Maggiore, Phys. Lett. B **304**, 65 (1993).
[19] F. Scardigli, Phys. Lett. B **452**, 39 (1999).
[20] R.J. Adler, D.I. Santiago, Mod. Phys. Lett. **A14**, 1371 (1999).
[21] A. Kempf, G. Mangano, R.B. Mann, Phys. Rev. D **52**, 1108 (1995).
[22] A.F.Ali, M.Faizal, M.Khalil, JCAP **09**, 025 (2015); M.Faizal, A.Ali, A.Nassar, Phys.Lett.B **765**, 238 (2017).
[23] F. Scardigli and R. Casadio, Class. Quantum Grav. **20**, 3915 (2003).
[24] W. Heisenberg, Zeitschrift für Physik, **43**, 172 (1927).
[25] G. Lambiase, F. Scardigli, *Lorentz violation and generalized uncertainty principle*, arXiv:1709.00637 [hep-th]
[26] F. Scardigli, Nuovo Cim. B **110**, 1029 (1995).
[27] R.J. Adler, P. Chen and D.I. Santiago, Gen. Rel. Grav. **33**, 2101 (2001).
[28] M. Cavaglia and S. Das, Class. Quant. Grav. **21**, 4511 (2004).
[29] M. Cavaglia, S. Das and R. Maartens, Class. Quant. Grav. **20**, L205 (2003).
[30] L. Susskind, J. Lindesay, *An Introduction to Black Holes, Information, and the String Theory Revolution* (World Scientific, Singapore, 2005). See chapter 10.
[31] K. Nouicer, Class. Quant. Grav. **24**, 5917 (2007).
[32] F. Scardigli, *Glimpses on the micro black hole planck phase*, [arXiv:0809.1832].
[33] F. Scardigli, G. Lambiase and E.C. Vagenas, Phys. Lett. B **767**, 242 (2017).
[34] G. Dvali, C. Gomez, *Black Hole's Quantum N-Portrait*, [arXiv:1112.3359]; R. Casadio, A. Giugno, A. Orlandi, Phys. Rev. D **91**, 124069 (2015); R. Casadio, A. Giugno, A. Giusti, Phys. Lett. B **763**, 337 (2016).
[35] S. Weinberg, *Gravitation and Cosmology*, John Wiley & Sons, New York (1972).
[36] P. Jizba, H. Kleinert, and F. Scardigli, Phys. Rev. D **81**, 084030 (2010).
[37] F. Scardigli, R. Casadio, Eur. Phys. J. C **75** 425 (2015).
[38] J.M. Weisberg, D.J. Nice, J.H. Taylor, Astrophysical Journal, **722**, 1030 (2010).
[39] J.H. Taylor and J.M. Weisberg, Astrophysical Journal, **345**, 434 (1989).
[40] I. Pikovski, C. Brukner, et al., Nat. Phys. **8**, 393 (2012).
[41] F. Marin, M. Cerdonio et al., Nat. Phys. **9**, 71 (2013).
[42] Z.-W. Feng, S.-Z. Yang, H.-L. Li, and X.-T. Zu, Phys. Lett. B **768**, 81 (2017).
[43] S. Ghosh, Class. Quantum Gravity 31, 025025 (2014).
[44] M. Bawaj, C. Biancofiore, F. Marin et al., Nat. Commun. **6**, 7503 (2015) [arXiv:1411.6410].
[45] S. Das, E.C. Vagenas, Phys. Rev. Lett. **101**, 221301 (2008). A.F. Ali, S. Das, E.C. Vagenas, Phys. Rev. D **84**, 044013 (2011). S. Das, E.C. Vagenas Can. J. Phys. **84**, 233 (2009).
[46] C. Quesne, V.M. Tkachuk, Phys. Rev. A **81**, 012106 (2010).
[47] D. Gao, M. Zhan, Phys. Rev. A **94**, 013607 (2016).
[48] H. Bondi, Rev. Mod. Phys. **29**, 423 (1957).
[49] C. Lammerzahl, Space Sci. Rev. **148**, 501 (2009). D.F. Barlett, D. van Buren, Phys. Rev. Lett. **57**, 21 (1986).
[50] M.J. Duff, Phys. Rev. D **9**, 1837 (1974).
[51] J.F. Donoghue, Phys. Rev. Lett. **72**, 2996 (1994); J.F. Donoghue, Phys. Rev. D **50**, 3874 (1994).
[52] I.B. Khriplovich, G.G. Kirilin, J.Exp.Theor.Phys. **95**, 981 (2002); N.E.J. Bjerrum-Bohr, J.F. Donoghue, B.R. Holstein, Phys.Rev.D **68**, 084005 (2003); Erratum-ibid.D **71**, 069904 (2005); N.E.J. Bjerrum-Bohr, J.F. Donoghue, B.R. Holstein, Phys.Rev.D **67**, 084033 (2003); Erratum-ibid.D **71**, 069903 (2005); A.Akhundov and A.Shiekh, EJTP 5, No. 17, 1 (2008).
[53] J. Steinhauer, Nature Physics **10**, 864 (2014). R. Cowen, Nature News, 12 October 2014, *Hawking radiation mimicked in the lab*, (doi:10.1038/nature.2014.16131).

IOP Conf. Series: Journal of Physics: Conf. Series **1071** (2018) 012020 doi:10.1088/1742-6596/1071/1/012020

Analytical solutions for the quantum parametric oscillator from corresponding classical dynamics via a complex Riccati equation

Dieter Schuch

Institut für Theoretische Physik, J.W. Goethe-Universität Frankfurt am Main, Max-von-Laue-Str. 1, D-60438 Frankfurt am Main, Germany

E-mail: `Schuch@em.uni-frankfurt.de`

Abstract. The time-dependent Schrödinger equation for quadratic Hamiltonians has Gaussian wave packets as exact solutions. For the parametric oscillator with frequency $\omega(t)$, the width of these wave packets must be time-dependent. This time-dependence can be determined via the solution of a complex nonlinear Riccati equation or an equivalent real nonlinear Ermakov equation. All quantum dynamical properties of the system can easily be constructed from these solutions, e.g., uncertainties of position and momentum, their correlations, tunnelling currents, ground state energies, etc. In addition, the link to the corresponding classical dynamics is supplied via linearization of the Riccati equation to a complex Newtonian equation, actually representing two equations of the same kind: one for the real and one for the imaginary part. If the solution of one part is known, the missing (linear independent) solution of the other can be obtained via a conservation law for the motion in the complex plane. Knowing these two solutions, the solution of the Ermakov equation can be determined immediately plus the explicit expressions for all the quantum dynamical properties mentioned above. Furthermore, these two solutions also provide the time-dependent Green function for the systems propagation for any given initial condition, not only for Gaussians. It is also possible to obtain the corresponding Wigner function and generalized creation and annihilation operators in this way. In searching for problems with analytical solutions, the crucial point is to identify systems that have either a wave packet solution with time-dependent width (in closed form) that obeys a Riccati equation, or a classical parametric oscillator with frequency $\omega(t)$ that still enables an analytical solution of the Newtonian equation. Comparing the wave packets with the solution of the diffusion equation – also a Gaussian with time-dependent width – leads to a real Riccati equation. A first attempt at "complexifying" this equation turns out to be problematic. Nevertheless, it provides a clue to time-dependent frequencies inversely proportional to time that lead to Newtonian equations with analytical solutions. Following a detailed analysis of the corresponding quantum dynamics and energetics, possible extensions of the method are indicated.

1. Introduction

One, if not THE, major difference between classical physics and quantum physics is that classical theories are essentially based on real quantities whereas complex quantities are in principle necessary for quantum theory. Take mechanics: one might think it sufficient to double the number of degrees of freedom by replacing one real variable with a complex variable consisting of a real and an imaginary part; not quite, in quantum theory! This is because these two parts are in this case not independent of each other but coupled via some conservation law. Taking the conventional view of quantum mechanics as a linear theory, this coupling is not at all obvious as real and imaginary parts of the complex variable fulfil the same (linear) equation and there seems to be no relation that couples these two components. A nonlinear perspective of quantum mechanics changes the situation completely; this is now possible via a formulation in terms of (quadratic) nonlinear (NL) complex Riccati equations. Squaring a complex number yields a new complex number where real and imaginary parts of the original one are uniquely interrelated. In the quantum mechanical context, this leads to a conservation law that enables the calculation of the missing part of the complex variable, providing one part (real or imaginary) is known.

Furthermore, the complex NL Riccati equation can be linearized to a complex Newtonian equation. Likewise, the conservation law mentioned previously makes it possible to calculate one part of the complex variable fulfilling the Newtonian equation, again, if the other part is known. This ultimately means it is sufficient to solve the classical Newtonian equation of motion for the classical system in order to obtain, via its complex version, the conservation law and corresponding NL complex Riccati equation, including all the information about the quantum dynamics of the system.

In Section 2 it is shown how the Gaussian wave packet (WP) solutions of the time-dependent (TD) Schrödinger equation (SE) for systems with exact analytic solutions (essentially Hamiltonians that are at most quadratic in position and momentum) can be traced back to the solution of the complex Riccati equation mentioned earlier. Different ways of solving this equation are discussed. From these solutions, all quantum dynamical properties – like uncertainties of position and momentum, their correlations, tunnelling currents, ground state energies, etc. – can be obtained. In addition, the TD Green function for the system's propagation can be constructed for any given initial condition, not only for Gaussians. Also the corresponding Wigner function and generalized creation and annihilation operators follow from this procedure.

The parametric oscillator with TD frequency $\omega(t)$ and potential $V = \frac{m}{2}\omega^2(t)x^2$ (in one dimension) is studied thoroughly in Section 3 as an example of a Hamiltonian quadratic in position variable with analytic solutions. An extension to more than one dimension is usually straightforward. It has been shown [1, 2] that for $\omega = \omega(t)$ the width of the Gaussian WP must be TD. It is also well-known that a solution of the diffusion equation is a Gaussian function with width spreading proportional to time t. Therefore, a first attempt is made to link the TDSE with the diffusion equation via the Riccati equation describing the time-dependence of the width of the Gaussian solutions that exist in both cases. However, the major difference is that the Riccati equation corresponding to the diffusion equation is a real one, whereas, in the quantum mechanical case it is a complex one and, essentially, the squaring of the complex variable provides the information linking the real and imaginary parts in a unique way.

A first "naive" attempt at "complexifying" the real Riccati equation using the specific form of real and imaginary parts that fulfil the complex Riccati equation proves not quite successful, but not futile. It implies that the frequency $\omega(t)$ should be inversely proportional to time t, i.e., $\omega(t) \propto \frac{1}{t}$. This finding, together with the above-mentioned information about the complex Riccati equation and the resulting linearized complex Newtonian equation (or its equivalent formulation in terms of a real NL Ermakov equation), makes it possible to solve the classical as well as the quantum mechanical problem of some particular parametric oscillators in closed form. The dynamical and energetic aspects of the quantum system are examined and possible extensions of the method indicated.

2. Quantum dynamics of Gaussian wave packets

There are only few cases where the SE can be solved analytically. Particularly for potentials that are at most quadratic in position variable, the TDSE possesses solutions that have the form of Gaussian WPs. A general ansatz for such a function can be written as [3, 4]

$$\Psi_{\mathrm{WP}}(x,t) \;=\; N(t)\exp\left\{\mathrm{i}\left[y(t)\,\tilde{x}^2 + \frac{\langle p\rangle}{\hbar}\,\tilde{x} + K(t)\right]\right\} \tag{1}$$

with the purely TD normalization factor $N(t)$ (and the TD function $K(t)$ that will not be relevant for the following discussion). The variable $\tilde{x}$ is defined as $x - \langle x\rangle$ where $\langle x\rangle$ is the mean value of position, calculated via $\langle x\rangle = \int_{-\infty}^{+\infty} dx \Psi^* x \Psi = \eta(t)$; similarly, $\langle p\rangle = m\dot{\eta}$ is the mean value of momentum. Important for the quantum mechanical aspect of the system's dynamics is the coefficient of $\tilde{x}^2$ in the exponent, $y(t)$. As the wave function Ψ is generally a complex function, the TD coefficient $y(t)$ is also assumed to be complex, i.e., $y(t) = y_{\mathrm{R}}(t) + \mathrm{i}y_{\mathrm{I}}(t)$. The imaginary part is related to the position uncertainty via $y_{\mathrm{I}} = \frac{1}{4\langle\tilde{x}^2\rangle}$ with the mean square deviation of position $\langle\tilde{x}^2\rangle = \langle x^2\rangle - \langle x\rangle^2$. The square root of $\langle\tilde{x}^2\rangle$ essentially determines the width of the Gaussian function. This function is completely determined by two parameters, the width and the maximum of the bell-shaped curve. In this quantum mechanical context, both parameters can be TD. The equations of motion that determine this time-dependence can be obtained by inserting the WP(1) into the TDSE,

$$\mathrm{i}\hbar\frac{\partial}{\partial t}\Psi_{\mathrm{WP}}(x,t) \;=\; \left\{-\frac{\hbar^2}{2m}\frac{\partial^2}{\partial x^2} + V(x,t)\right\}\Psi_{\mathrm{WP}}(x,t)\ , \tag{2}$$

in the following for $V(x,t) = \frac{m}{2}\omega^2 x^2$ with the general case $\omega = \omega(t)$ and the particular cases $\omega = \omega_0 = const.$ (harmonic oscillator) and $\omega = 0$ (fee motion). From the terms proportional to $\tilde{x} = x - \eta$, one obtains

$$\ddot{\eta} + \omega^2(t)\eta = 0\ , \tag{3}$$

i.e., the Newtonian equation of motion that determines the time-evolution of the maximum of the WP, located at $x = \eta(t)$ and thus following the classical trajectory. Defining the complex variable $\mathcal{C}(t)$ as $\mathcal{C}(t) = \frac{2\hbar}{m}y(t)$, the terms proportional to $\tilde{x}^2$ provide the complex nonlinear (NL) Riccati equation

$$\dot{\mathcal{C}} + \mathcal{C}^2 + \omega^2(t) \;=\; 0 \tag{4}$$

which determines the time-evolution of the WP-width, as $\mathcal{C}_{\mathrm{I}} = \frac{2\hbar}{m}y_{\mathrm{I}} = \frac{\hbar}{2m\langle\tilde{x}^2\rangle}$. Therefore, in order to obtain the information about the typical quantum mechanical aspects of the system's dynamics, this Riccati equation must be solved. One possible way of doing this [5, 6] is to transform the inhomogeneous Riccati equation (4) into the (still NL and complex) homogeneous Bernoulli equation for $\mathcal{V}(t)$,

$$\dot{\mathcal{V}} + 2\,\tilde{\mathcal{C}}\mathcal{V} + \mathcal{V}^2 = 0\ , \tag{5}$$

using, for the general solution of the Riccati equation, the ansatz

$$\mathcal{C} \;=\; \tilde{\mathcal{C}} \;+\; \mathcal{V}(t)\ , \tag{6}$$

where $\tilde{\mathcal{C}}$ is a particular solution of the Riccati equation that can be found in some cases (particularly for constant $\tilde{\mathcal{C}}$). The advantage of the Bernoulli equation is that it can be linearized to an inhomogeneous first order differential equation of the form

$$\dot{\kappa} - 2\,\tilde{\mathcal{C}}\,\kappa \;=\; 1 \tag{7}$$

with $\kappa = \frac{1}{\mathcal{V}(t)}$.

The general solution of the Riccati equation can then finally be written in the form

$$\mathcal{C}(t) \;=\; \tilde{\mathcal{C}} \;+\; \frac{d}{dt}\,\ln\left[\mathcal{I}(t) + \kappa_0\right] \tag{8}$$

where the (complex) parameter κ_0 is the initial value of $\kappa(t)$ and $\mathcal{I}(t)$ is an integral over exponential functions depending on the particular solution $\tilde{\mathcal{C}}$. For constant $\tilde{\mathcal{C}}$, the general solution can be written in the simple form

$$\mathcal{C}(t) \;=\; \tilde{\mathcal{C}} \;+\; \frac{e^{-2\tilde{\mathcal{C}}t}}{\frac{1}{2\tilde{\mathcal{C}}}\left(1 - e^{-2\tilde{\mathcal{C}}t}\right) \;+\; \kappa_0} \tag{9}$$

where $\tilde{\mathcal{C}}$ can be real, imaginary or complex. A different way of treating the Riccati equation (4) can be taken by introducing a new variable $\alpha(t)$ via

$$\mathcal{C}_{\rm I}(t) = \frac{1}{\alpha^2}\;. \tag{10}$$

From the imaginary part of Eq. (4) follows immediately

$$\mathcal{C}_{\rm R} = \frac{\dot{\alpha}}{\alpha}\;. \tag{11}$$

As $\alpha(t)$, via $\alpha = \sqrt{\frac{2m}{\hbar}\langle\tilde{x}^2\rangle}$, is directly proportional to the WP-width, only for constant width, i.e., $\dot{\alpha} = 0$, the real part $\mathcal{C}_{\rm R}$ vanishes. In all other cases, the Riccati variable $\mathcal{C}(t)$ is complex (particularly also always for the parametric oscillator with $\omega = \omega(t)$). From the real part of Eq. (4), one finally obtains the equation for $\alpha(t)$,

$$\ddot{\alpha} + \omega^2(t)\alpha = \frac{1}{\alpha^3}\;, \tag{12}$$

known in the literature as the Ermakov equation [7], but also associated with the names Steen [8, 9, 10], Milne [11], Pinney [12], Lewis [13, 14] and Riesenfeld [14].

A special feature of Eq. (12) describing the evolution of the WP-width is quite interesting for the dynamics of the system. Eliminating ω^2 using Eq. (3) that describes the evolution of the WP-maximum, leads to a dynamical invariant, the so-called Ermakov invariant

$$I \;=\; \frac{1}{2}\left[(\dot{\eta}\alpha - \eta\dot{\alpha})^2 \;+\; \left(\frac{\eta}{\alpha}\right)^2\right] \;=\; {\rm const}\;, \tag{13}$$

that is also a constant of motion for $\omega = \omega(t)$, although in this case the Hamiltonian (and thus the energy) no longer has this property.

The sum of two quadratic terms in (13) can also be re-written as the product of two linear terms (now expressed with $\mathcal{C}(t)$) as

$$I \;=\; \frac{1}{2}\,\alpha^2\left[(\dot{\eta} - \mathcal{C}_{\rm R}\eta)^2 \;+\; (\mathcal{C}_{\rm I}\eta)^2\right] \;=\; \frac{1}{2}\,\alpha^2\left[(\dot{\eta} - \mathcal{C}\eta)\,(\dot{\eta} - \mathcal{C}^*\eta)\right]\;. \tag{14}$$

This is already formally similar to the harmonic oscillator Hamiltonian written in terms of creation and annihilation operators as

$$H_{op} \;=\; \frac{1}{2m}\,p_{op}^2 \;+\; \frac{m}{2}\,\omega_0^2\,x^2 \;=\; \hbar\omega_0\,\left(a^+a \;+\; \frac{1}{2}\right) \tag{15}$$

with $p_{op} = \frac{\hbar}{\rm i}\frac{\partial}{\partial x}$, the number operator a^+a with the commutator $[a, a^+]_- = 1$ of the annihilation operator a and the creation operator a^+, defined as

$$a = \mathrm{i} \sqrt{\frac{m}{2\hbar\omega_0}} \left(\frac{p_{\mathrm{op}}}{m} - \mathrm{i}\, \omega_0\, x \right) , \tag{16}$$

$$a^+ = -\mathrm{i} \sqrt{\frac{m}{2\hbar\omega_0}} \left(\frac{p_{\mathrm{op}}}{m} + \mathrm{i}\, \omega_0\, x \right) . \tag{17}$$

For the harmonic oscillator with constant frequency ω_0 and constant width of the corresponding Gaussian function, $\mathrm{i}\, \omega_0 = \mathrm{i}\, \mathcal{C}_{\mathrm{I}} = \mathrm{i}\frac{1}{\alpha^2}$ with constant α is valid, i.e., $\mathrm{i}\, \omega_0$ is simply the imaginary part of the complex Riccati variable. The real part $\mathcal{C}_{\mathrm{R}} = \frac{\dot{\alpha}}{\alpha}$ vanishes, as $\dot{\alpha} = 0$. For the general case with TD width of the Gaussian, a generalization of the creation and annihilation operators [15] can be obtained by replacing $\eta \to x$, $\dot{\eta} \to \frac{p_{op}}{m}$ in the invariant (14) and taking into account the non-commutativity of x and p_{op}, leading to the corresponding operator

$$I_{op} = \frac{^-h}{m} \left[a^+(t) a(t) + \frac{1}{2} \right] \tag{18}$$

with the TD operators

$$a(t) = \mathrm{i} \sqrt{\frac{m}{2\hbar}}\, \alpha \left(\frac{p_{\mathrm{op}}}{m} - \mathcal{C}\, x \right) , \tag{19}$$

$$a^+(t) = -\mathrm{i} \sqrt{\frac{m}{2\hbar}}\, \alpha \left(\frac{p_{\mathrm{op}}}{m} - \mathcal{C}^*\, x \right) . \tag{20}$$

Finally, the Riccati equation (4) can also be linearized using the logarithmic derivative of a new complex variable $\lambda(t)$ via $\mathcal{C} = \frac{\dot{\lambda}}{\lambda}$ to yield the complex Newtonian equation

$$\ddot{\lambda} + \omega^2(t)\lambda = 0 \tag{21}$$

with

$$\lambda(t) = u + \mathrm{i}\, z = \alpha\, \mathrm{e}^{\mathrm{i}\varphi} . \tag{22}$$

Inserting the polar form into the above definition of $\mathcal{C}$ leads to

$$\mathcal{C} = \frac{\dot{\alpha}}{\alpha} + \mathrm{i}\, \dot{\varphi} . \tag{23}$$

Comparison with the definition as given in (10) and (11) shows that both definitions are consistent and the amplitude α of λ is identical to the Ermakov variable $\alpha(t)$, providing

$$\dot{\varphi} = \frac{1}{\alpha^2} \tag{24}$$

is valid. This can be proven easily by inserting the polar form of $\mathcal{C}$ into the imaginary part of the Riccati equation (4). Relation (24), written as a conserved quantity $\dot{\varphi}\, \alpha^2 = 1$, is at least formally similar to the conservation of angular momentum for the motion in a plane but here it is the motion of $\lambda(t)$ in the *complex* plane (for further details, see [1, 2, 3, 4, 6]).

Written in Cartesian coordinates, the conservation law (24) takes the form

$$\dot{z}u - \dot{u}z = 1 \tag{25}$$

which is equivalent to the Wronskian determinant of two linear independent solutions u and z of the complex Newtonian equation (21).

Knowing one solution (e.g., $z(t)$), it is possible to calculate the other via

$$u \;=\; -\, z \int^{t} \frac{1}{z^2}(t)\; dt' \;. \tag{26}$$

Knowing both solutions of the classical Newtonian equation immediately provides the solution for the dynamics of the WP width via

$$\alpha^2 \;=\; u^2 \;+\; z^2 \;. \tag{27}$$

If α is known, $\mathcal{C}_{\rm I}$ and $\mathcal{C}_{\rm R}$ can be determined and consequently all typical quantum mechanical properties like uncertainties and ground state energies:

$$\langle \tilde{x}^2 \rangle(t) = \frac{\hbar}{2m}\alpha^2(t) = \frac{\hbar}{2m}\frac{1}{\mathcal{C}_{\rm I}(t)} \tag{28}$$

$$\langle \tilde{p}^2 \rangle(t) = \frac{m\hbar}{2}\left[\dot{\alpha}^2(t) + \frac{1}{\alpha^2(t)}\right] = \frac{m\hbar}{2}\frac{\mathcal{C}_{\rm R}^2(t) + \mathcal{C}_{\rm I}^2(t)}{\mathcal{C}_{\rm I}(t)} \tag{29}$$

$$\langle [\tilde{x},\tilde{p}]_+ \rangle(t) = \langle \tilde{x}\tilde{p} + \tilde{p}\tilde{x} \rangle(t) = \hbar\; \alpha(t)\; \dot{\alpha}(t) = \hbar \frac{\mathcal{C}_{\rm R}(t)}{\mathcal{C}_{\rm I}(t)} \tag{30}$$

$$U \;=\; \langle \tilde{x}^2 \rangle \langle \tilde{p}^2 \rangle = \frac{\hbar^2}{4}\left\{1 + (\alpha(t)\;\dot{\alpha}(t))^2\right\} \;=\; \frac{\hbar^2}{4}\left\{1 + \left(\frac{\mathcal{C}_{\rm R}(t)}{\mathcal{C}_{\rm I}(t)}\right)^2\right\} \tag{31}$$

$$\tilde{E} \;=\; \frac{\langle \tilde{p}^2 \rangle}{2m} + \frac{m}{2}\omega^2 \langle \tilde{p}^2 \rangle = \frac{\hbar}{4}\left\{\dot{\alpha}^2(t) + \frac{1}{\alpha^2(t)} + \omega^2 \alpha^2(t)\right\} \;=\; \frac{\hbar}{4}\frac{1}{\mathcal{C}_{\rm I}}\left\{\mathcal{C}_{\rm R}^2(t) + \mathcal{C}_{\rm I}^2(t) + \omega^2\right\} \;. \tag{32}$$

So, with the solutions of the classical Newtonian equation it is possible to construct the solution of the Ermakov equation that describes the dynamics of the uncertainties and thus the quantum mechanical aspect. In other words, if the dynamics of the WP-maximum is known, the dynamics of the WP-width can be determined. This also works the other way round. Key point once more is the conservation law, this time not in the Cartesian form (25), but in the polar form (24). Again, as in (26), one integration is required to obtain the missing part, in this case $\varphi \;=\; \int^{t} \frac{1}{\alpha^2(t')}\, dt'$. Together with $\alpha(t)$ this yields $\lambda = e^{i\varphi} = u + i\, z = \alpha\cos\varphi + i\,\alpha\sin\varphi$, i.e., the two solutions of the Newtonian equation.

What is still missing is the information about which of the two Cartesian components is (up to a constant factor) identical to the classical trajectory and thus the maximum of the WP, u or z. This can be ascertained using the TD Green function (also called Feynman kernel) of the problem. With the help of this function, the wave function $\Psi(x,t)$ can be obtained starting from an initial wave function $\Psi(x',t')$ according to

$$\Psi_{\rm WP}(x,t) \;=\; \int dx'\; K(x,x',t,t'=0)\; \Psi_{\rm WP}\,(x',t'=0) \;. \tag{33}$$

Particularly for Gaussian wave functions it can be shown that the TD Green function is also a Gaussian function and can be expressed with the help of the TD parameters $u(t)$ and $z(t)$ according to [1, 2, 17]

$$K\,(x,x',t,t'=0) \;=\; \left(\frac{m}{2\pi i\hbar\alpha_0 z}\right)^{\frac{1}{2}} \exp\left\{\frac{im}{2\hbar}\left[\frac{\dot{z}}{z}x^2 - 2\frac{x}{z}\left(\frac{x'}{\alpha_0}\right) + \frac{u}{z}\left(\frac{x'}{\alpha_0}\right)^2\right]\right\} \;. \tag{34}$$

Inserting this kernel into (33) and choosing a Gaussian WP as initial state yields the TD Gaussian WP in the form

$$\Psi_{\mathrm{WP}}(x,t) = \left(\frac{m}{\pi \hbar}\right)^{\frac{1}{4}} \left(\frac{1}{u+\mathrm{i}z}\right)^{\frac{1}{2}} \exp\left\{\frac{\mathrm{i}m}{2\hbar}\left[\frac{\dot{z}}{z}x^2 - \frac{(x-\frac{p_0\alpha_0}{m}z)^2}{z(u+\mathrm{i}z)}\right]\right\}. \quad (35)$$

Comparison with the Gaussian WP as defined in (1) shows

$$z(t) = \frac{m}{\alpha_0\, p_0}\, \eta(t) \quad (36)$$

and, using $\lambda = u + \mathrm{i}\, z$ and the conservation law (25), the coefficient of the quadratic term in the exponent is $\frac{\mathrm{i}m}{2\hbar}\left(\frac{\dot{\lambda}}{\lambda}\right) = \frac{\mathrm{i}m}{2\hbar}\,\mathcal{C} = \mathrm{i}\ y(t)$. So, u and z are identical to real and imaginary parts of the complex variable $\lambda(t)$ and the imaginary part $z(t)$ is, up to a constant factor, identical to the classical trajectory $\eta(t)$.

To conclude this section, the Wigner function of the problem is introduced. This quantum mechanical phase space distribution function can be obtained from the wave function in position space via the transformation

$$W(x,p) = \frac{1}{2\pi\hbar}\int_{-\infty}^{+\infty} dq\ e^{\mathrm{i}pq/\hbar}\ \Psi^*_{\mathrm{WP}}(x+\frac{q}{2})\ \Psi_{\mathrm{WP}}(x-\frac{q}{2})\ . \quad (37)$$

Inserting our Gaussian WP into this Wigner transformation and using the relations between $\mathcal{C}_{\mathrm{I}}$, $\mathcal{C}_{\mathrm{R}}$ and the uncertainties as shown in (28)-(30), the Wigner function can be written as

$$W(x,p,t) = \frac{1}{\pi\hbar}\exp\left\{-\frac{2}{\hbar^2}\left[\langle\tilde{p}^2\rangle\tilde{x}^2 - \langle[\tilde{x},\tilde{p}]_+\rangle\tilde{x}\tilde{p} + \langle\tilde{x}^2\rangle\tilde{p}^2\right]\right\}\ . \quad (38)$$

Expressing the uncertainties in terms of α and $\dot{\alpha}$ and rearranging them in the exponent finally leads to

$$W(x,p,t) = \frac{1}{\pi\hbar}\exp\left\{-\frac{m}{\hbar}\left[\left(\dot{\alpha}\tilde{x} - \alpha\frac{\tilde{p}}{m}\right)^2 + \left(\frac{\tilde{x}}{\alpha}\right)^2\right]\right\}\ . \quad (39)$$

This means the exponent of the Wigner function is essentially the Ermakov invariant (multiplied by $-\frac{2m}{\hbar}$), only η is replaced by $\tilde{x} = x - \eta$ and $\dot{\eta}$ by $\frac{\tilde{p}}{m} = \frac{p}{m} - \dot{\eta}$. (For further details, see [16, 17].)

Figure 1 shows schematically the different treatments of the Riccati equation via the real Ermakov equation or the complex Newtonian equation and their corresponding interrelations.

3. Gaussian wave packets for the parametric oscillator

3.1. Gaussian functions with time-dependent width

It can be shown [1, 2] that a Gaussian WP solution of the TDSE with constant width exists for the harmonic oscillator with constant frequency $\omega = \omega_0$ only if this width is identical to the one of the corresponding ground state wave function. In all other cases, but especially for the parametric oscillator with $\omega = \omega(t)$, the WP width is TD. This time-dependence is obviously different for a different $\omega(t)$. For arbitrary $\omega(t)$, finding analytical solutions for the time-dependence of the WP width cannot be expected, nor for its maximum (obeying the corresponding classical Newtonian equation), as both quantities are not independent of each other, as demonstrated in the previous section. One might then ask if there are any cases of analytical solutions for the classical Newtonian and the corresponding Ermakov or Riccati equations. If there are, what are the corresponding frequencies $\omega(t)$?

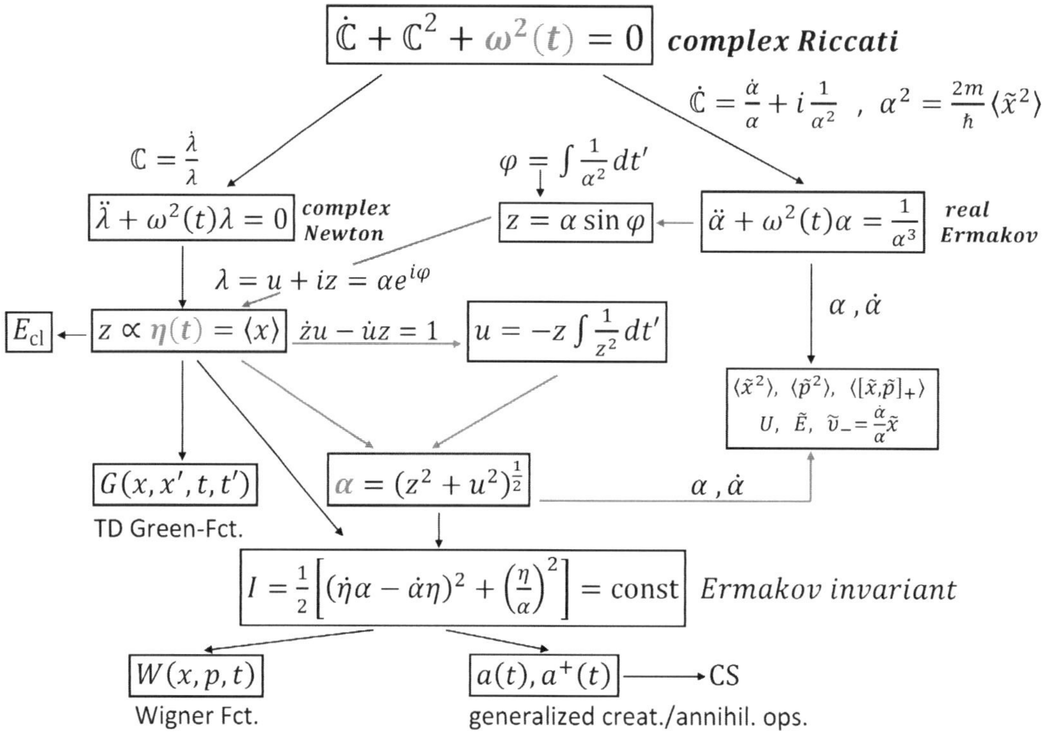

Figure 1: Different treatments of the complex Riccati equation for the TDSE.

In a first attempt at answering these questions, it might be useful to start out with a comparison of other systems that are described by Gaussian functions with TD width. The diffusion equation, for instance, has solutions with this property. The diffusion process is a classical one which means the corresponding Gaussian function is real. In our quantum mechanical case, it is complex. For comparison, however, it is possible to use the Gaussian WP that describes the probability density $\varrho_{\mathrm{S}}(x,t) = \Psi^*\Psi(x,t)$ where the subscript S indicates that ϱ_{S} is related to the solution $\Psi(x,t) \propto \exp\left\{\mathrm{i}\frac{\hbar}{2m}\mathcal{C}(t)\,\tilde{x}^2\right\}$ of the SE.

This probability density fulfils the continuity equation

$$\frac{\partial}{\partial t}\varrho_{\mathrm{S}} + \frac{\partial}{\partial x}(\varrho_{\mathrm{S}}\, v_-) = 0 \tag{40}$$

and has (up to a normalization factor) the form

$$\varrho_{\mathrm{S}}(x) \propto \exp\left\{-\frac{m}{\hbar}\frac{\tilde{x}^2}{\alpha^2}\right\} = \exp\left\{-\frac{\tilde{x}^2}{2\langle\tilde{x}^2\rangle_{\mathrm{S}}}\right\} . \tag{41}$$

The term $\varrho_{\mathrm{S}}v_-$ is a convection current with velocity field v_- and, for Gaussian WPs, is written in the form

$$v_- = \dot\eta + \mathcal{C}_{\mathrm{R}}\,\tilde{x} = \dot\eta + \frac{\dot\alpha}{\alpha}\,\tilde{x} , \tag{42}$$

i.e., the non-classical contribution, representing the tunnelling contribution to the current, is also completely determined by the solution of the complex Riccati or the equivalent Ermakov equation. In the diffusion equation, the convection current is replaced by the diffusion current $-D\,\frac{\partial}{\partial x}\varrho_{\rm D}$, leading to

$$\frac{\partial}{\partial t}\varrho_{\rm D} - D\,\frac{\partial^2}{\partial x^2}\varrho_{\rm D} = 0\,. \tag{43}$$

The corresponding diffusion distribution function $\varrho_{\rm D}$ is, again up to a normalization factor, given by

$$\varrho_{\rm D}(x) \propto \exp\left\{-Rx^2\right\} = \exp\left\{-\frac{x^2}{4Dt}\right\} = \exp\left\{-\frac{x^2}{2\langle\tilde{x}^2\rangle_{\rm D}}\right\}\,. \tag{44}$$

The coefficient R of the quadratic term is real and fulfils the Riccati/Bernoulli equation

$$\dot{R} + 4DR^2 = 0 \tag{45}$$

with the solution $R = \frac{1}{4D\,t}$.

Comparing the two mean-square deviations of position,

$$\langle\tilde{x}^2\rangle_{\rm S} = \frac{\hbar}{2m}\alpha^2 \tag{46}$$

and

$$\langle\tilde{x}^2\rangle_{\rm D} = 2D\,t \tag{47}$$

one arrives at

$$\alpha^2 = 2\,t \tag{48}$$

if $D = \frac{\hbar}{2m}$ is assumed; an assumption that can also be found in the literature [18, 19], e.g., when considering the free motion TDSE as a diffusion equation with imaginary diffusion coefficient.

In a first "naive" attempt at "*complexifying*" the real Riccati equation (45), one could assume that from

$$\mathcal{C}_{\rm I} = \frac{1}{\alpha^2} = \frac{1}{2\,t} \tag{49}$$

also the missing real part $\mathcal{C}_{\rm R}$ could be obtained via

$$\mathcal{C}_{\rm R} = \frac{\dot{\alpha}}{\alpha} = \frac{1}{2\,t}\,. \tag{50}$$

Inserting (49) and (50) into the real part of the complex Riccati equation (4) would finally provide the corresponding frequency $\omega(t)$,

$$\begin{aligned}\dot{\mathcal{C}}_{\rm R} + \mathcal{C}_{\rm R}^2 - \mathcal{C}_{\rm I}^2 + \omega_{\rm Ri}^2(t) &= \dot{\mathcal{C}}_{\rm R} + \omega_{\rm Ri}^2(t) = 0\\ &\rightarrow \omega_{\rm Ri}(t) = \frac{1}{\sqrt{2}\,t}\,.\end{aligned} \tag{51}$$

From Sect. 2 it is known that the complex Riccati equation can be linearized to the Newtonian equation

$$\ddot{\lambda} + \omega_{\rm Ne}^2(t)\,\lambda = 0 \tag{52}$$

where real and imaginary parts of λ must fulfil the classical equation of motion with the same frequency ω that also occurs in the Riccati equation.

An ansatz that fulfils Eq. (52) is $z = t^{\frac{1}{2}} \to \dot{z} = \frac{1}{2}\, t^{-\frac{1}{2}}$, leading to

$$\begin{aligned} \ddot{z} &= -\frac{1}{4}\, t^{-\frac{3}{2}} = -\frac{1}{4\, t^2}\, z \\ \to \omega_{\rm Ne}(t) &= \frac{1}{2\, t} \neq \omega_{\rm Ri}(t)\ . \end{aligned} \tag{53}$$

This obviously is in disagreement with (51).

Deaply related with this result is the question, if a Gaussian solution of the TDSE (or the corresponding continuity equation, respectively) for a parametric oscillator with specific $\omega(t)$ can be found that is also a solution of the diffusion equation. A Gaussian function fulfilling this requirement should obey the relation

$$\frac{\partial}{\partial x}\left(\varrho_{\rm S} v_-\right) = -D\, \frac{\partial^2}{\partial x^2}\varrho\ . \tag{54}$$

The first problem is due to the drift motion of the Gaussian obeying the continuity equation, where the maximum moves according to v_- with the classical velocity $\dot{\eta}(t)$, whereas the maximum of the Gaaussian describing the diffusion process is usually fixed. But even if this problem could be circumvended by, e.g., changing to some moving coordinate frame or similar procedures, still the relation between the quantities $\alpha(t)$ that describe the spreading of the Gaussians have to be matched. This means, the relation

$$D\frac{2\hbar}{m} - D\frac{4m^2}{\hbar^2}\frac{1}{\alpha^2}\tilde{x}^2 = \dot{\alpha}\alpha - \frac{2m}{\hbar}\frac{\dot{\alpha}}{\alpha}\tilde{x}^2 \tag{55}$$

has to be fulfilled.

For the choice $D = \frac{\hbar}{2m}$ (as assumed above) this leads to

$$1 - \frac{2m}{\hbar}\frac{1}{\alpha^2}\tilde{x}^2 = \dot{\alpha}\alpha - \frac{2m}{\hbar}\frac{\dot{\alpha}}{\alpha}\tilde{x}^2 \tag{56}$$

or

$$\dot{\alpha}\alpha = 1\ . \tag{57}$$

From the Gaussian solution of the diffusion equation follows

$$\alpha^2 = \frac{2m}{\hbar}\,\langle\tilde{x}^2\rangle = \frac{4m}{\hbar}Dt = 2t\ , \tag{58}$$

what is in agreement with (49), (50) and (57).

On the other side, the width of the Gaussian fulfilling the continuity equation corresponding to the TDSE of a parametric oscillator with frequency $\omega(t)$ has to fulfil the Ermakov equation (12), where $\omega(t)$ is the same frequency that also occurs in the classical Newtonian equation for $\eta(t)$.

From Eq. (57) follows that $\frac{\ddot{\alpha}}{\alpha} = -\frac{1}{\alpha^4}$, so the Ermakov equation (12) would turn into

$$\omega^2(t) = 2\left(\frac{1}{\alpha^4}\right) = 2\left(\frac{1}{2\, t}\right)^2 = \left(\frac{1}{\sqrt{2}\, t}\right)^2\ . \tag{59}$$

Therefore, the frequency of the Ermakov equation would be $\omega(t) = \frac{1}{\sqrt{2}\, t} = \omega_{\rm Ri}(t)$, like in the Riccati equation (see Eq. (51)) and different from the frequency $\omega_{\rm Ne}(t)$ that enters the classical Newtonian equation of the problem. So, there is no Gaussian solution of the parametric oscillator that fulfils the corresponding continuity equation and at the same time also the diffusion equation.

According to these results, the approach via the WP-width $\alpha(t)$ and the Riccati equation seems problematic. However, with $z(t)$, one solution of the Newtonian equation is already known. As shown in Section 2, with a second linear independent solution $u(t)$, the Ermakov variable can be obtained via (27).

3.2. Approach via the Newtonian equation of motion

For $\omega(t) = \frac{1}{2t}$, a first solution of the Newtonian equation is given by $z = t^{\frac{1}{2}}$. From (26), it then follows that $u(t) = -t^{\frac{1}{2}} \ln t$ and, thus [1] from (27),

$$\alpha^2 = u^2 + z^2 = t \left[1 + \ln^2 t\right] . \quad (60)$$

With this expression for α, now $\mathcal{C}_{\rm I}$ and $\mathcal{C}_{\rm R}$ can be determined as

$$\mathcal{C}_{\rm I} = \frac{1}{\alpha^2} = \frac{1}{t\left[1 + \ln^2 t\right]} \quad (61)$$

$$\mathcal{C}_{\rm R} = \frac{\dot{\alpha}}{\alpha} = \frac{1}{2} \frac{(1 + \ln t)^2}{t\left[1 + \ln^2 t\right]} = \frac{1}{2\,t} + \frac{\ln t}{t\left[1 + \ln^2 t\right]} \quad (62)$$

where $\frac{1}{2t}$ is actually also a particular solution $\tilde{\mathcal{C}}$ of the Riccati equation. A tricky aspect is still the divergence for $t \to 0$; but this can be resolved.

In the following, the modified TD frequency

$$\omega(t) = \frac{1}{2\,(t+b)} \quad (63)$$

is considered with $b = constant \neq 0$ and dimension *"time"*.

Two linear-independent solutions of the (complex) Newtonian equation

$$\ddot{\lambda} + \left(\frac{1}{2\,(t+b)}\right)^2 \lambda = 0 \quad (64)$$

are

$$z = (t+b)^{\frac{1}{2}} \quad (65)$$

and

$$u = -\,(t+b)^{\frac{1}{2}}\,\ln(t+b)\,, \quad (66)$$

leading to the WP width via

$$\alpha^2(t) = (t+b)\,\left[1 + \ln^2(t+b)\right] \quad (67)$$

with the well-behaved initial value $\alpha^2(0) = b\left[1 + \ln^2 b\right]$.

From (67), the imaginary and real parts of the Riccati variable can be obtained as

[1] In the following three equations it is assumed that $t \geq 0$, where the ln-terms diverge for $t = 0$

$$\mathcal{C}_{\rm I} = \frac{1}{(t+b)\,\left[1+\ln^2(t+b)\right]} \tag{68}$$

$$\mathcal{C}_{\rm R} = \frac{1}{2\,(t+b)} + \frac{\ln(t+b)}{(t+b)\,\left[1+\ln^2(t+b)\right]} \tag{69}$$

with initial values $\mathcal{C}_{\rm I}(0) = \frac{1}{b\,[1+\ln^2 b]}$ and $\mathcal{C}_{\rm R}(0) = \frac{1}{2\,b} + \frac{\ln b}{b\,[1+\ln^2 b]}$. The first term on the rhs of $\mathcal{C}_{\rm R}$ is again a particular solution of the Riccati equation, $\tilde{\mathcal{C}} = \frac{1}{2\,(t+b)}$. This (real) particular solution can also be utilized to prove the accuracy of (68) and (69) using the direct method of solving the Riccati equation outlined in Section 2; i.e., assuming that the general solution is given by $\mathcal{C} = \tilde{\mathcal{C}} + \mathcal{V}(t)$. In this case, $\mathcal{V}(t)$ must fulfil the Bernoulli equation

$$\dot{\mathcal{V}} + \frac{1}{(t+b)}\mathcal{V} + \mathcal{V}^2 = 0\,. \tag{70}$$

As $\tilde{\mathcal{C}}$ is real, the imaginary part of the Bernoulli equation is identical to the imaginary part of the general solution, $\mathcal{V}_{\rm I} = \mathcal{C}_{\rm I}$. With $\mathcal{C}_{\rm I}$ as given in (68), it follows from $\mathcal{V}_{\rm R} = -\frac{1}{2}\frac{\dot{\mathcal{V}}_{\rm I}}{\mathcal{V}_{\rm I}} - \frac{1}{2\,(t+b)}$ that

$$\mathcal{V}_{\rm R} = \frac{\ln(t+b)}{(t+b)\,\left[1+\ln^2(t+b)\right]}\,, \tag{71}$$

which is exactly the missing second term on the rhs of (69), fulfilling the real part of the Bernoulli equation,

$$\dot{\mathcal{V}}_{\rm R} + \frac{1}{(t+b)}\,\mathcal{V}_{\rm R} + \mathcal{V}_{\rm R}^2 - \mathcal{V}_{\rm I}^2 = 0\,. \tag{72}$$

With the expressions for α^2, $\mathcal{C}_{\rm I}$ and $\mathcal{C}_{\rm R}$, the analytical form of the quantum uncertainties, ground state energy and tunnelling contribution to the probability current can be given explicitly:

$$\langle\tilde{x}^2\rangle(t) = \frac{\hbar}{2m}\alpha^2(t) = \frac{\hbar}{2m}(t+b)\,\left[1+\ln^2(t+b)\right] \tag{73}$$

$$\langle\tilde{p}^2\rangle(t) = \frac{m\hbar}{2}\left[\dot{\alpha}^2(t) + \frac{1}{\alpha^2(t)}\right] = \frac{m\hbar}{2}\,\frac{1}{(t+b)}\left\{\frac{1}{4}[1 + \ln(t+b)]^2 + \frac{1}{2}\ln(t+b) + 1\right\} \tag{74}$$

$$\langle[\tilde{x},\tilde{p}]_+\rangle(t) = \hbar\,\alpha(t)\,\dot{\alpha}(t) = \frac{\hbar}{2}\,[1+\ln(t+b)]^2 \tag{75}$$

$$U(t) = \langle\tilde{x}^2\rangle\langle\tilde{p}^2\rangle = \frac{\hbar^2}{4}\left\{1+(\alpha\,\dot{\alpha})^2\right\} = \frac{\hbar^2}{4}\left\{1+\frac{1}{4}\,[1+\ln(t+b)]^4\right\} \tag{76}$$

$$\tilde{E}(t) = \frac{\langle\tilde{p}^2\rangle}{2m} + \frac{m}{2}\omega^2(t)\langle\tilde{x}^2\rangle = \frac{\hbar}{4}\left\{\dot{\alpha}^2 + \frac{1}{\alpha^2} + \omega^2(t)\alpha^2\right\} = \frac{\hbar}{4}\left\{\frac{1}{(t+b)}\left[\frac{1}{2}(1 + \ln(t+b))^2 + 1\right]\right\} \tag{77}$$

Table 1: Particular values of the quantum uncertainties, ground state energy and tunnelling contribution to the probability current for $t = 0$ and $t \to \infty$.

$$\tilde{v}_-(t) = v_- - \langle v \rangle = \frac{\dot{\alpha}}{\alpha}\,\tilde{x} = \left\{ \frac{1}{2(t+b)} + \frac{\ln(t+b)}{(t+b)\,\left[1 + \ln^2(t+b)\right]} \right\} \tilde{x}\,. \tag{78}$$

The particular values for $t = 0$ and $t \to \infty$ are given in Table 1.

The plots for $\langle \tilde{x}^2 \rangle(t) = \sigma_x^2$, $\langle \tilde{p}^2 \rangle(t) = \sigma_p^2$, $\frac{1}{2}\langle [\tilde{x}, \tilde{p}]_+ \rangle(t) = \sigma_{xp}, U(t) = \sigma_x^2 \sigma_p^2$ are given in Figures 2 and 3. The time-dependence of $\tilde{E}(t)$ and $\tilde{v}_-$ (essentially given by $\mathcal{C}_{\mathrm{R}} = \frac{\dot{\alpha}}{\alpha}$) is shown in Figure 4.

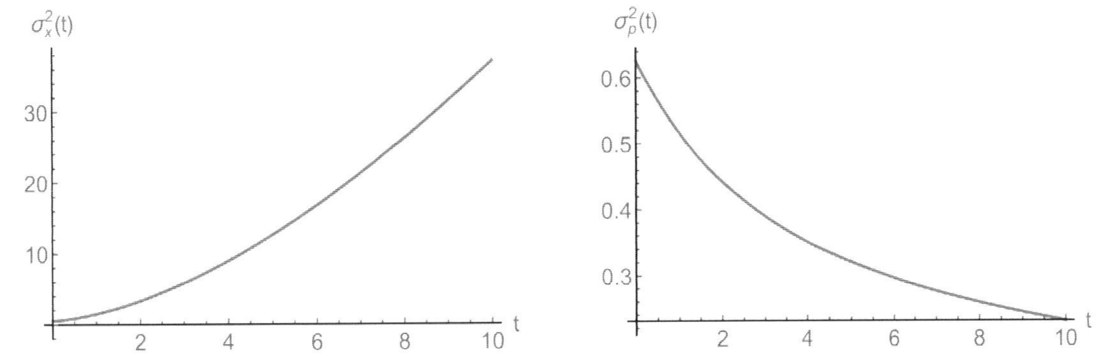

(a) (b)

Figure 2: Graphs of $\sigma_x^2(t) = \langle \tilde{x}^2 \rangle(t)$ and $\sigma_p^2(t) = \langle \tilde{p}^2 \rangle(t)$ as function of time.

Obviously, the position uncertainty and thus the WP width are diverging for $t \to \infty$ whereas the momentum uncertainty is approaching zero. The product of both is going to infinity for $t \to \infty$ while the energy contribution $\tilde{E}$ is going to zero in this case, as in a (classical) dissipative system (without fluctuating force).

3.3. Classical dynamics and energetics for this parametric oscillator

According to (36), $z(t)$ is proportional to the classical trajectory $\eta(t)$. Therefore, the classical Newtonian equation for this parametric oscillator takes the form

$$\ddot{\eta} + \left(\frac{1}{2(t+b)} \right)^2 \eta = 0 \tag{79}$$

with the solution

$$\eta(t) = v_0\, \tau^{\frac{1}{2}} (t+b)^{\frac{1}{2}} \tag{80}$$

with $\eta_0 = v_0 \tau^{\frac{1}{2}} b^{\frac{1}{2}}$ where the constant τ with dimension time has been introduced to get the correct dimension for $\eta(t)$. The corresponding velocity $\dot{\eta}$ is then given by

$$\dot{\eta}(t) = \frac{1}{2}\, v_0\, \tau^{\frac{1}{2}} (t+b)^{-\frac{1}{2}} = v_0\, \frac{\tau^{\frac{1}{2}}}{2\,(t+b)^{\frac{1}{2}}} \tag{81}$$

IOP Conf. Series: Journal of Physics: Conf. Series **1071** (2018) 012020 doi:10.1088/1742-6596/1071/1/012020

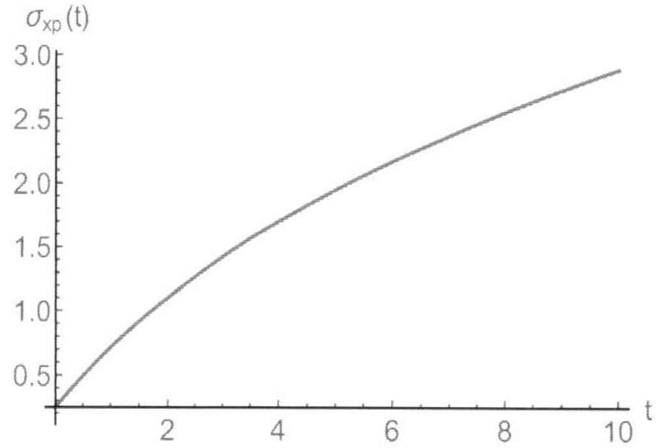

(a)

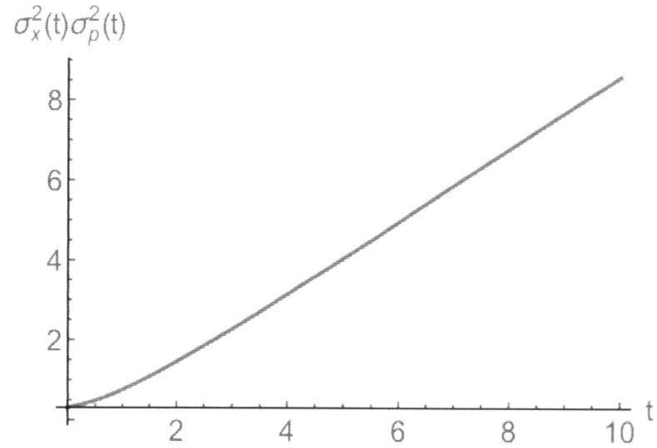

and $\tilde{E}(t)$

(b)

Figure 3: Graphs of $\sigma_{xp}(t) = \langle \frac{1}{2}[\tilde{x}, \tilde{p}]\rangle(t)$ and $U(t) = \sigma_{\tilde{x}}^2(t)\sigma_{\tilde{p}}^2(t)$ as function of time.

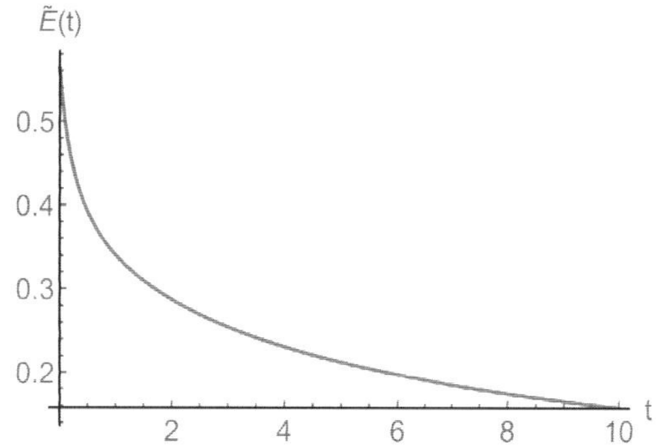

(a)

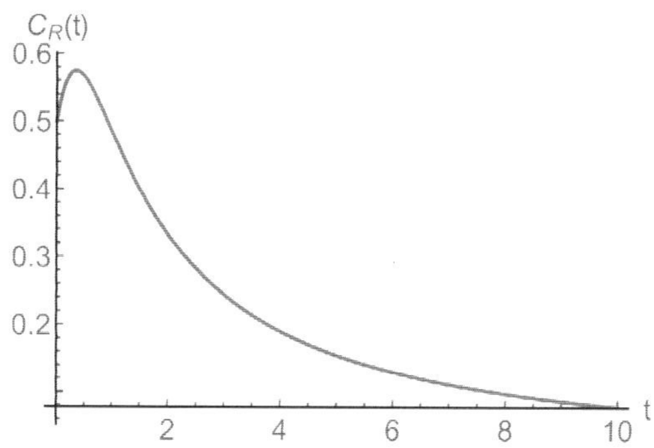

(b)

Figure 4: Graphs of $\tilde{E}(t) = \frac{1}{2m}\sigma_{\tilde{p}}^2(t) + \frac{m}{2}\left(\frac{1}{2\,(t+b)}\right)^2 \sigma_{\tilde{x}}^2(t)$ and $\mathcal{C}_{\mathrm{R}} = \frac{\dot{\alpha}}{\alpha}$ as function of time.

with $\dot{\eta}_0 = v_0\, \frac{\tau^{\frac{1}{2}}}{2\, b^{\frac{1}{2}}}$. From $\dot{\eta}_0 = v_0$ follows $b = \frac{\tau}{4}$.

The classical energy of the system then acquires the form

$$\begin{aligned} E &= T + V = \frac{m}{2}\,\dot{\eta}^2 + \frac{m}{2}\left(\frac{1}{2(t+b)}\right)^2 \eta^2 \\ &= \frac{m}{2}\, v_0^2\, \frac{\tau}{2(t+b)} \propto \frac{1}{t} \end{aligned} \tag{82}$$

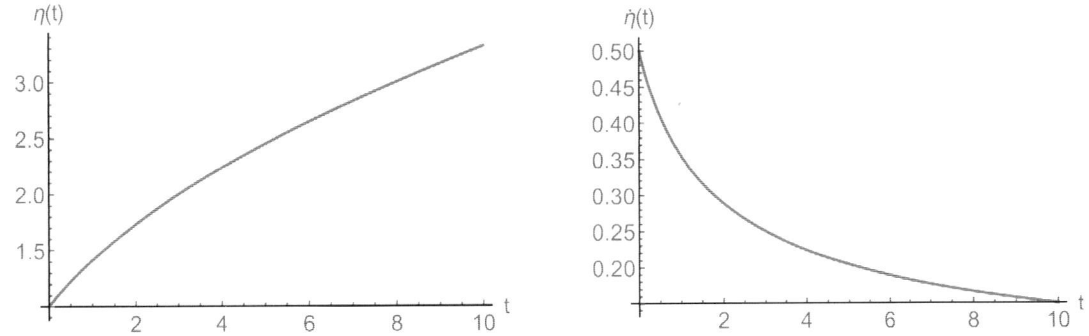

(a) (b)

Figure 5: Graphs of the classical trajectory $\eta(t)$ and the classical velocity $\dot{\eta}(t)$.

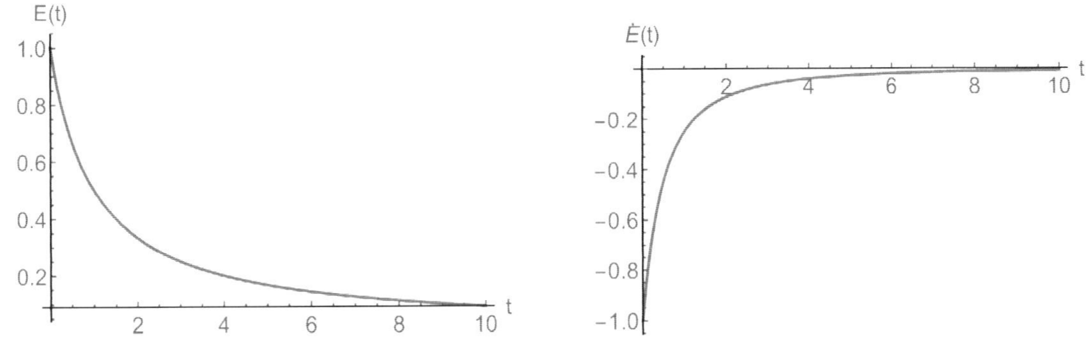

(a) (b)

Figure 6: Graphs of the classical energy $E(t)$ and its change in time, $\dot{E}(t)$, as function of time.

and decreases inversely proportional to time t. The initial energy E_0 is given by $E_0 = \frac{m}{2}\, v_0^2\, \frac{\tau}{2b} = \frac{m}{2}\, v_0^2$, the final energy for $t \to \infty$ approaches zero, $E_\infty \to 0$, as in a dissipative system.

The decrease in energy is given by

$$\dot{E} \;=\; -\,\frac{m}{2}\, v_0^2\, \frac{\tau}{2} \left(\frac{1}{t+b}\right)^2 \;=\; -2\omega(t)\, E \;=\; -\,4\,\omega(t)\, T \;<\; 0\;. \tag{83}$$

Comparison with a dissipative system with a friction force linear proportional to velocity, i.e., the equation of motion

$$\ddot{\eta} + \gamma\, \dot{\eta} + \omega^2 \eta = 0\;, \tag{84}$$

shows that, in this case, the decrease in energy is

$$\dot{E} \;=\; -\,2\gamma\, T \;=\; -\,4\frac{\gamma}{2}\, T\;. \tag{85}$$

So, the time-dependent frequency would correspond to half of the friction parameter γ,

$$\omega(t) \,\hat{=}\, \frac{\gamma}{2}\;. \tag{86}$$

The plots of $\eta(t)$, $\dot{\eta}(t)$, $E(t)$ and $\dot{E}(t)$ are shown in Figures 5 and 6.

4. Conclusions

The TDSE for potentials that are at most quadratic in position variable possesses exact solutions that have the form of Gaussian WPs. These functions are uniquely determined by two parameters, the maximum and the width. In this case, both parameters are TD. The time-dependence of the maximum obeys the classical Newtonian equation of motion for the given potential; the time-evolution of the width is determined by a complex NL Riccati equation or, equivalently, by a real NL Ermakov equation. For the free motion ($\omega = 0$) and the harmonic oscillator ($\omega = \omega_0 =$ const.), both equations can be solved in closed form. For TD frequency $\omega = \omega(t)$, it is the explicit time-dependence of $\omega(t)$ that determines whether analytic solutions exist or not.

Independent of the particular $\omega(t)$, the solutions of the Riccati or Ermakov equations provide the complete information about the quantum dynamics of the system. These solutions also provide the necessary information to describe the dynamics in terms of TD Green functions or Feynman kernels, generalized creation/annihilation operators or Wigner functions.

Furthermore, the complex Riccati equation can also be linearized to a complex Newtonian equation where real and imaginary parts of the complex variable $\lambda(t)$ fulfilling this equation are two linear-independent solutions of the second-order Newtonian differential equation. However, these two solutions are not independent of each other but coupled via a kind of conservation law that is related to an invariant that can be obtained from the coupled pair of differential equations provided by the Newtonian and the Ermakov (or Riccati) equation of the problem. Essentially, the solution of the classical Newtonian equation makes it possible to obtain a second linear-independent solution with both providing the solution of the Ermakov equation and thus all information about the quantum dynamics of the system.

On the other hand, the interrelations between the classical and quantum aspects of the system's dynamics also lead to the solution of the classical dynamics, once the solution of the quantum dynamics describing the time-dependence of the WP width is known.

In looking for a TD frequency $\omega(t)$ that allows analytic solutions for the WP maximum and width, a comparison with the time-dependence of the Gaussian function that is a solution of the diffusion equation suggests a frequency that is inversely proportional to time, $\omega(t) \propto \frac{1}{t}$.

One problem is that the width of the Gaussian function describing the classical diffusion process obeys a real Riccati equation, whereas the quantum dynamics is determined by a complex Riccati equation. Finding the missing imaginary part of the Riccati equation when only the real part is known is not trivial. A more promising attempt is to solve the corresponding classical Newtonian equation of motion for the given $\omega(t)$, construct the second linear-independent solution via the above-mentioned conservation law which finally leads to the solution of the real Ermakov or complex Riccati equation.

This procedure proved quite successful for $\omega(t) = \frac{1}{2t}$ except for the problematic aspect of divergencies for $t \to 0$ and $t < 0$ (already on the classical level; but could be avoided by replacing t with $|t|$). The divergence for $t \to 0$ can be circumvented by changing $\omega(t)$ to $\omega(t) = \frac{1}{a(t+b)}$.

In particular, the case for $a = 2$ was considered in this paper. The dynamics and energetics of the parametric oscillator with frequency $\omega(t) = \frac{1}{2(t+b)}$ for the cases $b = 0$ and $b \neq 0$ can be compared with that of a damped harmonic oscillator with linear friction.

For $b = 0$, time-reversal seems not to affect the equation of motion at first sight, i.e., Eq. (79) with $b = 0$, as in this case $\omega^2(t) = \omega^2(-t)$. However, Eq. (80) for the classical trajectory and Eq. (81) for the classical velocity would, for $t < 0$, lead to purely imaginary values. This seemingly unphysical behaviour could be resolved by replacing t with its absolute value $|t|$ for $t < 0$. Despite this apparent reversibility of the motion, the energy still decreases inversely-proportional to time and approaches zero, like in the dissipative, frictionally-damped system, i.e., the system displays dissipation without irreversibility.

For $b \neq 0$, the situation changes concerning the equation of motion as, in this case,

$\omega(t) \neq \omega(-t)$. So now, also the equation of motion is no longer invariant under time-reversal since the addition of the constant b breaks the time-reversal symmetry! Still, $\dot{E} < 0$, i.e., the energy decreases similarly to the situation in a dissipative system. Regarding the trajectory and velocity, the same essentially applies as in the case $b = 0$, i.e., replacing t with $|t|$ prevents the problem of imaginary values.

Comparison of the frequency $\omega(t)$ with $\frac{\gamma}{2}$ in (86) is analogous to the aperiodic limit of the damped harmonic oscillator, i.e., the case $\omega_0 = \frac{\gamma}{2}$. Indeed, it can be shown that there are analytical solutions for the parametric oscillator with the more general frequency

$$\omega(t) = \frac{1}{a(t+b)} \tag{87}$$

for arbitrary constant b. However, the systems display qualitatively different behaviour for different values of a. The case of $a = 2$ has been discussed in detail in this work and is formally similar to the aperiodic limit of the damped harmonic oscillator, i.e., $\omega_0 = \frac{\gamma}{2}$.

The case $a < 2$ resembles the harmonic oscillator with undercritical damping, $\omega_0 < \frac{\gamma}{2}$ and exhibits oscillations with damped amplitudes of the velocity or momentum (contrary to the position!).

Finally, the case $a > 2$ behaves like an overcritically damped harmonic oscillator without oscillatory motion and $\omega_0 > \frac{\gamma}{2}$. The latter two cases will be discussed in detail in a forthcoming paper.

References

[1] Schuch D and Moshinsky M 2006 Connection between quantum-mechanical and classical time evolution via a dynamical invariant. *Phys. Rev.* A **73** 063111
[2] Schuch D 2008 Riccati and Ermakov Equations in Time-Dependent and Time-Independent Quantum Systems. *SIGMA* **4** 043
[3] Schuch D 1989 Connections between Newton- and Schrödinger-type equations in the description of reversible and irreversible dynamics. *Int. J. Quant. Chem., Quant. Chem. Symp.* **36 S23** 59-72
[4] Schuch D 1992 On the complex relations between equations describing the dynamics of wave and particle aspects. *Int. J. Quant. Chem.* **42** 663-683
[5] Schuch D 2014 Nonlinear Riccati equations as a unifying link between linear quantum mechanics and other fields of physics. *J. Phys. Conf. Ser.* **504** 012005
[6] Cruz H, Schuch D, Castaños O and Rosas-Ortiz O 2015 Time-evolution of quantum systems via a complex nonlinear Riccati equation I. Conservative systems with time-independent Hamiltonians. *Ann. Physics* **360** 44-60
[7] Ermakov V P 1880 Second-Order Differential Equations: Conditions of Complete Integrability. *Univ. Jzv. Kiev* **20 No 9** 1-25; Harin A O (English translation) 2008 *Appl. Anal. Discrete Math.* **2** 123-145
[8] Steen A 1874 Om Formen for Integralet af den lineaere Differentialligning af anden Orden. *Overs. over d. K. Danske Vidensk. Selsk. Forh.* **(1874)** 1-12
[9] Redheffer R 1999 Steen's equation and its generalizations. *Aequationes Math.* **58** 60-72
[10] Redheffer R and Redheffer I 2001 Steen's 1874 paper: historical survey and translation. *Aequationes Math.* **61** 131-150
[11] Milne W E 1930 The Numerical Determination of Characteristic Numbers. *Phys. Rev.* **35** 863-867
[12] Pinney E 1950 The nonlinear differential equation $y" + p(x)y + cy^{-3} = 0$. *Proc. Amer. Math. Soc.* **1** 681
[13] Lewis Jr. H R 1967 Classical and Quantum Systems with Time-Dependent Harmonic-Oscillator-Type Hamiltonians. *Phys. Rev. Lett.* **18** 510; 1967 Erratum, Classical and Quantum Systems with Time-Dependent Harmonic-Oscillator-Type Hamiltonians. *Phys. Rev. Lett.* **18** 636
[14] Lewis Jr. H R and Riesenfeld W B 1969 An Exact Quantum Theory of the Time-Dependent Harmonic Oscillator and of a Charged Particle in a Time-Dependent Electromagnetic Field *J. Math. Phys.* **10** 1458-1473
[15] Castaños O, Schuch D and Rosas-Ortiz O 2013 Generalized coherent states for time-dependent and nonlinear Hamiltonian operators via complex Riccati equations. *J. Math. Phys. A: Math. Theor.* **46** 075304
[16] Schuch D 2005 On the relation between the Wigner function and an exact dynamical invariant. *Phys. Lett. A* **338** 225-231

[17] Schuch D 2018 *Quantum Theory from a Nonlinear Perspective – Riccati Equations in Fundamental Physics* (Heidelberg: Springer)
[18] Grössing G, Mesa Pasasio J and Schwabl H 2011 A Classical Explanation of Quantization. *Found. Phys.* **41**, 1437-1453
[19] Schuch D 2012 Nonlinear quantum mechanics, complex classical mechanics and conservation laws for closed and open systems. *J. Phys.: Conf. Ser.* **361** 012020

IOP Conf. Series: Journal of Physics: Conf. Series **1071** (2018) 012021 doi:10.1088/1742-6596/1071/1/012021

Low-energy antimatter experiments at the antiproton decelerator at CERN: Testing CPT invariance and the WEP

Ingmari C. Tietje

Experimental Physics Department, CERN, 1211 Geneva 23, Switzerland
Centre of Astronomy and Astrophysics, TU Berlin, 10623 Berlin, Germany

E-mail: ingmari.tietje@cern.ch

Abstract. The riddle of the baryon asymmetry, i.e. the matter antimatter imbalance in the universe can be addressed by comparing matter particles with their antimatter counterparts. At the antiproton decelerator (AD) at CERN several antimatter experiments investigate whether CPT (charge-parity-time reversal) invariance and the WEP (weak equivalence principle) hold. The systems probed are antihydrogen ($\bar{H}$), antiprotonic helium and individual antiprotons ($\bar{p}$). This article is meant to give an overview of the experiments located at the AD, discuss some commonly used experimental techniques and point out what the different experimental approaches entail. The research done on low-energy antimatter systems can be seen as complementary to the high energy research carried out at CERN and elsewhere: It provides bounds on CPT invariance and directly addresses the question of whether the WEP holds for antimatter. It is noted that the AD - at the moment - is the only low-energy antiproton source on earth.

1. Introduction

Despite the striking success in deciphering the inner workings of matter in particle physics, there is so far no answer to a simple question stemming from cosmology - the matter-antimatter asymmetry. In the standard model of the Big Bang, matter and antimatter are supposed to be created in equal amounts. While the cosmic microwave background photon density - the remnant of the annihilation of most matter with antimatter - is a bit less than 10^9 photons per cubic metre, there is, on average, less than 1 matter particle per cubic metre in the universe giving rise to a baryon to photon ratio of $\eta = (6.5^{+0.4}_{-0.3}) \times 10^{-10}$(see e.g. [1], p. 7, 11). So far, we have not seen any primordial antimatter in the universe. An excess of about one baryon, i.e. a ratio of 1 000 000 001 : 1 000 000 000 (baryons : antibaryons) could therefore account for the asymmetry.

In 1967 Sakharov proposed that there could have been an imbalance in the baryon to antibaryon number in case three conditions were fulfilled: (1) CP violation in the laws of physics, (2) Baryon number violation and (3) a period of the early universe, that was out of thermodynamic equilibrium, see [2] and [3], p. 98. Sakharov's argumentation is based on an expanding universe, that requires a superdense intial state of matter and therefore seems to outrule macroscopic compartments of matter and antimatter in the universe.

A theory of a fragmented or patchwork universe with volumes of matter adjoining volumes of antimatter could still survive assuming the domains are of the size of our visible universe [4, 5]. Exploring the possibility of antigravity [6], i.e. a gravitational repulsion of matter and antimatter, it is even imaginable that the fragments/domains could have coexisted being spacially separated.

The nature of all these considerations about the mystery of the missing antimatter is speculative. Thorough experimental tests of charge-parity-time reversal (CPT) symmetry and weak equivalence principle (WEP) are, step by step, shedding light on the various speculations that have emerged since Dirac developed the notion of and postulated antimatter in the years 1928-1931 [7], p. 46.

At present the antiproton decelerator (AD) at the European Organisation for Nuclear Research (CERN) is the only place that provides low-energy antiprotons worldwide. 6 collaborations that all focus on low-energy antimatter physics are operating or are setting up their experiments.

The ALPHA, ASACUSA, ATRAP and BASE experiments (when listed, the names of the experiments occur in alphabetical order) are focussing on precision tests of CPT symmetry comparing antimatter with matter. Any deviation in a property of antimatter from the same property in matter, e.g. the resonance frequency of the 1S to 2S spectroscopy in (anti)hydrogen, would immediately indicate CPT symmetry violation.

The AE$\bar{g}$IS, ALPHAg and GBAR experiments pursue tests of the WEP by determining the sign and absolute value of the gravitational acceleration $\bar{g}$ of antimatter in Earth's gravitational field. So far no conclusive direct test of the WEP has been carried out, so it cannot be definitely excluded that an antibaryon-antilepton system like antihydrogen might fall up. However, under the assumption of exact CPT symmetry a comparison of the cyclotron-frequency measurements of (anti)protons (as well as positrons and electrons) lead to constraints on the possible deviation of the local acceleration for matter g and antimatter $\bar{g}$ [8]: For an effective gravitational potential U for matter, $\alpha_g U$ gives the effective gravitational potential for antimatter. The parameter α_g expresses the possible gravitational deviation. The most recent comparison of the antiproton-to-proton charge-to-mass ratio with the BASE experiment gave a constraint of $|\alpha_g - 1| < 8.7 \times 10^{-7}$ [9] assuming scalar gravitational interactions.

The systems investigated to probe CPT symmetry range from individual antiprotons (ATRAP, BASE), antihydrogen atoms (ALPHA, ASACUSA, ATRAP) to antiprotonic helium (ASACUSA). Since antihydrogen is uncharged and fully comprised of antimatter (lepton-baryon system) all WEP experiments aim at making use of the antiatom in their experimental schemes (AE$\bar{g}$IS, ALPHAg and GBAR).

The prerequesites to testing either CPT symmetry or WEP in the domain of low-energy antimatter physics are to efficiently trap and cool antiprotons in a Penning-Malmberg trap or to form a beam of antiprotons cold enough so it can be more easily manipulated (e.g. focused to a smaller diameter) and used in an experimental protocol.

If antihydrogen formation is required, i.e. in all collaborations at the AD except for BASE, there are two different ways to go: The older and therefore well-tested way - which is commonly refered to as 'mixing' - is to form antihydrogen via three-body collisions of antiprotons and positrons [10, 11]. The other method is a charge-exchange reaction between an antiproton and a Rydberg positronium (i.e. a blown-up electron-positron quasiatom) in which the antiproton 'steals' the positron of the positronium to form antihydrogen and sets the electron free [12].

Since the experimental methods of trapping and cooling antiprotons, positron preparation as well as antihydrogen formation are shared across many of the experiments, we will give a brief overview of the methods in Section 2.

In Section 3 we will first describe the individual experiments testing CPT symmetry. Afterwards, the experiments testing the WEP will be explained (like above the experiments will be listed according to topic, then alphabetically). Section 4 comprises a summary of the experiments presented and Section 5 gives an outlook on near-future and future low-energy antimatter physics.

For a reminder of WEP and CPT theorem please have a look at [13] and [14]. Note that this article does not discuss subexperiments at the AD that do not focus directly on CPT or WEP.

2. Trapping of antiprotons & antihydrogen formation

As addressed in the introduction a precise investigation of an entity involving spectroscopy or free fall requires to cool it down and localise it in a suitable trap/environment. Since the antiprotons mostly determine the antihydrogen formation temperature they need to be cold. After their creation they are therefore decelerated/cooled down by the AD. They are further electrostatically decelerated and/or trapped by an experiment and further cooled down within a trap.

In most experiments (ALPHA, AE$\bar{g}$IS, ASACUSA and ATRAP) positrons are obtained from a radioactive isotope of sodium. The energetic positrons coming from the Na^{22} source are moderated, buffer-gas cooled and accumulated.

Antihydrogen formation can be achieved via 'mixing' positrons with antiprotons at sufficiently high positron densities (typically achieved in Penning-Malmberg traps) or through a charge-exchange reaction between an antiproton and a positronium 'quasi'-atom.

An insight into the cooling and trapping of antiprotons, the preparation of positrons from a radioactive source (for positron production via a small linear accelerator, see Section 3.7) as well as the conditions necessary for the two different antihydrogen production methods are given in the following.

2.1. *Cooling and trapping*

The antiprotons are first created in the target region of the AD via 26 GeV/c protons impinging on a target leading to pair production. After their creation the antiprotons are extracted via a magnetic horn (also known as Van der Meer horn) from the target region [15, 16] and injected into the AD ring. There they are first stochastically and then electron-cooled from 3.57 GeV/c to 100 MeV/c before they are injected into the experiment beam lines after roughly 110s, see Fig. (1).

Most experiments then use a degrader foil that lowers the energy of parts of the beam to the order of 10keV/c, but also annihilates and scatters a great part of the impinging antiprotons. The limit in trapping efficiency in the commonly-used Penning-Malmberg traps is about 1% partly due to the limited trapping voltages of $5 - 9$keV and the impinging antiproton bunch kinetic energy, partly because of the losses in and due to the scattering in the degrader foil [17].

Another decelerator - the Extra Low Energy Antiproton (ELENA) ring - is currently in its comissioning phase. It will further lower the antiproton bunch energy and increase the trapping efficiency of the experiments by a factor of ~ 100 [18].

2.2. *Penning-Malmberg traps*

Penning-Malmberg traps, see Fig. (2), are composed of static magnetic and electric fields. They confine charged particles radially via an $E \times B$ drift and axially via electric potential wells. The majority of the AD experiments use Penning-Malmberg traps due to the long trapping times (static E and B field) and the high number of particles that can be stored. They also offer a wide range of applications spanning from various cooling techniques (e.g. electron cooling, resistive cooling) to precision single-particle frequency measurements [19].

The trap configurations used in the AD experiments are often composed of several tens of cylindrical electrodes placed in the homogeneous field of a superconducting magnet.

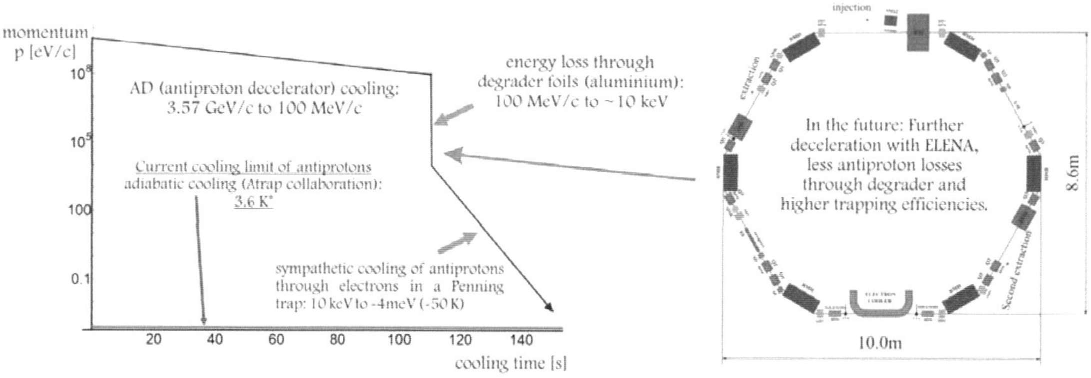

Figure 1. On the left: Kinetic energy of antiprotons in the AD ring, energy drop due to degrader foil and due to sympathetic cooling via electrons in the trap systems as a function of time. On the right: Sketch of the Extra Low ENergy Antiproton (ELENA) ring - an upgrade to the AD which will increase the cooling of the beam and thus permit much better antiproton trapping efficiencies in the experiments. *The present cooling limit of 3.6 K of an antiproton plasma is reached through adiabatic cooling [20].

Commonly shared across the great part of experiments is the trapping and cooling which is carried out as follows: Before the antiprotons are caught electrons are loaded into the trap and cool down via the emission of cyclotron radiation until they approach the equilibrium temperature (of the order of 40K) of the cryogenic environment. Antiprotons are caught in between two high-voltage electrodes and cool down via collisions with the electrons. The lowest temperature measurement of an antiproton plasma measured at the AD according to the author's knowledge is 3.6K and involved an adiabatic expansion after electron-cooling the antiproton ensemble in the trap [20].

2.3. Positron moderation, cooling and accumulation

In most experiments at the AD positrons are obtained as a pulsed beam from a radioactive source (for a different method see Sec. 3.7 about the GBAR experiment). The sodium isotope Na^{22} which is typically used emits high energy positrons in a reverse-β decay with a maximum energy of 543 keV.

In an experimental setup the Na^{22} source is located at the bottom of a parabolic or conical opening of a cryogenic copper holder. The positrons are emitted into 4π, however by means of a titanium block a fraction of the positrons otherwise lost is back-reflected [21]. The positrons are preferentially directed toward the opening of the holder where they thermalise while they pass through a solid neon moderator grown at a temperature of about $5-9$ K on the holder surface [22]. Due to the positive work function of solid neon [23] the thermalised positrons exit the moderator with a well-defined energy and form a monoenergetic positron beam.

The moderated positrons are then guided along magnetic field lines into a buffer-gas Penning-Malmberg trap often referred to as Surko-type trap [24, 25]. Initially, the positrons cool via collisions with N_2 molecules ($e^+ + N_2 \rightarrow e^+ + N_2^*$), a process that at sufficiently low pressures competes with positron annihilation through positronium formation ($e^+ + N_2 \rightarrow Ps + N^+{}_2$). The cooling of positrons by means of a specific buffer-gas is limited. At too cold temperatures positrons annihilate through the positronium channel rather than cool. A second buffer gas such as CF_4 or SF_6 is often employed in a different trap or part of the trap to cool the antiparticles to lower temperatures than possible with N_2.

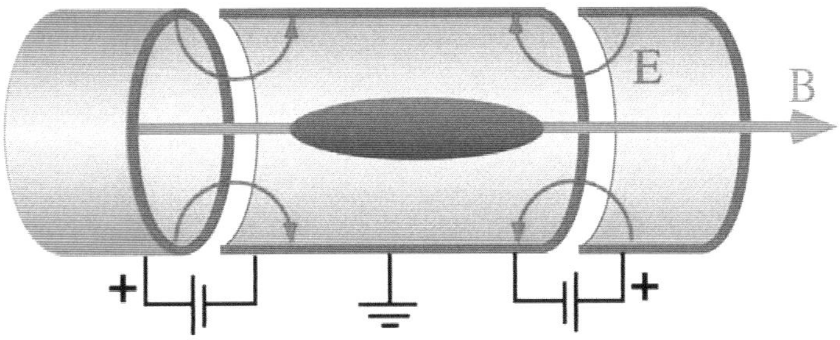

Figure 2. Charged particles are confined in a Penning-Malmberg trap via strong magnetic and electric fields. The magnetic field (in the AD experiments it is typically 1T to 5T) continually diverts the particles and forces them to orbit around the trap center as well as to undergo a cyclotron motion [19]. The electric field applied by the ring electrodes impedes the particles from exiting on axis. In short - they are trapped.
The aforementionend cyclotron motion causes the trapped particles to emit synchrotron radiation, more so for lighter particles. As a consequence electrons and positrons emit so much radiation that they approach equilibrium with the (cryogenic) environment while antiprotons do not cool by themselves. Image courtesy of [26].

The positrons are repeatedly released from the trap. Varying from one experimental setup to the next the particles are either accumulated in another Penning-Malmberg trap where they can be further buffer-gas cooled or retrapped in a stronger magnetic field with a lower background pressure where they cool through cyclotron radiation. With the techniques described above positrons can be stacked to high numbers of about 10^8 and classify as non-neutral plasmas. As has been shown in the Surko group positrons can also be stored in a multi-cell trap from which they can be released in greater numbers $\sim 10^{15}$ [27].

2.4. Antihydrogen formation via 'mixing'
'Mixing' is a three-body recombination in which two positrons and one antiproton collide at small relative velocities. One positron binds with the antiproton to form an antihydrogen atom, while the other positron carries away the binding energy.

$$\bar{p} + e^+ + e^+ \rightarrow \bar{H}^* + e^+ \tag{1}$$

Since it is a three-body collision the production rate of antihydrogen is proportional to n_e^2, where n_e is the positron density. The antihydrogen atom is formed in an excited state $n > 100$, where n refers to the principal quantum number.

In Fig. (3) it is illustrated how the two species 'mix' in a so-called nested Penning trap. Oppositely charged, the positrons are trapped in the center well, whereas the antiprotons are trapped in the double well surrounding the positrons, hence the name nested trap. Just like electrons, positrons are light enough to cool by means of their cyclotron radiation to the cryogenic temperatures of the environment. They are therefore anticipated to cool the antiprotons sympathetically once the plasmas overlap.

The recombination rate of an antiproton with a positron is determined by their relative velocity. The antiproton-positron mass ratio is roughly 1800, i.e. the antiproton is a factor of

~ 42 slower than the positron at the same kinetic energy. As illustrated in [16], however, a 2eV antiproton is still slow enough to combine with a 15K positron and due to the mass ratio the formed antihydrogen will keep the energy of the initial antiproton. To facilitate the formation of low-energy antihydrogen needed for spectroscopy the antiprotons can be carefully mixed with the positrons by slowly manipulating the nested trap potentials instead of simply releasing the antiproton cloud into the positron plasma like in earlier experiments [10, 11].

The formed antihydrogen is de-excited to lower quantum numbers via collisions with positrons [28]. Antihydrogen in an excited state too close to the ionisation threshold is field-ionised when it exits the field free bulk of the plasma [29].

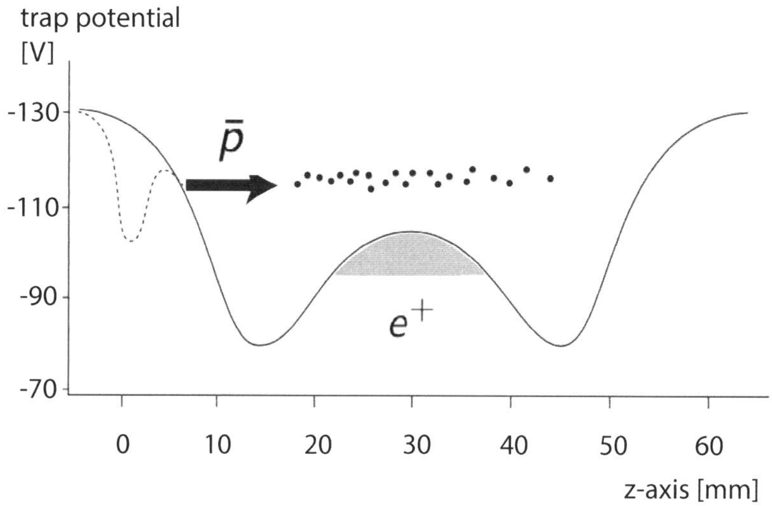

Figure 3. The antiprotons are released into an axial double well, where they cool and collide with positrons waiting in the middle of the so-called nested trap. The positrons are light enough to cool by themselves emitting cyclotron radiation into the cryogenic environment. Once the antiprotons have cooled down to energies low enough such that the relative velocities between antiprotons and positrons are small antihydrogen is formed. To form cold antihydrogen through 'mixing' it is crucial to carefully manipulate the overlap of antiproton and positron plasmas, see text. Note, that the voltage configuration depicted here is for positively charged particles. Adapted from [16].

2.5. Antihydrogen formation via charge-exchange

In the charge-exchange reaction of (Rydberg) positronium Ps^* and antiprotons $\bar{p}$

$$\bar{p} + Ps^* \rightarrow \bar{H}^* + e^- \tag{2}$$

the electron in the positronium 'quasi'-atom is replaced by an antiproton to form (Rydberg) antihydrogen. The cross section σ of the charge-exchange reaction is of the order of $a_0 n_{Ps}^4$, where a_0 is the Bohr radius and n_{Ps} the principal quantum number of positronium [30]. The highest cross section σ for a set principal quantum number n_{Ps} will be obtained for $k = \frac{v_{Ps}}{v_{orb}} < 1$, where v_{Ps} is the positronium velocity and v_{orb} the positron angular velocity [31].

Charge-exchange for antihydrogen production was demonstrated by the ATRAP collaboration in 2004 [32]. In this first demonstration Cs* atoms were used for positronium formation whereas - in the current experiments nanoporous silica targets are utilised. These targets are undergoing a lot of improvement and exist in reflection [33] and in transmission mode [34, 35]. The transmission mode positronium targets offer a much improved solid angle for positronium production as can be easily seen from Fig. (4). They however tend to be very fragile. Also the converters in transmission geometry have been proven to work with efficiency at best of about 25% of that of converters in reflection geometry.

Charge-exchange, unlike 'mixing', permits a pulsed antihydrogen production and provides antihydrogen with a more narrow quantum distribution.

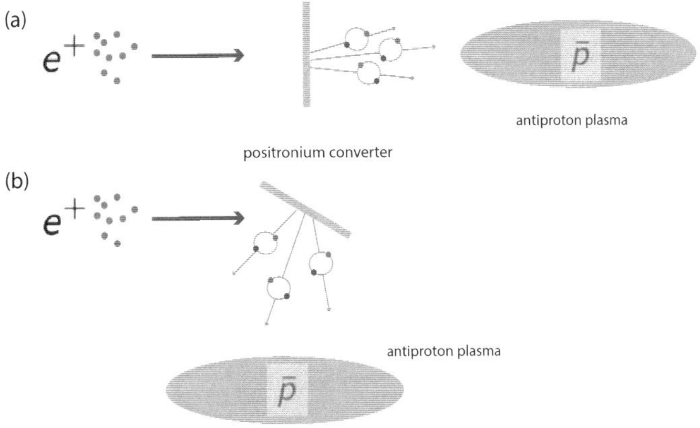

Figure 4. In scenarios (**a**) and (**b**) a bunch of positrons is released onto a transmission target and a reflection target, respectively. The solid angle for antihydrogen production is much improved in the transmission geometry compared to the reflection geometry.

3. AD Experiments

3.1. ALPHA — Antihydrogen Laser PHysics Apparatus

The ALPHA collaboration focusses on tests of CPT invariance by comparing antihydrogen transition frequencies to those precisely measured for hydrogen [36]. In 2017 they observed the 1S - 2S transition in antihydrogen trapped in a magnetic trap to a relative precision of about 2×10^{-10} [37].

The antiatoms trapped by means of their small magnetic moment in their low-field seeking states were observed in three different scenarios: (1) with the 243-nm laser light on resonance, (2) with the 243-nm laser light on, but detuned by 200kHz below the resonance and (3) with the laser switched off. Each scenario lasted about 600s and was concluded by slowly (1.5s) ramping the magnetic trap down to detect the remaining antihydrogen atoms.

The numbers detected for (2) and (3) were consistent and about twice as high as the numbers for (1) indicating that the laser on resonance had removed about 58% $\pm 6\%$ of the trapped antiatoms by an ionisation event or a transition to a high-field seeking state, i.e. spin-flip that would not be contained in the trap.

Other endeavours of the ALPHA collaboration include setting an experimental limit on the charge of antihydrogen [38, 39] and probing resonant quantum transitions in trapped antihydrogen atoms [40] with a similar method as described above. Recent measurements have benefitted from an increased trapping rate of antihydrogen ($\sim$ 50 antiatoms per trial). The ALPHA collaboration is also building a new experiment ALPHAg, which will focus on a test of the gravitational acceleration $\overline{g}$ antimatter experiences in Earth's gravitational field, see [41] as well as Section 3.6.

3.2. ASACUSA - Atomic Spectroscopy And Collisions Using Slow Antiprotons

The ASACUSA collaboration has two research objectives that focus on directly testing CPT invariance: (1) Spectroscopy on the ground state (GS) hyper-fine splitting (HFS) of antihydrogen and hydrogen, respectively. (2) Probing the level structure of a three-body exotic atom - antiprotonic helium [42].

ASACUSA intends to measure the GS HFS of a polarised beam of antihydrogen. This presents a different approach to the ALPHA or ATRAP collaborations, who focus on the trapping of antihydrogen, see Section 3.1 and Section 3.3.

Laser spectroscopy of antiprotonic helium allows for a precise determination of the antiproton-to-electron mass ratio. If CPT invariance holds it should be exactly the same as the proton-to-electron mass ratio.

ASACUSA also measures nuclear cross sections of antiprotons [43] at low ($<$ 100keV) energies which will not be covered here.

3.2.1. Spectroscopy of the HFS of the GS of antihydrogen and hydrogen: Antihydrogen is formed via 'mixing' in a CUSP trap made of superconducting coils in anti-Helmholtz configuration and a stack of electrodes. While antihydrogen in its high field seeking (hfs) state is defocused in the CUSP trap and annihilates, antihydrogen in its low field seeking (lfs) state is preferentially focussed along the axis of the trap. This discrimiation leads to the formation of a polarised beam of antihydrogen.

Once outside the trap the beam passes through a Rabi module which drives the hyperfine transition(s). To discriminate the antihydrogen atoms which have made a transition from a lfs state to a hfs the beam passes through a sextupole magnet, that (like the CUSP trap) defocusses antihydrogen atoms in their hfs state. The lfs antihydrogen atoms in the beam are not deflected and counted by means of a BGO calorimeter and a 2 layer hodoscope [42].

Recently, ASACUSA measured two HFS hydrogen transitions - π_1 and σ_1 - and determined the zero-field value with the same method as described above, but for a beam of hydrogen [42]. To measure the π_1 and σ_1 transitions simultaneously the Rabi module/RF cavity was placed at 45 degrees (for the π transition the RF field needs to be oriented perpendicular to the magnetic field) and a magnetic field was oriented perpendicular to the propagation axis of the hydrogen beam [42].

As a milestone towards a comparable measurement for antihydrogen the collaboration demonstrated beam formation of antihydrogen in 2014 [44].

3.2.2. Laser spectroscopy of antiprotonic helium: Laser spectroscopy of antiprotonic helium allows for a precise determination of the antiproton-to-electron mass ratio.

The experimental procedure is as follows [45] A low-energy beam of antiprotons is being stopped in a cryogenic helium target. About 3% of the antiprotons are automatically trapped as constituents of so-called antiprotonic helium atoms. These metastable neutral systems have a lifetime of more than a microsecond and consist of a helium nucleus with an antiproton replacing the second electron.

The longevity occurs for large n ($\sim$ 38) and L ($\geq$ 35) when the orbit of the antiproton is near-circular.

While being in a circular Rydberg state that shields the bound antiprotons from premature annihilation the exotic three-body atoms cool down to 1.5 - 1.7 K via collisions with helium atoms.

Once at cryogenic temperatures, the exotic atoms are probed. Let us assume an antiprotonic helium is in the metastable state $(n, L) = (35, 33)$. A 372.6nm laser can transfer it to the $(n, L) = (34, 32)$ state, which is short lived and Auger-decays to an ionic state $(n_i, L_i) = (30, 29)$ within les than 10ns. The ionic state in turn is very quickly destroyed by Stark collisions in the buffer gas. The annihilation signal of the now unshielded antiproton is detected [45].

In order to obtain the antiproton-to-electron mass ratio many transition frequencies have to be measured and compared to precise three-body QED calculations of the level structure of antiprotonic helium.

The cryogenic temperatures allow for a precision of the antiproton-to-electron mass ratio measurement of 1836.1526736(23) [46] which is well in agreement with the proton-to-electron mass ratio value that is known to a similar precision [47].

3.3. ATRAP — The Antihydrogen TRAP Experiment

The ATRAP collaboration dates back much longer than the other collaborations at the AD. In 2002 ATRAP was among the first two collaborations to report production of slow antiprotons via the 'mixing' technique, see Section 2.4 (the other collaboration was ATHENA which split up into the ALPHA and AE$\bar{g}$IS collaborations later on). Its predecessor was the TRAP collaboration which pioneered antiproton trapping and manipulation techniques used throughout all AD experiments.

ATRAP has measured charge-to-mass ratios of antiprotons and protons to 90 ppt [48] which was the most precise result of CERN's antiproton program before the AD. The collaboration demonstrated the feasibility of the charge-exchange reaction (see Section 2.5) to form antihydrogen [32] and successfully trap antihydrogen atoms in an Ioffe trap [49]. In 2013 ATRAP was the first experiment to measure the magnetic moment of a single antiproton [50].

The ATRAP collaboration's objective is to trap antihydrogen and perform spectroscopy on it in order to test CPT invariance. The experiment consists of building blocks very similar to those in the ALPHA apparatus. Probably the most striking difference is that the traps are stacked upon each other vertically similar to the ALPHAg but different from the ALPHA design.

As mentioned above both experiments use the 'mixing' technique to produce antihydrogen and a superimposed Ioffe trap to confine the antihydrogen atoms via their magnetic moment. The ATRAP Ioffe trap has a similar trap depth of 0.5K - confining only the few antihydrogen atoms created at very slow velocities. The main difference between the experiments is that ALPHA goes through the production and trapping cycle of antihydrogen more frequently while ATRAP makes use of higher antiproton and positron numbers for antihydrogen production[51].

3.4. BASE — The Baryon Antibaryon Symmetry Experiment

The BASE collaboration is focussing on precision measurements of the proton and antiproton magnetic moments as well as comparisons of the proton-to-antiproton charge-to-mass ratios.

They recently measured the magnetic moment of the antiproton to a parts-per-billion precision [52]. BASE also performed a high-precision comparison of the antiproton-to-proton charge-to-mass ratio with a fractional precision of 69 parts in a billion which constitutes the to date most precise test of CPT invariance in the baryon sector [53].

The experimental setup is comprised of several Penning traps at cryogenic temperatures. One of the traps serves as a catching and storage trap which confines antiprotons for more than a year [54]. The reservoir of antiprotons makes the BASE experiment fairly independent from the beam time schedules of the AD. A second trap has the purpose of efficiently cooling the particle degrees of freedom via resistive cooling.

The other two traps are precision traps used to measure the particle frequencies in two different scenarios which is refered to as the double Penning trap method, see [55]. The so-called precision trap is used to measure the proton/antiproton frequencies and the trap refered to as analysis trap is designed for proton/antiproton spin state analysis using the continuous Stern-Gerlach effect [56]. Highly-sensitive electronics and the cryogenic temperatures of the trap electrodes allow to pick up the image current signal of a single antiproton/proton in the trap. This sensitivity permits to determine the small magnetic moment of the particle by measuring the ratio of the cyclotron frequency ν_c and the spin precession frequency ν_L, which gives access to the g-factor $g = 2\nu_L/\nu_c$.

3.5. AE$\bar{g}$IS — Antimatter Experiment: Gravity, Interferometry,Spectroscopy

The AE$\bar{g}$IS collaboration aims at measuring the gravitational acceleration of antihydrogen $\bar{g}$ in Earth's gravitational field [57]. The experimental idea is to measure the deflection due to gravity of a pulsed $\overline{H}$ beam. The precise determination of $\bar{g}$ presents a test of the WEP - fundamental to special and general relativity.

Antiprotons are first captured in a Penning-Malmberg trap (4.46T magnetic field), where they are electron-cooled and compressed with the so-called rotating wall technique. Afterwards they are transfered to another trap region in a 1T field, recooled and recompressed in a trap with partially perforated electrodes.

At the same time positrons from a Na^{22} source are moderated by means of a cold head and accumulated in a Surko-type trap system that after accumulation stores, cools and compresses them [58]. The positrons are released from the trap and injected into a nanoporous positronium target which sits above the perforated electrodes. The fraction of the positronium formed in the target which lives long-enough to escape from the nano-pores is laser excited into Rydberg states, see e.g. [59]. A fraction of these long-lived Rydberg quasi-atoms passes through the perforated electrodes into the trap and forms antihydrogen through charge-exchange as explained in Section 2.5. AE$\bar{g}$IS is currently working on the antihydrogen production.

Once Rydberg antihydrogen atoms are produced they can be formed into a beam with a Stark accelerator. In a beam the Rydberg antiatoms must be deexcited into their GS before they pass through a Moiré deflectometer: two microscopic (however classical) gratings that are suitable to measure the deflection of the antihydrogen atoms over a distance of roughly 1m [60]. The antiparticles are detected by means of an emulsion [61] and/or a time-pix silica detector [62].

3.6. ALPHAg

The ALPHA collaboration is building a second experiment which is designed to test the WEP via a statistical method [63]. The experiment under construction is meant to first measure the sign of the acceleration of antihydrogen in the terrestrial gravitational field $\bar{g}$ and later make a precision measurement [64, 65].

The experimental protocol is very similar to the protocol in the ALPHA experiment, see Section 3.1. Antihydrogen is created via mixing, see Section 2.4, and the antihydrogen atoms are confined in an Ioffe trap. Unlike in the ALPHA apparatus, the trap axis of the Penning-Malmberg trap is vertically oriented.

The gravitational measurement comprises opening the Ioffe trap on axis and at both ends, which in the vertical geometry point away and toward Earth. The atoms are no longer trapped and they escape through either opening and annihilate on the Penning-Malmberg trap wall.

The gravity measurement is thus essentially converted into a counting experiment that compares the numbers of antiatoms that escaped upwards to those that escaped downwards.

3.7. GBAR — Gravitational Behaviour of Antihydrogen at Rest

The GBAR collaboration aims at testing the WEP by directly measuring the free fall acceleration of neutral antihydrogen atoms at μK temperatures [66, 67]. Their experimental scheme relies on the onset of the new ELENA decelerator which started operation in 2017, see Fig. (1). As discussed in Section 2.1 the low-energy antiprotons from ELENA permit a 100-fold (or 10-fold for ASACUSA) increased antiproton trapping efficiency, i.e. of the order of 10^7 antiprotons per bunch instead of 10^5. Via an electrostatic decelerator GBAR intends to slow the antiproton beam coming from ELENA down even further to a few keV.

To scale up in positron numbers (compared to the other experiments) GBAR is operating a small linear accelerator that produces a 10MeV electron beam that hits a Tungsten target. The energy released leads to pair production of electrons and positrons. The positrons are moderated and subsequently 'harvested' with a magnetic separator [68]. This offers a much higher flux in positrons than the flux that can be obtained with a radioactive Na^{22} source used in the other experiments that produce or aim at producing antihydrogen.

The positron beam is accumulated and further moderated in a buffer gas trap (N_2) and then transfered into a 5T Penning-Malmberg trap where the particles (just like electrons) cool down to a few meV via synchroton radiation until they reach the cryogenic temperature of their environment.

Once the energy spread is small enough the positrons are injected into a porous silica target (cf. Section 3.5) located in a reaction target where some fraction of the positrons binds with electrons to form positronium. The quasi atom gets reflected from the walls and thus exits through the nanochannels of the material. Once outside - the positronium is excited to a higher energy state (possibly $n = 2$ or $n = 3$) to improve its lifetime and enhance the probability of charge-exchange reactions with the slow (keV) antiproton beam that crosses through the chamber, see Eq. (2). A second charge-exchange reaction with already produced $\overline{H}$ atoms leads to $\overline{H}^+$ production [69]

$$\overline{H} + Ps \rightarrow \overline{H}^+ + e^- \ . \tag{3}$$

The yield of antihydrogen ions $\overline{H}^+$ will be sympathetically cooled with $^9Be^+$ ions to a few μK using two Paul traps - the first one containing a few thousand of Doppler cooled Be^+ ions ($\sim$mK), the second one to perform ground state cooling of a $Be^+/\overline{H}^+$ ion pair.

A photodetachment of the extraneous positron sets the antihydrogen atom free. The free fall of the antiatoms (i.e. the gravitational acceleration of antimatter $\overline{g}$) can then be measured by a set of annihilation detectors surrounding the chamber of the atom cloud. This presents a direct test of the WEP [70].

4. Discussion

From a theoretical and experimental point of view antimatter particles are the 'perfect' mirror objects. We can investigate and compare them to their matter counterparts in the search for a (minimal) difference that would indicate CPT or WEP violation. The experiments at the AD presented in Section 3 investigate relatively simple (anti)matter systems — antiprotons, antihydrogen and antiprotonic helium.

ASACUSA, see Section 3.2, aims at doing spectroscopy on antihydrogen and investigates a very exotic mixed antimatter-matter system. Antiprotonic helium lends itself to compare the antiproton-to-electron with the proton-to-electron mass ratio. ALPHA and ATRAP concentrate their efforts on antihydrogen trapping and spectroscopy while BASE focusses on the fundamental properties - magnetic moment and charge-to-mass ratio - of antiprotons and their counterparts. Complementary to these experiments designed to test CPT invariance AE$\overline{g}$IS, GBAR and ALPHAg (part of the ALPHA collaboration) all aim at measuring $\overline{g}$: (1) AE$\overline{g}$IS with (anti)matter interferometry performed on a beam of antihydrogen, (2) GBAR with ultra-cold sympathetically cooled antihydrogen ions whose temperatures are small enough that, once ionised - they can simply be dropped to measure $\overline{g}$ (3) the ALPHA collaboration aims at using a statistical method in their ALPHAg experiment which turns the gravity measurement of the sign of $\overline{g}$ into a simple counting experiment.

When looking at all the experiments in detail it becomes evident that they have a lot of experimental techniques in common. Most experiments use (or are going to use) the same trapping and cooling techniques for antiprotons and similar ones to prepare positrons. Since the antiprotons, positrons and electrons (used for electron-cooling, see Section 2.1) are often stored as non-neutral plasmas there are a number of methods borrowed from non-neutral plasma physics. One method used to compress the particles which is commonly refered to as the rotating wall (RW) technique [71] helps to store large numbers of charged particles for long times. RW also eases the preparation of the transport of plasmas from one to the other end of the long Penning-Malmberg traps used in the AD experiments.

The methods of antihydrogen production are so far two-fold: 'Mixing', see Section 2.4, is the more established technique. It has been used since the ATHENA and ATRAP collaboration synthesized cold antihydrogen atoms in 2002. Charge-exchange between an antiproton and a (Rydberg) positrium quasi atom, see Section 2.5, is less well-tried.

In 'mixing' (see Section 2.4) the positron and antiproton populations are mixed in a trap and form antihydrogen through three-body collisions. [28]. The formation rate per antiproton is proportional to the positron density squared n_e^2, which permits a high yield of antihydrogen with regard to typical densities of non-neutral plasmas inside Penning-Malmberg traps.

The antihydrogen atoms are formed at high quantum states $n > 100$. To leave the field-free plasma without being ionised by the trap potential gradient, the antiatoms have to have undergone enough collisions to relax into a lower state below the ionisation threshold [29].

The yield of antihydrogen atoms through the charge-exchange reaction (see Section 2.5) is determined by the cross section of the reaction and is proportional to the number of antiproton-positronium encounters. The cross section scales with n_{Ps}^4, where n_{Ps} is the principal quantum number of the positronium 'quasi'-atom, and critically depends on the ratio of the positron orbital velocity and the relative velocity between antiproton and positronium [30]. Charge-exchange permits a sharp onset of antihydrogen production suitable for pulsed antihydrogen beam formation, see Section 3.5. Further, the antihydrogen atoms that are formed show a more narrow and well-defined quantum state distribution than those formed through 'mixing'.

The geometry and positron to positronium conversion rate in a nanoporous silica target is very important for the charge-exchange. A lot of research is being done on how to produce stable, highly efficient transmission positronium targets. Both AE$\overline{g}$IS and GBAR collaboration use targets in reflection geometry. GBAR will work with very high particle numbers, see Section 3.7.

One of the driving forces in the advancement of low-energy antimatter physics are (as in most high precision research) newly developed methods, that often have in common that they allow higher control over the particle ensemble properties, see e.g. [72]. The precision of free fall experiments on antimatter and spectroscopy on antihydrogen as well as antiprotonic helium critically depend on the number of antiatoms available. Constantly developing as well as newly invented methods determine the experimental precision and the bounds that can be set.

Bounds on CPT invariance and WEP violation help outrule models which are part of the frameworks going beyond the standard model like the theories of the super symmetric framework (SUSY). In providing such bounds the low-energy antimatter experiments can be regarded as being complementary to the high-energy experiments at CERN.

5. Outlook

The number of available antihydrogen and antiprotonic helium atoms matters for the precision of spectroscopy experiments performed on them. In 2017 the ELENA ring started operation and will - in the future - supply the experiments with a higher number of antiprotons (a factor of about 100) at a higher rate, see Section 2.1. More antiprotons available for antihydrogen production lead to higher numbers of antihydrogen atoms that can be probed and thus higher precision in less time. The increased numbers will provide momentum to the great part of the AD experiments. In particular, the experimental schemes relying on the charge-exchange reaction (AE$\bar{g}$IS and GBAR) will benefit from the increased numbers, since the formation of antihydrogen through charge-exchange critically depends on the absolute number of antiprotons. In addition, GBAR cranks up the positron number for positronium formation via operating a linear electron accelerator instead of using a Na^{22} positron source.

In all experiments the mass of the (anti)particles involved is either measured directly or expressed in the experimental outcome. A thought suggested by Antonino Zichichi is whether we can regard (anti)hydrogen as representative of all (anti)matter since the nucleus is a single antiproton [73]. The question comes to mind whether antideuteron/deuteron would be a suitable 'mirror' pair and (anti)matter candidate instead, since the nuclei mass incorporates a nuclear binding mass term.

Seen in this light the tests of CPT and WEP could possibly have a different outcome for an antideuterion atom than for antihydrogen.

No doubt interesting could also be to perform a WEP test on protonium, also known as antiprotonic hydrogen - which is a purely baryonic system. Cold antiprotonic hydrogen production in vacuum was reported in the ATHENA collaboration in 2006 [74].

The same would be the case for the leptonic system positronium. Unlike antiprotons positrons are available without a particle decelerator. A WEP test with positronium could therefore be carried out as a table-top experiment [75].

6. Acknowledgement

I am indebted to the AD community for illuminating discussions (and suggestions), in particular to Michael Doser, Joel Fajans, Stefan Ulmer, Patrice Perez, Laurent Hilico, Will Bertsche, Barbara Latacz, Celeste Carruth, Benjamin Rienacker, Helga Holmestad, James Harrington and Steve Jones. Many of the discussions took place at the Winter school on Trapped Charged Particles in Les Houches, some of them at CERN. I am also very thankful for the invitation of the organizers of the symposium and grateful for the constructive criticism of the referees. Partial financial support by the German Federal Ministry of Education and Research is thankfully acknowledged.

References

[1] D. N. Spergel et al., arXiv:astro-ph/0302209v3 (2003)
[2] A. D. Sakharov, ZhETF Pis'ma 5, No. 1, 32-35, 1January 1967
[3] A singularly unfeminine profession, Mary K. Gaillard, World Scientific Publishing Co. Pte. Ltd. (2015)
[4] Laurent Canetti et al 2012 New J. Phys. 14 095012
[5] A. G. Cohen et al., Astrophys. J. **495**, pp. 539-549 (1998)
[6] A. Benoit-Lévy et al, A& A 537, A78 (2012)
[7] Antimatter, Frank Close, Oxford University Press Inc., New York (2009)
[8] R. J. Hughes and M. H. Holzscheiter, PRL **66**, pp. 854-857, (1991)
[9] S. Ulmer et al., Nature **524**, pp. 196-199 (2015)
[10] M. Amoretti et al., Nature **419**, 456–459 (2002)
[11] G. Gabrielse et al., PRL **100**, 113001 (2002)
[12] Production of Antihydrogen via double charge exchange: PhD thesis, A. Müllers, Johannes Gutenberg-Universität Mainz (2012)
[13] C. C. Speake et al., Class. Quantum Grav. **29**, 180301 (2012)
[14] Background information on Lorentz and CPT violation: http://www.physics.indiana.edu/ kostelec/faq.html
[15] https://espace.cern.ch/AD-site/default.aspx
[16] N. Madsen, Phil. Trans. R. Soc. A, **368**, 3671-3682 (2010)
[17] D. Krasnický et al., AIP Conf. Proc. **1521**, 144 (2013)
[18] ELENA decelerator: https://espace.cern.ch/elena-project/sitepages/home.aspx

[19] Physics of Nonneutral Plasmas: R. C. Davidson, Imperial College Press and World Scientific Publishing Co. Pte. Ltd. (2001)
[20] Trapped antihydrogen in its ground state: PhD thesis, P. J. Richerme, Harvard University (2012)
[21] R. Krause-Rehberg et al., Nucl. Instr. and Meth. in Phys. Res. B **221**, 165-167, (2004)
[22] R. G. Greaves et al., Can. J. Phys. **51**, 445-448 (1996)
[23] A. P. Mills and E. M. Gullikson, Appl. Phys. Lett. **49**, 1121 (1986)
[24] The Positron Trap - A New Tool for Plasma Physics: C. M. Surko, et al., Positron Studies of Solids, Surfaces, and Atoms, A. P. Mills, et al. eds. World Scientific, p. 222-233 (1984)
[25] C. M. Surko et al., PRL **62**, 901 (1989)
[26] By Dhdpla at English Wikipedia, CC BY-SA 3.0, https://commons.wikimedia.org/w/index.php?curid=61934417
[27] C. M. Surko and R. G. Greaves, Radiation Physics and Chemistry **68**, 419-425 (2003)
[28] G. Gabrielse et al., Physics Letters A **129** 38-42 (1988)
[29] P. O. Fedichev, Physics Letters A **226**, 289-292 (1997)
[30] Prop. for the AE$\bar{g}$IS Exp. at the CERN antiproton decelerator, SPSC-P-334. CERN-SPSC-2007-017
[31] D. Krasnický et al., Phys. Rev. A **94**, 022714 (2016)
[32] C. H. Storry et al., PRL 93, 263401 (2004)
[33] S. Mariazzi et al., PRL **104** , 243401 (2010)
[34] S. Aghion et al., Nuclear Instruments and Methods in Physics Research **B 407**, 55-66 (2017)
[35] S. L. Andersen et al., J. Phys. B, **48**, 204003 (2015)
[36] C. G. Parthey et al., PRL **107**, 203001 (2011)
[37] M. Ahmadi et al., Nature, **541**, 506-512 (2017)
[38] C. Amole et al., Nat. Comm. **5**, 3955 (2014)
[39] M. Ahmadi et al., Nature **529**, 373-376 (2016)
[40] C. Amole et al., Nature **483**, 439-443 (2012)
[41] The ALPHA Collaboration & A. E. Charman, Nat. Comm., DOI: 10.1038/ncomms2787 (2013)
[42] C. Malbrunot et al., The ASACUSA antihydrogen and hydrogen program : results and prospects, arXiv:1710.03288v1 [physics.atom-ph] 9 Oct 2017
[43] H. Aghai-Khozani et al., Nuclear Physics A **970**, 366–378 (2018)
[44] N Kuroda et al., Nat. Comm., **5** (2014)
[45] R. S. Hayano, Proc. Jpn. Acad., Ser. B 86 (2010)
[46] M. Hori et al., Science **354**, 6312, pp. 610-614 (2016)
[47] M. Hori et al., Nature **484**, 475 (2011)
[48] G. Gabrielse et al., PRL **82**, 3198 (1999)
[49] G. Gabrielse et al., PRL **108**, 113002 (2012)
[50] J. DiSciacca, PRL **110**, 130801 (2013)
[51] G. Gabrielse, Physics Today 63, 3, 68 (2010)
[52] C. Smorra et al., Nature, **550**, 371 (2017)
[53] S. Ulmer et al., Nature **524**, 196 (2015)
[54] https://home.cern/about/updates/2017/10/more-precise-measurement-antimatter-matter
[55] https://www.mpi-hd.mpg.de/blaum/gfactor/silicon/index.en.html
[56] C. Smorra, Eur. Phys. J. Special Topics **224**, 3055–3108 (2015)
[57] M. Doser et al., Class. Quantum Grav. **29**, 18 (2012)
[58] C. M. Surko, Accumulation, storage and manipulation of large numbers of positrons in traps I., Proceedings of the International School of Physics "Enrico Fermi" , Course CLXXIV
[59] S. Aghion et al., Phys. Rev. A **94**, 012507 (2016)
[60] S. Aghion et al., Nat. Comm. **5**, 4538 (2014)
[61] C. Amsler et al., arXiv:1211.1370v2 [physics.ins-det] 27 Dec 2012
[62] Development and data analysis of a position detector for AE$\bar{g}$IS (Antimatter Experiment: Gravity, Interferometry, Spectroscopy): PhD thesis, A. Gligorova, University of Bergen (2014)
[63] ALPHA collaboration, Nature comm., DOI: 10.1038/ncomms2787 (2013)
[64] private communication with W. Bertsche, Les Houches 2018
[65] Addendum to the ALPHA Prop., The ALPHA-g App., CERN-SPSC-2016-031 / SPSC-P-325-ADD-1
[66] P. Pérez and Y. Sacquin, Class. Quantum Grav. **29** 184008 (2012)
[67] P. Pérez et al., Hyperfine Interact. **233**, pp.21-27 (2015)
[68] L Liszkay et al 2013 J. Phys.: Conf. Ser. 443 012006
[69] P. Comini et al., Hyperfine Interact. **228**, pp.159-165 (2014)
[70] gBAR webpage: https://gbar.web.cern.ch/GBAR/public/
[71] E. M. Hollmann, F. Anderegg, and C. F. Driscoll, PHYSICS OF PLASMAS **7**, 7 (2000)
[72] M. Ahmadi et al., PRL **120**, 025001 (2018)

[73] Antimatter, past present and future, Antonino Zichichi
[74] N. Zurlo et al., PRL **97**, 153401 (2006)
[75] D. B. Cassidy and S. D. Hogan, International Journal of Modern Physics: conference proceedings - Antimatter and Gravity (WAG 2013) **30**, 1460259 (2014)

On Clifford groups in quantum computing

J Tolar [1]

[1] Department of Physics
Faculty of Nuclear Sciences and Physical Engineering
Czech Technical University in Prague
Břehová 7, 115 19 Prague 1, Czech Republic

E-mail: jiri.tolar@fjfi.cvut.cz

Abstract. The term Clifford group was introduced in 1998 by D. Gottesmann in his investigation of quantum error-correcting codes. The simplest Clifford group in multiqubit quantum computation is generated by a restricted set of unitary Clifford gates - the Hadamard, $\pi/4$-phase and controlled-X gates. Because of this restriction the Clifford model of quantum computation can be efficiently simulated on a classical computer (the Gottesmann-Knill theorem). However, this fact does not diminish the importance of the Clifford model, since it may serve as a suitable starting point for a full-fledged quantum computation.

In the general case of a single or composite quantum system with finite-dimensional Hilbert space the finite Weyl-Heisenberg group of unitary operators defines the quantum kinematics and the states of the quantum register. Then the corresponding Clifford group is defined as the group of unitary operators leaving the Weyl-Heisenberg group invariant. The aim of this contribution is to show that our comprehensive results on symmetries of the Pauli gradings of quantum operator algebras – covering any single as well as composite finite quantum systems – directly correspond to Clifford groups defined as quotients with respect to U(1).

1. Introduction

In quantum mechanics of single N-level systems in Hilbert spaces of finite dimension N, the basic operators are the *generalized Pauli matrices*. They generate the finite *Weyl-Heisenberg group* (defining quantum kinematics and the states of the quantum register) as a subgroup of the unitary group $\mathrm{U}(N)$ [1, 2, 3, 4].

Its normalizer within the unitary group $\mathrm{U}(N)$, or in other words the largest subgroup of the unitary group having the Weyl-Heisenberg group as a normal subgroup, is in the papers on quantum information conventionally called the *Clifford group* [5]. Since this normalizer necessarily contains the continuous group U(1) of phase factors, some authors adopt an alternative definition of the Clifford group as the quotient of the normalizer with respect to U(1) [6]. In this paper we call it the *Clifford quotient group.*

Our results on symmetries of the Pauli gradings of quantum operator algebras [7, 8, 9] account for all possible Clifford quotient groups corresponding to arbitrary single or composite quantum systems. This contribution presents a brief review of their description. In Sect. 2 the Weyl-Heisenberg groups of N-level quantum systems are defined as subgroups of $\mathrm{U}(N)$ for $N = 2, 3, \ldots$. In Sect. 3 the corresponding Clifford quotient groups are described and in Sect. 4 the Clifford quotient groups for arbitrary composite quantum systems are shortly introduced.

2. Weyl-Heisenberg groups of single N-level systems

In finite-dimensional quantum mechanics of a single N-level system the N-dimensional Hilbert space $\mathcal{H}_N = \mathbb{C}^N$ has an orthonormal basis $\mathcal{B} = \{|0\rangle, |1\rangle, \ldots |N-1\rangle\}$. The basic unitary operators Q_N, P_N are defined by their action on the basis [1, 2]

$$Q_N|j\rangle = \omega_N^j|j\rangle, \quad (1)$$
$$P_N|j\rangle = |j-1 \pmod N\rangle, \quad (2)$$

where $j = 0, 1, \ldots, N-1$, $\omega_N = \exp(2\pi i/N)$. This is the well-known *clock-and-shift representation* of the basic operators Q_N, P_N. In the canonical or computational basis $\mathcal{B}$ the operators Q_N and P_N are represented by the *generalized Pauli matrices*

$$Q_N = \operatorname{diag}\left(1, \omega_N, \omega_N^2, \cdots, \omega_N^{N-1}\right) \quad (3)$$

and

$$P_N = \begin{pmatrix} 0 & 1 & 0 & \cdots & 0 & 0 \\ 0 & 0 & 1 & \cdots & 0 & 0 \\ 0 & 0 & 0 & \cdots & 0 & 0 \\ \vdots & & & \ddots & & \\ 0 & 0 & 0 & \cdots & 0 & 1 \\ 1 & 0 & 0 & \cdots & 0 & 0 \end{pmatrix}. \quad (4)$$

Their commutation relation

$$P_N Q_N = \omega_N Q_N P_N \quad (5)$$

expresses the minimal non-commutativity of the operators Q_N and P_N. Further, they are of the order N,

$$P_N^N = Q_N^N = I, \quad \omega_N^N = 1.$$

The *Pauli group* Π_N of order N^3 is generated by ω_N, Q_N and P_N [7]

$$\Pi_N = \left\{\omega_N^l Q_N^i P_N^j | l, i, j = 0, 1, 2, \ldots, N-1\right\}. \quad (6)$$

Elements of $\mathbb{Z}_N = \{0, 1, \ldots N-1\}$ label the vectors of the canonical basis $\mathcal{B}$ with the physical interpretation that $|j\rangle$ is the (normalized) eigenvector of position at $j \in \mathbb{Z}_N$. In this sense the cyclic group $\mathbb{Z}_N$ plays the role of the *configuration space for an N-level quantum system.*

Now the Clifford group to be constructed should contain the corresponding *Weyl-Heisenberg group* as its normal subgroup. The phase factors emerging in the Clifford group lead to the necessity to define the Weyl-Heisenberg groups $H(N)$ in even dimensions N by doubling the Pauli group so that $H(N)$ for even N contains the Pauli group Π_N as its subgroup. For this purpose the phase factor $\tau_N = -e^{\frac{\pi i}{N}}$ is introduced such that $\tau_N^2 = \omega_N$ [10, 11, 12] (see also [13]). Then the Weyl-Heisenberg groups

$$H(N) = \Pi_N = \left\{\omega_N^l Q_N^i P_N^j \mid l, i, j = 0, 1, \ldots, N-1\right\} \quad \text{for odd } N,$$

$$H(N) = \left\{\tau_N^l Q_N^i P_N^j \mid i, j = 0, 1, \ldots, N-1, l = 0, 1, \ldots, 2N-1\right\} \quad \text{for even } N,$$

are of orders N^3, $2N^3$, respectively. Note that for $N = 2$ we have $H(2) = \langle \tau_2 I_2, Q_2, P_2\rangle$, where $\tau_2 = -i$, $Q_2 = \sigma_z$, $P_2 = \sigma_x$; $H(2)$ is of order 16.

The *center* $Z(H(N))$ *of the Weyl-Heisenberg group* is the set of all those elements of $H(N)$ which commute with all elements in $H(N)$. For odd N we have

$$Z(H(N)) = \{\omega_N, \omega_N^2, \cdots, \omega_N^N = 1\} = \{\tau_N, \tau_N^2, \cdots, \tau_N^N = 1\},$$

while for even N

$$Z(H(N)) = \{\tau_N, \tau_N^2, \cdots, \tau_N^{2N} = 1\}.$$

Since the center is a normal subgroup, one can go over to the quotient group

$$\mathcal{P}_N = H(N)/Z(H(N)).$$

Its elements are the cosets labeled by pairs of exponents (i,j), $i,j = 0,1,\ldots,N-1$. The quotient group $\mathcal{P}_N$ is usually identified with the *finite phase space* $\mathbb{Z}_N \times \mathbb{Z}_N$. Namely, denoting the cosets corresponding to elements (i,j) of the phase space by $Q^i P^j = \left\{\tau_N^l Q_N^i P_N^j\right\}$, the correspondence

$$\Phi : H(N)/Z(H(N)) \to \mathbb{Z}_N \times \mathbb{Z}_N : Q^i P^j \mapsto (i,j), \tag{7}$$

is an *isomorphism of Abelian groups*, since

$$\Phi\left(\left(Q^i P^j\right)\left(Q^{i'} P^{j'}\right)\right) = \Phi\left(\left(Q^i P^j\right)\right)\Phi\left(\left(Q^{i'} P^{j'}\right)\right) = (i,j) + (i',j') = (i+i', j+j'). \tag{8}$$

3. Clifford groups of single N-level quantum systems

As already mentioned, in quantum information the term *Clifford group* means the group of symmetries of the Weyl-Heisenberg group in the following sense [10, 11]:

Definition

The Clifford group comprises all unitary operators $X \in \mathrm{U}(N)$ for which the Ad*-action preserves the subgroup $H(N)$ in* $\mathrm{U}(N)$, *i.e. the Clifford group is the normalizer $N_{\mathrm{U}(N)}(H(N))$.*

In this sense the Clifford group consists of all those $X \in \mathrm{U}(N)$ such that their Ad-action leaves $H(N)$ invariant,

$$\mathrm{Ad}_X H(N) = X H(N) X^{-1} = H(N).$$

But $H(N)$ is generated by τ_N, Q_N and P_N, so the Clifford group consists of all $X \in \mathrm{U}(N)$ such that

$$\mathrm{Ad}_X Q_N = X Q_N X^{-1} \in H(N) \quad \text{and} \quad \mathrm{Ad}_X P_N = X P_N X^{-1} \in H(N).$$

Some authors use the notion of *Clifford operations* [12]. They are the one-step unitary evolution operators acting on the nodes of a quantum register and are physically realized by so-called *Clifford gates*.[1]

Since the subgroup $H(N)$ is a normal subgroup of the normalizer, one can formally write a short exact sequence of group homomorphisms

$$1 \to H(N) \to N_{\mathrm{U}(N)}(H(N)) \to N_{\mathrm{U}(N)}(H(N))/H(N) \to 1.$$

[1] Note that the original name Clifford group is connected with Clifford algebras. For instance, the Clifford algebra generated by four Dirac matrices γ^μ is the real Dirac algebra D of dimension $2^4 = 16$. The corresponding Clifford group is the universal covering of the Lorentz group $O(3,1)$ of isometries of the real vector space $\mathbb{R}^4$ with an indefinite inner product, spanned by γ^0, γ^1, γ^2 and γ^3 such that

$$S\gamma^\mu S^{-1} - \Lambda_\nu^\mu \gamma^\nu.$$

The real Pauli algebra is the even part of D of dimension $2^3 = 8$ and its isometries form the universal covering of the orthogonal group $O(3)$.

It turns out that the full structure of the normalizer $N_{U(N)}(H(N))$ for arbitrary N is complicated by phase factors and rather difficult to describe in general [12]. In order to get insight into the structure of the normalizer *up to arbitrary phase factors* we turn to the definition of the *Clifford quotient group* $C(N)$ as the quotient [6]

$$C(N) = N_{U(N)}(H(N))/U(1).$$

Its elements are the cosets $\{e^{i\alpha}X\}$. The following lemma is crucial for our alternative view of the Clifford quotient group.

Lemma

Let $X, Y \in \mathrm{U}(N)$*. Then the equality* $\mathrm{Ad}_X A = \mathrm{Ad}_Y A$ *holds for all* $A \in H(N)$ *if and only if* $X = e^{i\alpha}Y$*.*

The proof of the converse implication is trivial. For the proof of the direct implication one writes down the first identity in the form $Y^{-1}XA = AY^{-1}X$, i.e. $Y^{-1}X$ commutes with all $A \in H(N)$. But the elements of $H(N)$ form an irreducible set, hence by Schur's lemma $Y^{-1}X$ is proportional to the unit matrix, $Y^{-1}X = e^{i\alpha}I_N$.□

In accordance with the Lemma, instead of the cosets $\{e^{i\alpha}X\}$ one can equivalently consider Ad-actions induced by the elements X of the normalizer. Generally, the mappings $\mathrm{Ad}_X : A \rightarrow XAX^{-1}$, where $A \in \mathrm{GL}(N, \mathbb{C})$, are inner automorphisms of $\mathrm{GL}(N, \mathbb{C})$ induced by elements $X \in \mathrm{GL}(N, \mathbb{C})$. In our case we need the subgroup $\mathcal{M}_N$ of $\mathrm{Int}(\mathrm{GL}(N, \mathbb{C}))$ generated by unitary operators, $\mathcal{M}_N = \{\mathrm{Ad}_X | X \in \mathrm{U}(N)\}$. In fact, if inner automorphisms Ad_X transform unitary operators A in unitary operators $A' = XAX^{-1}$, then to each $X \in \mathrm{GL}(N, \mathbb{C})$ there exists a unitary operator $U \in \mathrm{U}(N)$ such that $\mathrm{Ad}_U = \mathrm{Ad}_X$ and U is unique up to a phase factor.

In this equivalent approach the Clifford quotient group $C(N)$ can be studied as a subgroup of $\mathcal{M}_N$ – the subgroup of those Ad-actions which preserve the Weyl-Heisenberg group $H(N)$. But $C(N)$, consisting of the cosets $\{e^{i\alpha}X\}$ leaving $H(N)$ invariant, contains the cosets $\{e^{i\alpha}A\}$ of operators $A \in H(N)$ as a subgroup, and this subgroup is isomorphic to the group of Ad-actions Ad_A, $A \in H(N)$. Note that Ad-actions of unitary operators commute $\mathrm{Ad}_A\mathrm{Ad}_B = \mathrm{Ad}_B\mathrm{Ad}_A$ if and only if there exists $q \in \mathbb{C}^*$ such that $AB = qBA$ and $q^N = 1$. Hence in this picture the group of Ad-actions Ad_A, $A \in H(N)$, is generated by the commuting Ad-actions Ad_{Q_N} and Ad_{P_N},

$$\{\mathrm{Ad}_{Q_N^i P_N^j} | i, j = 0, 1, \ldots, N-1\} \cong \mathbb{Z}_N \times \mathbb{Z}_N \cong \mathcal{P}_N.$$

Proposition

The Clifford quotient group $C(N)$ *is isomorphic to the subgroup of those inner automorphisms in* $\mathcal{M}_N$ *which preserve* $\mathcal{P}_N$*, i.e.* $C(N)$ *is the normalizer of* $\mathcal{P}_N$ *in* $\mathcal{M}_N$*,*

$$C(N) \cong N_{\mathcal{M}_N}(\mathcal{P}_N).$$

We will show now that the short exact sequence of homomorphisms of subgroups of $\mathcal{M}_N$

$$1 \rightarrow \mathcal{P}_N \rightarrow N_{\mathcal{M}_N}(\mathcal{P}_N) \rightarrow N_{\mathcal{M}_N}(\mathcal{P}_N)/\mathcal{P}_N \rightarrow 1 \tag{9}$$

can be fully decoded [7].

Obviously, Ad-actions of elements of $\mathcal{P}_N$ leave $\mathcal{P}_N$ invariant. Then the elements of the quotient group $N_{\mathcal{M}}(\mathcal{P}_N)/\mathcal{P}_N$ are the cosets corresponding to possibly non-trivial transformations of $\mathcal{P}_N$ forming a symmetry (or Weyl) group. Let us consider the Ad-actions $\mathrm{Ad}_X(A) = XAX^{-1}$, where $X \in U(N)$, on elements $A \in H(N)$, which induce permutations of cosets in

IOP Conf. Series: Journal of Physics: Conf. Series **1071** (2018) 012022 doi:10.1088/1742-6596/1071/1/012022

$H(N)/Z(H(N)$. We consider them to be equivalent if, for each pair $(i,j) \in \mathbb{Z}_N \times \mathbb{Z}_N$, they define the same transformation of cosets in $H(N)/Z(H(N))$:

$$\mathrm{Ad}_Y \sim \mathrm{Ad}_X \quad \Leftrightarrow \quad YQ^iP^jY^{-1} = XQ^iP^jX^{-1}.$$

We have seen that the group $\mathcal{P}_N$ has two generators, Ad_{Q_N} and Ad_{P_N}, corresponding to cosets Q and P. Hence if Ad_Y induces a permutation of elements in $\mathcal{P}_N$, then there must exist $a,b,c,d \in \mathbb{Z}_N$ such that

$$YQY^{-1} = Q^aP^b \quad \text{and} \quad YPY^{-1} = Q^cP^d.$$

It follows that to each equivalence class of Ad-actions Ad_Y a quadruple (a,b,c,d) of elements in $\mathbb{Z}_N$ is assigned. In matrix notation

$$\mathrm{Ad}_Y \begin{pmatrix} 1 \\ 0 \end{pmatrix} = \begin{pmatrix} a \\ b \end{pmatrix} = \begin{pmatrix} a & c \\ b & d \end{pmatrix} \begin{pmatrix} 1 \\ 0 \end{pmatrix},$$

$$\mathrm{Ad}_Y \begin{pmatrix} 0 \\ 1 \end{pmatrix} = \begin{pmatrix} c \\ d \end{pmatrix} = \begin{pmatrix} a & c \\ b & d \end{pmatrix} \begin{pmatrix} 0 \\ 1 \end{pmatrix}.$$

Now inserting the relations

$$\mathrm{Ad}_Y Q_N = \mu Q_N^a P_N^b, \qquad \mathrm{Ad}_Y P_N = \nu Q_N^c P_N^d,$$

into the basic commutation condition

$$\mathrm{Ad}_Y(P_N Q_N) = \omega_N \mathrm{Ad}_Y(Q_N P_N),$$

we find

$$\omega_N^{ad-1} = \omega_N^{bc} \quad \text{i.e.} \quad ad - bc = \det \begin{pmatrix} a & c \\ b & d \end{pmatrix} = 1 \pmod{N}.$$

This result can also be stated as the value in $\mathbb{Z}_N$ of a bilinear alternating non-degenerate form (symplectic form) on the finite phase space $\mathcal{P}_N = \mathbb{Z}_N \times \mathbb{Z}_N$. Namely, the symplectic form is the bilinear mapping $q \,:\, \mathcal{P}_N \times \mathcal{P}_N \longrightarrow \mathbb{Z}_N$, where

$$q \,:\, ((i,j)(i',j')) \mapsto i'j - ij' = (i,j) \begin{pmatrix} 0 & -1 \\ 1 & 0 \end{pmatrix} \begin{pmatrix} i' \\ j' \end{pmatrix} = \det \begin{pmatrix} i' & j' \\ i & j \end{pmatrix}.$$

Theorem [7]
For integer $N \geq 2$ there is an isomorphism Φ between the set of equivalence classes of Ad*−actions* Ad_Y *which induce permutations of cosets, and the group* $\mathrm{SL}(2,\mathbb{Z}_N)$ *of 2×2 matrices with determinant equal to* $1 \pmod{N}$,

$$\Phi(\mathrm{Ad}_Y) = \begin{pmatrix} a & c \\ b & d \end{pmatrix}, \qquad a,b,c,d \in \mathbb{Z}_N. \tag{10}$$

The action of these automorphisms on $\mathcal{P}_N$ is given by (left) action of $\mathrm{SL}(2,\mathbb{Z}_N)$ *on elements $(i,j)^T$ of the phase space $\mathcal{P}_N = \mathbb{Z}_N \times \mathbb{Z}_N$,*

$$\mathrm{Ad}_Y \begin{pmatrix} i \\ j \end{pmatrix} = \begin{pmatrix} i' \\ j' \end{pmatrix} = \begin{pmatrix} a & c \\ b & d \end{pmatrix} \begin{pmatrix} i \\ j \end{pmatrix}. \tag{11}$$

It follows from the above results that the groups entering (9) are isomorphic to

$$\mathcal{P}_N \cong \mathbb{Z}_N \times \mathbb{Z}_N, \tag{12}$$
$$N_{\mathcal{M}_N}(\mathcal{P}_N)/\mathcal{P}_N \cong \mathrm{SL}(2,\mathbb{Z}_N), \tag{13}$$
$$C(N) = N_{\mathcal{M}_N}(\mathcal{P}_N) \cong (\mathbb{Z}_N \times \mathbb{Z}_N) \rtimes \mathrm{SL}(2,\mathbb{Z}_N), \tag{14}$$

where $\rtimes$ denotes a semidirect product. Here $\mathcal{P}_N$ is a normal subgroup of the normalizer $N_{\mathcal{M}_N}(\mathcal{P}_N)$ with two generators Ad_{Q_N}, Ad_{P_N}.

Summarizing, the *Clifford quotient group* $C(N)$ is isomorphic to the normalizer of the Abelian subgroup $\mathcal{P}_N$ in the group of unitary inner automorphisms $\mathcal{M}_N$. Since it contains all inner automorphisms transforming the phase space into itself, it necessarily contains $\mathcal{P}_N$ as an Abelian semidirect factor. The symmetry (or Weyl) group is then isomorphic to the quotient group of the normalizer with respect to $\mathcal{P}_N$.

The generators of the normalizer $N_{\mathcal{M}_N}(\mathcal{P}_N)$ are Ad_{Q_N}, Ad_{P_N} and Ad_{S_N}, Ad_{D_N} defined below [7]: The unitary Sylvester matrix S_N is the matrix of the discrete Fourier transformation (for $N = 2$ the Hadamard gate):

$$(S_N)_j k = \frac{\omega_N^{jk}}{\sqrt{N}}. \tag{15}$$

It acts on Q_N and P_N according to

$$S_N Q_N S_N^{-1} = P_N^{-1} \quad S_N P_N S_N^{-1} = Q_N, \tag{16}$$

The unitary operator D_N (for $N = 2$ the phase gate) is diagonal,

$$D_N = \mathrm{diag}\ (d_0, d_1, \ldots, d_{N-1}),$$

where $d_j = \tau_N^{j(1-j)}$ if N is odd, $d_j = \tau_N^{j(N-j)}$ if N is even. It acts on Q_N and P_N according to

$$D_N Q_N D_N^{-1} = Q_N, \qquad D_N P_N D_N^{-1} = \alpha_N Q_N P_N$$

where $\alpha_N = 1$ for N odd and $\alpha_N = \tau_N^{N+1}$ for N even.

Given the prime decomposition of $N = \prod_{i=1}^{r} p_i^{k_i}$, the general formula for the number of elements of $\mathrm{SL}(2,\mathbb{Z}_N)$ is the following multiplicative function of number theory:

$$|\mathrm{SL}(2,\mathbb{Z}_N)| = N^3 \prod_{i=1}^{r} \left(1 - \frac{1}{p_i^2}\right).$$

Some cardinalities are given in the table:

N	$\lvert\mathcal{P}_N\rvert$	$\lvert\mathrm{SL}(2,\mathbb{Z}_N)\rvert$	$\lvert\mathcal{P}_N \rtimes \mathrm{SL}(2,\mathbb{Z}_N)\rvert$
2	4	6	24
3	9	24	216
4	16	48	768
5	25	120	3000
6	36	144	5184
7	49	336	16464
8	64	384	24576

Example $N = 2$ [14]: The phase space consists of 4 elements $(0,0)$, $(1,0)$, $(0,1)$, $(1,1)$. The group $\mathrm{SL}(2,\mathbb{Z}_2)$ has 6 elements

$$\begin{pmatrix} 1 & 0 \\ 0 & 1 \end{pmatrix}, \begin{pmatrix} 0 & 1 \\ 1 & 0 \end{pmatrix}, \begin{pmatrix} 1 & 0 \\ 1 & 1 \end{pmatrix}, \begin{pmatrix} 1 & 1 \\ 0 & 1 \end{pmatrix}, \begin{pmatrix} 1 & 1 \\ 1 & 0 \end{pmatrix}, \begin{pmatrix} 0 & 1 \\ 1 & 1 \end{pmatrix},$$

and acts transitively on the orbit $\{(1,0),(0,1),(1,1)\}$. Unitary operators S_2 and D_2 are

$$S_2 = \frac{1}{\sqrt{2}} \begin{pmatrix} 1 & 1 \\ 1 & -1 \end{pmatrix}, \quad D_2 = \begin{pmatrix} 1 & 0 \\ 0 & -i \end{pmatrix}.$$

The finite Clifford group generated by S_2 and D_2 has $24 \times 8 = 192$ elements, since $(S_2 D_2)^3 = \eta I_2$ is of order 8:

$$\begin{pmatrix} 1 & 0 \\ 0 & \alpha \end{pmatrix}, \begin{pmatrix} 0 & 1 \\ \alpha & 0 \end{pmatrix}, \frac{1}{\sqrt{2}} \begin{pmatrix} 1 & \beta \\ \alpha & -\alpha\beta \end{pmatrix}, \eta^\nu \begin{pmatrix} 1 & 0 \\ 0 & 1 \end{pmatrix},$$

where $\eta = \exp(i\frac{\pi}{4})$, $\nu = 0, 1, \ldots, 7$, $\alpha, \beta \in \{1, i, -1, -i\}$. Note that $Q_2 = \sigma_3$, $P_2 = \sigma_1$ as well as their products and powers are generated by S_2 and D_2.

In this example the finite Clifford group is defined neither as the whole normalizer of the Heisenberg-Weyl group nor as its quotient with respect to $U(1)$, but a certain finite subgroup of unitary operators in the normalizer [15].

In mathematical terms the motivation for this construction comes from Cartan's method for simple Lie algebras. There in the first step the maximal commutative subalgebra containing semi-simple elements is identified. This Cartan subalgebra generates a commutative subgroup (complex torus) in the corresponding complex Lie group or the torus in the corresponding compact Lie group. Then the eigenspaces of its Ad-action are the root subspaces of the Lie algebra. They are permuted by the Weyl group of symmetries of the root subspaces.

The construction of the *finite Clifford group* can then be seen as an analogue of the *Demazure-Tits finite group* for simple Lie algebras [16]. In dimension N the finite Clifford group is a special finite subgroup of the normalizer $N_{U(N)}(H(N))$ completely defined by its generators τ_N, Q_N, P_N and S_N, D_N. Moreover, it was shown in [15] that the finite Clifford group for odd N is generated only by S_N and D_N.

4. Clifford quotient groups of multipartite systems

Our further results concern detailed description of groups of symmetries of finite Weyl-Heisenberg groups for *finitely composed quantum systems consisting of subsystems with arbitrary dimensions.* We have fully described these symmetries on the level of Ad-actions. In our notation [9] the most general *symmetry (or Weyl) groups* in the sense of [17] are $\mathrm{Sp}_{[n_1,\ldots,n_k]}$, where the indices denote arbitrary dimensions of the constituent Hilbert spaces.

More in detail, let the Hilbert space of a composite system be the tensor product $\mathcal{H}_{n_1} \otimes \ldots \otimes \mathcal{H}_{n_k}$ of dimension $N = n_1 \ldots n_k$, where $n_1, \ldots, n_k \in \mathbb{N}$. For the *composite system*, quantum phase space is the Abelian subgroup of $\mathrm{Int}(\mathrm{GL}(N,\mathbb{C}))$ defined by

$$\mathcal{P}_{(n_1,\ldots,n_k)} = \{\mathrm{Ad}_{M_1 \otimes \cdots \otimes M_k} |\ M_i \in H(n_i)\}.$$

The *Clifford quotient group*, or the normalizer of this Abelian subgroup in the group of unitary inner automorphisms of $\mathrm{GL}(N,\mathbb{C})$, contains all unitary inner automorphisms transforming the phase space into itself, hence necessarily *contains* $\mathcal{P}_{(n_1,\ldots,n_k)}$ *as an Abelian semidirect factor.* The symmetry (or Weyl) group is then given by the quotient group of the normalizer with respect to this Abelian subgroup.

The generating elements of $\mathcal{P}_{(n_1,\ldots,n_k)}$ are the inner automorphisms

$$e_j := \mathrm{Ad}_{A_j} \quad \text{for} \quad j = 1, \ldots, 2k,$$

IOP Conf. Series: Journal of Physics: Conf. Series **1071** (2018) 012022 doi:10.1088/1742-6596/1071/1/012022

where (for $i = 1, \ldots, k$)

$$A_{2i-1} := I_{n_1 \cdots n_{i-1}} \otimes P_{n_i} \otimes I_{n_{i+1} \cdots n_k}, A_{2i} := I_{n_1 \cdots n_{i-1}} \otimes Q_{n_i} \otimes I_{n_{i+1} \cdots n_k}.$$

The normalizer of $\mathcal{P}_{(n_1,\ldots,n_k)}$ in $\mathrm{Int}(\mathrm{GL}(n_1 \cdots n_k, \mathbb{C}))$ will be denoted

$$\mathcal{N}(\mathcal{P}_{(n_1,\ldots,n_k)}) := N_{\mathrm{Int}(\mathrm{GL}(n_1 \cdots n_k, \mathbb{C}))}(\mathcal{P}_{(n_1,\ldots,n_k)}),$$

We need also the normalizer of $\mathcal{P}_n$ in $\mathrm{Int}(\mathrm{GL}(n, \mathbb{C}))$,

$$\mathcal{N}(\mathcal{P}_n) := N_{\mathrm{Int}(\mathrm{GL}(n,\mathbb{C}))}(\mathcal{P}_n),$$

and

$$\mathcal{N}(\mathcal{P}_{n_1}) \times \cdots \times \mathcal{N}(\mathcal{P}_{n_k}) := \{\mathrm{Ad}_{M_1 \otimes \cdots \otimes M_k} |\ M_i \in \mathcal{N}(\mathcal{P}_{n_i})\} \subseteq \mathrm{Int}(\mathrm{GL}(N, \mathbb{C})).$$

Further,

$$\mathcal{N}(\mathcal{P}_{n_1}) \times \cdots \times \mathcal{N}(\mathcal{P}_{n_k}) \subseteq \mathcal{N}(\mathcal{P}_{(n_1,\ldots,n_k)}).$$

Now the *symmetry group* $\mathrm{Sp}_{[n_1,\ldots,n_k]}$ is defined in several steps. First let $\mathcal{S}_{[n_1,\ldots,n_k]}$ be a set consisting of $k \times k$ matrices H of 2×2 blocks

$$H_{ij} = \frac{n_i}{\gcd(n_i, n_j)} A_{ij}$$

where 2×2 matrices $A_{ij} \in \mathrm{M}_2(\mathbb{Z}_{n_i})$ for $i, j = 1, \ldots, k$. Then $\mathcal{S}_{[n_1,\ldots,n_k]}$ is (with the usual matrix multiplication) a monoid. Next, for a matrix $H \in \mathcal{S}_{[n_1,\ldots,n_k]}$, we define its adjoint $H^* \in \mathcal{S}_{[n_1,\ldots,n_k]}$ by

$$(H^*)_{ij} = \frac{n_i}{\gcd(n_i, n_j)} A_{ji}^T.$$

Further, we need a skew-symmetric matrix

$$J = \mathrm{diag}(J_2, \ldots, J_2) \in \mathcal{S}_{[n_1,\ldots,n_k]}$$

where

$$J_2 = \begin{pmatrix} 0 & 1 \\ -1 & 0 \end{pmatrix}.$$

Then the *symmetry group* is defined as

$$\mathrm{Sp}_{[n_1,\ldots,n_k]} := \{H \in \mathcal{S}_{[n_1,\ldots,n_k]} |\ H^* J H = J\}$$

and is a finite subgroup of the monoid $\mathcal{S}_{[n_1,\ldots,n_k]}$.

Our first theorem states the group isomorphism:

Theorem 1 [9]

$$\mathcal{N}(\mathcal{P}_{(n_1,\ldots,n_k)})/\mathcal{P}_{(n_1,\ldots,n_k)} \cong \mathrm{Sp}_{[n_1,\ldots,n_k]}.$$

Our second theorem describes the generating elements of the normalizer:

Theorem 2 [9]

The normalizer $\mathcal{N}(\mathcal{P}_{(n_1,\ldots,n_k)})$ *is generated by*

$$\mathcal{N}(\mathcal{P}_{n_1}) \times \cdots \times \mathcal{N}(\mathcal{P}_{n_k}) \quad \textit{and} \quad \{\mathrm{Ad}_{R_{ij}}\},$$

where (for $1 \le i < j \le k$*)*

$$R_{ij} = I_{n_1 \cdots n_{i-1}} \otimes \mathrm{diag}(I_{n_{i+1} \cdots n_j}, T_{ij}, \ldots, T_{ij}^{n_i - 1}) \otimes I_{n_{j+1} \cdots n_k}$$

and

$$T_{ij} = I_{n_{i+1}\cdots n_{j-1}} \otimes Q_{n_j}^{\frac{n_j}{\gcd(n_i,n_j)}}.$$

An important special case is

Corollary [9]

If $n_1 = \ldots = n_k = n$, *i.e.* $N = n^k$, *the symmetry group is* $\mathrm{Sp}_{[n,\ldots,n]} \cong \mathrm{Sp}_{2k}(\mathbb{Z}_n)$.

These cases are of particular interest, since they uncover *symplectic symmetry* of k-partite systems composed of subsystems with the same dimensions. This circumstance was found, to our knowledge, first by [18] for $k = 2$ under additional assumption that $n = p$ is prime, leading to $\mathrm{Sp}(4, \mathrm{F}_p)$ over the field F_p. We have generalized this result [8] to bipartite systems with arbitrary n (non-prime) leading to $\mathrm{Sp}(4, \mathbb{Z}_n)$ over the modular ring, and also to multipartite systems [9]. The corresponding result was independently obtained by [19] who studied symmetries of the tensored Pauli grading of the Lie algebra $\mathrm{sl}(n^k, \mathbb{C})$ (see also [20]).

5. Illustrative examples

Consider a *bipartite system* created by coupling two single multi-level subsystems with arbitrary dimensions n, m, i.e.

$$G = \mathbb{Z}_n \times \mathbb{Z}_m \quad \text{and} \quad \mathcal{H} = \mathcal{H}_n \otimes \mathcal{H}_m.$$

The corresponding finite Weyl-Heisenberg group is embedded in $\mathrm{U}(N)$, $N = nm$. Via *inner automorphisms* it induces an Abelian subgroup in $\mathrm{Int}(\mathrm{GL}(N, \mathbb{C}))$.

The *Clifford quotient group*, or the normalizer of this Abelian subgroup in the group of inner automorphisms of $\mathrm{GL}(N, \mathbb{C})$, contains all unitary inner automorphisms transforming the phase space into itself, hence necessarily *contains* $\mathcal{P}_{(n,m)}$ *as an Abelian semidirect factor*. The symmetry is then given by the quotient group of the normalizer with respect to this Abelian subgroup.

According to the Corollary the case $n = m$, $N = n^2$, corresponds to the symmetry group

$$\mathrm{Sp}_{[n,n]} \cong \mathrm{Sp}(4, \mathbb{Z}_n).$$

If $N = nm$, n, m coprime, the symmetry group is

$$\mathrm{Sp}_{[n,m]} \cong \mathrm{SL}(2, \mathbb{Z}_n) \times \mathrm{SL}(2, \mathbb{Z}_m) \cong \mathrm{SL}(2, \mathbb{Z}_{nm}).$$

Further, if $d = gcd(n, m)$, $n = ad$, $m = bd$, the finite configuration space can be further decomposed under the condition that a, b are both coprime to d,

$$G = \mathbb{Z}_n \times Z_m = Z_{ad} \times Z_{bd} \cong (Z_a \times Z_d) \times (Z_b \times Z_d).$$

Thus the symmetry group is reduced to the direct product

$$\mathrm{Sp}_{[n,m]} \cong \mathrm{Sp}_{[a,b]} \times \mathrm{Sp}(4, \mathbb{Z}_d)$$

For instance, if $n = 15$, $m = 12$, then $d = 3$ is coprime to both $a = 5$ and $b = 4$, and also a and b are coprime, hence the symmetry group is reduced to the standard types SL and Sp,

$$\mathrm{Sp}_{[n,m]} \cong \mathrm{SL}(2, \mathbb{Z}_a) \times \mathrm{SL}(2, \mathbb{Z}_b) \times \mathrm{Sp}(4, \mathbb{Z}_d).$$

Consider a still more general situation, for instance let $n = 180$ and $m = 150$. Then $d = 30$, $a = 180/30 = 6$ and $b = 150/30 = 5$, hence a divides d and also b divides d. In this case the reduction to solely standard groups Sp and SL is not possible. One has to break down the composite system consisting of two single systems into its *elementary building blocks* [21]. We decompose each of the finite configuration spaces

$$\mathbb{Z}_{180} \times Z_{150} = (\mathbb{Z}_{2^2} \times Z_{3^2} \times Z_5) \times (\mathbb{Z}_2 \times Z_3 \times Z_{5^2}),$$

and take notice of coprime factors 2.2^2, 3.3^2 and 5.5^2 leading to the factorization of the symmetry group in agreement with the elementary divisor decomposition

$$\mathrm{Sp}_{[180,150]} \cong \mathrm{Sp}_{[2,2^2]} \times \mathrm{Sp}_{[3,3^2]} \times \mathrm{Sp}_{[5,5^2]}.$$

Thus if attention is paid to the elementary building blocks of finite quantum systems (quantal degrees of freedom) [21], then the symmetries can be reduced – in accordance with the elementary divisor decomposition – to direct products of finite groups of the types $\mathrm{SL}(2, \mathbb{Z}_n)$, $\mathrm{Sp}(2k, \mathbb{Z}_n)$, and $\mathrm{Sp}_{[p^k,p^l,\ldots]}$. The last type of symmetry groups corresponds to a *new class of Clifford quotient groups of composite systems*. These symmetry groups like $\mathrm{Sp}_{[p^k,p^l]}$ with indices given by different powers of the same prime p deserve to be added to the standard types Sp and SL.

6. Conclusion

We have considered three definitions of the Clifford group. As the most interesting seems not the first one – the whole normalizer of the Heisenberg-Weyl group, but the second – its quotient with respect to $U(1)$ or even the third – a certain finite subgroup of unitary operators in the normalizer like that presented in the example in dimension $N = 2$.

Acknowledgements

The author is grateful to M. Havlíček, M. Korbelář and P. Novotný for fruitful discussions on the subject of the paper. Partial support of the Ministry of Education of Czech Republic (from the research plan RVO:68407700) is gratefully acknowledged.

References

[1] Weyl H 1931 *The Theory of Groups and Quantum Mechanics* (New York: Dover) pp 272–280
[2] Schwinger J 1960 Unitary operator bases *Proc. Nat. Acad. Sci. U.S.A.* **46** 570–579, 1401–1415
[3] Šťovíček P and Tolar J 1984 Quantum mechanics in a discrete space-time *Rep. Math. Phys.* **20** 157–170
[4] Vourdas A 2004 Quantum systems with finite Hilbert space *Rep. Progr. Phys.* **67** 267–320
[5] Gottesman D 1998 Theory of fault-tolerant quantum computation *Phys. Rev. A* **57** 127–137
[6] Koenig R and Smolin J A 2014 How to efficiently select an arbitrary Clifford group element *J. Math. Phys.* **55** 122202
[7] Havlíček M, Patera J, Pelantová E and Tolar J 2002 Automorphisms of the fine grading of $sl(n, \mathbb{C})$ associated with the generalized Pauli matrices *J. Math. Phys.* **43** 1083-1094; arXiv: math-ph/0311015
[8] Korbelář M and Tolar J 2010 Symmetries of the finite Heisenberg group for composite systems *J. Phys. A: Math. Theor.* **43** 375302 (15pp); arxiv: 1006.0328 [quant-ph]
[9] Korbelář M and Tolar J 2012 Symmetries of finite Heisenberg groups for multipartite systems *J. Phys. A: Math. Theor.* **45** 285305 (18pp); arXiv: 1210.6167 [quant-ph]
[10] Appleby D M, Bengtsson I, Brierley S, Grassl M, Gross D, Larsson J-A 2012 The monomial representations of the Clifford group *Quantum Information and Computation* **12** No 5&6 0404-0431
[11] Appleby D M 2005 SIC-POVMs and the extended Clifford group *J. Math. Phys.* **46** 052107
[12] Hostens E, Dehaene J and De Moor B 2005 Stabilizer states and Clifford operations for systems of arbitrary dimensions, and modular arithmetic *Phys. Rev. A* **71** 042315 (11pp)
[13] Bengtsson I and Zyczkowski K 2017 *Geometry of Quantum States. An Introduction to Quantum Entanglement* 2nd Edition (Cambridge: Cambridge University Press) Sect. 12.6
[14] Rains E M and Sloane N J A 1998 Self-dual codes In: *Handbook of Coding Theory* Pless V S and Huffman W C eds 177-294; arXiv: math/0208001 Sect. 6.2
[15] Bartoňová P 2017 *Group Theory and Quantum Computing* Bachelor Thesis at Czech Technical University in Prague (in Czech)

[16] Michel L, Patera J and Sharp R S 1988 The Demazure-Tits subgroup of a simple Lie group *J. Math. Phys.* **29** 777-796
[17] Patera J and Zassenhaus H 1989 On Lie gradings I *Lin. Alg. Appl.* **112** 87–159
[18] Pelantová E, Svobodová M and Tremblay J 2006 Fine grading of sl($p^2, \mathbb{C}$) generated by tensor product of generalized Pauli matrices and its symmetries *J. Math. Phys.* **47** 5341–5357
[19] Han G 2010 The symmetries of the fine gradings of sl($n^k, \mathbb{C}$) associated with direct product of Pauli groups *J. Math. Phys.* **51** 092104
[20] Vourdas A and Banderier C 2010 Symplectic transformations and quantum tomography in finite quantum systems *J. Phys. A: Math. Theor.* **43** 042001 (9pp)
[21] Tolar J 2014 A classification of finite quantum kinematics *J. Phys.: Conf. Series* **538**, 012020 (12pp)